Origin of the Earth

Edited by

Horton E. Newsom
University of New Mexico

John H. Jones
NASA Johnson Space Center

Associate Editors

M. Bruce Fegley Jr.
Massachusetts Institute of Technology and Max-Planck-Institut für Chemie

John H. Jones
NASA Johnson Space Center

William M. Kaula
University of California, Los Angeles

Horton E. Newsom
University of New Mexico

David J. Stevenson
California Institute of Technology

Oxford University Press, New York
Lunar and Planetary Institute, Houston
1990

Oxford University Press

Oxford New York Toronto
Delhi Bombay Calcutta Madras Karachi
Petaling Jaya Singapore Hong Kong Tokyo
Nairobi Dar es Salaam Cape Town
Melbourne Auckland

and associated companies in
Berlin Ibadan

Published by Oxford University Press, Inc.,

200 Madison Avenue, New York, New York 10016

Library of Congress Cataloging-in-Publication Data
Origin of the earth / edited by Horton E. Newsom and John H. Jones.
p. cm.
Based on a conference entitled "The Origin of the earth,"
held in Houston, Tex., Dec. 1988.
Includes bibliographical references and indexes.
ISBN 0-19-506619-7
1. Earth—Origin—Congresses. I. Newsom, Horton E.
II. Jones, John H.
QB632.075 1990 525—dc20 90-38335

135798642
Printed in the United States of America
on acid-free paper

PREFACE

This book is an outgrowth of the LPI Conference on the Origin of the Earth, which was held in Berkeley, California in December 1988. When we began planning this conference/book project in 1986, our idea was to build on two previous meetings: the Workshop on the Early Earth and the Origin of the Moon Conference. We envisioned a meeting that included dynamical and cosmochemical aspects of the origin of the Earth-Moon system. But we wanted to emphasize questions about the origin of the Earth: Was the Earth essentially molten? How did the major geochemical reservoirs in the Earth form? What happened to the primordial atmosphere of the Earth? We hoped for a book that would provide an opportunity to reevaluate the origin of the Earth in light of the recent advances in many fields.

The conference was a tremendous success, with a great deal of excitement and communication across disciplines. As a result, we offer a book that summarizes relevant results from many fields and that is written to an interdisciplinary audience. While the book is not representative of our total knowledge about the early Earth, it does contain a spectrum of opinion on the implications for the origin and earliest history of the Earth. Suggested additional readings are offered in the introductions to each section.

A strong theme of both the conference and many of this book's papers is the Giant Impact Hypothesis. Even the papers that do not specifically address the Giant Impact are influenced by that hypothesis. If the conference had a single conclusion, it was the recognition that the origin of the Moon by giant impact has serious implications for the early history of the Earth. It seems unavoidable that a collision of a Mars-sized body would have vaporized a significant amount of the Earth and would have totally melted the rest. How is it then that there are samples from the Earth's mantle that appear to have undergone minimal processing? This is the paradox that the authors in this book wrestle with but do not solve (at least not to the satisfaction of all).

The Giant Impact hypothesis has been presented by some as a revolution and panacea, similar to the advent of plate tectonics, that can explain the long-standing problems of the angular momentum of the Earth-Moon system, the depletion of iron metal in the Moon, and, quite possibly, the difference in chemical composition between the Earth and Moon. As illustrated by the papers in the volume, however, the implications of this theory for the Earth have not led to the unanimous agreement that giant impacts (connected with the origin of the Moon or not) are consistent with all geophysical and geochemical observations. What we need is for all the pieces of the puzzle to fit without serious whittling, carving or shoving. As of now, they don't. We hope that this book will help set the stage for the next generation of investigation and for fitting some pieces of the puzzle together.

The book is divided into six sections. The first three sections deal with accretion and giant impact and the next three discuss the establishment of chemical and physical reservoirs. The book begins with a discussion of terrestrial accretion within the more general context of the origin of the solar nebula. The discussions of accretion then focus on the effects of a giant impact and whether there was a terrestrial magma ocean. The final three sections deal with the formation of the reservoirs that we are familiar with today: the core, the continental crust, and the atmosphere/hydrosphere. We have tried to group papers into unifying topics rather than by discipline, but some papers could fit into any one of several categories and our arrangement is necessarily somewhat artificial.

There are many people that we would like to thank for their hard work and encouragement. In particular, the conference and this book would not have been successful without the help and cooperation of the program committee, the associate editors, and the LPI logistical staff. We also appreciate the support of Kevin Burke (former Director) and David Black (Director), Lunar and Planetary Institute.

The program committee for the LPI Conference on the Origin of the Earth consisted of R. Arculus, L. Ashwal, K. Burke, M. Drake, B. Fegley, J. Jones, M. Kurz, H. Newsom, D. Stevenson, and G. Wetherill. Assistance at the conference was provided by D. Musselwhite, P. Noll, K. Righter, and K. Wirth. Summarizers for the technical sessions were C. Agee, B. Fegley, S. Galer, P. Warren, and Q. Williams. Logistics for the conference were ably handled by Pamela Jones and LeBecca Simmons.

Our scientific quality control for *Origin of the Earth* was the responsibility of our Associate Editors: B. Fegley, J. Jones, W. Kaula, H. Newsom, and D. Stevenson. Our publications quality control was handled by the staff of the LPI Publications Services Department, particularly Stephanie Tindell, Renee Dotson, and Ronna Hurd. Finally, we wish to thank Gary Larson and Universal Press Syndicate for their views on the origin of the Earth.

Horton Newsom, University of New Mexico, Albuquerque
John Jones, NASA Johnson Space Center, Houston

In God's kitchen

CONTENTS

THE EARLIEST STAGES: THE SOLAR NEBULA AND ACCRETION OF PLANETESIMALS

There are four papers that address aspects of condensation and accretion from the solar nebula:

A. P. Boss: *3D Solar Nebula Models: Implications for Earth Origin*
M. J. Gaffey: *Thermal History of the Asteroid Belt: Implications for Accretion of the Terrestrial Planets*
S. R. Taylor and M. D. Norman: *Accretion of Differentiated Planetesimals to the Earth*
W. M. Kaula: *Differences Between the Earth and Venus Arising from Origin by Large Planetesimal Infall*

Boss emphasizes the importance of possible asymmetries during the formation of the solar nebula and predicts that the inner solar nebula was rather strongly heated. His calculation is consistent with the observation that the terrestrial planets of the inner solar system are depleted in volatiles relative to the jovian planets and the sun. However, it is not presently clear how such a hot, turbulent nebula could preserve the heterogeneities in oxygen isotopic composition that we observe today in meteorites.

Gaffey argues on the basis of spectral observation of asteroids that many of the planetesimals that accreted to form the Earth had been strongly heated and that temperatures in these objects were sufficient to initiate differentiation—e.g., core formation and the production of basalts. Planetesimals may have still been hot at the time of accretion. Taylor and Norman take this conclusion one step further, speculating that some of the chemical signatures of the Earth's mantle may have been determined prior to accretion.

Kaula has attempted to explain the difference in rotation rates between the Earth and Venus (which are otherwise rather similar in terms of total mass and distance from the sun) as being due to the influence of large, late-stage impacts. In so doing, he sets the stage for the next section of this book, which is mainly devoted to the effects of a giant impact on the Earth.

There is little discussion in this section on the general nature of the accretion process. We still do not know if accretion was rather quiescent and "localized" or whether accretion was violent and dominated by collisions of large, late-stage planetesimals having highly eccentric orbits. These two endmember hypotheses have dramatically different implications for the thermal state of the early Earth.

The newest concepts in accretion theory—hot planetesimals, giant impacts, and thermal insulation by the protoatmosphere—all point in the direction of a hot early Earth. Even without considering the tremendous amount of heat released during core formation, it may be difficult to avoid the formation of a terrestrial magma ocean. However, as will be seen in subsequent sections, chemical analysis of mantle-derived materials does not necessarily support the magma ocean hypothesis; this is a paradox that must be resolved.

The collection of papers in *The Formation and Evolution of Planetary Systems* (H. L. Weaver and L. Danly, eds., Cambridge, 1989) summarizes the latest views on the formation of solar systems. For a current review of the accretion process, the interested reader is encouraged to seek out Wetherill (1990) Formation of the Earth. In *Annual Reviews of Earth and Planetary Science, vol. 18.*

3D SOLAR NEBULA MODELS: IMPLICATIONS FOR EARTH ORIGIN

A. P. Boss

Department of Terrestrial Magnetism, Carnegie Institution of Washington, 5241 Broad Branch Road, N.W., Washington, DC 20015

The physical properties of the early solar nebula determined much of the subsequent course of terrestrial planet formation. A survey of the parameter space for three-dimensional (3D) models of early phases of the solar nebula has been completed recently; the implications of these models for Earth formation are presented here. The models show that nebula evolution may have been dominated by angular momentum transport through gravitational torques, even during only mildly nonaxisymmetric phases, implying that the turbulent diffusion and mixing associated with viscous solar nebula models might be obviated. Little or no evidence is found for planet formation through giant gaseous protoplanet instability; rather, Earth formation through planetesimal accumulation and impacts is indicated. Surface densities at 1 A.U. are typically quite adequate for Earth formation in a minimum mass ($\sim$0.05 $M_{\odot}$) nebula. The models also imply that the inner solar nebula may have been heated initially to temperatures on the order of 1500 K (or even higher) through the compressional heating associated with nebula formation. Temperatures at 1 A.U. may have been regulated to values around 1500 K by a thermostatic effect produced by the vaporization of iron grains; the thermostatic effect minimizes the thermal gradient in the inner nebula and hence suppresses production (through subsequent condensation) of a strong compositional gradient in the terrestrial planet region. A hot inner nebula is consistent with the gross depletion of volatiles on the Earth compared to solar abundances.

INTRODUCTION

In order to develop a complete theory of Earth origin, an understanding of the physics and chemistry of the solar nebula is required. Physical conditions in the solar nebula must have affected the earliest phases of dust grain coagulation and hence the beginning of the process of accumulating submicron-sized dust particles into kilometer-sized planetesimals and eventually into the Earth and the other terrestrial planets (e.g., *Weidenschilling,* 1988). Indeed, the solar nebula defined the initial conditions for planet formation in the grossest sense, by determining the global distribution of gases and solids and the spatially and temporally varying thermal profiles that controlled the possible condensation, melting, and/or evaporation of solids. Thus, while the solar nebula may seem only remotely related to the present-day Earth, the properties of the solar nebula largely determined the dynamical mode of terrestrial planet formation and the early thermodynamical history of the matter incorporated into the planets.

Solar nebula models have been amply reviewed recently by *Cameron* (1988), *Wood and Morfill* (1988), and *Boss et al.* (1989). The main focus of these reviews was on two-dimensional, axisymmetric (symmetric about the nebula's rotation axis) models, because the great majority of solar nebula models have been based on the concept of the viscous accretion disk, for which the evolution can be determined in axisymmetry given an effective viscosity. The most promising mechanism for producing a large effective viscosity is turbulence associated with convective instability in the direction perpendicular to the midplane of the nebula (*Lin and Papaloizou,* 1980). Viscous stresses act in precisely the direction desired for solar nebula evolution: Angular momentum is transported outward to the preplanetary region, while mass moves inward to be accreted onto the growing protosun. This evolution is desired because the sun contains nearly all of the mass of the solar system but relatively little of the angular momentum. Viscous accretion disk models, with a few exceptions (*Cassen and Moosman,* 1981; *Cassen and Summers,* 1983), do not treat the phase of evolution where the nebula is forming through the collapse of a dense interstellar cloud, and hence they assume that the compressional energy produced by the collapse has already been completely radiated away. Instead, the temperature of the nebula is determined by a steady-state balance between the heat produced by viscous friction in the nebula and the heat lost by radiation from the nebula surface; the typical result is a warm nebula (~1000 K; *Wood and Morfill,* 1988) at 1 A.U. with a strong temperature gradient throughout the terrestrial planet region.

Collapse models of solar nebula evolution are directed toward understanding the actual formation of the nebula through the self-gravitational collapse of a dense, rotating interstellar cloud. Because of the approximate conservation of angular momentum during the rapid, dynamical collapse, an initially spherical cloud will become strongly flattened by rotation by the time the cloud collapses down to the size of the solar nebula. The flattening process can be well studied in axisymmetry, and axisymmetric presolar nebula collapse models (including radiative transfer and full gas thermodynamics) have been calculated by *Tscharnuter* (1978, 1987), *Boss* (1984a), and *Bodenheimer et al.* (1988). While calculations including the collapse and formation phases are evidently more realistic at early times than steady-state nebula models, collapse models are hindered by the tremendous range in length and timescales involved, and hence are often limited to the earliest stages of nebula formation. *Tscharnuter* (1978, 1987) has pushed axisymmetric models forward the farthest in time, while Bodenheimer and colleagues (personal communication, 1989) have used a "multiple collapse" approach to divide the problem into a series of more manageable calculations.

Boss et al. (1989) also described recent work on fully three-dimensional (3D) solar nebula models. In a 3D collapse calculation, the possibility exists for the cloud to fragment into a binary protostellar system (e.g., *Boss,* 1988b). Considering that the sun is not thought to be a member of a binary system, binary fragmentation must be avoided, and this can be accomplished either by starting the collapse from a strongly centrally condensed cloud (*Boss,* 1987) or from a slowly rotating cloud (specific angular momentum $J/M < 10^{20}$ cm^2 sec^{-1}; *Boss,* 1985, 1986).

Cassen et al. (1981) calculated several models of the nonaxisymmetric evolution (i.e., allowing variations in the density around the rotation axis) of a nebula with an assumed initial surface density structure stabilized by a central protosun. The calculations employed a type of N-body code to model the dynamics of an infinitely thin, isothermal solar nebula. *Cassen et al.* (1981) found that relatively cool (~100 K) nebula models, containing at least as much mass as the central protosun, could rapidly break up into a number of massive protoplanets ("giant gaseous protoplanets," e.g., *Cameron,* 1978), or else produce a well-defined spiral arm structure. *Cassen and Tomley* (1988) are using this N-body technique to study the gravitational stability of solar nebula models with simulated thermal gradients.

This paper describes the implications for the origin of the Earth of the first major survey of 3D models of solar nebula formation (*Boss,* 1989). Following a brief discussion of the numerical techniques and assumptions employed in the models, the dynamical and thermodynamical implications of these models are discussed and compared with the results of previous axisymmetric solar nebula models.

3D SOLAR NEBULA MODELS

Solar nebula formation is an initial value problem: Given the physical conditions in the dense, rotating interstellar cloud that collapsed to form our solar system, the equations of protostellar formation can be used to predict the intermediate phases through which the solar nebula passed. While this approach sidesteps the need to make assumptions about, e.g., the steady-state nebula structure, we unfortunately do not know very much about the physical conditions in the presolar cloud. Observations of dense molecular clouds in current regions of low-mass star formation provide indications about the possible structure of the presolar cloud, but there is a large degree of uncertainty within these constraints. Solar nebula formation models thus need to be calculated over a range of possible initial conditions in order to try to isolate the type of initial conditions that could have led to the formation of our solar system. Of course, one hopes that a unique set of initial conditions will emerge, but this is yet to be demonstrated.

The models to be described here were produced by calculating the collapse of initially spherical clouds, with specified density and rotation profiles, onto protosuns of varied mass (*Boss,* 1988a, 1989). The calculations involve numerical solutions of the 3D equations of hydrodynamics, gravitation, and radiative transfer in the diffusion approximation. This set of six partial differential equations for the nine dependent variables (density, three components of velocity, specific internal energy, gravitational potential, pressure, opacity, and temperature) is closed by specifying how the specific internal energy, pressure, and opacity depend on the density and temperature (*Boss,* 1984a). The equations of state employed in these specifications include all the physics thought to be important in the density and temperature regime of the solar nebula, such as dust grain opacities (*Pollack et al.,* 1985) and molecular hydrogen thermodynamics. The partial differential equations are solved using explicit finite differences on a spherical coordinate grid, with variable grid spacing in the radial and colatitude directions to provide enhanced resolution of compact and thin structures. The fact that radiative transfer effects are included means that the temperature of the nebula is self-consistently determined.

One important modification compared to the previous 3D solar nebula models of *Boss* (1985) was the introduction of a central sink cell to simulate the gravitational attraction of the growing protosun. The central cell accepts mass and angular momentum from the collapsing nebula, and its mass is added into the gravitational potential. The detailed evolution of the matter within this sink cell is ignored, however, and in particular the neglect of radiation from the central protosun implies that the nebular temperature distributions thereby obtained are *lower bounds.* If protostellar heating had been included, significantly higher temperatures could be produced in models where the protosun mass is large (cf. the results in the section on compressional heating with those of *Tscharnuter,* 1987, whose axisymmetric model included heating from the protosun and produced temperatures exceeding 1600 K out to ~5 A.U.). The sink cell artifice allows advanced phases of solar nebula evolution, where most of the mass lies within the protosun, to be simulated without having to worry about the detailed evolution of the protosun itself. Clearly though, future models at least should include an approximate treatment of central protosun heating.

Because of the limitation on the size of the time step required for numerical stability of this solution method, the initial cloud densities ($\rho_i \sim 10^{-13}$ g cm^{-3}) were chosen to be considerably higher than those typically observed in the densest molecular clouds ($\rho \sim 10^{-18}$ g cm^{-3}). While we do not know at precisely what density interstellar clouds begin to collapse, the critical density is probably closer to the dense molecular cloud value. The effect of starting from abnormally high initial densities is to speed up the collapse process, which greatly reduces the computer time required, but also means that the cloud may not be allowed to radiate away as much of its compressional energy as would be the case if lower initial densities had been used. *Boss* (1989) argued that starting from these high densities may not have a deleterious effect on the dynamics or thermodynamics, but this assumption will clearly tend to *overestimate* nebula temperatures. Combined with the neglect of protosun heating mentioned previously, it must be admitted that the thermal structure of the present 3D models cannot be considered definitive—it would be fortuitous indeed if these two systematic errors (with opposite sign) conspired to cancel each other exactly. However, recent axisymmetric solar nebula models by *Tscharnuter* (1987) and Bodenheimer and colleagues (personal communication, 1989), including protostellar heating and starting from lower (but still relatively high) initial densities, have produced temperature profiles similar to those found in the present 3D models.

Several different parameters were varied in the models calculated by *Boss* (1989): the initial mass of the protosun (0.0 to 1.0 $M_\odot$), the initial mass of the nebula (0.10 to 1.0 $M_\odot$), and the initial specific angular momentum (J/M) of the nebula (2.0×10^{18} to 6.9×10^{19} cm^2 sec^{-1}). The initial density profile was either uniform density or a power law profile with $\rho_i \propto r^{-1}$, where r is the spherical coordinate radius. The initial angular velocity was taken to be either uniform (solid body rotation) or varying as $\Omega_i \propto R$, where R is the cylindrical radius; the latter dependence was chosen as an extreme case, in order to try to enhance surface densities in the outer nebula (see section on formation by planetesimal accumulation). At the termination of the calculations, the 3D models had protosun masses ranging from 0.001 to 1.8 $M_\odot$ and nebula masses ranging from 0.002 to 1.02 $M_\odot$, with most models having the properties of a minimum mass nebula (e.g., *Safronov,* 1969; *Weidenschilling,* 1977): a protosun mass $\sim$1 $M_\odot$ and a nebula mass $\sim$0.05 $M_\odot$. It should be emphasized that even the minimum mass nebula models depict possible nebula properties *immediately after formation,* i.e., the models are restricted to early phases of nebula evolution, probably prior to significant planetesimal accumulation.

DYNAMICAL IMPLICATIONS

Nebula Evolution

If the presolar cloud had very little angular momentum initially (e.g., J/M < 10^{18} cm^2 sec^{-1}), then nearly all of the cloud would have collapsed directly onto the protosun. However, for larger initial angular momenta (J/M > 10^{19} cm^2 sec^{-1}), a substantial amount of the matter would have become centrifugally supported in the nebula. Measurements of rotation rates in dense molecular clouds (*Goldsmith and Arquilla,* 1985), the preponderance of binary stars (*Abt,* 1983), and the ubiquity of relatively large-scale disks associated with young stellar objects (e.g., *Bally,* 1982) all suggest that the larger values of J/M are to be expected. If this was the case for the presolar cloud, then the solar nebula must have evolved in such a way as to transport the excess mass inward to the protosun. Substantial nebula evolution also appears to be necessary in order to produce the angular momentum distribution of the solar system from a presolar cloud with roughly uniform J/M (*Bodenheimer et al.,* 1988).

Three processes have been suggested for transporting mass and angular momentum in the solar nebula: viscous stresses, magnetic stresses, and gravitational torques. While the effects of viscous and magnetic stresses can be profitably investigated in axisymmetric calculations, 3D (or at least nonaxisymmetric) calculations are required to study angular momentum transport by gravitational torques. Gravitational torques require the presence of significant nonaxisymmetry, such as spiral density waves (*Larson,* 1984 and references therein), large-scale bars formed during collapse (*Boss,* 1985), or an ellipsoidal protosun (*Yuan and Cassen,* 1985).

An analytical estimate of the strength of gravitational torques (*Boss,* 1984b) showed that a mildly nonaxisymmetric nebula could evolve by gravitational torques just as rapidly as an efficient viscous accretion disk. This may be roughly demonstrated by dimensional analysis: The units of the time rate of change of the specific angular momentum (d(J/M)/dt) are $cm^2 sec^{-1}$, i.e., $d(J/M)/dt \propto R^2 \tau^{-2}$, where R and τ are characteristic lengths and times, respectively. The characteristic time for oscillations of a self-gravitating body is $\tau \propto (G\rho)^{-1/2}$, where G is the gravitational constant and ρ is the density. This reasoning suggests that $d(J/M)/dt \sim G\rho R^2$, which is actually a good approximation to the analytical result of *Boss* (1984b). For a nebula with $(J/M) = 10^{19}\, cm^2 sec^{-1}$, $\rho = 10^{-8}\, g\, cm^{-3}$, and R = 2 A.U., the timescale for angular momentum transport (J/M)/[d(J/M)/dt] is $\sim 10^7$ years.

Figure 1 shows the midplane density structure for one of the models from *Boss* (1989) that formed a minimum mass nebula. Mild nonaxisymmetry is apparent; in terms of a Fourier expansion of the density in the angle ϕ around the rotation axis, the amplitude of the $\cos(2\phi)$ term is ~0.1. The nonaxisymmetry in this model is apparently caused by nonlinear coupling with the ongoing collapse of the nebula envelope; in other models, nonaxisymmetry can be excited primarily by rotational instability or by gravitational instability of a massive nebula. Model C (Fig. 1) started from strongly centrally condensed initial conditions that tend to suppress the growth of nonaxisymmetry (*Boss,* 1987). Even still, the nonaxisymmetry that does arise is sufficiently large to imply a timescale for angular momentum transport in the inner regions of this model on the order of 10^6 years; this timescale is similar to the timescale for collapse of the presolar cloud, which may be important if nonaxisymmetry is driven primarily through coupling to the collapse motions of the initial cloud.

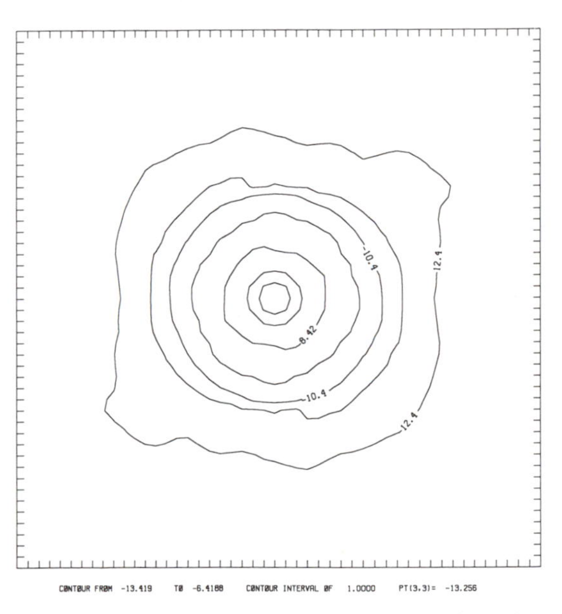

Fig. 1. *Density contours in the midplane of a solar nebula model (C) formed by collapse of an initial $1.02\, M_\odot$ cloud with $\rho_i \propto r^{-1}$, $\Omega_i = c$ onto a protosun with initial mass $0.01\, M_\odot$ (Boss, 1989). At the time shown in this figure and Figs. 2–4, the protosun has grown to $0.96\, M_\odot$ and the nebula mass has dropped to $0.07\, M_\odot$. Region shown is 20 A.U. across. The rotation axis (and protosun) is at the center of the plot; each contour corresponds to a factor of 10 change in density. Contours are labeled with logarithmic (base 10) densities in $g\, cm^{-3}$. Mild nonaxisymmetry is apparent in this minimum mass nebula; no tendency is exhibited for the nebula to breakup into giant gaseous protoplanets.*

Other 3D models from *Boss* (1989) demonstrate that gravitational torques can be even more efficient at transporting angular momentum in the solar nebula. Models with massive nebulae ($\sim 1\ M_{\odot}$) and low-mass protosuns often form well-defined, trailing spiral arm patterns, a result consistent with the models of *Cassen et al.* (1981). Assuming that the instantaneous transport rates from the 3D models can be maintained indefinitely, timescales for angular momentum transport for strongly nonaxisymmetric models can be as short as $\sim 10^3$ years, while timescales of $\sim 10^6$-10^7 years typify moderately nonaxisymmetric models. These timescales are comparable to or less than inferred ages of young solar-type stars that no longer show any evidence for the presence of a gaseous, dusty nebula (*Walter,* 1988), implying that gravitational torques can be efficient enough to account for the removal of the bulk of the nebula gases (any residual gases are presumably removed by the enhanced early solar wind). In the context of Earth formation, the most important implication of the efficiency of gravitational torques is that they remove the apparent need to ascribe nebula evolution to viscous stresses, and thus remove the necessity that the nebula was well mixed and possibly homogenized by diffusion and turbulent motions. The nebula might well have been turbulent, at least locally, but global turbulence was not a prerequisite for nebula evolution if gravitational torques were effective.

Giant Gaseous Protoplanet Formation

Cameron (1978) proposed forming both the terrestrial and giant planets through a gravitational instability of the gaseous portion of the solar nebula; the result would be the formation of fairly massive ($>0.007\ M_{\odot}$), giant gaseous protoplanets (GGPP). *Cameron* (1978) hypothesized that a sufficiently massive nebula would break up into a number of rings, and that each of the rings would subsequently break up into a small number (perhaps 3 to 6) of GGPP. As previously noted, *Cassen et al.* (1981) found that a massive nebula could break up directly into a small number (~ 4) of GGPP, though the absence of a massive central protosun means that such models perhaps more closely resemble the formation of a multiple stellar system. While *Cameron* (1988) no longer feels that the GGPP theory is promising, it is worthwhile to examine the evidence for GGPP formation in the *Boss* (1989) models.

The 3D models showed very little tendency to break up into GGPP; none of the models collapsed and directly formed GGPP. One large-scale (~ 15 A.U. in major radius) bar formed, as well as several smaller-scale bars, but none of the bars began to fragment. A few rings were observed, but these were all weak, low-mass, transient rings associated with the buildup of the nebula through accretion. Furthermore, the total mass in these ring models was quite low ($\sim 0.2\ M_{\odot}$ or less), implying that at least $0.8\ M_{\odot}$ of matter still had to be processed through the nebula to produce a solar system, and survival of a GGPP through this subsequent evolutionary phase is doubtful (see comments in following subsection on proto-Jupiter survival).

These results are somewhat different from those of *Cassen et al.* (1981); the differences appear to result from several improvements in the *Boss* (1989) models. First, the inclusion of compressional heating during formation has led to considerably higher nebula temperatures than those used by *Cassen et al.* (1981), and the thereby enhanced thermal pressure forces resist breaking up the nebula into GGPP. Partially as a result of these higher temperatures, surface densities in the models generally fall below the critical values necessary for a thin, equilibrium disk to be subject to axisymmetric gravitational instability (the $Q > 1$ stability criterion of *Toomre,* 1964); surface densities did exceed the critical values in a model that experienced binary fragmentation, in excellent agreement with the predictions of the Toomre criterion. Second, because of the gradual formation of the nebula through collapse of a dense cloud in these new models, strongly unstable disks did not have to be used as initial conditions.

The relatively coarse spatial resolution (32 grid points in ϕ) in the *Boss* (1989) models means that only the formation of a small number (say 2 or 4) of GGPP at any given radius can be simulated; the calculations are unable to simulate the formation of a large number (>8) of GGPP at a given radius. However, considering that *Cameron's* (1978) prediction was for a small number of GGPP, and also considering the results of *Cassen et al.* (1981), the present results seem to severely limit the possibilities for GGPP formation in an evolving solar nebula.

The inability of the 3D models to give any support for the formation of giant gaseous protoplanets suggests that planet formation probably occurred through the only alternative mechanism: The coagulation of dust grains followed by the accumulation of planetesimals through collisions.

Formation by Planetesimal Accumulation

As reviewed by *Safronov* (1969) and *Wetherill* (1980), the most likely means for planet formation is through the accumulation of planetesimals. While there is much that we do not yet know about the planetary accumulation process, particularly in the earliest phases (*Weidenschilling,* 1988), it is fairly certain that planet formation by accumulation cannot occur if there is insufficient solid matter present to account for the masses of the present-day planets. *Weidenschilling* (1977) reconstituted the planets to solar abundances, spread the resulting planetary masses out over a thin disk, and thereby derived rough lower bounds on the amount of mass (quantified by the surface density σ in g cm^{-2}) necessary to produce the planets. Hence a very basic criterion for solar system formation is achieving surface densities in the solar nebula at least as great as those of *Weidenschilling* (1977).

Safronov (1969) and *Goldreich and Ward* (1973) showed that a gravitational instability of a very thin dust subdisk could greatly speed up the accumulation process. Based on the same reasoning as *Weidenschilling* (1977), *Goldreich and Ward* (1973) used a dust surface density at 1 A.U. of $\sigma_d \sim 7.5$ g cm^{-2} for their study of the gravitational instability. Using a gas to dust ratio of 200:1 at 1 A.U., this critical surface density corresponds to a critical gas surface density of 1500 g cm^{-2} at 1 A.U.

Giant planet formation is also relevant for Earth origin for two reasons: (1) in order to see if we can construct a complete theory of solar system formation, and (2) because the relative timing of giant planet formation may have a profound effect on the gravitational perturbations experienced by the planetesimals in the terrestrial planet zone. In order to explain their composition, forming the giant planets prior to removal of the gaseous portion of the solar nebula is necessary, but this is a perennial problem in planetary accumulation theories (e.g., *Safronov,* 1969). Formation on a timescale of less than 10^7 years seems to be required by observations of young solar-type stars (*Walter* 1988). *Safronov* (1969) proposed two solutions to this sort of problem: (1) a more massive nebula than is indicated by the present planetary masses, and (2) small relative velocities between planetesimals, leading to enhanced collisional cross-sections because of gravitational focusing of orbits. *Levin* (1978) pointed out that runaway accretion, where one body accretes all of its neighbors, leads to the desired small relative velocities in the planetesimals. *Lissauer* (1987) has suggested that runaway accretion of icy-rock planetesimals coupled with a nebula more massive than the minimum possible (e.g., $\sigma_d > 15$ g cm^{-2} at 5 A.U.) could produce $\sim 15\ M_\oplus$ cores on a timescale of $\sim 10^6$ years. These cores would then accrete gaseous envelopes from the nebula on an even shorter timescale, thereby accounting for rapid giant planet formation. Using a gas to dust ratio of 50:1 at 5 A.U. (lower compared to 1 A.U. because of the condensation of water and other ices), Lissauer's surface density corresponds to a gas surface density of 750 g cm^{-2} at 5 A.U. This estimate assumes that once a runaway planetesimal accumulates all bodies within its sphere of influence, the planetesimal is effectively isolated and growth stops. Ensuring that the isolation mass is large enough to yield a ~15 $M_{\oplus}$ core at 5 A.U. is the motivation for increasing σ in Lissauer's scenario. However, protoplanets whose orbits are evolving because of tidal interactions with the nebula may continue to sweep through enough material to prevent isolation (*Ward,* 1989); in this case, the need for significantly enhanced surface densities in the outer nebula may be removed.

The models of *Boss* (1989) produce direct estimates of nebula surface densities immediately following nebula formation. In the 3D models, the dust grains are assumed to be moving with the gas during nebula formation, so a comparison of gas surface densities should be adequate. Figure 2 shows a typical result, in this case for the same model as in Fig. 1 (model C). The 3D models have surface densities in the inner solar nebula that are nearly always sufficient to account for terrestrial planet formation; in fact, the terrestrial region surface densities may be somewhat high. On the other hand, surface densities in the outer solar nebula are substantially less than the required amounts in a minimum mass nebula. Only if the nebula is quite massive (~1 $M_{\odot}$) does the surface density exceed the minimum values. However, any proto-Jupiter that is formed in such a massive nebula is likely to be lost during subsequent nebula evolution, either because the protoplanet will clear a gap in the gas and will be transported into the protosun with the bulk of the nebula (*Hourigan and Ward,* 1984; see also *Lin and Papaloizou,* 1986), or because tidal torques between the protoplanet and the nebula will result in rapid orbital decay of the protoplanet onto the protosun (*Ward,* 1986).

The 3D results on surface densities hold even for rather large variations in the assumptions about the initial cloud, such as the initial density profile and initial angular velocity profile. Furthermore, these results appear to be in agreement with those produced by a semi-analytic model of the collapse with conserved angular momentum of axisymmetric clouds to infinitely thin disks (*Stemwedel et al.,* 1987, 1990).

While terrestrial planet formation through planetesimal accumulation is thus strongly supported by these models, giant planet formation remains something of a problem, and to that extent Earth origin must also be considered somewhat uncertain. However, giant planet formation might well be simply explained as a result of nebula evolution subsequent to the formation phase studied by *Boss* (1989). That is, a nebula evolving through gravitational torques may very well act

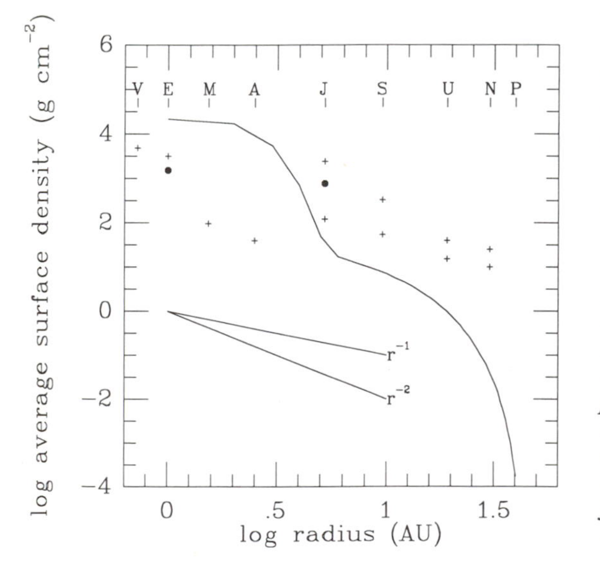

Fig. 2. *Azimuthally averaged surface density profile for model C. Filled circles represent σ thought necessary for Earth (Goldreich and Ward, 1973) and rapid Jupiter (Lissauer, 1987) formation. Representative power-law profiles are shown, as well as planetary locations. Plus signs are σ obtained by reconstituting all planets to solar composition (Weidenschilling, 1977). Model C (solid line) evidently has sufficient surface density to account for terrestrial planet formation, but insufficient surface density in the giant planet region.*

like a viscous accretion disk (*Lin and Pringle,* 1987); viscous nebulae do tend to increase their outer region surface densities (e.g., *Lissauer,* 1987). The former possibility is yet to be demonstrated, however. Diffusive redistribution of water vapor, suggested by *Stevenson and Lunine* (1988), could lead to preferential condensation of water ice at 5 A.U. and thus help account for rapid Jupiter formation, but this process would not aid in the formation of the other giant planets.

THERMODYNAMICAL IMPLICATIONS

Compressional Heating

The 3D models of *Boss* (1989) focus on the nebula formation phase, when the envelope of the progenitor dense interstellar cloud is still collapsing onto the nebula. The collapsing gas is compressed by the collapse and therefore heated. Because of the relatively high optical depths in this phase, little radiation escapes from the nebula and the nebular material undergoes significant heating above the initial cold interstellar cloud temperature of 10 K. Midplane temperatures typically reach 1500 K in the inner solar nebula (Fig. 3) during this formation phase (*Boss,* 1988a). Temperatures (and densities) fall off strongly with distance from the midplane, so that a typical temperature at a height of ~0.1 A.U. above the nebula midplane would be ~500 K. Cooling of the nebula by radiative losses (in a dusty envelope with an effective temperature ~20 K) to significantly lower temperatures should occur within about 10^5 years after nebula formation (*Boss,* 1988a, 1989), once a well-defined accretion shock is formed and is capable of cooling the nebula, or once accretion has stopped altogether. Cooling to temperatures well below 1500 K is thus likely (and probably *necessary*) prior to significant growth of planetesimals in the inner nebula.

Similar or even higher inner early nebula temperatures have been obtained in the axisymmetric collapse models of solar nebula formation of *Tscharnuter* (1987) and *Bodenheimer et al.* (1988, and personal communication, 1989). *Morfill* (1988) has recently presented models of axisymmetric, steady-state, viscous solar nebula models with temperatures as high as 1500 K at 1 A.U. for certain model parameters.

Maximum temperatures on the order of 1500 K imply that at least some portion of the equilibrium condensation sequence (e.g., *Grossman,* 1972) is likely to have been applicable to the earliest phases of the inner solar nebula as it cooled over a timescale on the order of 10^5 years. High temperatures in the inner solar nebula also could account for gross depletion of volatile species (e.g., C, N, H_2O) in the terrestrial planets by factors on the order of 10^4 or more compared to solar abundances (*Prinn and Fegley,* 1989). At these temperatures, the volatiles remain gaseous and would have been removed from

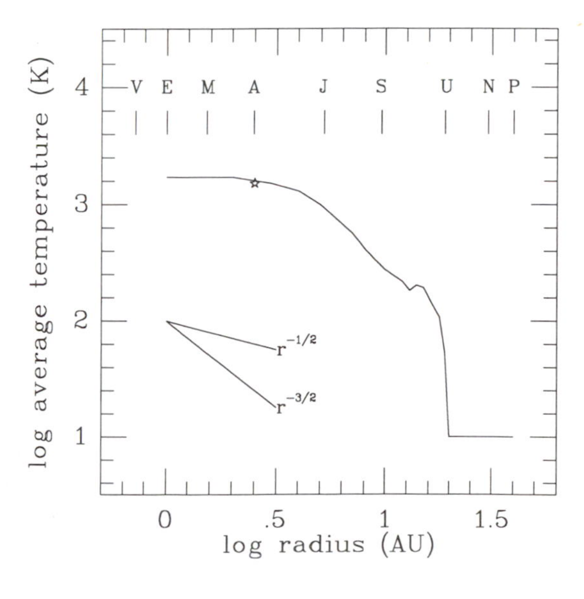

Fig. 3. *Azimuthally averaged midplane temperature profile for model C. The star denotes 1500 K at 2.5 A.U. Representative power-law profiles and planetary locations are noted. The thermostatic effect of the iron grain opacity is evident in the inner solar nebula, leading to temperatures close to 1500 K.*

the terrestrial planet region either through subsequent evolution of the hydrogen-rich nebula onto the protosun or through the enhanced early solar wind. Temperatures around 1500 K also imply that equilibrium chemistry is likely to have characterized the inner solar nebula and that thermochemistry was the dominant energy source (*Prinn and Fegley,* 1989).

If the nebula became strongly bar-like, then the temperature distribution would have been bar-like also. A mildly nonaxisymmetric 3D model is shown in Fig. 4. *Boss* (1988a) presented a strongly nonaxisymmetric model where the temperature at 1 A.U. varied by ~600 K with azimuthal angle, and suggested the possibility of periodic thermal cycling of solids. This might occur because solid bodies that are large enough to be decoupled from the gas will move on Keplerian orbits, while the pressure-supported gas (and the thermal distribution it defines) will rotate at a slightly slower rate. The cycling time would be on the order of $\sim 10^3$ years or more. *Morfill et al.* (1988) have recently suggested that thermal cycling on a much shorter timescale (~ hours) might result from a sequence of dust grain evaporation and recondensation.

Thermostatic Effect

Nebula temperatures are strongly dependent on the nebula opacity, which controls nebula cooling through radiative losses. The opacity in a hot inner solar nebula is controlled by iron dust grains, which vaporize around 1420 K (*Pollack et al.,* 1985). At higher temperatures (~1500 K), the absence of iron grains means that the opacity drops by a factor of ~100 (molecular hydrogen and other molecular and atomic species become the primary sources of opacity). Reduced opacity leads to increased radiative losses and hence increased cooling. Conversely, when temperatures drop below ~1500 K, the iron grains reform, the opacity rises again, radiative losses drop, and the temperature tends to increase. Thus the iron grains serve as an effective thermostat for the inner solar nebula, regulating the temperature to ~1500 K. The thermostatic effect is especially obvious in Fig. 5, where the adiabats for three quite different 3D solar nebula models all converge at inner nebula pressures. *Morfill et al.* (1988) have also argued for the existence of this thermostatic tendency.

While a hot inner solar nebula implies that sequential condensation of solids may have occurred, the thermostatic effect may have prevented the maximum nebula temperature from exceeding ~1500 K, and so may have led to the preservation of the most refractory relict interstellar grains (cf. *Huss,* 1988). Equilibrium condensation sequences may have to be reinterpreted in view of the likely survival of the most refractory material. Furthermore, equilibrium condensation need not have led to a strong compositional gradient in the terrestrial planet region,

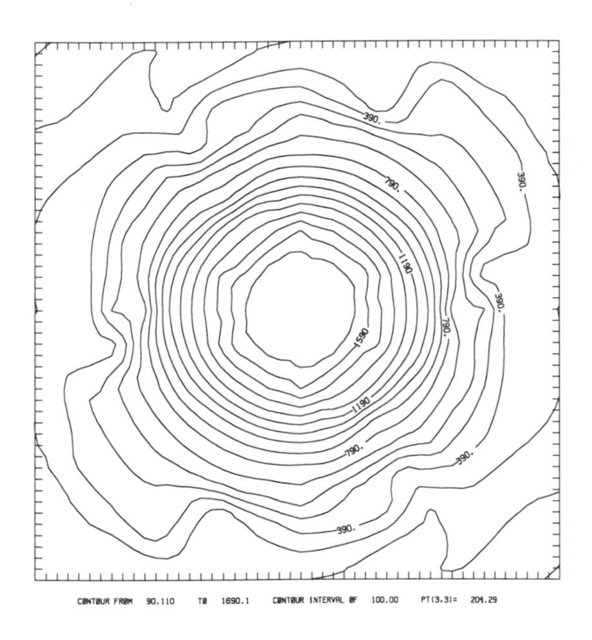

Fig. 4. *Temperature contours in the midplane of model C. Region shown is 20 A.U. across. Each contour corresponds to a change in temperature of 100 K. Contours are labeled with temperatures in K. Contours are missing from the innermost 0.5 A.U. because of the neglect of protosun heating in these 3D models. Mild thermal nonaxisymmetry is evident.*

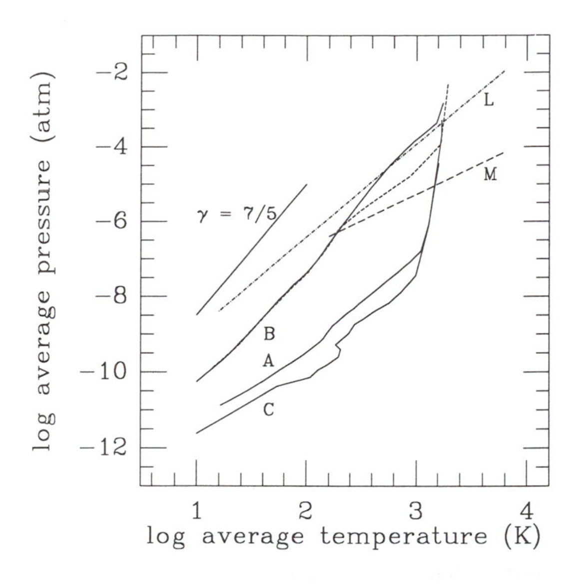

Fig. 5. *Pressure as a function of midplane temperature for three 3D models of the solar nebula. Model A (Boss, 1988a) has a protosun of mass 1.0 $M_\odot$ and a nearly axisymmetric nebula of mass 0.06 $M_\odot$. Model B (Boss, 1988a) has not formed a central protosun yet; the nebula is strongly bar-like with a mass of 1.0 $M_\odot$, though most of the mass (0.75 $M_\odot$) lies within 2 A.U. Model C is described in more detail in the captions to the previous figures. Pressures and temperatures are averaged in azimuthal angle except for the dashed segment of model B, which shows the variation along the bar. An arbitrary adiabat for molecular hydrogen ($\gamma = 7/5$) is shown, as well as the Larson (1969; L) adiabat and a generic viscous accretion disk profile (Wood and Morfill, 1988; M). The 3D models all converge at ~1500 K because of the thermostatic effect of the iron grain opacity.*

because the thermostatic effect suppresses any radial thermal gradient in the inner solar nebula, implying similar condensation sequences throughout the inner solar nebula.

CONCLUSIONS

The new 3D models of solar nebula formation calculated by *Boss* (1989) are consistent with the prevailing view that the Earth formed through the collisional accumulation of planetesimals. Planetary formation through giant gaseous planet formation does not appear to be possible in these models. While we thus seem to be converging on a generally acceptable model for Earth origin, it must be noted that Jupiter formation remains an outstanding problem in the accumulation theory, and until we believe that we also understand Jupiter's origin, we must feel somewhat uneasy about our current understanding of the formation of the Earth.

Considering the various approximations made in these 3D models, it must be emphasized that there is some uncertainty about the thermodynamical predictions of the models; improved numerical models are necessary before these predictions can be considered to be definitive. However, the available evidence suggests that the entire terrestrial planet region of the early solar nebula may well have experienced temperatures of 1500 K or more, with obvious implications for processes such as condensation during nebula cooling and volatile depletions. The high temperatures would probably be restricted to the earliest phases of nebula evolution (the first ~10^5 years), when significant infall of nebula material was still ongoing. In the absence of a complete chronology for planetesimal accumulation, the implications of these models for specific solar system objects must remain uncertain. Hot solar nebula models were first considered by *Cameron* (1962) in one of the pioneering solar nebula models, but have long since fallen into disfavor (e.g., *Wood and Morfill,* 1988; *Wood,* 1988), so it will be interesting to see whether high temperatures in the early inner solar nebula can be successfully resurrected and shown to be of importance for terrestrial planet formation.

Acknowledgments. I thank P. Cassen, B. Fegley, Jr., W. Ward, and G. Wetherill for valuable discussions about solar nebula models, and P. Cassen and W. Ward for detailed reviews leading to many improvements in the manuscript. This research was partially supported by the NASA Planetary Geology and Geophysics Program under grant NAGW-1410.

REFERENCES

Abt H. A. (1983) Normal and abnormal binary frequencies. *Annu. Rev. Astron. Astrophys., 21,* 343-372.

Bally J. (1982) Energetic activity in a star-forming molecular cloud core: a disk constrained bipolar flow in NGC 2071. *Astrophys. J., 261,* 558-568.

Bodenheimer P., Różyczka M., Yorke H. W., and Tohline J. E. (1988) Collapse of a rotating protostellar cloud. In *Formation and Evolution of Low Mass Stars* (A. K. Dupree and M. T. V. T. Lago, eds.), pp. 139-151. Kluwer, Dordrecht.

Boss A. P. (1984a) Protostellar formation in rotating interstellar clouds. IV. Nonisothermal collapse. *Astrophys. J., 277,* 768-782.

Boss A. P. (1984b) Angular momentum transfer by gravitational torques and the evolution of binary protostars. *Mon. Not. R. Astron. Soc., 209,* 543-567.

Boss A. P. (1985) Three dimensional calculations of the formation of the presolar nebula from a slowly rotating cloud. *Icarus, 61,* 3-9.

Boss A. P. (1986) Protostellar formation in rotating interstellar clouds. V. Nonisothermal collapse and fragmentation. *Astrophys. J. Suppl., 62,* 519-552.

Boss A. P. (1987) Protostellar formation in rotating interstellar clouds. VI. Nonuniform initial conditions. *Astrophys. J., 319,* 149-161.

Boss A. P. (1988a) High temperatures in the early solar nebula. *Science, 241,* 505-628.

Boss A. P. (1988b) Binary stars: formation by fragmentation. *Comm. Astrophys., 12,* 169-190.

Boss A. P. (1989) Evolution of the solar nebula. I. Nonaxisymmetric structure during nebula formation. *Astrophys. J., 345,* 554-571.

Boss A. P., Morfill G. E., and Tscharnuter W. M. (1989) Models of the formation and evolution of the solar nebula. In *Planetary and Satellite Atmospheres: Origin and Evolution* (S. K. Atreya and J. B. Pollack, eds.), pp. 35-77. Univ. of Arizona, Tucson.

Cameron A. G. W. (1962) The formation of the sun and planets. *Icarus, 1,* 13-69.

Cameron A. G. W. (1978) Physics of the primitive solar accretion disk. *Moon and Planets, 18,* 5-40.

Cameron A. G. W. (1988) Origin of the solar system. *Annu. Rev. Astron. Astrophys., 26,* 441-472.

Cassen P. M. and Moosman A. (1981) On the formation of protostellar disks. *Icarus, 48,* 353-376.

Cassen P. M. and Summers A. (1983) Models of the formation of the solar nebula. *Icarus, 53,* 26-40.

Cassen P. and Tomley L. (1988) The dynamical behavior of gravitationally unstable solar nebula models (abstract). *Bull. Am. Astron. Soc., 20,* 815.

Cassen P. M., Smith B. F., Miller R. H., and Reynolds R. T. (1981) Numerical experiments on the stability of preplanetary disks. *Icarus, 48,* 377-392.

Goldreich P. and Ward W. R. (1973) The formation of planetesimals. *Astrophys. J., 183,* 1051-1061.

Goldsmith P. F. and Arquilla R. (1985) Rotation in dark clouds. In *Protostars and Planets II* (D. C. Black and M. S. Matthews, eds.), pp. 137-149. Univ. of Arizona, Tucson.

Grossman L. (1972) Condensation in the primitive solar nebula. *Geochim. Cosmochim. Acta, 36,* 597-619.

Hourigan K. and Ward W. R. (1984) Radial migration of preplanetary material: Implications for the accretion timescale problem. *Icarus, 60,* 29-39.

Huss G. R. (1988) The role of presolar dust in the formation of the solar system. *Earth, Moon, and Planets, 40,* 165-211.

Larson R. B. (1969) Numerical calculations of the dynamics of a collapsing proto-star. *Mon. Not. R. Astron. Soc., 145,* 271-295.

Larson R. B. (1984) Gravitational torques and star formation. *Mon. Not. R. Astron. Soc., 206,* 197-207.

Levin B. J. (1978) Relative velocities of planetesimals and the early accumulation of planets. *Moon and Planets, 19,* 289-296.

Lin D. N. C. and Papaloizou J. (1980) On the structure and evolution of the primordial solar nebula. *Mon. Not. R. Astron. Soc., 191,* 37-48.

Lin D. N. C. and Papaloizou J. (1986) On the tidal interaction between protoplanets and the protoplanetary disk. III. Orbital migration of protoplanets. *Astrophys. J., 309,* 846-857.

Lin D. N. C. and Pringle J. E. (1987) A viscosity prescription for a self-gravitating accretion disk. *Mon. Not. R. Astron. Soc., 225,* 607-613.

Lissauer J. J. (1987) Timescales for planetary accretion and the structure of the protoplanetary disk. *Icarus, 69,* 249-265.

Morfill G. E. (1988) Protoplanetary accretion disks with coagulation and evaporation. *Icarus, 75,* 371-379.

Morfill G. E., Goertz C. K., and Havnes O. (1988) Thermal cycling and fluctuations in the protoplanetary nebula. *Icarus, 76,* 391-403.

Pollack J. B., McKay C. P., and Christofferson B. M. (1985) A calculation of the Rosseland mean opacity of dust grains in primordial solar system nebulae. *Icarus, 64,* 471-492.

Prinn R. G. and Fegley B. Jr.(1989) Solar nebula chemistry: Origin of planetary, satellite, and cometary volatiles. In *Origin and Evolution of Planetary and Satellite Atmospheres* (S. K. Atreya, J. B. Pollack, and M. S. Matthews, eds.), pp. 78-136. Univ. of Arizona, Tucson.

Safronov V. S. (1969) *Evolution of the Protoplanetary Cloud and Formation of the Earth and the Planets.* Nauka, Moscow. Translated by the Israel Program for Scientific Translation (1972).

Stemwedel S. W., Yuan C., and Cassen P. (1987) A model for self-gravitating disks resulting from the collapse of uniformly rotating clouds (abstract). *Bull. Am. Astron. Soc., 19,* 1092.

Stemwedel S. W., Yuan C., and Cassen P. (1990) Equilibrium models for self-gravitating inviscid disks resulting from the collapse of rotating clouds. *Astrophys. J.,* in press.

Stevenson D. J. and Lunine J. I. (1988) Rapid formation of Jupiter by diffusive redistribution of water vapor in the solar nebula. *Icarus, 75,* 146-155.

Toomre A. (1964) On the gravitational stability of a disk of stars. *Astrophys. J., 139,* 1217-1238.

Tscharnuter W. M. (1978) Collapse of the presolar nebula. *Moon and Planets, 19,* 229-236.

Tscharnuter W. M. (1987) A collapse model of the turbulent presolar nebula. *Astron. Astrophys., 188,* 55-73.

Walter F. M. (1988) Implications for planetary formation timescales from the nakedness of the low mass PMS stars. In *Formation and Evolution of Planetary Systems* (H. A. Weaver, F. Paresce, and L. Danly, eds.), pp. 71-78. Space Telescope Science Institute, Baltimore.

Ward W. R. (1986) Density waves in the solar nebula: Differential Lindblad torque. *Icarus, 67,* 164-180.

Ward W. R. (1989) Disc tides and the formation of giant planet cores (abstract). In *Lunar and Planetary Science XX,* pp. 1175-1176. Lunar and Planetary Institute, Houston.

Weidenschilling S. J. (1977) The distribution of mass in the planetary system and solar nebula. *Astrophys. Space Sci., 51,* 153-158.

Weidenschilling S. J. (1988). Formation processes and timescales for meteorite parent bodies. In *Meteorites and the Early Solar System* (J. F. Kerridge and M. S. Matthews, eds.), pp. 348-371. Univ. of Arizona, Tucson.

Wetherill G. W. (1980) Formation of the terrestrial planets. *Annu. Rev. Astron. Astrophys., 18,* 77-113.

Wood J. A. (1988) Chondritic meteorites and the solar nebula. *Annu. Rev. Earth Planet. Sci., 16,* 53-72.

Wood J. A. and Morfill G. E. (1988) A review of solar nebula models. In *Meteorites and the Early Solar System* (J. F. Kerridge and M. S. Matthews, eds.), pp. 329-347. Univ. of Arizona, Tucson.

Yuan C. and Cassen P. (1985) Protostellar angular momentum transport by spiral density waves. *Icarus, 64,* 435-447.

THERMAL HISTORY OF THE ASTEROID BELT: IMPLICATIONS FOR ACCRETION OF THE TERRESTRIAL PLANETS

M. J. Gaffey

Department of Geology, West Hall, Rensselaer Polytechnic Institute, Troy, NY 12180-3590

Asteroid surface compositions derived from the analysis of spectral and other remote-sensing data indicate that the planetesimals in the inner portions of the asteroid belt underwent strong postaccretionary heating that produced extensive melting and magmatic differentiation. The intensity of the heating process—or the susceptibility of the bodies to that process—fell off steeply with increasing heliocentric distance, from 100% igneous bodies at 2.0 A.U. to 0% at 3.5 A.U. Radiometric ages of igneous meteorites and the limitations on the heating of small bodies make it probable that the heating of these planetesimals occurred during the first few million years of solar system history. It may have been produced by electrical induction during the T-Tauri phase of solar evolution or perhaps by decay of the shortlived radioisotope ^{26}Al. Heating took place after the growth of these planetesimals to sizes comparable with those of the present asteroids. It seems probable that most—if not all—of inner solar system planetesimals underwent a similar strong early heating. Models of terrestrial planet formation, thermal evolution, and core-mantle segregation that involve accretion times of more than a few million years from a population of asteroid-sized (or larger) planetesimals should assume that the accreting population was composed primarily of differentiated objects with chondritic bulk compositions and hot interiors.

INTRODUCTION

Models for the origin and early thermal evolution of the Earth and other terrestrial planets have generally assumed that these bodies accreted from a population of chondritic planetesimals (e.g., *Hanks and Anderson,* 1969; *Toksöz et al.,* 1978; *Stevenson,* 1981; *Abe and Matsui,* 1986). This "chondritic" model actually includes several distinct and independent assumptions concerning the nature of the population of accreting objects. In the strictest sense, "chondritic" refers only to the bulk composition and stipulates relative abundances of the nonvolatile elements similar to those of the sun. However, the term "chondritic" also invokes the textural and genetic properties characteristic of undifferentiated meteorite assemblages.

Based upon our knowledge of the bulk composition of the Earth, the assumption that it formed from a reservoir with chondritic chemistry appears justified (e.g., *Anderson,* 1989). However, this should not be taken a priori to indicate that the accreting population also shared those chondritic textural and genetic properties. An intact differentiated planetesimal will have essentially the same bulk composition as its chondritic precursor, although its internal structure and properties will have been greatly changed by the melting and gravitational segregation of phases associated with the magmatic differentiation processes.

Models for the formation of the terrestrial planets have not generally distinguished between the explicit compositional assumption and the implicit textural/temperature assumptions involved in the accretion of "chondritic" planets. With a few exceptions (e.g., *Smith,* 1979; *Stevenson,* 1981; *Hutchison,* 1982) these models have assumed that the accreting population is composed exclusively of undifferentiated planetesimals. Although the differentiated or undifferentiated nature of the accreting bodies should have no significant effect on the major element composition of the final planet, it could have profound effects on the inventory of volatile elements and upon the isotopic patterns in trapped noble gases. Also, unless a significant cooling time elapses prior to accretion into the terrestrial bodies, such differentiated planetesimals would have hot interiors and could contribute significantly to the thermal energy budget of the growing planet, especially during the early stages of accretion.

Two general types of terrestrial accretion models have arisen based on assumptions concerning the retention of accretional energy within the growing planet. The earlier "iron catastrophe"-type endmember model (no significant retention of accretional energy) assumes that at the end of accretion the interior of the body is composed of an intimate metal-silicate mixture at subsolidus temperatures (e.g., *Hanks and Anderson,* 1969). Subsequent heating from the decay of long-lived radionuclides eventually raised the internal temperature above the Fe-FeS eutectic, producing a dense melt phase. The downward migration of this dense liquid converted the gravitational potential energy of dense phases initially located at shallow depths into heat through frictional interaction with the underlying layers. Models suggest that this would have been a positive feedback process that rapidly raised the temperature of the Earth's interior by 1500-2000 K. In this type of model the Earth was "turned on"—i.e., transformed from solid planet to a molten one—approximately 0.5-2.0 G.y. after its formation, producing core-mantle segregation and initiating outgassing as a significant contributor to the atmosphere and oceans.

The more recent "accretionary heating"-type endmember models assume that after the planetary nucleus had grown to a critical size, the retention of accretional energy was sufficient to melt or partially melt the near-surface regions, resulting in a primordial Earth with a cold chondritic core beneath a hot and perhaps differentiated mantle. Outgassing would have begun during the late stages of accretion, providing an early ocean and atmosphere. In these models, core formation occurred subsequent to the late stages of accretion, perhaps well after the end of accretion, as the gravitationally unstable density distribution was overturned to form the core-mantle system (e.g., *Stevenson,* 1981; *Ida et al.,* 1987).

In both classes of models, the "chondritic" assumption has generally been taken to indicate cool planetesimals whose internal heat did not contribute significantly to the thermal budget of the accreting planet. Based upon the properties of the planetesimal population inferred from petrologic studies of the meteorites, this was both a reasonable and a plausible assumption. However, recent asteroid studies indicate that most—if not all—of inner solar system planetesimals underwent strong heating in early short-lived thermal events. If the Earth and other terrestrial planets underwent accretion on timescales longer than a few million years, they would postdate the early transient heat sources and would not be directly affected. However, models that invoke terrestrial planet accretion from a population of asteroid-sized (or larger) bodies should assume that these bodies consisted primarily of differentiated planetesimals of chondritic bulk compositions with hot interiors resulting from the action of the early transient heat source or sources.

METEORITES, ASTEROIDS, AND THE EARLY SOLAR SYSTEM

The meteorites and the asteroids are generally presumed to be the surviving remnants of the inner solar system planetesimal population from which the terrestrial planets accreted. It has long been known that one or more episodic thermal events occurred in the early solar system. The meteorites exhibit the effects of a wide range of very early heating events. While some meteorites escaped any significant heating (e.g., $T_{max} < 200°C$), others were heated sufficiently to produce total melting of the metal and silicate phases ($T > 1600°C$; *Takahashi,* 1983). The spatial distribution (e.g., heliocentric distances) of early solar heating events based upon meteoritic evidence is quite model-dependent and is therefore uncertain. The meteorites do, however, provide excellent chronologies of those events, indicating that they occurred very early in the history of the solar system, probably within the first few million years. For detailed reviews, see *Tilton* (1988), *Podosek and Swindle* (1988), and the references therein.

A number of igneous meteorites, including many of the basaltic achondrites, have radiometric ages very close (< 20 m.y.) to the age of the solar system derived from the study of chondritic meteorites. Their mineralogies are the result of the differentiation of solid and liquid components by density-controlled gravitational segregation. The antiquity of such differentiated meteorites requires that planetesimals of significant size (at least tens to hundreds of kilometers) existed at the time of the heating event in order to provide gravitational forces sufficient to segregate the phases with different densities. The cooling times to the closure of these radiometric systems—the point at which the sample is cool enough to begin to retain the relevant isotope or daughter product—even in the relatively near-surface regions characteristic of such lithologies may be a few million years or more. Therefore, the timescale of the early heating event and of planetesimal growth must have been very short. Only a very limited set of heating mechanisms (e.g., short-lived isotopes such as ^{26}Al or electrical induction heating during a T-Tauri phase of solar evolution) can be plausibly invoked to produce the early thermal episodes recorded in the meteorites (e.g., *Lee et al.,* 1976, 1977; *Herbert and Sonett,* 1979; *Sonett and Reynolds,* 1979; *Herbert,* 1989).

Meteorites are generally presumed to be samples of some subset of the present asteroid population. The meteorite evidence therefore suggests that the asteroids suffered a range of thermal events. Since the igneous meteorites appear to have originated in the asteroid belt (see, for example, *Wetherill and Chapman,* 1988 and references therein) and since the asteroids and the meteorite parent bodies appear to have comparable sizes, the chronology of meteorite parent body heating events is assumed to also apply to any igneous asteroids.

COMPOSITION AND THERMAL HISTORY OF THE ASTEROID TAXONOMIC CLASSES

More than 600 of the larger asteroids have been classified into a variety of types based upon color and albedo data (e.g., *Bowell et al.,* 1978; *Tholen,* 1984; *Barucci et al.,* 1987; *Tedesco et al.,* 1989; see also the review by *Tholen and Barucci,* 1989). Analysis of visible and near-infrared spectral reflectance data obtained for individual members of each type has produced general mineralogical characterizations for these asteroid classes from which the probable thermal history can be inferred. This work has been reviewed in *Gaffey et al.* (1989). The probable asteroid compositions and their meteoritic analogues for each taxonomic type are summarized in Table 1.

Although the larger asteroid classes are likely to include a diverse range of assemblages and to have experienced a variety of thermal histories (e.g., primitive C3 assemblages as a subtype within the predominantly igneous S-type asteroids; *Bell et al.,* 1987, 1989; *Bell,* 1988), useful generalizations can be made concerning the compositional nature of each asteroid class. Although indirect, these remote-sensing characterizations are relatively unambiguous concerning the existence of a number of strongly heated asteroids (e.g., *Gaffey et al.,* 1989; *Bell et al.,* 1989; and references therein). Since these small bodies have large surface-to-volume ratios and little accretional

TABLE 1. *General mineralogy and meteoritic analogues for the asteroid taxonomic types.*

Taxonomic Types	Surface Mineralogy	Meteoritic Analogues
A [4]	Olivine or olivine-metal	Olivine achondrite or pallasite (igneous)
B, C, F, G [112]	Hydrated silicates + carbon/organics/opaques	CI1-CM2 assemblages and assemblages produced by aqueous alteration and/or metamorphism of CI/CM precursor materials
D, P [49]	Carbon/organic-rich silicate (anhydrous?) assemblages	Organic-rich cosmic dust grains? CI1-CM2 plus organics? C3 plus organics? (primitive)
E [8]	Enstatite or possibly other iron-free silicates	Enstatite achondrites (igneous)
M [21]	Nickel-iron metal possibly with trace silicates	Iron meteorites (igneous) Enstatite chondrites? (primitive)
Q [1]	Olivine + pyroxene + metal	Ordinary chondrites (primitive)
R [1]	Pyroxene + olivine	Pyroxene-olivine achondrite (igneous)
S [144]	Metal → olivine → pyroxene	Pallasites with accessory pyroxene; olivine-dominated stony-irons, ureilites and primitive achondrites (igneous); small component of CV/CO chondrites and possibly ordinary chondrites (primitive)
K [7]	Olivine + pyroxene + carbon(?)	C3 chondrites
V [1]	Pyroxene + feldspar	Basaltic achondrites (igneous)
T [4]	Anhydrous silicates(?) Carbon/ organic compounds(?)	Organic-rich cosmic dust grains? C3 plus organics? (primitive) Metamorphosed CI-CM?

Asteroid classification from *Tholen* (1984). The number in brackets is the number of classified asteroids classified as that type by *Tholen* (1984). Number of K-type asteroids is from *Tedesco et al.* (1989). Mineralogy and meteoritic analogues are from *Gaffey et al.* (1989) and references therein.

energy release, exotic heat sources are generally required for any significant heating. Similar conditions existed for the meteorite parent bodies, which indicates the plausibility of such exotic heat sources even if the genetic relationships between individual meteorite types and asteroids remain obscure.

The strongly heated taxonomic types are identified by the presence of surface materials produced by igneous processes (melting and segregation of different phases) within their parent planetesimals. The surface assemblages of these asteroids are analogous to the meteorites formed by igneous processes (i.e., the irons, stony-irons, and achondrites). Among the 15 asteroid types, 6 are predominantly differentiated: types A, V, R, M, E, and S. The igneous natures of asteroid types A, E, V, M, and R are widely—and to an extent, uncritically—accepted by the asteroid community, even though the strength of the evidence varies considerably between these different classes. Moreover, there may be assemblages with diverse histories within a single class. Objects such as asteroid 4 Vesta (type V) with its basaltic surface assemblage and 16 Psyche (type M) with a metallic spectrum and extremely high radar reflectivity (*Ostro et al.,* 1985) can only plausibly be produced by igneous processes. The relatively low radar reflectivity of some other M-type asteroids as well as their photometric properties suggests assemblages more similar to the undifferentiated enstatite chondrites (e.g., *Lupishko and Belskaya,* 1989).

The igneous interpretation of the S-type asteroids has generated significant controversy. Of all the asteroid types inferred to have igneous affinities, the S-type asteroids are the most common, dominating the inner belt population, and will have the greatest impact on models of the thermal history of the asteroid belt. The spectral data for S-type asteroids have been interpreted by some investigators as indicating primitive assemblages analogous to the ordinary chondritic meteorites, and have been interpreted by others as indicating igneous assemblages analogous to the stony-iron meteorites. *Gaffey* (1984) and *Gaffey et al.* (1989) have reviewed the history of this controversy. Since this interpretation strongly affects the subsequent discussion, a brief review of the S-type asteroids is in order.

THE NATURE OF THE S-TYPE ASTEROIDS

The reflectance spectra of S-type asteroids indicate surface assemblages composed of NiFe metal, olivine, and pyroxene. This is not diagnostic of evolutionary history since both ordinary chondritic (undifferentiated) and stony-iron (differentiated) meteoritic assemblages exist with these phases present. While there are meteoritic and dynamical reasons to prefer that the S-type asteroids are similar to the ordinary chondrites (e.g., *Wetherill,* 1985), the telescopic investigations of specific main belt S-type bodies have to date produced no cases where the data require or even prefer the ordinary chondritic interpretation. Conversely, as discussed below, most of the specific data on S-type asteroids favor—and in some cases, require—the presence of differentiated surface assemblages.

A few of the S-type asteroids have been subjected to detailed spectral study (e.g., 8 Flora: *Gaffey,* 1984; 15 Eunomia: *Gaffey and Ostro,* 1987). The igneous nature of these bodies has been established by the nature of their rotational spectral variations. The spectral reflectance curve of the largest S-type asteroid, 15 Eunomia, is shown on Fig. 1. The absorption features near 1 and 2 μm arise from crystal field transitions of iron in the mafic minerals, pyroxene and olivine. The wavelength position of these features is controlled by the compositions and relative abundance of both minerals in the case of the 1 μm feature, and by the composition of the pyroxene in the case of the 2 μm feature.

The individual points on Fig. 2 show the wavelength position of the 2 μm absorption feature relative to the ratio of the areas (total absorbance) for the 1 and 2 μm absorption features in the reflectance spectrum of 15 Eunomia as the object rotated to expose different portions of its surface. The values obtained for sequential rotational aspects are connected by the light lines. The heavy line represents the error-weighted, linear least-squares fit to this set of points. The stippled area is the observed region for nebular

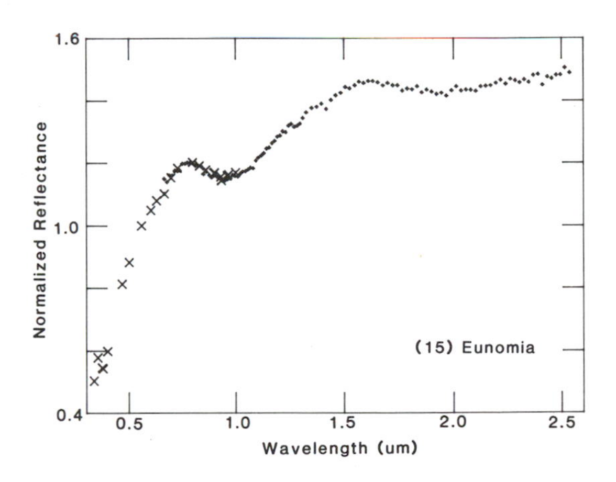

Fig. 1. *Spectral reflectance curve for the S-type asteroid 15 Eunomia (Gaffey and Ostro, 1987). The spectral curve has been normalized to unity at 0.56 μm. Errors are plotted for all points, but are often smaller than the symbols. The overall increase in reflectance toward longer wavelengths is attributed to NiFe metal, while the broad absorption features near 1 μm and 2 μm are due to olivine and pyroxene.*

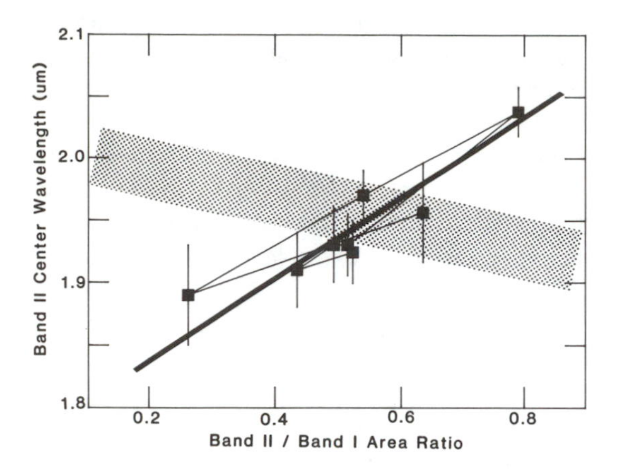

Fig. 2. *Relative variations in the spectral parameters of 15 Eunomia with rotation. Sequential rotational aspects are joined with a light line. The linear least-squares fit to these points is shown by the heavy solid line. Band II (2 μm) center is primarily a function of pyroxene composition. Band II/Band I area ratio is a function of the relative abundances of pyroxene and olivine. The negative slope of the stippled area (the same parameters for ordinary chondritic assemblages) is an intrinsic consequence of the undifferentiated nature of the chondrites. The positive slope for the observed variations of Eunomia rules out an undifferentiated surface assemblage.*

anhydrous silicate assemblages, such as the chondritic meteorites, that have not undergone melting and differentiation.

Among the chondrites, the abundance of olivine increases at the expense of pyroxene and metallic iron by a reaction under oxidizing conditions that can be represented as

$$\text{Fe [metal]} + \text{O} \rightarrow \text{FeO and}$$
$$\text{FeO} + (\text{FeO,MgO})\text{SiO}_2 \text{ [pyroxene]} \rightarrow$$
$$(\text{FeO,MgO})_2\text{SiO}_2 \text{ [olivine]}$$

Under reducing conditions the reaction proceeds in the opposite direction. Thus within the chondritic suite, increasing olivine content correlates strongly with decreasing pyroxene and metal contents and with increasing iron content in the mafic silicates (e.g., *Gaffey,* 1984). Although all chondritic materials do not have a single bulk composition, their range of variations [e.g., Fe/Si(atomic) = 0.66-0.98; Mg/Si = 0.73-1.06; *Mason,* 1979] is much smaller than that seen among achondritic and differentiated meteorites that are products of open magmatic systems (i.e., major elements and compounds can be gained or lost by local regions). The compositions and relative abundances of the mafic silicate minerals and metal in differentiated (nonchondritic) assemblages are not correlated with each other in any particular manner.

For any assemblage with a chondritic composition, there will exist a general functional relationship between the abundances and the compositions of the mafic and metallic phases that will produce correlated variations between the spectral parameters of such assemblages. In the spectra of olivine-orthopyroxene mixtures, the band area ratio increases with increasing pyroxene abundance (e.g., *Cloutis et al.,* 1986), while the 2 μm band center shifts toward longer wavelength with increasing iron content (e.g., *Adams,* 1974). In Fig. 2 the correlation between chondritic mafic silicate abundance and compositions produces the negative slope (stippled region) observed for these assemblages. By contrast the slope of the Eunomia distribution is positive. While calibrational uncertainties, systematic errors in the observational data, and artifacts of interpretive methodologies may all cause inaccuracies in the spectral parameters, none of these can materially affect the slope of the distribution. The presence of a chondritic slope for the observed points would be necessary, but not sufficient, to identify an undifferentiated assemblage. A slope with the wrong sign is sufficient to unambiguously eliminate a chondritic assemblage.

The most plausible means of producing the observed slope is by magmatic processes within the parent body of Eunomia. In such a system, mineral species and melt phases move toward the core or the surface depending upon their density and upon the viscosity or permeability of the surrounding medium.

The addition or removal of the associated elements and compounds in the various radial zones within the parent object results in an open system that is no longer constrained to follow the chondritic pattern of mineral abundance and mineral composition.

Almost any slope or distribution of points is possible due to the open system behavior of the processes that produced the differentiated assemblages. It is a fortuitous consequence of the scale and distribution of surface exposures of different igneous lithologies that the Eunomia points lie along a relatively straight line. The distribution of points for asteroid 8 Flora shows a quite different pattern (*Gaffey,* 1984), indicating a different distribution of surface exposures. This is analogous to the much greater mineralogic and chemical diversity seen in the differentiated meteorites than in the chondrites.

This type of detailed study of individual S-type asteroids is a very slow process. Seven less sensitive but more widely available diagnostic tests (Table 2; *Gaffey et al.,* 1990) can be applied to the approximately 150 S-type asteroids. Each of these either favors or is consistent with most S-type asteroids being differentiated assemblages. Although several of these tests are consistent with the alternative interpretation, none favor it.

For example, an olivine and metal-rich surface is readily produced by the exposure of the core-mantle boundary zone within a differentiated body, but cannot be reconciled

TABLE 2. *Diagnostic criteria for differentiated or undifferentiated S-type asteroids.*

Criteria	Implications	Reliability	Indication
High spectral abundance of metal (NiFe)	Metal abundance above chondritic range	Moderate	S-types mostly differentiated unless some surface or regolith process enriches or spectrally enhances apparent metal abundance
High olivine/pyroxene ratio (Ol/Px)	Generally above chondritic range with some overlap	Low	Generally favors S-types as differentiated but ambiguous between high olivine chondrites and achondrites
Iron content of pyroxene (Fs, ferrosilite)	Generally above chondritic range with some overlap	Moderate	S-types mostly differentiated but some chondritic assemblages may be present
Simultaneously high NiFe, Ol/Px and Fs	Mutually inconsistent for chondrites	Strong	S-types mostly differentiated
Relative variation in apparent metal and olivine abundances with heliocentric distance	Inverse of pattern for simple nebula [$T,P \propto r^{-n}$]	Strong	S-types mostly differentiated or large nebula contained large temperature and/or redox inversions
Systematically high 12/25 μm infrared flux ratios	High surface metal abundances	Moderate	S-types mostly differentiated
S-type asteroid radar albedos are generally intermediate between the C- and M-types	Relatively abundant surface metal	Low	S-type asteroids mostly differentiated but interpretation is very model dependent

with an undifferentiated body without invoking *ad hoc* regolith processes. Similarly, the metal abundance is apparently high at many scales (the submicron scale of visible light and the centimeter scale of the radar data). This is very difficult to reconcile with the low physical abundance and small grain sizes of metal in chondrites. Although all available direct evidence on the S-type asteroids indicates that they are predominantly differentiated, the presence of a minor component of undifferentiated bodies within the S-type population cannot be eliminated by present data (e.g., the new K class proposed by *Bell,* 1988; *Bell et al.,* 1989; *Tedesco et al.,* 1989). The most conservative approach is to infer that the bulk of the S-type asteroid population must have undergone a strong postaccretionary heating episode.

THE HELIOCENTRIC DISTRIBUTION OF IGNEOUS ASTEROIDS

It has been known for some time that there are systematic trends in the spatial distribution of asteroidal taxonomic types (e.g., *Chapman et al.,* 1975; *Zellner and Bowell,* 1978; *Bowell et al.,* 1978; *Tholen,* 1984; *Gradie et al.,* 1989). A number of attempts have been made to use the distribution of asteroid types to constrain the nature and processes of the late solar nebula and early solar system (e.g., *Gradie and Tedesco,* 1982; *Tholen,* 1984; *Zellner et al.,* 1985a,b; *Bell,* 1986; *Gradie et al.,* 1989; *Bell et al.,* 1989). The conclusions of such analyses are critically dependent upon the particular assumptions that were made concerning the compositional nature of each taxonomic type.

The distribution of asteroid classes with heliocentric distance has been interpreted as the signature of both a nebular compositional gradient and of early postaccretionary heating events. *Gradie and Tedesco* (1982) assumed that the S-type asteroids were undifferentiated chondritic assemblages and interpreted the change from an S-dominated inner belt to a C-dominated outer belt as the fossil signature of the nebular compositional gradient, with increasing oxidation state and lower temperatures at greater heliocentric distances across this interval.

Bell (1986; see also *Bell et al.,* 1989) defined three asteroid superclasses—primitive, metamorphic, and igneous—based upon the processes required to produce the inferred asteroid surface assemblages. The igneous superclass (taxonomic types V, R, S, A, M, E) include those objects whose surface mineralogy require melting and magmatic differentiation within their parent planetesimal. The metamorphic superclass (types B, G, F, T) exhibit surface assemblages indicating subsolidus aqueous alteration and thermal metamorphism. The primitive superclass (types D, P, C, K, Q) are consistent with unaltered nebular (chondritic) material. The absence of hydrated silicate minerals in the P and D types of the outer belt is interpreted as an indication that these objects never achieved postaccretionary temperatures sufficient to melt any included water ice (*Jones,* 1988; *Lebofsky et al.,* 1989, 1990).

Bell (1986) and *Bell et al.* (1989) showed the relative abundance of his superclass objects binned into seven intervals of heliocentric distance between 1.9 and 5.2 A.U. The relative abundance of the igneous superclasses decreased in a nearly linear manner from approximately 100% at 1.9 A.U to 0% at 3.4 A.U. The relative abundance of the primitive superclass increased from approximately 0% to 100% over the same interval, reaching 100% at 4.0 A.U. The metamorphic superclass showed a broad low distribution peaking at about 10% near 3.0 A.U. It was concluded that this pattern indicated a planetesimal heating mechanism whose intensity declined with increasing heliocentric distance.

Gaffey (1988) divided the asteroid types classified by *Tholen* (1984) into igneous (those having surface assemblages resulting from at least partial melting) and nonigneous types based upon spectral analysis of a few members of each type. The igneous group included asteroid types S, M, A, E, V, and R. The relative proportions of the igneous objects among 423 classified asteroids in 11 dynamically defined asteroid zones (*Zellner et al.,* 1985b) are shown vs. heliocentric distance on the upper half of Fig. 3. The solid line is a linear least squares fit to the points between 1.9 and 3.5 A.U. For comparison, the lower portion of the figure shows a histogram of total asteroid number density vs. heliocentric distance. From 1.9 to 3.5 A.U., across the interval of the main belt, the proportion of objects that underwent strong (T > 1000°C) postaccretionary heating decreases linearly from 100% to 0%, with a slope of approximately 67% per A.U. Based upon plausible heating mechanisms for such small bodies, the chronology of meteorite heating events, and the size requirements for density-controlled magmatic segregation by gravitation forces within the planetesimals, these temperatures seem to have been attained very early in solar system history (the first few million years) but after the parent planetesimals had grown to the present sizes or somewhat larger.

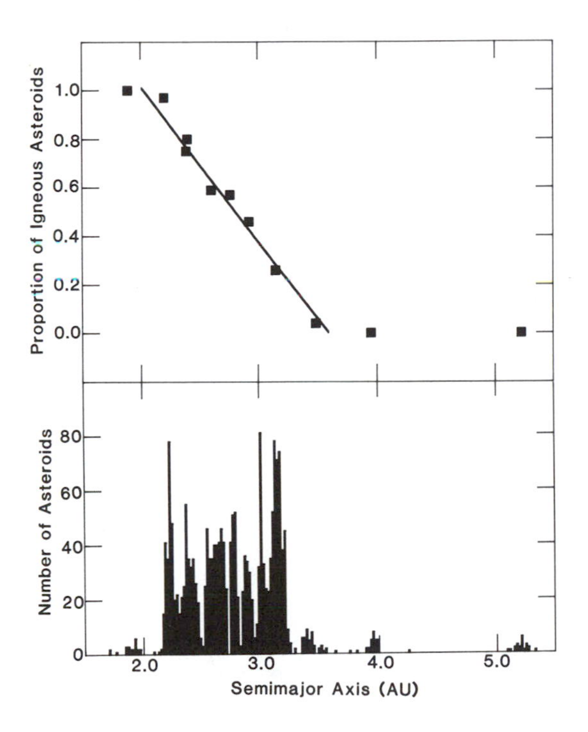

Fig. 3. Top: *The relative proportion of igneous bodies (types A, E, M, S, R, V) among 423 classified asteroids (Tholen, 1984) as a function of semimajor axis for the 11 asteroid belt zones defined by Zellner et al. (1985b). The solid line is the linear least squares fit for the nine points at 3.5 A.U. or less.* **Bottom:** *The distribution of asteroids nos. 1–2016 as a function of heliocentric distance binned into 0.02 A.U. intervals (Gaffey, 1988).*

The asteroid compositional variation with heliocentric distance appears to be primarily a function of the intensity of (or susceptibility to) a very early heating mechanism, perhaps modified by a radial gradient in the composition of the initial material preserved from the solar nebula. The simplest model would invoke a sun-related heat source with a strong dependence on heliocentric distance. A non-sun-centered heat source such as ^{26}Al or other short-lived radioisotopes can be made consistent with the asteroidal pattern, but would require the additional complexity of an accretionary wave moving outward through the early solar system, such that the inner belt bodies accreted early and incorporated a large amount of the heating element while the outer belt bodies accreted several halflives later and incorporated little or none of the heating element. This seems less probable than a single-stage process. It is also possible that both processes were operative to some extent.

THERMAL EVOLUTION OF INNER SOLAR SYSTEM PLANETESIMALS

Unless accretion of the planetesimals in the inner solar system was delayed until after the exotic heating episode had ceased or unless the heating was not active at smaller heliocentric distances, it seems probable that all planetesimals formed inside 2 A.U. were subjected to this strong heating event and that they underwent differentiation. Based upon the nature of plausible heat sources in small bodies and the chronology of the differentiated meteorites, this occurred during the first few million years of solar system history. If accretion of the terrestrial planets took place from a population of asteroid-sized (or larger) planetesimals on a timescale of 10^7-10^8 years as commonly proposed (see the review by *Wetherill,* 1980), then those planetesimals would have undergone a strong thermal episode prior to accretion. Since the timescale for significant internal cooling of planetesimals in this size range is long (e.g., $>10^8$ years; *Lazarewicz and Gaffey,* 1980; *Gaffey and Lazarewicz,* 1990) compared to the proposed duration of planetary accretion, the bulk of the material contributed by the planetesimals will be near or above 1000°C, and substantial fractions (10-90% depending upon planetesimal size and time of accretion) of the accreting material should be in the form of magma and liquid metal.

Models for the accretion and early evolution of the terrestrial planets should consider hot planetesimals as the baseline case. The presence of high internal temperatures in the growing terrestrial planets is not dependent on their reaching a size where retention of impact heat is significant. The presence of significant amounts of dense molten metallic phases should permit gravitational core-mantle segregation to begin with the aggregation of the first two planetesimals to form the Earth's embryo. The internal thermal energy of the planetesimal population should dominate the internal heat budget of the growing Earth until it reaches a size where retention of heat deposited by the impacting projectiles becomes significant.

Most of the accreting material will have undergone at least some degree of degassing prior to incorporation into the terrestrial planets, and the exposure of planetesimal magmas during impacts onto the terrestrial planets during the latter stages of accretion may incorporate or lose particular suites of gases from or to any ambient atmosphere. The volatile and rare gas inventory of the Earth thus may represent the net results of several different processes. These consequences of the accretion of hot, differentiated planetesimals remain to be explored.

Acknowledgments. Aspects of this research were supported by NSF (Solar System Astronomy) grant AST-8616634 and by NASA (Planetary Geology and Geophysics) grant NAGW-642.

REFERENCES

Abe Y. and Matsui T. (1986) Early evolution of the Earth: Accretion, atmosphere formation, and thermal history. *Proc. Lunar Planet. Sci. Conf. 17th,* in *J. Geophys. Res., 91,* E291-E302.

Adams J. B. (1974) Visible and near-infrared diffuse reflectance spectra of pyroxenes as applied to remote sensing of solid objects in the solar system. *J. Geophys. Res., 79,* 4829-4836.

Anderson D. L. (1989) Composition of the Earth. *Science, 20,* 367-370.

Barucci M. A., Capria M. T., Coradini A., and Fulghignoni M. (1987) Classification of asteroids using G-mode analysis. *Icarus, 72,* 304-324.

Bell J. F. (1986) Mineralogical evolution of the asteroid belt (abstract). In *Meteoritics, 21,* 333-334.

Bell J. F. (1988) A probable asteroidal parent body for the CV or CO chondrites (abstract). In *Meteoritics, 23,* 256-257.

Bell J. F., Hawke B. R., and Owensby P. D. (1987) Carbonaceous chondrites from S-type asteroids (abstract). In *Bull. Am. Astron. Soc., 19,* 841.

Bell J. F., Davis D. R., Hartmann W. K., and Gaffey M. J. (1989) Asteroids: The big picture. In *Asteroids II* (R. P. Binzel, T. Gehrels, and M. S. Matthews, eds.), pp. 921-945. Univ. of Arizona, Tucson.

Bowell E., Chapman C. R., Gradie J. C., Morrison D., and Zellner B. (1978). Taxonomy of asteroids. *Icarus, 35,* 313-335.

Chapman C. R., Morrison D., and Zellner B. (1975) Surface properties of asteroids: A synthesis of polarimetry, radiometry, and spectrophotometry. *Icarus, 25,* 104-130.

Cloutis E., Gaffey M. J., Jackowski T. L., and Reed K. L. (1986) Calibrations of phase abundance, composition, and particle size distribution for olivine-orthopyroxene mixtures from reflectance spectra. *J. Geophys. Res., 91,* 11641-11653.

Gaffey M. J. (1984) Rotational spectral variations of asteroid (8) Flora: Implications for the nature of the S-type asteroids and for the parent bodies of the ordinary chondrites. *Icarus, 60,* 83-114.

Gaffey M. J. (1988) Thermal history of the asteroid belt: Implications for accretion of the terrestrial planets (abstract). In *Lunar and Planetary Science XIX*, pp. 369-370. Lunar and Planetary Institute, Houston.

Gaffey M. J. and Lazarewicz A. R. (1990) The thermal, tectonic, and mineralogical evolution of small planetary objects. *Icarus,* in press.

Gaffey M. J. and Ostro S. J. (1987) Surface lithologic heterogeneity and body shape for asteroid (15) Eunomia: Evidence from rotational spectral variations and multi-color lightcurve inversions (abstract). In *Lunar and Planetary Science XVIII,* pp. 310-311. Lunar and Planetary Institute, Houston.

Gaffey M. J., Bell J. F., and Cruikshank D. P. (1989) Reflectance spectroscopy and asteroid surface mineralogy. In *Asteroids II* (R. P. Binzel, T. Gehrels, and M. S. Matthews, eds.), pp. 98-127. Univ. of Arizona, Tucson.

Gaffey M. J., Bell J. F., Brown R. H., and Burbine T. (1990) Mineralogical variations within the S-asteroid population (abstract). In *Lunar and Planetary Science XXI,* pp. 399-400. Lunar and Planetary Institute, Houston.

Gradie J. and Tedesco E. (1982) Compositional structure of the asteroid belt. *Science, 216,* 1405-1407.

Gradie J. C., Chapman C. R., and Tedesco E. F. (1989) Distribution of taxonomic classes and the compositional structure of the asteroid belt. In *Asteroids II* (R. P. Binzel, T. Gehrels, and M. S. Matthews, eds.), pp. 316-335. Univ. of Arizona, Tucson.

Hanks T. C. and Anderson D. L. (1969) The early thermal history of the Earth. *Phys. Earth Planet. Inter., 2,* 19-29.

Herbert F. (1989) Primordial electrical induction heating of asteroids. *Icarus, 78,* 402-410.

Herbert F. and Sonett C. P. (1979) Electromagnetic heating of minor planets in the early solar system. *Icarus, 40,* 484-496.

Hutchison R. (1982) Meteorites—evidence for the interrelationships of materials in the solar system of 4.55Ga ago. *Phys. Earth Planet. Inter., 29,* 199-208.

Ida S., Nakagawa Y., and Nakazawa K. (1987) The Earth's core formation due to Rayleigh-Taylor instability. *Icarus, 69,* 239-248.

Jones T. D. (1988) An infrared reflectance study of water in outer belt Asteroids: Clues to composition and origin. Ph.D. thesis, Univ. of Arizona, Tucson. 281 pp.

Lazarewicz A. R. and Gaffey M. J. (1980) Thermal, stress and mineralogic evolution of small planetary objects (abstract). In *Lunar and Planetary Science XI,* pp. 616-620. Lunar and Planetary Institute, Houston.

Lebofsky L. A., Jones T. D., and Herbert F. (1989) Asteroid volatile inventories. In *Origin and Evolution of Planetary and Satellite Atmospheres* (S. K. Atreya, J. B. Pollack, and M. S. Matthews, eds.), pp. 192-229. Univ. of Arizona, Tucson.

Lebofsky L. A., Jones T. D., Owensby P. D., Feierberg M. A., and Consolmagno G. J.(1990) The nature of low albedo asteroids from 3-μm spectrophotometry. *Icarus, 83,* 16-26.

Lee T., Papanastassiou D. A., and Wasserburg G. J. (1976) Demonstration of ^{26}Mg excess in Allende and evidence for ^{26}Al. *Geophys. Res. Lett., 3,* 109-112.

Lee T., Papanastassiou D. A., and Wasserburg G. J. (1977) ^{26}Al in the early solar system: Fossil or fuel? *Astrophys. J. Lett., 228,* L93-L98.

Lupishko D. F. and Belskaya I. N. (1989) On the surface composition of the M-type asteroids. *Icarus, 78,* 395-401.

Mason B. (1979) Chapter B. Cosmochemistry, Part 1. Meteorites. In *Data of Geochemistry,* 6th edition (M. Fleischer, ed.), pp. B29-B31 and B46-B48. U.S. Geol. Surv. Prof. Pap. 440-B-1.

Ostro S. J., Campbell D. B., and Shapiro I. I. (1985) Mainbelt asteroids: Dual-polarization radar observations. *Science, 229,* 442-446.

Podosek F. A. and Swindle T. D. (1988) Extinct radionuclides. In *Meteorites and the Early Solar System* (J. F. Kerridge and M. S. Matthews, eds.), pp. 1093-1113. Univ. of Arizona, Tucson.

Smith J. V. (1979) Mineralogy of the planets: A voyage in space and time. *Mineral. Mag., 43,* 1-89.

Sonett C. P. and Reynolds R. T. (1979) Primordial heating of asteroidal parent bodies. In *Asteroids* (T. Gehrels and M. S. Matthews, eds.), pp. 822-848. Univ. of Arizona, Tucson.

Stevenson D. J. (1981) Models of the Earth's core. *Science, 214,* 611-619.

Takahashi E. (1983) Melting of a Yamato L3 chondrite (Y-74191) up to 30 Kbar. In *Proceedings of the Eighth Symposium on Antarctic Meteorites,* pp. 168-180. Natl. Inst. of Polar Res., Tokyo.

Tedesco E. F., Williams J. G., Matson D. L., Veeder G. J., Gradie J. C., and Lebofsky L. A. (1989) A three-parameter asteroid taxonomy. *Astron. J., 97,* 580-606.

Tholen D. J. (1984) Asteroid taxonomy from cluster analysis of photometry. Ph.D. thesis, Univ. of Arizona, Tucson. 150 pp.

Tholen D. J. and Barucci M. A. (1989) Asteroid taxonomy. In *Asteroids II* (R. P. Binzel, T. Gehrels, and M. S. Matthews, eds.), pp. 298-315. Univ. of Arizona, Tucson.

Tilton G. R. (1988) Age of the solar system. In *Meteorites and the Early Solar System* (J. F. Kerridge and M. S. Matthews, eds.), pp. 259-275. Univ. of Arizona, Tucson.

Toksöz M. N., Hsui A. T., and Johnston D. H. (1978) Thermal evolution of the terrestrial planets. *Moon and Planets, 18,* 281-320.

Wetherill G. W. (1980) Formation of the terrestrial planets. *Annu. Rev. Astron. Astrophys., 18,* 77-113.

Wetherill G. W. (1985) Asteroidal sources of ordinary chondrites. *Meteoritics, 20,* 1-22.

Wetherill G. W. and Chapman C. R. (1988) Asteroids and meteorites. In *Meteorites and the Early Solar System* (J. F. Kerridge and M. S. Matthews, eds.), pp. 35-67. Univ. of Arizona, Tucson.

Zellner B. and Bowell E. (1978) Asteroid compositional types and their distributions. In *Comets, Asteroids, Meteorites—Interrelations, Evolution, and Origins* (A. H. Delsemme, ed.), pp. 185-197. Univ. of Toledo, Toledo, Ohio.

Zellner B., Tholen D. J., and Tedesco E. F. (1985a) The eight-color asteroid survey: Results for 589 minor planets. *Icarus, 61,* 355-416.

Zellner B., Thirunagari A., and Bender D. (1985b) The large-scale structure of the asteroid belt. *Icarus, 62,* 505-511.

ACCRETION OF DIFFERENTIATED PLANETESIMALS TO THE EARTH

S. R. Taylor[1] and M. D. Norman

Research School of Earth Sciences, Australian National University, Canberra, Australia 2601

It is argued here that the Earth and the other planets of the inner solar system accreted predominantly from volatile-depleted planetesimals that were already differentiated into metallic cores and silicate mantles. Metal-sulfide-silicate partitioning seems to have been present in the early dust components of the solar nebula, prior to chondrule formation, and proceeded efficiently during growth of planetesimals and accretion of the planets. Metal-sulfide-silicate equilibrium was thus mainly established at low pressures within relatively small planetesimals. Rapid core-mantle separation in the Earth of precursor metal and silicate makes their reequilibration at higher pressures less likely. Thus, the trace and minor element composition of the mantle and core will reflect the low pressures and oxygen fugacities of the small precursor bodies, rather than the megabar pressures of the deep terrestrial mantle. In this model, core formation on Earth was not a catastrophe affecting a fully formed planet, but more likely occurred during accretion of the previously differentiated planetesimals. Since metal-sulfide-silicate equilibrium was accomplished predominantly at low pressures, troilite will be the main accreting phase for sulfur, which thus remains a viable candidate for the light element in the Earth's core.

INTRODUCTION

A fundamental question about the origin of the Earth concerns the state of the precursor material prior to planet formation, and the mode of accretion of the planets. Did the terrestrial planets form directly from the dispersed dust and gas of the nebula or were they built up brick by brick from planetesimals? In the latter model, were the precursor bodies composed of primitive aggregates of undifferentiated solar nebula matter, or were at least the larger ones differentiated into metallic cores and silicate mantles? If much of the accreting material were already present as metal, sulfide, and silicate that had separated into cores and mantles in planetesimals at low pressures, then this provides a major constraint on core formation and metal-silicate equilibration at higher pressures in the early Earth.

[1]Also at the Lunar and Planetary Institute, 3303 NASA Road 1, Houston, TX 77058

Although the concept that planets accreted directly from the fine-grained dust and gas of the solar nebula was popular for many years, several observations suggest that they are actually the end products of a hierarchical accretionary process that first produced a large number of planetesimals that were later accreted to form the larger planets (*Safronov,* 1969; *Wetherill,* 1986). Some of these observations in support of large precursor objects include the ubiquity of ancient cratered terrains across the solar system (*Basaltic Volcanism Study Project,* 1981; *Chapman and McKinnon,* 1986; *Melosh,* 1989), the tilts of the axes of rotations of most planets relative to the ecliptic, and the striking chemical heterogeneity of the planets and asteroids (*Wasson,* 1985, pp. 112-113; *Taylor,* 1988).

The oldest material available for direct study of the early conditions of the nebula, the accretionary processes that were active then, and the timing of these processes, is found in meteorites. In this paper first we review the meteoritic evidence for conditions in the early solar system, then examine the meteorites as analogs for the terrestrial precursor planetesimals, and conclude with a discussion of the implications of planetesimal accretion for the compositions of the inner planets. We will focus this discussion on the Earth, although it is impossible to consider the origin of Earth in isolation from that of the Moon and the other planets of the inner solar system.

THE METEORITIC EVIDENCE

Meteorites provide tantalizing evidence about the early solar system, our current understanding being aptly summarized by the comment made in 1492 about the fall of the Ensisheim, Alsace, meteorite: "Many know much about this stone, everyone knows something, but no one knows quite enough" (*Marvin,* 1986). Meteorites are conveniently divided into two classes: those from primitive, undifferentiated parent bodies (chondrites), and those from parent bodies that were apparently fractionated into at least silicate mantles and metallic cores (eucrites, irons, SNCs, etc.). Each of these two classes of meteorites holds unique information about conditions in the early solar system.

Primitive Bodies

The composition of CI carbonaceous chondrites is very close to that of the solar photosphere for nongaseous elements, and so is probably the best estimate available for the composition of the earliest condensed material from the solar nebula. However, it should be recalled that the CI meteorites have undergone low temperature aqueous alteration (*Zolensky and McSween,* 1988), and that the primitive mineralogy of the nebula was anhydrous and reduced. Although this compositional similarity between the solar photosphere and the CI chondrites implies thorough mixing within the early nebula, there are some differences between the solar composition and that of the CI meteorites that belie a completely homogeneous nebula. For example, heterogeneities among the meteorites and the planets, such as those recorded in their oxygen isotopes (*Thiemens,* 1988) and K/U ratios (*Taylor,* 1988), suggest either incomplete physical mixing within the nebula, or some fractionation within the nebula superimposed on an initially homogeneous condition.

Recent studies also suggest that the sun may contain 30-40% more iron than do the CI meteorites (*Anderson,* 1989). Is the discrepancy real? There is a long history of difficulty in measuring the iron abundance in the sun and for several years before 1965, the calculated solar value was only 20% of the meteoritic value, until the oscillator strength of the spectral lines was revised (*Urey,* 1967). Now the best photospheric estimate is 7.67 ± 0.03 relative to $\log N_H = 12.00$, compared to the CI value of $7.51 = 0.01$ on the same scale (*Anders and Grevesse,* 1989). However, there is still considerable uncertainty in the solar value.

The problem is summarized by *Anders and Grevesse* (1989), who record photospheric values from various workers, relative to $\log N_H = 12.00$, ranging from 7.56 to 7.72, compared to the well established meteoritic value of 7.51 on the same scale. In addition to the solar photospheric data, values are obtainable from the corona and the solar wind (7.53 ± 0.27) or solar energetic particles (SEP) (*Anders and Grevesse,* 1989). The SEP value agrees with the higher solar photospheric value, but these coronal abundances are too subject to fractionation by little understood processes to be considered a reliable second opinion.

The position is adopted here that too much uncertainty surrounds the solar Fe values at present to justify moving from the meteoritic value for the overall solar system Fe abundance, in agreement with the assessment of *Anders and Grevesse* (1989) and E. Anders (personal communication, 1989), despite the siren-like attractions of the higher Fe abundance (e.g., *Anderson,* 1989).

Chondrules

Chondrules are millimeter-sized silicate spherules that make up the bulk of ordinary chondritic meteorites (e.g., *Grossman et al.,* 1988; *Hewins,* 1988). They are depleted in siderophile and chalcophile elements, and apparently contain a record of some of the earliest events of the solar system. They formed, in the words of *Sorby* (1877) over 100 years ago, as "molten drops in a fiery rain." Our understanding of the place and conditions of their origin has advanced only a little since that time.

Chondrules form by melting of preexisting dust in the nebula. Volcanic and impact origins can be ruled out (*Taylor et al.,* 1983). The preservation of local isotopic anomalies in the refractory inclusions (CAI) and the ubiquitous oxygen isotopic variations (*Thiemens,* 1988) also rule out condensation from a hot vaporized nebula; all would be homogenized in a hot nebula. Cooling times of chondrules are very fast, on a timescale of hours (*Hewins,* 1988). It is not possible to cool molten drops so rapidly in a hot nebula, so the process must have been highly localized in an overall cool environment (*Wood,* 1987).

What was the nature of the chondrule precursor material? One possibility is that the chondrules were melted from oxidized preexisting dust and that reduction of iron, depletion of siderophiles, chalcophiles, and some volatiles occurred during this stage. The alternative is that metal, sulfide, and silicate phases existed already in the early nebula, either as interstellar dust grains or condensed from nebular gas (*Grossman et al.,* 1988). The latter scenario is preferred for a number of reasons.

Silicate dust was melted selectively during the chondrule-forming event, with some minor reduction and loss of iron, and minor depletion of sulfides and volatile elements. Thus, although some reduction of iron and depletion of siderophiles, chalcophiles, and some volatiles may have occurred during chondrule formation (*Grossman et al.,* 1988), the major separation of these phases and elements apparently occurred prior to the chondrule melting event within the dispersed material of the solar nebula.

How was the silicate dust melted preferentially, without involving the metal and sulfide phases to more than a minor extent? Perhaps silicate dust was separated from metal and sulfide, either by differential gravitational settling or magnetically in the case of the metal, or perhaps the silicates stuck together more efficiently. As the dust settled to the mid-plane, physical and perhaps magnetic separation of the dust occurred, possibly accounting for the complementary relationship between chondrules and the more iron-rich (denser?) matrix as well (*Scott et al.,* 1988).

Many explanations have been offered for the origin of chondrules in the past (e.g., *King,* 1982), but none has accounted for all the observations. The most plausible scenario is that chondrules form by melting of silicate dust by nebular flares, analogous to solar flares, high above the mid-plane of the nebula (*Sonett,* 1979; *Grossman,* 1988; *Levy and Araki,* 1989). In this scenario, the separation of metal from silicate might be accomplished by the preferential gravitational settling of metal to the mid-plane of the nebula, while the existence of strong magnetic fields associated with the nebula flares could account for the magnetic signature in chondrites (*Sugiura and Strangway,* 1988).

In summary, the chondritic meteorites preserve evidence of the very early existence of metal, sulfide, and silicate phases. These are, however, intimately mixed in the chondritic parent bodies that were never melted. Chondrules, which comprise over half the mass of ordinary chondrites, however, record the preferential rapid melting of mainly silicate phases as one of the earliest events in the nebula.

DIFFERENTIATED BODIES

Chondritic meteorites were derived from undifferentiated parent bodies that preserve some memory of primitive solar nebula compositions (*McSween,* 1987). However, other meteorites, such as the eucrites and irons, attest to episodes of metal-sulfide-silicate partitioning within parent bodies.

Eucrites

Good evidence for early solar system fractionation is provided by the eucrites. Apparently derived by partial melting of a silicate mantle in a small asteroid, these meteorites have low K/U ratios (~3100) and are depleted in siderophile elements (e.g., Ni is depleted by a factor of 10^4 relative to CI abundances). Samarium-neodymium and $^{207}Pb/^{206}Pb$ data indicate crystallization ages of 4.54-4.56 aeons (*Tilton,* 1988). Two possibilities exist to explain the occurrence of these highly differentiated igneous rocks so early in the history of the solar system: (1) The eucrite parent body was accreted from metal-free, volatile depleted precursors, or (2) the eucrites were derived from the mantle of an asteroid that was depleted in volatiles and in siderophile elements that may have segregated into a metallic core.

In either alternative, metal-silicate partitioning was a *fait accompli* and K and Rb had already been depleted relative to CI abundances of U and Sr (see later discussion of possible alternatives).

Iron Meteorites

Iron meteorites, perhaps the ultimate in differentiated objects, come from at least 60 separate bodies (*Wasson,* 1985, pp. 112-113), providing evidence for many episodes of low-pressure metal-silicate differentiation in precursor bodies prior to planetary accretion. One possible model is that the irons melted and solidified in small (1-10 km) bodies, which were then reaccreted into larger planetesimals where they were slowly cooled through 800 K (*Wood,* 1987). This could preserve some compositional heterogeneity within the core of a single body, and would be consistent with the planetesimal hypothesis. There is a caveat, however. The reaccretion of these bodies must have been at low velocities, otherwise the iron core would have been disrupted. This adds another argument for accretion of planetesimals from localized zones, rather than from a random mixture from widely separated zones when high approach velocities would be common (*Wetherill,* 1985).

It is also of much interest that these differentiated iron meteorites were almost never remixed with silicates. Clearly little physical or chemical mixing occurred following the differentiation events, which implies rather quiet conditions in the asteroid belt following the differentiation of the asteroids.

Overriding all the meteoritic evidence is an uncertainty. Are meteorites a representative sample of the material that went up to make the terrestrial planets, or are they even an adequate sample of the present asteroid belt? For example, there is no spectral match among the asteroids for the ordinary chondrites that are the most common terrestrial falls. However, the evidence seems clear that differentiation events were common and widespread in the early solar system before the accretion of the terrestrial planets.

VOLATILE ELEMENT DEPLETION IN THE SOLAR NEBULA

It has been clear for some time that although the inner planets are broadly chondritic in their major element composition, they cannot be matched to any specific meteorite class, and differ from the CI meteorites principally in their depletion of elements that are volatile below about 1000 K. This depletion of volatile elements relative to the presumed initial composition of the solar nebula is not only a feature of the inner planets, but also extends to several types of meteorites, and so must be a reflection of some fundamental processes acting in the early solar system.

This depletion of volatile elements is well shown by the relative abundances of K (a moderately volatile element) and U (a refractory element) (e.g., *Taylor,* 1988, Fig. 7.8.6). Since these elements are both gamma-ray emitters, K and U determinations can be made remotely and are two of the few geochemical measurements available for Earth, Venus, and Mars, as well as for the meteorites. Potassium and U, although distinctly different in chemical properties, ionic radius, and valency, nevertheless share a common characteristic which makes them useful as geochemical indices of planetary compositions: They are both excluded from the common rock-forming minerals in basalts (i.e., they are "incompatible" elements), and so are concentrated together in residual melts during crystallization of basaltic silicate melts. Accordingly, they tend to preserve their bulk planetary ratios during planetary differentiation. Thus, their measurement can provide information about the depletion of volatile relative to refractory elements in their parent body.

The primordial K/U ratio, as given by the CI meteorites, is appproximately 60,000, while terrestrial ratios are around 10,000 (*Mason,* 1979). Potassium is also depleted, relative to U, in the surface rocks of Mars and Venus. The K/U ratios for the SNC meteorites, almost certainly derived from Mars (*Bogard et al.,* 1984), appear to be slightly higher (1.5×10^4), suggesting a somewhat higher volatile content for that planet, but still a factor of 4 lower than primordial values.

In addition to the chemical evidence, the Rb/Sr isotopic systematics also indicate that the Earth is depleted in volatile Rb relative to refractory Sr (*Basaltic Volcanism Study Project,* 1981, Chapter 7). Samples derived from the mantle indicate that the mantle Rb/Sr ratio has always been much lower than the primordial nebula values given by the CI meteorites. Since Rb has closely similar geochemical properties to K, it is likely that neither K nor Rb are present in the mantle in their primordial solar nebula concentrations and that Rb and the other volatile elements were depleted in the precursor planetesimals from which the inner planets accumulated.

Formation of Meteorite Parent Bodies

Meteorites contain unique and crucial information about the dates of early events in the solar system. The oldest reliably dated objects in meteorites are the millimeter- to centimeter-sized refractory calcium-aluminum inclusions (CAI) (*Tilton,* 1988). These give an age of 4559 ± 4 m.y., which is taken here as T_0 (*Tilton,* 1988). The ordinary chondritic meteorites (H, L, LL, and E classes) all have similar whole rock ages, the best estimate being 4555 ± 4 m.y. (*Tilton,* 1988). Samarium-neodymium ages for phosphates from chondrites yield ages of 4550 ± 45 m.y. (*Brannon et al.,* 1987). The whole-rock ages appear to be marginally younger than the ages recorded by the CAI inclusions.

Formation of igneous rocks in asteroidal bodies began within a few million years of T_0 (*Tilton,* 1988). Angra dos Reis (ADOR) is an igneous meteorite with an age of 4551 ± 2 m.y. The basaltic achondrites are true basalts similar to lunar basalts. They have an average crystallization age of 4539 ± 4 m.y., notionally only 20 m.y. younger than the oldest material dated by the Allende CAI. Clearly the achondrites reveal that planetary type differentiation and the eruption of basalts was occurring on a small but extensive scale in asteroids very close to T_0.

Timing of Volatile Depletion

The meteorites provide us with the time of volatile depletion in the inner nebula since the Pb-Pb and Rb-Sr ages give the time of separation and depletion of volatile Pb and Rb relative to refractory U and Sr from the primordial solar nebula values. All the meteorite data give effectively the same age for this event, indicating that this depletion in volatile elements occurred at about 4560 m.y. Accordingly, volatile depletion in meteorites, and presumably also in the inner nebula closer to the sun, in the region of the terrestrial planets, occurred effectively at T_o.

The age and initial Sr isotopic data from meteorites (*Tilton,* 1988) indicate that they record a single massive loss of volatile Rb relative to refactory Sr effectively at T_o, so that this was a nebula-wide event rather than being connected with isotopic evolution in individual parent bodies. This depletion must have occurred very early, since very low initial $^{87}Sr/^{86}Sr$ ratios are observed in the achondritic meteorites. They record igneous events at about 4560 m.y. and indicate a major separation of Rb from Sr at this early stage of solar system formation. The terrestrial Rb/Sr ratio of 0.03 is nearly a factor of 10 lower than the CI ratio (*Basaltic Volcanism Study Project,* 1981), implying a massive preterrestrial loss of volatile elements, with the Earth accreting from already depleted planetesimals. Loss of heavy elements from massive bodies is precluded, since the temperatures required for Jeans escape from the Earth of elements such as Rb are several thousand Kelvin.

The most primitive initial $^{87}Sr/^{86}Sr$ ratio is that given by CAI in Allende (*Gray et al.,* 1973). This appears to represent early Rb loss in isolated heating events and to record the earliest events we can trace in the nebula. A little later the massive nebula-wide loss of Rb and K relative to Sr occurred. Such a widespread event must be connected with the early violent solar activity, which in this scheme occurs a little later than the production of CAI.

Some further loss of volatiles may occur during heating of material in high energy collisions. Such processes may occur during the formation of the Moon by a single large impact. Thus the lunar ratios, both for K/U and Rb/Sr are lower than terrestrial. If the terrestrial ratios of $K/U = 10^4$ and $Rb/Sr = 0.03$ are typical of the inner planetary regions and of Mars-sized impactors as discussed later, then it is possible that the major separation of Rb from Sr occurred at T_o, with a secondary, lesser depletion during lunar formation. Such scenarios appear to be required by the low lunar initial $^{87}Sr/^{86}Sr$ (LUNI = 0.69895 ± 3), which is close to BABI, the initial Sr ratio of the basaltic achondrites (0.69898 ± 3; *Nyquist,* 1977). New measurements on ADOR show that it has an initial Sr ratio identical to that of BABI (*Nyquist et al.,* 1989; *Lugmair et al.,* 1989). These data are consistent with a loss of volatile Rb relative to refractory Sr in a widespread early nebula event rather than from individual planetesimals during local heating episodes.

In summary, the depletion of volatile relative to refractory elements (as measured by K/U ratios) appears to have been a first-order nebular event affecting planetary material from at least Venus out to the asteroid belt. All the meteorite data give effectively the same age for this event, about 4560 m.y. (*Tilton,* 1988). Accordingly, volatile depletion in meteorites, and presumably also closer to the sun, in the region of the inner planets, occurred effectively at T_o.

Processes of Volatile Depletion

It is sometimes argued that volatile loss occurs from condensed bodies, during explosive igneous processes (e.g., *Wasson,* 1985, p. 112; *Mittlefeldht,* 1987). However, there are several problems with such scenarios. Among these are that molten bodies freeze over very quickly, and even a thin surface skin is sufficient to prevent volatile loss. The very early date for the depletion event makes it unlikely that molten planetesimals were common; the pervasive nature of the volatile depletion makes a nebula-wide event more likely. Thus if the achondrites had retained primordial Rb/Sr ratios, much higher $^{87}Sr/^{86}Sr$ ratios would be present than are observed. However, the very depleted nature of the eucrites, which have lunar-like K/U ratios, may require additional events, the most likely being volatile loss by vaporization during massive impacts.

Apart from these essentially minor processes, the view is adopted here that the depletion of the volatile elements occurred from dispersed phases in the nebula, and that the inner nebula was cleared of gas and depleted in volatile elements very early, probably within about 1 m.y. of the arrival of the sun on the main sequence. What mechanisms were responsible? Two alternatives seem possible. In the first, temperatures in the inner nebula were too hot for the volatiles to condense. Either they were swept into the early sun with the instreaming H and He, or swept out along with the other noncondensed gases into the outer nebula by early strong solar winds. The cause of this clearing of the inner regions of the nebula is most probably connected with early intense solar activity, with strong stellar winds, and flare outbursts of the type observed with the very young T-Tauri and FU Orionis stars, as the sun settled onto the main sequence.

In summary, differentiated and volatile-depleted bodies existed within a few million years of T_0. This is much earlier than the projected time for the accretion of the Earth, for which timescales of 10^7-10^8 years are required (*Wetherill,* 1986, 1988) so that differentiated objects were available for incorporation into the growing Earth.

IMPLICATIONS FOR PLANETARY ACCRETION

Planetesimals

In assessing theories for the origin of the Earth, the current view is that planets accrete from a hierarchy of planetesimals. Accretion times for the inner planets appear to be on the order of 10^7-10^8 years, with the final population of objects that were swept up into the terrestrial planets being 10-20% of the mass of the planet (*Wetherill,* 1986).

What sort of evidence do we have for these now vanished objects? There are several different converging lines of thought.

The asteroids are left-over planetesimals. The absence of a planet in the asteroid belt (*Binzel et al.,* 1989), in which over 3700 small bodies (the largest is 1 Ceres, 933 km in diameter) have been numbered (apart from countless smaller ones), is probably due to the influence of massive Jupiter, which swept up or ejected many of the bodies originally present and may have prevented the asteroids from accreting into a fully formed planet. The small size of Mars is probably due to a similar cause, i.e., starvation caused by massive Jupiter. Phobos, one of the martian satellites, appears to be a primitive object and may be a captured asteroid. Thus, it may provide us with an analog for a planetesimal.

The second direct piece of evidence for the former existence of planetesimals comes from the observation that all of the older preserved surfaces on planets and satellites are saturated with craters. The lunar surface is the classic example (*Wilhelms,* 1987), but from Mercury, close to the sun, out to the satellites of Uranus (*Chapman and McKinnon,* 1986) and Neptune, the photographs reveal a massive bombardment that struck the planets and their satellites. Craters of all sizes are present, from micron-sized pits on lunar samples due to impact of tiny grains, up to giant ringed basins over 1000 km in diameter. Like the smile of the Cheshire Cat in *Alice in Wonderland,* these craters record the previous existence of now-vanished objects—in this case, the planetesimals.

The extent of this bombardment on the Moon after it had reached its present size *and after the lunar crust had formed* is revealed by the 80 basins with diameters greater than 300 km and 10,000 craters in the 30-300 km size range that formed before the main bombardment ceased (*Wilhelms,* 1987). The last major ringed basin, Mare Orientale, formed on the Moon at about 3800 m.y. ago (*Taylor,* 1982). A similar but probably more intense barrage must have struck the Earth, probably accounting for the absence of identifiable rocks older than that age. Between 4400 and 3800 m.y. (i.e., following the accretion of the Earth) an estimated 200 ringed basins with diameters greater than 1000 km formed on the Earth due to the impact of bodies a few hundred kilometers in diameter. On the Earth, in contrast to the Moon, the evidence has been removed, since the smashed up breccias would have been easy prey for the terrestrial agents of erosion.

The next major piece of evidence comes from the tilt or inclination of the planets to their axis of rotation. The largest impact is required to account for Uranus. Calculations show that a body the size of the Earth, crashing into that planet, would be needed to tip it through 90° (*Benz and Cameron,* 1989). Smaller collisions are needed to account for the tilt of the other planets, but a few very large objects must have been responsible, since the impacts of many small bodies will average out (*Benz et al.,* 1989).

A large impact is probably responsible for the strange fact that Mercury has such a small rocky mantle and such a large iron core, and an inclined orbit so close to the sun. Two explanations are current. The first proposes that the silicate was boiled away in some early high-temperature event connected with early solar activity (the surface temperature on the present sunlit side of Mercury is 425°C, hot enough to melt lead). However, extremely high temperatures of several thousand degrees are required to boil off the rocky mantle. The alternative, and currently favored explanation, is that Mercury was struck by a body about 1/6 of its mass at a late stage in its accretion. The collision fragmented the planet with most of the silicate lost to space but the iron core surviving to reaccrete with a depleted silicate mantle. Simulations in three dimensions on supercomputers have revealed the details of such a collision (*Benz et al.,* 1988).

Finally, the long-standing problem of the origin of the Moon is resolved by the impact of an already differentiated massive (0.14 Earth mass) body with the Earth, the material making up the Moon being mostly (80%) derived from the silicate mantle of the impactor (*Benz et al.,* 1989).

In summary, there is ample evidence for the existence of large precursor bodies, or planetesimals, in the early solar system.

How Many Objects and How Big Were They?

How many bodies were involved and what were the relative sizes of the accreting planetesimals? Recent calculations (*Stevenson et al.,* 1986; *Wetherill,* 1986) assume that 100 objects of lunar mass (7.35×10^{25} gm), 10 with masses exceeding that of Mercury (3.19×10^{26} gm), and several exceeding the mass of Mars (6.42×10^{26} gm) were present in the final stages of the accretion of the terrestrial planets. About one-third of these objects would have accreted to the Earth, thus providing a total of 50-75% of present Earth mass from these massive planetesimals, with the remainder coming from a multitude of small bodies.

Fractionation in Precursor Planetesimals

What was the history of the planetesimals prior to their incorporation into the inner planets? Some of the largest, the size of Mars, would have made respectable planets in their own right if fate had taken a different course. Were they already differentiated into silicate mantles and metallic cores before they came to a violent end as they were swept up into Earth or Venus?

Based on evidence from meteorites, even some relatively small planetesimals underwent internal differentiation into metallic cores and silicate mantles quite early in their history. The larger planetesimals almost certainly had already gone through at least one intraplanetary melting episode, with core formation occurring before they were accreted by the inner planets. Such bodies, of course, may have been broken up by collisions and reaccreted in differing proportions of metal and silicate fractions, so that much diversity of composition among the accreting bodies can be expected.

The arrival of Moon- and Mars-sized objects on the accretionary scene presents another facet of the scenario. In the large impact hypothesis for lunar origin, a crucial parameter is that the Mars-sized body has already formed a metallic core and a silicate mantle, from which the Moon is essentially derived. The iron core accretes to the Earth, adding perhaps 10% of core volume. If the Earth were put together from a large number of such objects, then metal-silicate fractionation would have already occurred in relatively low-pressure environments in precursor bodies before final accretion, and the siderophile element signature of the resulting bulk planetary mantle would not necessarily reflect equilibrium with the core.

Is there independent evidence that such melting and fractionation occurred? Perhaps the most dramatic information is provided by the achondritic and iron meteorites, which indicate that melting and differentiation had occurred by about 4550 m.y. in many small parent bodies (e.g., *Bell,* 1988; *Bell et al.,* 1990; *Gaffey et al.,* 1990). 4 Vesta, a potential source of the eucrites, may represent only a small population of similar objects, as not all differentiated asteroids would proceed to the stage of lava effusion on their surfaces, as apparently happened on Vesta. However, most meteorites record fractionation on at least a small scale. Even the ordinary chondrites contain reduced metal, sulfide, and silicate phases, and have in addition lost some volatile elements.

Heat Sources

What were the sources of heat for melting small bodies in the solar system? This is most likely distinct from the volatile depletion event, connected with early solar activity, since direct solar heating cannot melt planetesimals, and in any event, it is unlikely that large planetesimals were around so early. This question of heat supply for early planetesimal metamorphism and melting is essentially unresolved. Two principal mechanisms. are currently discussed. If ^{26}Al ($t_{1/2} = 730{,}000$ yr) was present in the early solar system (*Podosek and Swindle,* 1988), it could have constituted an important heat source. The second possibility is by inductive heating during the early intense T-Tauri and FU Orionis stages of solar activity. Both of these mechanisms encounter difficulty and early planetesimal heating may be the result of processes not presently understood (J. A. Wood, personal communication, 1989).

A reasonably firm constraint on the heat source is that it varied with distance from the sun. The evidence for this comes from the zonal nature of the asteroid belt, which shows a steep change from differentiated to primitive character with increasing distance (*Bell,* 1988). Heating appears to have fallen off rapidly beyond 2 A.U. and does not seem to have been effective beyond about 3 A.U. This provides strong support for the sun as the source of the energy; ^{26}Al would be expected to be more uniformly distributed in the nebula. Although the amount of ice in the outer reaches of the nebula might affect the amount of melting, the temperature gradient across the asteroid belt seems too steep for this explanation to be valid.

Problems with a solar source revolve about the timescale involved. Such electromagnetic inductive heating can occur only during the early T-Tauri and FU Orionis stages, which last only for about 10^6 years or perhaps less. At this stage, the planetesimals have not grown to an appreciable size in most versions of the planetesimal hypothesis. But it should also be recalled that the evidence for live ^{107}Pd ($t_{1/2} = 6.5 \times 10^6$ yr) from the presence of anomalous ^{107}Ag/^{109}Ag ratios in iron meteorites (*Chen and Wasserburg,* 1983) indicates that segregations of metallic iron were present at a very early stage of solar system history.

This compression of the timescale does not alter the sequence of events, so that, for example, the gas was gone well before the accretion of the inner planets. However, it does raise some interesting possibilities. If the heating is restricted to very short timescales, a requirement for both currently discussed mechanisms, then most planetesimals were still very small, less than 10 km in diameter. These objects could then be accreted into larger bodies, possibly with some accretional heating causing metamorphic effects. Thus, two periods of heating, the first due to ^{26}Al, inductive heating, or some more exotic mechanism, and the second milder heating due to accretion, might explain the record. Cooling rates of the meteorite parent bodies may also be consistent with this scenario.

Width of Planetary Feeding Zones

A crucial question for the terrestrial planets is the width of the feeding zones from which they accumulated (*Wetherill,* 1985). Equilibrium condensation models predict a strict radial zoning outwards from the sun, with accretion of the planets from distinct compositional zones. Such regularity of planetary composition is not observed, and the evidence seems mostly against simple condensation models related to heliocentric distance for the formation of the inner planets. Rather, the inner planets seem to have accreted from feeding zones of differentiated planetesimals in a system in which stochastic processes played a leading role.

Classic models of planetesimal accretion (e.g., *Safronov,* 1969) suggest that a planet will grow almost exclusively from a narrow zone in which the planet's gravitational attraction is predominant. *Wetherill's* (1988) models of planetesimal accretion are consistent with relatively narrow feeder zones for the bulk of existing planetary mass, but they also show that each of the four major planets of the inner solar system could have received at least some material from all heliocentric distances in the region from 0.6 to 1.2 A.U. In this model, Mercury, Earth, and Mars preserve a fairly strong bias in their provenance with half of Mercury apparently coming from the region inside 0.6 A.U., two-thirds of Earth built from material from within 0.9-1.1 A.U., and over half of Mars built from material outside 1.0 A.U. Venus, on the other hand, may represent the best average of material originally in the region from 0.6-1.2 A.U. (*Wetherill,* 1988).

Compositional Zoning in the Inner Solar System

In attempting to constrain the width of planetary feeding zones, a comparison of planetary and meteoritic compositions may be useful, for if the planets grew from a well-mixed population of planetesimals, then there should be a high degree of compositional similarity not only among the inner planets, but also among the planets and the only surviving population of planetesimals—the asteroids.

Although the inner planets obviously share some broad compositional similarities such as depletion in volatile elements, it is also clear that the inner planets differ in some fundamental compositional parameters. The oxygen isotope data show that, except for the EH and EL groups, there is no overlap of meteoritic values with terrestrial values. Likewise, the SNC data, assumed to represent martian values, occupy a field distinct from the common chondrites, and also separate from Earth and Moon. No presently known meteorite group has appropriate K/U ratios to compose suitable building blocks for the Earth (*Taylor,* 1988). These compositional differences between the planets and asteroids imply considerable heterogeneity within the inner solar system and relatively narrow feeding zones for the major planets, and rule against nebula-wide mixing or homogenization during accretion of the planets.

The present population of meteorites, except for the rare lunar and SNC (martian) examples, are derived from 2-4 A.U. There is really no indication that any of them were closer to the sun at an early stage. The zonal structure in the asteroid belt is probably a primary feature, imposed on the asteroid belt due to early heating, and not disturbed very much since that time, a fact that argues for little large-scale lateral mixing in the solar nebula.

Thus, although the compositions of the inner planets are chondritic in a general sense, it does not appear possible to build the Earth and the other terrestrial planets out of the building blocks as supplied by the currently sampled population of meteorites. Our sampling of meteorite types is probably incomplete, with several samples (e.g., Kakangari, ALHA 85085) falling outside the main chondrite groups. Even so, substantial compositional differences apparently exist between the inner asteroid belt and the zone sunward of about 2 A.U., in which the terrestrial planets accumulated. Apart from the oxygen isotopes, the most significant difference between these regions appears to have been a greater depletion of the volatile elements in the region in which the terrestrial planets now reside.

Although we thus lack samples of the planetesimals that contributed to the accretion of the terrestrial planets, the structure of the asteroid belt, with its strong concentration of differentiated bodies in the inner belt, strongly supports the notion that those planetesimals sunward of the inner belt were more strongly heated and, thus, also differentiated. In this context, *Wasson and Rubin* (1985, p. 168) argue that "more than 90% of the bodies in heliocentric orbits having radii near 1 A.U. were differentiated."

In summary, it is reasonable to suppose that planetesimals within the feeding zone for the Earth were not only depleted in volatiles, but also differentiated.

Core-Mantle Equilibrium

The "predestination" scenario (*Taylor,* 1983) in which the terrestrial planets accrete from planetesimals that were already mostly differentiated into metallic, silicate, and sulfide phases implies little further reaction between metal and silicate once these bodies accreted to the Earth. Core formation is expected to occur in the planetesimals once melting temperatures are reached, with metallic iron in the planetesimals sinking rapidly through their silicate mantle, even in the relatively low gravitational fields. The addition of such differentiated bodies to the Earth adds a complicating factor to our present limited understanding of terrestrial core-mantle relationships (*Jones and Drake,* 1986; *Treiman,* 1986; *Treiman et al.,* 1986, 1987), since in this view much of the core and mantle arrived prepackaged.

One constraint on terrestrial core formation seems firm, which is that the present upper mantle did not achieve equilibrium with the core. There are high abundances in the upper mantle of the Earth of Re, Au, Ni, Co, and the platinum group elements (PGE = Ru, Rh, Pd, Os, Ir, Pt), with the distribution of the PGE rather uniform, and present in approximately primordial (CI) proportions, although depleted relative to Fe (*Arculus and Delano,* 1981; *Delano,* 1986; *Newsom and Palme,* 1984; *Newsom,* 1986). These highly siderophile elements would have been efficiently extracted into the metal phase under equilibrium conditions so that the present upper mantle was apparently never in equilibrium with the core.

Late accretion of CI planetesimals rich in PGE is a common explanation for their overabundance in the upper mantle. A similar scenario of terminal accretion of volatile-rich planetesimals is often invoked to account for the volatile (e.g., H_2O) inventory of the Earth. The addition of the metallic core of the impactor responsible, in the single impact hypothesis, for the origin of the Moon (*Benz et al.,* 1989) is another possible source of material. A cometary influx might be an equally viable source, although the high impact velocities of comets derived from the outer solar system may cause removal rather than addition of material. The late infall of planetesimals to planets, which are essentially complete, is like adding icing to a cake; the decoration may give little insight about the composition of the interior.

We have argued that the terrestrial depletion in K and other volatile elements was an inherent feature of the incoming planetesimals. Are there other explanations? Possibly some volatile loss occurred through vaporization in massive collisional events. If core-mantle equilibrium were attained in an initially molten Earth, then pressures are high enough for K to enter the core. However, Mars also appears, from the SNC data, to have low K/U ratios, although a little higher than the Earth. The central pressure in Mars is only 400 kbar, too low to permit K to enter the martian core (L.-J. Liu, personal communication, 1988) so that the depletion in K on that planet must have another explanation. In summary, no compelling reasons appear to require high-pressure core-mantle equilibrium during accretion and core-mantle separation.

A further consequence may be noted. The metallic core of the Earth contains about 10% of a light element. The two current contenders are O and S. Although meteorites are not a perfect analog for the terrestrial precursor planetesimals, they do tell us that elemental and mineralogical fractionation was endemic in the early nebula. If silicate, sulfide, and metal phases, formed under low-pressure equilibrium, were already present in the accreting planetesimals, separation of these phases may occur concomitantly with accretion and thus there may be little high-pressure equilibration between core and mantle in the Earth. Simulation of Mars-sized impacts indicates that the cores of such bodies penetrate the terrestrial mantle (*Benz et al.,* 1989). Accretion of large objects likewise causes mantle melting, independent of the large Moon-forming event, so that metal is likely to sink on very short timescales. Whether these scenarios apply to the accretion of smaller planetesimals is less certain. However, it is a tenet of the planetesimal hypothesis that such small bodies are swept up early, so that pressures within the growing planet will be low. The arrival of large planetesimals, from which the Earth acquires most mass and which induce melting, occurs later, when higher pressures will prevail in the interior. Thus it seems likely that melting, and hence rapid metal segregation, will occur concomitantly with planet growth.

If high-pressure core-mantle equilibrium were not attained in the early Earth, then it seems unlikely that O entered the core, since this requires megabar pressures, as is the case for K. Sulfur then becomes the most viable candidate for the light element in the Earth's core. The often-repeated argument that the Earth must be depleted in S, since that element is more volatile than K, fails to recognize that most of the S accreting to the Earth in planetesimals is stabilized by combination with Fe as troilite (FeS).

CONCLUSIONS

1. The Earth and the other planets of the inner solar system were built from precursor planetesimals that had already experienced at least one episode of melting, differentiation, and volatile depletion.

2. These planetesimals were composed mostly of separate metal, sulfide, and silicate phases. The larger ones at least had already formed metallic cores and silicate mantles.

3. The differentiated planetesimals formed within a few million years of T_0, well before the timescales of 10^7-10^8 years required for the accretion of the Earth.

4. Planetesimals within the feeding zone for the Earth were mostly differentiated.

5. Separation of metallic phases from silicates was a continuous process, beginning in the early dust of the solar nebula and continuing through chondrule formation, meteorite and planetesimal accretion, and planet formation.

6. Because this metal-sulfide-silicate fractionation occurred largely at low pressures, the Earth's core and mantle may not record the expected effects of high-pressure equilibration; the geochemistry of the core and mantle may

instead be dominated by the low-pressure equilibria established in the precursor planetesimals.

Acknowledgments. Careful and perceptive reviews by J. T. Wasson, who has advanced similar views on planetary accretion, A. Treiman, and J. Jones materially improved this paper. S.R.T. thanks H. Newsom and J. Jones for the opportunity to participate in the Origin of the Earth conference, and for encouraging us to submit this paper.

REFERENCES

Anders E. and Grevesse N. (1989) Abundances of the elements: Meteoritic and solar. *Geochim. Cosmochim. Acta, 53,* 197-214.

Anderson D. L. (1989) Composition of the Earth. *Science, 243,* 367-370.

Arculus R. J. and Delano J. W. (1981) Siderophile element abundances in the upper mantle: Evidence for a sulfide signature and equilibrium with the core. *Geochim. Cosmochim. Acta, 45,* 1331-1343.

Basaltic Volcanism Study Project (1981) *Basaltic Volcanism on the Terrestrial Planets,* Chapter 8. Pergamon, New York. 1286 pp.

Bell J. F. (1988) The 52-color asteroid survey: Final results and interpretation (abstract). In *Lunar and Planetary Science XIX,* pp. 57-58. Lunar and Planetary Institute, Houston.

Bell J. F., Davis D. R., Hartmann W. K., and Gaffey M.J. (1989) Asteroids: The big picture. In *Asteroids II* (R. P. Binzel, T. Gehrels, and M. S. Matthews), pp. 921-945. Univ. of Arizona, Tucson.

Benz W. and Cameron A. G. W. (1989) Tilting Uranus in a giant impact (abstract). *Bull. Am. Astron. Soc., 21,* 916.

Benz W., Cameron A. G. W., and Melosh H. J. (1988) The origin of the Moon: Further studies of the giant impact (abstract). In *Lunar and Planetary Science XIX,* pp. 61-62. Lunar and Planetary Institute, Houston.

Benz W., Cameron A. G. W., and Melosh H. J. (1989) The origin of the moon and the single-impact hypothesis III. *Icarus, 81,* 113-131.

Binzel R. P., Gehrels T., and Matthews M. S., eds. (1989) *Asteroids II.* Univ. of Arizona, Tucson.

Bogard D. D., Nyquist L. E., and Johnson P. (1984) Noble gas contents of shergottites and implications for the martian origin of SNC meteorites. *Geochim. Cosmochim. Acta, 48,* 1273-1739.

Brannon J. C., Podosek F. A., and Lugmair G. W. (1987) Initial $^{87}Sr/^{86}Sr$ and Sm-Nd chronology of chondritic meteorites. *Proc. Lunar Planet. Sci. Conf. 18th,* pp. 555-564.

Chapman C. R. and McKinnon W. B. (1986) Cratering of planetary satellites. In *Satellites* (J. A. Burns and M. S. Matthews, eds.), pp. 492-580. Univ. of Arizona, Tucson.

Chen J. H. and Wasserburg G. J. (1983) The isotopic composition of silver and lead in two iron meteorites: Cape York and Grant. *Geochim. Cosmochim. Acta, 47,* 1725-1737.

Delano J. W. (1986) Abundances of cobalt, nickel, and volatiles in the silicate portion of the Moon. In *Origin of the Moon* (W. K. Hartmann, R. J. Phillips, and G. J. Taylor, eds.), pp. 231-247. Lunar and Planetary Institute, Houston.

Gaffey M. J., Bell J. F., and Cruikshank D. P. (1989) Reflectance spectroscopy and asteroid surface mineralogy. In *Asteroids II* (R. P. Binzel, T. Gehrels, and M. S. Matthews, eds.), pp. 98-127. Univ. of Arizona, Tucson.

Gray C. M., Papanastassiou D. A., and Wasserburg G. J. (1973) The identification of early condensates from the solar nebula. *Icarus, 20,* 213-219.

Grossman J. N. (1988) Formation of chondrules. In *Meteorites and the Early Solar System* (J. F. Kerridge and M. S. Matthews, eds.), pp. 680-696. Univ. of Arizona, Tucson.

Grossman J. N., Rubin A. E., Nagahara H., and King E. A. (1988) Properties of chondrules. In *Meteorites and the Early Solar System* (J. F. Kerridge and M. S. Matthews, eds.), pp. 619-659. Univ. of Arizona, Tucson.

Hewins R. (1988) Experimental studies on chondrules. In *Meteorites and the Early Solar System* (J. F. Kerridge and M. S. Matthews, eds.), pp. 660-679. Univ. of Arizona, Tucson.

Jones J. H. and Drake M. J. (1986) Geochemical constraints on core formation in the Earth. *Nature, 322,* 221-228.

King E. A., ed. (1982) *Chondrules and their Origins.* Lunar and Planetary Institute, Houston. 377 pp.

Levy E. H. and Araki S. (1989) Magnetic reconnection flares in the protoplanetary nebula and the possible origin of meteorite chondrules. *Icarus, 81,* 74-91.

Lugmair G. W., Galer S. J. G., and Loss R. (1989) Rb-Sr and other isotopic studies of the angrite LEW 86010 (abstract). In *Lunar and Planetary*

Science XX, pp. 604-605. Lunar and Planetary Institute, Houston.

Marvin U. B. (1986) Meteorites, the moon and the history of geology. *J. Geol. Educ., 34,* 140-165.

Mason B. (1979) Data of Geochemistry, Chapter B, Cosmochemistry, Part 1. Meteorites. *U.S. Geol. Surv. Prof. Pap. 440-B-1.* 132 pp.

McSween H. Y. Jr. (1987) *Meteorites and their Parent Planets.* Cambridge, New York. 237 pp.

Melosh H. J. (1989) *Impact Cratering: A Geologic Process.* Oxford, New York. 245 pp.

Mittlefehldt D. W. (1987) Volatile degassing of basaltic achondrite parent bodies: Evidence from alkali elements and phosphorus. *Geochim. Cosmochim. Acta, 51,* 267-278.

Newsom H. E. (1986) Constraints on the origin of the Moon from the abundance of molybdenum and other siderophile elements. In *Origin of the Moon* (W. K. Hartmann, R. J. Phillips, and G. J. Taylor, eds.), pp. 203-220. Lunar and Planetary Institute, Houston.

Newsom H. E. and Palme H. (1984) The depletion of siderophile elements in the Earth's mantle: New evidence from molybdenum and tungsten. *Earth Planet. Sci. Lett., 69,* 354-364.

Nyquist L. E. (1977) Lunar Rb-Sr chronology. *Phys. Chem. Earth, 10,* 103-142.

Nyquist L. E., Wiesmann H., Bansal B., and Shih C.-Y. (1989) Rb-Sr age of an eucrite clast in the Bholghati howardite and initial Sr composition of the Lewis Cliff 86010 angrite (abstract). In *Lunar and Planetary Science XX,* pp. 798-799.

Podosek F. A. and Swindle T. D. (1988) Extinct radionuclides. In *Meteorites and the Early Solar System* (J. F. Kerridge and M. S. Matthews, eds.), pp. 1093-1113. Univ. of Arizona, Tucson.

Safronov V. (1969) *Evolution of the Protoplanetary Cloud and Formation of the Earth and Planets.* Nauka, Moscow. Translated by the Israel Program for Scientific Translation (1972).

Scott E. R. D., Barber D. J., Alexander C. M., Hutchison R., and Peck J. A. (1988) Primitive material surviving in chondrites: Matrix. In *Meteorites and the Early Solar System* (J. F. Kerridge and M. S. Matthews, eds.), pp. 718-745. Univ. of Arizona, Tucson.

Sonett C. P. (1979) On the origin of chondrules. *Geophys. Res. Lett., 6,* 677-680.

Sorby H. C. (1877) On the structure and origin of meteorites. *Nature, 15,* 495-498.

Stevenson D. J., Harris A. W., and Lunine J. I. (1986) Origins of satellites. In *Satellites* (J. A. Burns and M. S. Matthews, eds.), p. 58. Univ. of Arizona, Tucson.

Sugiura N. and Strangway D. W. (1988) Magnetic studies of meteorites. In *Meteorites and the Early Solar System* (J. F. Kerridge and M. S. Matthews, eds.), pp. 595-615. Univ. of Arizona, Tucson.

Taylor G. J., Scott E. R. D., and Keil K. (1983) Cosmic setting for chondrule formation. In *Chondrules and Their Origins* (E. A. King, ed.), pp. 262-278. Lunar and Planetary Institute, Houston.

Taylor S. R. (1982) *Planetary Science: A Lunar Perspective.* Lunar and Planetary Institute, Houston. 481 pp.

Taylor S. R. (1983) Element fractionation in the solar nebula and planetary compositions: A "predestination" scenario (abstract). In *Lunar and Planetary Science XIV,* pp. 779-780. Lunar and Planetary Institute, Houston.

Taylor S. R. (1988) Planetary compositions. In *Meteorites and the Early Solar System* (J. F. Kerridge and M. S. Matthews, eds.), pp. 512-534. Univ. of Arizona, Tucson.

Thiemens M. H. (1988) Heterogeneity in the nebula: Evidence from stable isotopes. In *Meteorites and the Early Solar System* (J. F. Kerridge and M. S. Matthews, eds.), pp. 899-923. Univ. of Arizona, Tucson.

Tilton G. W. (1988) Age of the solar system. In *Meteorites and the Early Solar System* (J. F. Kerridge and M. S. Matthews, eds.), pp. 259-275. Univ. of Arizona, Tucson.

Treiman A. H. (1986) The parental magma of the Nakhla achondrite: Ultrabasic volcanism on the shergottite parent body. *Geochim. Cosmochim. Acta, 50,* 1061-1070.

Treiman A. H., Drake M. J., Janssens M.-J., Wolf R., and Ebihara M. (1986) Core formation in the Earth and shergottite parent body (SPB): Chemical evidence from basalts. *Geochim. Cosmochim. Acta, 50,* 1071-1091.

Treiman A. H., Jones J. H., and Drake M. J. (1987) Core formation in the shergottite parent body and comparison with the Earth. *Proc. Lunar Planet. Sci. Conf. 17th,* in *J. Geophys. Res., 92,* E627-E632.

Urey H. C. (1967) The abundance of the elements with special reference to the problem of the iron abundance. *Q. J. R. Astron. Soc., 8,* 23-47.

Wasson J. T. (1985) *Meteorites.* Freeman, New York. 267 pp.

Wasson J. T. and Rubin A. E. (1985) Formation of mesosiderites by low-velocity impacts as a natural consequence of planet formation. *Nature, 318,* 168-170.

Wetherill G. W. (1985) Occurrence of giant impacts during the growth of the terrestrial planets. *Science, 228,* 877-879.

Wetherill G. W. (1986) Accumulation of the terrestrial planets and implications concerning lunar origin. In *Origin of the Moon* (W. K. Hartmann, R. J. Phillips, and G. J. Taylor, eds.), pp. 519-550. Lunar and Planetary Institute, Houston.

Wetherill G. W. (1988) Accumulation of Mercury from planetesimals. In *Mercury* (F. Vilas, C. R. Chapman, and M. S. Matthews, eds.), pp. 670-691. Univ. of Arizona, Tucson.

Wilhelms D. E. (1987) *The Geologic History of the Moon.* U.S. Geol. Surv. Prof. Pap. 1347.

Wood J. A. (1987) Was chondritic material formed during large-scale, protracted nebular evolution or by transient local events in the nebula? (abstract). In *Lunar and Planetary Science XVIII,* pp. 1100-1101. Lunar and Planetary Institute, Houston.

Zolensky M. and McSween H. Y. Jr. (1988) Aqueous alteration. In *Meteorites and the Early Solar System* (J. F. Kerridge and M. S. Matthews, eds.), pp. 114-143. Univ. of Arizona, Tucson.

DIFFERENCES BETWEEN THE EARTH AND VENUS ARISING FROM ORIGIN BY LARGE PLANETESIMAL INFALL

W. M. Kaula

Department of Earth and Space Sciences and Institute of Geophysics and Planetary Physics, University of California, Los Angeles, CA 90024

Two differences between the Earth and Venus must be primordial: (1) their rotations and (2) their retentions of inert gases. These differences are so big as to appear to require catastrophic events, such as differences in magnitude of the largest impacts in the terminal phases of formation. This hypothesis is testable by numerical simulations. Four protoplanets were assumed, each with 81% of the final mass of the terrestrial planets, and orbits coplanar with the same semimajor axes and eccentricities as the final planet orbits. Starting spins were selected randomly from a Gaussian distribution about a prograde rate proportionate to $M^{1/6}$, and obliquities were selected from a distribution about zero. Also assumed were planetesimal populations of 50 members, with randomly selected distributions of the form $f(p) = kp^{-q}$ for four parameters p: mass, semimajor axis, eccentricity, and inclination. These conditions were selected to resemble the late stages of the accretion simulations of Wetherill (1986). Sixty Monte Carlo runs were made of interactions of such randomly selected sets of 50 planetesimals with the 4 planets, using an Öpik algorithm (modified to take into account eccentricity of the planets' orbits). The outcomes for proto-Venus and proto-Earth yielded retrograde spins in 30% of cases, speed-up of spins to LODs less than 8 hr in 15% of cases, and energy dissipations ranging from 1000 to 20,000°K equivalent mean temperature rise. Seven cases combined fast prograde rotation (LOD < 9.5 hrs) in one planet with retrograde rotation for the other. Three of these cases had much higher energy dissipation in the fast prograde rotator. Hence the marked differences of Earth and Venus in spin plus satellite and noble gas retention are plausible, though not probable, outcomes of formation from large planetesimals.

INTRODUCTION

Among all the planets, the Earth and Venus are by far the most similar to each other in their primary properties. The two planets differ by less than 20% in mass, mean density, and equilibrium black-body temperature with respect to solar radiation. However, they differ extremely in many secondary properties such as surface temperature, atmospheric water abundance, magnetic field intensity, and topography. The aforelisted differences are all explicable as consequences of evolution (although there are debates as to just how). But there are two great differences between the Earth and Venus that must depend on formation.

First are the coupled phenomena of rotation and satellite: The Earth rotates prograde rather rapidly and has a large satellite, while Venus rotates retrograde very slowly and has no satellite. The absence of a satellite around Venus is attributable to that planet's slow rotation, since tidal friction would cause the decay of any satellite orbit, prograde or retrograde (*Counselman,* 1973). The evolution and maintenance of Venus's very slow spin and extreme obliquity (177°) are attributable to dissipative processes (*Dobrovolskis and Ingersoll,* 1980), provided that Venus initially had a retrograde rotation: i.e., an obliquity greater than 90°. Given an obliquity greater than 90°, tidal dissipation would have driven Venus's obliquity toward 180° (*Goldreich and Peale,* 1970). Slowness of spin, more than degree of retrogradeness, facilitates the process, which is complicated enough that it is not easy to state a minimum 1/Q requirement.

Second is the difference in atmospheric abundances of primordial noble gases, particularly $^{36+38}$Ar, of which Venus has 80 times as much as the Earth (*Donahue and Pollack,* 1983). Attempts to explain the Ar difference by nebula or solar wind effects (*Pollack and Black,* 1982; *Wetherill,* 1981) have remained unpersuasive. In particular, these hypotheses forbid mixing between planetary zones in late-stage planet/planetesimal interaction.

The obvious hypothesis for these two primordial differences between Earth and Venus is the late-stage impact history, as suggested by *Cameron* (1983): The Earth was hit by a big body, which knocked off both the protolunar matter and nearly all of its primordial atmosphere, while Venus was not hit by anything nearly as large. As discussed below, "knocking off" is inappropriate terminology for the atmospheric loss; more probable is a planetary wind from the extreme heating caused by the impact.

Hence a problem is whether such a large difference in late-stage impact history could evolve. What was the population of bodies—masses and orbits—characterizing the penultimate stage of terrestrial planet formation? What constraints can be placed on formation by the spin and noble gas differences between Venus and Earth (as well as the primary properties of all four planets)? *Wetherill* (1985, 1986) showed that, given a suitable confinement in semimajor axis of a starting swarm of 500 planetesimals, about 4 planets could be obtained, with the final stages entailing impacts of bodies up to 0.3 $M_{\oplus}$ in mass at velocities of about 9 km/sec. Hence his Monte Carlo models support *Cameron's* (1983) conjecture, but Wetherill did not calculate the spins resulting from collisions in the model, and did not explore in any detail the implications of the observed outcomes in spin and volatile retention for the planetesimal population.

Earlier discussions of the implications of planet spin rates and obliquities (*Safronov and Zvjagina,* 1969; *Harris,* 1977) concluded that they limited the size of the forming planetesimals. Indeed, *Safronov's* (1969) main observational argument for the runaway growth of the largest body in a zone was the size of the obliquities. This problem has been reexamined by *Hartmann and Vail* (1986) with a model defined by distributions in mass, approach velocity, and offset of impactors, with the conclusion that impactors could have had masses as much as 20% of the impacted planets.

Hence, it seems appropriate to explore a random set of possible late-stage planetesimal populations to see what is the plausible range of properties and their consequences for the terrestrial planets. I take as the starting point the results of *Wetherill* (1986) at 81% completion, plus assumptions of random distributions in spin rates and obliquities about prograde rotations. These assumptions obviously limit the possible outcomes. However, they are sufficient to test qualitatively the hypothesis that the differences between the Earth and Venus are stochastic outcomes of formation from sizeable planetesimals. Extension from the Wetherill model gives a consistent model against which to test variations. The assumptions as to spin rates and obliquities probably bias the results on the conservative side.

ALGORITHM

General

I adapted an Öpik algorithm developed for another purpose and applied it to this problem. The main difference between this algorithm and Wetherill's is that four bodies are *a priori* designated as "planets," the remainder, as "planetesimals"; and *only* planet/planetesimal interactions are taken into account. Planetesimal/planetesimal collisions are not allowed. This omission is also a conservative bias, since it removes a mechanism by which orbital momentum is transferred to spin angular momentum, and by which bodies are heated. My computation is essentially a combination of the later stages of the modeling by *Wetherill* (1986) with the impact/spin modeling of *Hartmann and Vail* (1986). In addition to spin orientation and magnitude, energy dissipation upon collision is calculated, since variation therein pertains to the hypothesis of an impact-generated volatile difference (*Cameron,* 1983).

Input

Determinate input parameters that are the same for all runs of a Monte Carlo set are (1) planet properties (starting masses, radii, and orbital elements); (2) the total mass, number, and bodily density of the planetesimals; (3) the mean values and standard deviations for the spin rates and obliquities of the planets; and (4) the defining statistical parameters of the planetesimal population. These statistical parameters were the frequency exponent q, minimum p_m, and maximum p_x for the distribution of each parameter p, for a distribution density of the form kp^{-q} (see Appendix A for statistical definitions and derivations). The parameter p may be mass m, or semimajor axis a, or eccentricity e, or inclination I. In addition, correlations among parameters were prescribed.

Numerical values used for the starting planet set are given in Table 1, and those for the planetesimal population in Table 2. These values approximate those of the 31-m.y., or 81%-accreted, stage of *Wetherill* (1986, p. 537): 4 planets, 81% as massive as the final, plus 50 planetesimals of masses m randomly selected from a distribution density $f(m) = k\,m^{-0.5}$ (see Appendix A), with a maximum not more than 10 times the minimum, and a total mass 23.5% of the total mass of the planets. The only nonzero correlations assumed were negative correlations of -0.3 of the eccentricities and inclinations with masses.

The prograde starting spins have mean magnitudes in accord with the empirical rule that the angular-momentum densities were proportionate to $M^{5/6}$ (*Goldreich and Soter,* 1966), equivalent to spin proportionate to $M^{1/6}$ for homogeneous bodies. *Burns* (1975) found $M^{2/3}$ to be a better fit to the angular-momentum density, but the difference is slight for as narrow a mass range as Mercury to Earth; the resulting lengths-of-day (LODs) were 30 to 19 hr. The standard deviations for the Gaussian distributions about the mean were 50% of the mean spin rate, giving, for example, standard deviation LODs for Earth of 12.7 hr and 38 hr. The obliquities were assumed to have standard deviations varying inversely with the mean spin rate, from ±0.3, or ±17°, for Earth to ±28° for Mercury.

TABLE 1. *Starting planet properties.*

Proto-Planet	Mass $M_\oplus$	Radius $R_\oplus$	Spin Rev./Hr.	Obliquity ° Arc	a A.U.	e
"Mercury"	0.0444	0.3546	(1±0.5)/30.	0°±27°	0.387	0.21
"Venus"	0.6529	0.8825	(1±0.5)/19.	0°±17°	0.723	0.01
"Earth"	0.8000	0.9283	(1±0.5)/19.	0°±17°	1.000	0.02
"Mars"	0.0861	0.4911	(1±0.5)/27.	0°±24°	1.524	0.09

For spin and obliquity, the ± quantities are standard deviations of Gaussian distributions from which values were randomly selected for each simulation.

TABLE 2. *Starting planetesimal properties.*

Property p	Units	p_m	p_x	q
Mass m	—	0.100	1.000	0.5
Semimajor Axis	AU	0.400	1.800	0.5
Eccentricity e		0.060	0.700	0.8
Inclination I		0.030	0.350	0.8

Nonzero prescribed correlations:	M, e: -0.3 M, I: -0.3
Ratio of total mass to summed planetary mass:	0.25
Maximum allowable ratio of single planetesimal mass to total planetary mass:	0.10
Planetesimal density:	3.0 gm/cm³

Planetesimal Population Setup

In each Monte Carlo run, four random numbers x, $0 < x < 1$, were picked for each of the 50 planetesimals, to define parameters p = m, a, e, I, as described in Appendix A.

Encounter Probability Calculation

Since one test of the model is to produce orbital elements of the planets comparable to those observed, it is necessary to take into account the eccentricity of planetary orbits, as well as of planetesimal orbits, in calculating the probability of encounter between a given planetesimal and planet. Hence, the procedure given by *Öpik* (1976, pp. 35-50) was modified to allow for this noncircularity (see Appendix A). As in Öpik's work, a distance from the planet must be prescribed to define an encounter. The customary radius-of-influence proportionate to $a_p m_p^{1/3}$, where m_p is the mass and a_p is the semimajor axis of the planet, was used. In addition, probabilities per revolution were modified to probabilities per unit time, to obtain relative probabilities of encounter among all possible planet-planetesimal pairs, as discussed in Appendix A.

Given a complete set of relative probabilities, a specific planet and then a planetesimal are selected randomly, and three more random numbers are then selected to define the distance from the sun, the offset, the azimuth angle, and directions (inbound or outbound) of the two bodies in each encounter. Thus, seven random numbers are normally used for each encounter.

After an encounter, the probabilities for the encountering planetesimal are adjusted, but not for the other planet-planetesimal pairs, until the change in eccentricity of a planet orbit has exceeded a specified limit—0.001 was used—upon which all probabilities are recalculated.

Encounter Dynamics

The customary hyperbolic orbits in a reference frame fixed at the center-of-mass of the planet plus planetesimal were calculated (see Appendix B). Cases where the approach velocity was so slow as to make this approximation inaccurate, according to the criteria of *Wetherill and Cox* (1984), virtually never occurred.

For encounters not resulting in collision, the energy and angular momentum transfers resulting from the hyperbolic orbit were calculated, and both planetesimal and planet orbits adjusted. Planetesimals going into hyperbolic orbits in the heliocentric frame are dropped from the calculation; this virtually never happened (see Appendix B).

For encounters resulting in collision, the bodies were assumed to merge, and the angular momentum of the merged body calculated to get the revised rotation rate and obliquity. The energy dissipation consequent upon merger was also calculated.

The sums of energy dissipations under the foregoing assumption of merger were always more than the binding energy of Mercury; sometimes more for Mars; but never for Earth or Venus. In addition, the smaller bodies, but not the larger ones, spun up to rotational instability. The implications of these extreme outcomes for Mars and Mercury were not pursued (see Appendix B).

The runs were terminated for one of three conditions: (1) no crossings of planet and planetesimal orbits remained; (2) 3000 encounters were calculated; or (3) the calculated elapsed time reached 10^9 yr. In the 60 cases, termination (1) never happened; termination (2) occurred in about 20% of the runs; and termination (3) occurred in 80% of the runs. The elapsed time of 10^9 yr is artificially long because of the lack of long-range effects and the small ratio to influence radius used to define close (0.1) approaches in order not to use up too much computer time to get a majority of the planetesimals impacting. With the parameters used for run time and influence radii, typically 15 planetesimals were left at the end, half of them not intersecting planet orbits.

RESULTS

Sixty Monte Carlo runs were made. Histograms of the outcomes in energy dissipated, LOD, and obliquity for the four planets are given in Figs. 1-4.

While the averages of the 60 outcomes for Venus and Earth did not differ significantly from each other, the conjecture that a terminal phase of large impacts leads to a wide range of possible outcomes is supported. In particular, relevant to the actual outcomes for these two planets, there were (1) retrograde spins for 30% of the cases; (2) speed-ups of spins to angular momenta of the Earth and Moon (i.e., LOD of less than six hours) for 3% of the cases; (3) summed energy dissipations/unit mass ranging from 10^6 to 2×10^7 J/kg (equivalent to 1000 to 20,000°C temperature rise); (4) positive correlations (about 0.3) between spin rates and energy dissipations for both planets; and (5) orbit eccentricities ranging from 0.001 to 0.074. Furthermore, these effects may be too low because the $m^{-0.5}$ formula for masses gave maxima too low: 0.01 to 0.02 $M_\oplus$, less than suggested by *Wetherill's* (1986) figure.

It can be argued that giving outcomes separately, as in the foregoing paragraph, is deceptive, since in a given Monte Carlo run the results are coupled. There tends to be one body appreciably bigger than the rest; if Earth is hit by it, Venus is not, or vice-versa. Table 3 gives the coupling of events, a "fast prograde" rotation being defined as at least a doubling over the mean initial spin, i.e., an LOD less than 9.5 hr. It does show 7 of the 60 cases being combinations of fast prograde rotation for one planet and retrograde rotation for the other. Table 4 gives the energy dissipations for these cases. Of the seven, five obtain higher energy dissipation in the fast prograde spinner than in the retrograde spinner. Therefore the Earth and Venus appear to be quite possible, but not very probable, outcomes of origin by great planetesimal infall.

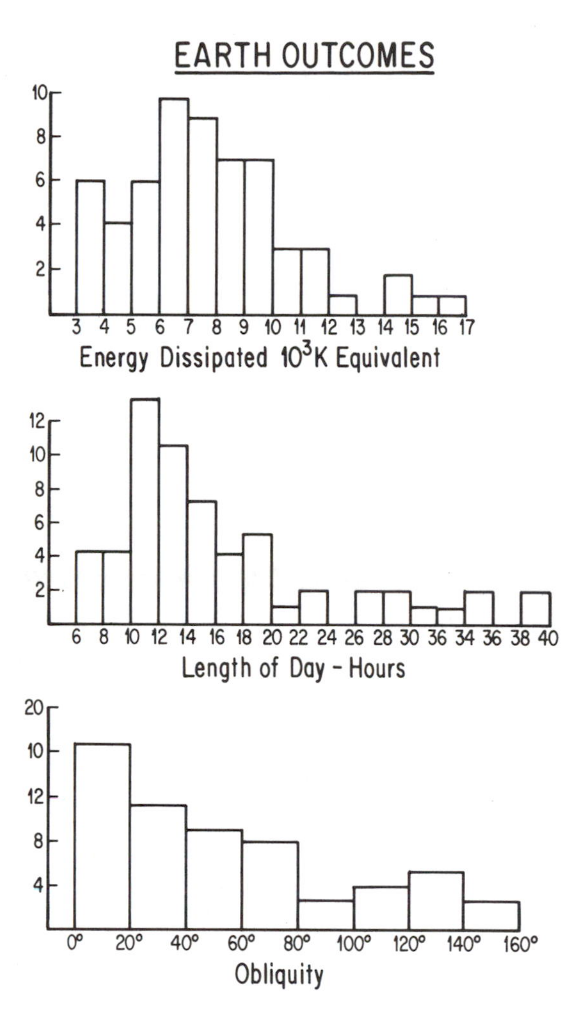

Fig. 1. *Histogram of "Earth" outcomes.*

The smaller outliers received considerably more impacts. The results for Mercury were fragmentation in all cases, i.e., energy dissipations more than twice binding (*Benz et al.*, 1988), equivalent to about 22,000°K. Mars, while having much greater dissipation per unit mass than Earth, did not reach this limit in any run.

DISCUSSION AND CONCLUSIONS

The modelings are thus consistent with the hypotheses that the spin plus satellite states of Venus and Earth, the greater loss of volatiles by the Earth, the stunted growths of Mars and Mercury, and the predominantly iron composition of Mercury are all results of the late stages of their growths being impacts by planetesimals with masses ranging up to well above the Moon's.

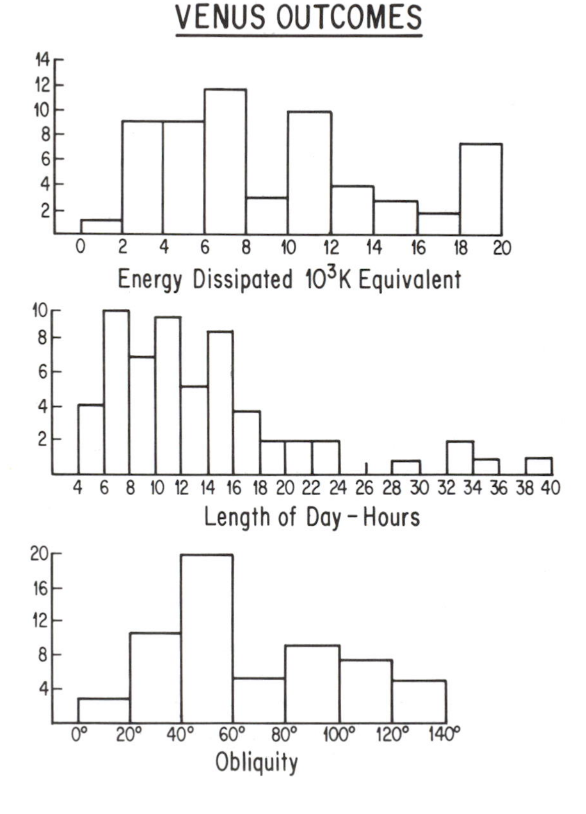

Fig. 2. *Histogram of "Venus" outcomes.*

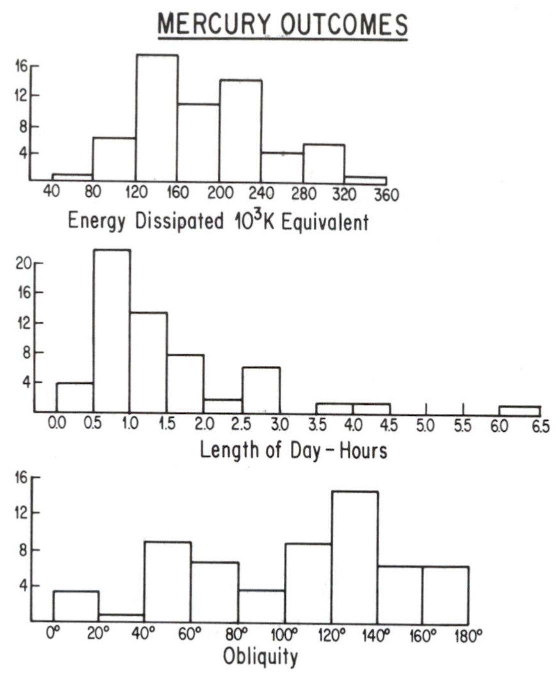

Fig. 3. *Histogram of "Mercury" outcomes.*

There remain several questions affecting differences between the Earth and Venus, which I discuss as (1) constraints from Mercury and Mars; (2) implications of heating by great impacts for volatiles; and (3) influences from the outer solar system.

Constraints from Mercury and Mars

The consequences in the model for proto-Mercury and proto-Mars were usually more violent than appears to have actually occurred for these planets. Starting conditions of the model that may be incorrect include the following. (1) 81% of final mass for Mercury and Mars, as well as Earth and Venus. It could be argued that Mars and Mercury accomplished more of their growth earlier, before they were subject to high-velocity impacts by bodies scattered by the larger planets. They might better be regarded as survivors, or fragments, of a larger population, rather than dominators of their zones (*Wetherill*, 1986). (2) The same distribution of mass and orbital elements of planetesimals in all zones. Plausibly, the eccentricities and inclinations in a zone should be positively correlated with the mass of the planet in the zone. Such a model could possibly lead to faster growth of the smaller planets, so that they could withstand impacts of planetesimals scattered by the larger planets.

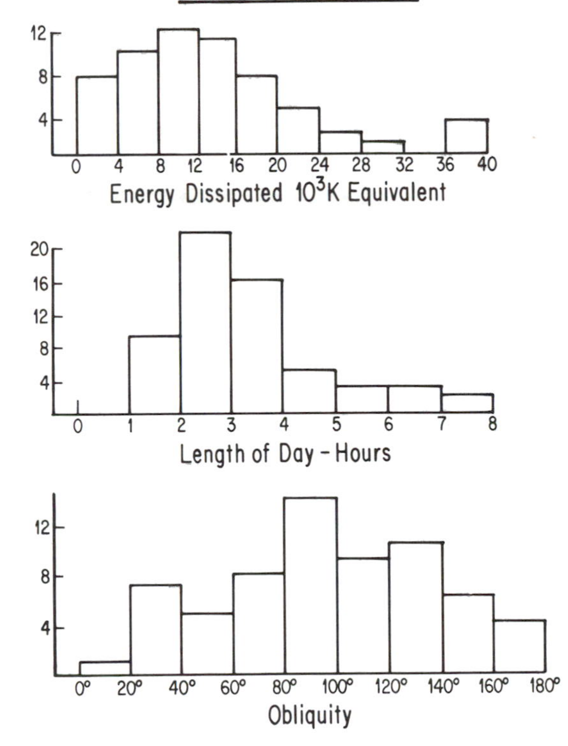

Fig. 4. *Histogram of "Mars" outcomes.*

Implications of Heating by Great Impacts for Volatiles

The differences in energy dissipated in some cases (Table 4) seem more than enough to cause great differences in loss by blowoff of primordial atmospheres. However, in a great impact only a minor part of an atmosphere is lost in the impact event itself because (1) most of the energy of an impact goes into the incompressible solids directly involved therein, rather than into the compressible gases, and (2) the compressive wave in the atmosphere to the sides of the impact travels faster in the upper atmosphere, thus driving the bulk of the atmosphere back toward the planet (*Walker,* 1986; *Melosh and Kipp,* 1989; *Benz and Cameron,* 1990). However, these characteristics act to make the atmospheric loss more chemically selective. The energy delivered is more than enough to make a magma ocean by convective transfer of heat soon after impact, indeed, enough to even vaporize most of the mantle, according to Benz and Cameron. This melted and vaporized mantle would have been hot enough to cause blowoff of atmosphere by a "planetary wind," similar to a stellar wind. In the phase where the mantle has cooled enough that a magma ocean exists in thermodynamic equilibrium with the atmosphere, the volatiles lost in a wind would have tended to be those that are *not* soluble in silicate melts. The solubility of argon is only 1/700 that of H_2O (*Zhang and Zindler,* 1989). Hence in this manner the Earth could possibly suffer severe loss of its noble gases while retaining its water. This

TABLE 3. *Coupling of Earth and Venus outcomes.*

		"Earth" Fast Prograde	"Earth" Slow Prograde	"Earth" Retrograde
"Venus"	Fast Prograde	0	6	5
"Venus"	Slow Prograde	3	19	9
"Venus"	Retrograde	2	15	1

TABLE 4. *Outcomes combining retrograde and fast prograde rotation.*

"Venus"			"Earth"		
Temp. °K	LOD Hrs.	Obliquity ° Arc	Temp. °K	LOD Hrs.	Obliquity ° Arc
14,000	5	44	6500	23	88
14,300	5	22	8100	33	132
18,500	8	91	9100	7	31
10,300	6	50	8300	19	123
10,500	8	119	4750	9	21
8,100	8	38	6800	44	157
19,300	7	6	3200	22	148

process should also result in appreciable loss of CO_2, which is 1/100 as soluble as H_2O. But the loss of CO_2 may have been much less than that of noble gases, since in planetesimal impacts the former was acquired mostly as carbonates, much of which was incorporated in the solid (or liquid) parts of the protoplanet, while much more of the noble gases was immediately outgassed to the atmosphere at these earlier impacts. As a consequence of iron plunging through to the core in a great impact, nearly all the mantle would be involved in a considerable stirring. However, if the distribution of impact-generated heat were concentrated more toward the outer layers, then much material may not have been circulated to the rock atmosphere (later the magma ocean) on top.

Influences from the Outer Solar System

A few runs were made to incorporate Jupiter, as suggested by *Kaula and Bigeleisen* (1975). However, these additional runs made little difference in the dynamics, partly because of the longer timescale at jovian distance, and partly because the energy required for Jupiter to scatter a body as far as the Earth is greater than for the more probable event of ejection from the solar system (*Weidenschilling,* 1975). Jupiter obviously did affect Mars, and might have played a significant role in sorting out the inner solar system through its effect on the secular rates (*Ward,* 1981). However, a more important implication of scattering from the outer solar system would have been its effect on volatiles. It has long been popular to account for the enrichment of volatiles in the outer Earth, in disequilibrium with the core, by a "late veneer" added from the outer solar system (*Hunten et al.,* 1988). The problem is how to keep these late impacts from adding noble gases as they add H_2O, CO_2, etc. The processes by which noble gases are incorporated in solids in laboratory experiments do not appear to be less effective at low temperatures (*Swindle,* 1988), so that comets should not have lower contents thereof than carbonaceous chondrites. But these laboratory processes appear to be quite inadequate to account for the abundances of "planetary" noble gases observed in chondrites. An alternative hypothesis is that noble gases are lost much more severely immediately upon the higher-velocity impacts of icy bodies from the outer solar system. However, laboratory experiments indicate that rather high temperatures are required to outgas noble gases from their sites, apparently adsorbed into carbon (*Swindle,* 1988). Hence, the primordial noble gas scarcity in the Earth may be a significant restriction on the addition of outer solar system matter, and possibly a second stochastic event was a large icy impact into Venus (*Hunten et al.,* 1988), but this event was not nearly as energetic as the impact into the Earth that created the Moon.

Clearly, there are several requirements to obtain more comprehensive and meaningful results: an earlier starting stage, with more bodies; a greater variety of starting conditions; more realistic modeling of collisions; and modeling of Jupiter effects. The work presented here required a 1400-line FORTRAN program and consumed considerable time on a Sun-3 work station. The results support the hypothesis that differences between the Earth and Venus could be stochastic outcomes of formation from large planetesimals. Perceptible refinement would be an appreciable programming and computational effort.

APPENDIX A: STATISTICS

Parameter Distributions

Many probability distributions of a parameter p can be defined by a density f proportionate to a power of p (see, e.g., *Safronov,* 1969)

$$f(p) = kp^{-q}, \; p_m < p < p_x \qquad \text{(A1)}$$

where p is a parameter with the prescribed density f between a minimum p_m and a maximum p_x, and the density f is subject to the condition $F(p_x) = 1$, where

$$F(p) = \int_{p_m}^{p} f(p')\, dp' \qquad \text{(A2)}$$

whence

$$\left.\begin{aligned} k &= (1 - q)/(p_x^{1-q} - p_m^{1-q}), \; q \neq 1 \\ &= 1/[\ln(p_x) - \ln(p_m)], \quad q = 1 \end{aligned}\right\} \qquad \text{(A3)}$$

Given a random number x, $0 < x < 1$, then by setting $F(p) = x$, a value of the parameter p can be recovered

$$\left.\begin{aligned} p &= [p_m^{1-q} + x(p_x^{1-q} - p_m^{1-q})]^{1/(1-q)}, \; q \neq 1 \\ &= \exp\,[x \ln(p_x) + (1-x)\ln(p_m)], \; q = 1 \end{aligned}\right\} \qquad \text{(A4)}$$

If a set of parameters p_i have correlations, as hypothesized for eccentricity and inclination with mass in this exercise, then an uncorrelated set of random numbers y_i, $0 < y_i < 1$, is converted to a correlated set x_i as follows. Let

$$f_i = x_i - 0.5, \; g_i = y_i - 0.5 \qquad \text{(A5)}$$

Then from the condition for the correlation coefficients r_{ij}

$$r_{ij} = \langle f_i f_j \rangle / \langle f_i^2 \rangle^{1/2} \langle f_j^2 \rangle^{1/2} \qquad \text{(A6)}$$

there are obtained for the transformation coefficients A_{ik}

$$f_i = A_{ik}\, g_k, \text{ summing over k} \qquad \text{(A7)}$$

$$\begin{aligned} A_{ik} &= r_{ik}/2 + r_{ik}^3/16 + \dots, \; i = k, \\ A_{ii} &= [1 - \textstyle\sum_{j \neq i} A_{ij}^2]^{1/2} \end{aligned} \qquad \text{(A8)}$$

A_{ik} is a unit strain matrix with zero dilatation. Hence sometimes an uncorrelated random number y_i close to 0 or 1 will produce a correlated number x_i outside these limits. However, for the modest correlations of -0.3, the resulting forays outside prescribed parameter limits are negligible in effect compared to differences of the assumptions from reality. By empirical test, calculating the r_{ij} from the f_i obtained in Monte Carlo runs, the approximate formula (A8) gives a reasonable fluctuation from one run to another.

Encounter Probabilities

Öpik (1976) and *Arnold* (1964) were concerned about the probabilities per unit revolution of close approach by a planetesimal with respect to a planet assumed to be in a circular orbit. In the present problem, there are needed probabilities per unit time for planetesimal orbits with respect to several elliptic planet orbits. Hence Öpik's equation (73), p. 37, $p = s^2U/\pi \sin I\, U_x$, where s is the ratio of target radius to planet orbit radius, U and U_x are total and radial velocities, and I is mutual inclination, must be modified to

$$P = s^2\, F\, F_p\, n\, n_p/\pi^2 \sin I \qquad \text{(A9)}$$

where n and n_p are mean motions of the planetesimal and planet respectively, and F and F_p are the portions of the time that the planetesimal and planet spend between the minimum and maximum radii that their orbits can intersect. Thus, setting S_i for sin E_i

$$F(E_i) = \int_{M_m}^{M_m} dM/\pi = \int_{E_m}^{E_i} (1 - e \cos E)\, dE/\pi$$

$$= [E_i - E_m - e (S_i - S_m)]/\pi \qquad (A10)$$

where the subscript m denotes the minimum radius at which intersection can occur, and the limits of M, the mean anomaly, and E, the eccentric anomaly, are 0 and π. The subscript x denotes the maximum radius. If the maximum is aphelion, then E_x is π; if the minimum is perihelion, then E_m is 0. If $E_x < \pi$ or $E_m > 0$, then the standard formulae for radius as a function of eccentric anomaly is inverted

$$E = \cos^{-1} [(1 - r/a)/e], 0 < E < \pi \qquad (A11)$$

But in the Monte Carlo study, what is given is not the radius r, but rather a random number x that corresponds to radius r and eccentric anomalies E_i (planetesimal) and E_p (planet)

$$x = F(E)F_p(E_p)/G$$
$$= F(r)F_p(r)/G, r_m < r < r_x \qquad (A12)$$

where

$$G = F(r_x)\, F_p(r_x) \qquad (A13)$$

and r_m is the greater of the two perihelia and r_x is the lesser of the two aphelia. Equation (A12) must be solved iteratively for r. The following procedure was found most efficient.

Let A_i, A_j be the two aphelia, and P_i, P_j be the two perihelia. Then if $A_j < A_i$ and $P_j > P_i$, the initial approximation is based on orbit j

$$E_j0 = x\, F_j + E_{jm} - e_j\, S_{jm}$$

$$E_{j1} = E_{j0} + e_j\, S_{j0} \qquad (A14)$$

If, e.g., $A_j < A_i$ and $P_j < P_i$, then the calculation in equation (A14) is based on orbit j for $x > 0.5$ and orbit i for $x < 0.5$. Next

$$r_1 = a_j [1 - e_j \cos (E_{j1})] \qquad (A15)$$

Then, using equation (A11) to get E_i from the r_1 given by equation (A15) and equation (A10) for F, and setting n = 1 to start

$$y_n = F_i(r_n)\, F_j(r_n)/G \qquad (A16)$$

If $x\text{-}y_n < 0$, set $r_x = r_n$; if $x\text{-}y_n > 0$, set $r_m = r_n$ for purposes of iteration (*not* for G recalculation). Then

$$dr/dy = G\, S_i\, S_j/r\, [a_i^2 e_i S_i F_i + a_j^2 e_j S_j F_j] \qquad (A17)$$

and

$$r_{n+1} = r_n + a_c\, (x - y_n)\, dr/dy \qquad (A18)$$

where a_c is an acceleration factor set at 1.0 to start, but reduced by a factor such as 0.8 if equation (A18) gives an $r_{n+1} < r_m$ or $> r_x$. The calculation then uses equation (A11) to solve for new E_i and E_j, and thence F_i and F_j by equation (A10), to use in equation (A16). For a limit of 0.3×10^{-3} on $x\text{-}y_n$, this procedure usually converged in three or four steps. If one of the orbits is driven to perihelion or aphelion, dr/dy = 0, and the second derivative must be used

$$d^2r/dy^2 = (\pi G/r)^2\, (\cos E_k) e_k\, (a/F_k)^2/2 \qquad (A20)$$

where k corresponds to the body closer to perihelion or aphelion. If $F_k = 0$ (as occurs at perihelion), then the previous d^2r/dy^2 is used; even an incorrect value prevents hangup (since d^2r/dy^2 is always positive at perihelion). Then

$$r_{n+1} = r_n + (x\text{-}y_n)^2\, d^2r/dy^2/2 \qquad (A21)$$

and return to equations (A11), (A10), (A16), etc.

The random number generator used was RAN1 in *Press et al.* (1986, p. 196).

APPENDIX B: DYNAMICS

Given planetesimal and planet orbit elements, plus approach offset d_o and azimuth A with respect to the sun-planet line both *in the plane of the planet orbit,* and whether an out- or inbound encounter for both bodies, then the two-body hyperbolic orbit in a frame defined by the planet orbit and the center-of-mass of the two bodies at time of planetesimal intersection of the plane can be calculated, as described by *Öpik* (1976, pp. 33-50), *Arnold* (1964), and *Kaula* (1968, pp. 244-247). Next, either (1) in a close pass, a transformation is made back to the heliocentric reference frame of the changed orbits or (2) in a collision, assuming merger, the resulting orbit, mass, rotation vector, and energy dissipation of the merged body are calculated. The dynamics are not novel, but are given here tersely as a record of what was done. Opportunity is also taken to correct an error in *Kaula* (1968), equation 5.3.40, p. 246, whose derivation I am unable to reconstruct [see equation (B9) below]. Planet parameters are subscripted p throughout, while planetesimal parameters are unsubscripted.

The planetesimal orbit radius is

$$r = r_p + d_o \cos A \qquad \text{(B1)}$$

Define q as

$$q = \dot{r}/r\,\dot{f}\,\sin I \qquad \text{(B2)}$$

where the radial and true anomaly rates, $\dot{r}$ and $\dot{f}$, are calculated from a, e, r by standard elliptic orbit formulae, the sign of $\dot{r}$ depends on whether an out- or inbound pass for the planetesimal, and I is inclination of the planetesimal orbit with respect to the planet's orbital plane. Then the rectangular coordinates of the unperturbed offset are

$$x_1 = r d_o \cos A/(1+q^2),\ x_2 = d_o \sin A,\ x_3 = -q x_1 \qquad \text{(B3)}$$

while the unperturbed relative velocity components are

$$u_1 = \dot{r} - \dot{r}_p,\ u_2 = r\dot{f}\cos I - r_p\dot{f}_p,\ u_3 = r\dot{f}\sin I \qquad \text{(B4)}$$

the out- or inboundness of the planet coming through $\dot{r}_p$.

The approach offset distance squared d^2 of the planetesimal from the planet (rather than planet orbit) is

$$d^2 = x_2^2 + x_1 d_o \cos A \qquad \text{(B5)}$$

The collision offset squared is

$$d_c^2 = (R+R_p)^2 + 2G(M+M_p)(R+R_p)/u^2 \qquad \text{(B6)}$$

where R, R_p and M, M_p are radii and masses of the bodies, and $u^2 = u_1^2 + u_2^2 + u_3^2$, the u_i from equation (B4).

If $d^2 > d_c^2$, then the deflection angle g of the hyperbolic orbit in the plane defined by the relative velocity and position vectors is obtained from

$$A = du^2,\ B = G(M+M_p)$$
$$\cos g = (A^2 - B^2)/(A^2 + B^2)$$
$$\sin g = 2\,\mathrm{sign}(u_2)\,\mathrm{sign}(x_1)\,AB/(A^2 + B^2) \qquad \text{(B7)}$$

The angles g and I, plus the mass ratio $m = M_p/(M+M_p)$, are then used to obtain the changes in relative velocity components du_i of the planetesimal referred to the planet orbit plane and radial direction by rotation from the hyperbolic orbit velocity deflection

$$R = \begin{Bmatrix} c_g - 1, & s_g c_I, & s_g s_I \\ -s_g c_I, & c_I^2 c_g + s_I^2 - 1, & s_I c_I (c_g - 1) \\ -s_g s_I, & s_I c_I (c_g - 1), & s_I^2 (c_g - 1) \end{Bmatrix} \qquad \text{(B8)}$$

where $c_g = \cos g$, $s_g = \sin g$, $c_I = \cos I$, and $s_I = \sin I$. Finally

$$\mathbf{du} = m\,\mathbf{R}\,\mathbf{u} \qquad \text{(B9)}$$

Next, the du_i and x_i are used to get the changes in components of angular momentum per unit mass, and the new planetesimal velocity components referred to the sun, which are used to get a new kinetic energy per unit mass. The new angular momenta and kinetic energy are then used to get new a, e, I; opposite changes therein get the new a_p, e_p, and I_p.

If $d^2 < d_c^2$, then the velocity components of the merged body are obtained by conservation of linear momentum, and the spin of the merged body by conservation of angular momentum about the mutual center of mass. The bodies are assumed spherical and homogeneous before and after collision. The energy dissipation D per unit mass of the merged body then works out to be

$$D = M M_p u^2/2(M+M_p)^2 + 3GM^2 \cdot [(1 + M_p/M)R^{5/3} - (1 + (M/M_p)^{5/3})]/5 R (M + M_p) \quad (B10)$$

Acknowledgments. This work was partially supported by NASA grant 05-007-002. The paper has been appreciably improved by a thorough and insightful review by S. J. Weidenschilling, which stimulated a complete redoing of the computations after program modification.

REFERENCES

Arnold J. R. (1964) The origin of meteorites as small bodies. In *Isotopic and Cosmic Chemistry* (H. Craig, S. L. Miller, and G. J. Wasserburg, eds.), pp. 347-364. North-Holland, Amsterdam.

Benz W. and Cameron A. G. W. (1990) Terrestrial effects of the giant impact. In *Origin of the Earth*, this volume.

Burns J. L. (1975) The angular momenta of solar system bodies: Implications for asteroid strengths. *Icarus, 25,* 545-554.

Cameron A. G. W. (1983) Origin of the atmospheres of the terrestrial planets. *Icarus, 56,* 195-201.

Counselman C. C. III (1973) Outcomes of tidal evolution. *Astrophys. J., 180,* 307-314.

Dobrovolskis A. R. and Ingersoll A. P. (1980) Atmospheric tides and the rotation of Venus: I. Tidal theory and the balance of torques. *Icarus, 41,* 1-35.

Donahue T. M. and Pollack J. B. (1983) Origin and evolution of the atmosphere of Venus. In *Venus* (D. M. Hunten, L. Colin, T. M. Donahue, and V. I. Moroz, eds.), pp. 1003-1036. Univ. of Arizona, Tucson.

Goldreich P. and Peale S. J. (1970) The obliquity of Venus. *Astron. J., 75,* 273-284.

Goldreich P. and Soter S. (1966) Q in the solar system. *Icarus, 5,* 375-389.

Harris A. W. (1977) An analytical theory of planetary rotation rates. *Icarus, 31,* 168-174.

Hartmann W. K. and Vail S. M. (1986) Giant Impactors: Plausible sizes and populations. In *Origin of the Moon* (W. K. Hartmann, R. J. Phillips, and G. J. Taylor, eds.), pp. 551-566. Lunar and Planetary Institute, Houston.

Hunten D. M., Pepin R. O., and Owen T. C. (1988) Planetary atmospheres. In *Meteorites and the Early Solar System* (J. F. Kerridge and M. S. Matthews, eds.), pp. 565-591. Univ. of Arizona, Tucson.

Kaula W. M. (1968) *An Introduction to Planetary Physics: The Terrestrial Planets.* Wiley, New York. 490 pp.

Kaula W. M. and Bigeleisen P. E. (1975) Early scattering by Jupiter and its collision effects in the terrestrial zone. *Icarus, 25,* 18-33.

Melosh H. J. and Kipp M. E. (1989) Giant impact theory of the Moon's origin: First 3-D hydrocode results (abstract). In *Lunar and Planetary Science XX,* pp. 685-686. Lunar and Planetary Institute, Houston.

Öpik E. J. (1976) *Interplanetary Encounters: Close-Range Gravitational Interactions.* Elsevier, Amsterdam. 155 pp.

Pollack J. B. and Black D. C. (1982) Noble gases in planetary atmospheres: Implications for the origin and evolution of atmospheres. *Icarus, 51,* 169-198.

Press W. H., Flannery B. P., Teukolsky S. A., and Vetterling W. T. (1986) *Numerical Recipes: The Art of Scientific Computing.* Cambridge Univ., Cambridge. 818 pp.

Safronov V. S. (1969) *Evolution of the Protoplanetary Cloud and Formation of the Earth and the Planets.* Nauka, Moscow. Translated by the Israel Program for Scientific Translation (1972).

Safronov V. S. and Zvjagina E. V. (1969) Relative sizes of the largest bodies during the accumulation of the planets. *Icarus, 10,* 109-115.

Swindle T. D. (1988) Trapped noble gases in meteorites. In *Meteorites and the Early Solar System* (J. F. Kerridge and M. S. Matthews, eds.), pp. 535-564. Univ. of Arizona, Tucson.

Walker J. G. C. (1986) Impact erosion of planetary atmospheres. *Icarus, 68,* 87-98.

Ward W. R. (1981) Solar nebula dispersal and the stability of the planetary system. I. Scanning secular resonance theory. *Icarus, 47,* 234-264.

Weidenschilling S. J. (1975) Mass loss from the region of Mars and the asteroid belt. *Icarus, 26,* 361-366.

Wetherill G. W. (1981) Solar wind origin of ^{36}Ar on Venus. *Icarus, 46,* 70-80.

Wetherill G. W. (1985) Occurrence of giant impacts during the growth of the terrestrial planets. *Science, 228,* 877-879.

Wetherill G. W. (1986) Accumulation of the terrestrial planets and implications concerning lunar origin. In *Origin of the Moon* (W. K. Hartmann, R. J. Phillips, and G. J. Taylor, eds.), pp. 519-550. Lunar and Planetary Institute, Houston.

Wetherill G. W. and Cox L. P. (1984) The range of validity of the two-body approximation in models of terrestrial planet accumulation. I. Gravitational perturbations. *Icarus, 60,* 40-55.

Zhang Y. and Zindler A. (1989) Noble gas constraints on the evolution of the Earth's atmosphere. *J. Geophys. Res., 94,* 13719-13737.

THE GIANT IMPACT AND EARTH-MOON RELATIONSHIPS

The Giant Impact hypothesis for the origin of the Moon has gained great popularity since the LPI conference on the Origin of the Moon. In short, the model hypothesizes that a Mars-sized planetesimal collided with the Earth after core formation was essentially complete. The vaporized ejecta then coalesced in orbit about the Earth and formed the Moon. This model could explain the absence or near-absence of a metallic core in the Moon and could also explain the high angular momentum of the Earth-Moon system.

The physical processes involved in the Giant Impact hypothesis require more extensive study before strong constraints can be placed on the chemical composition of the Moon. In principle, the Moon could (1) resemble the Earth, (2) resemble the impactor, or (3) have a composition that is dominated by complex vaporization processes. Distinguishing between the possibilities may be difficult and should provide ample opportunities for further research.

The three papers in this section are closely related to the Giant Impact hypothesis:

W. Benz and A. G. W. Cameron: *Terrestrial Effects of the Giant Impact*
H. J. Melosh: *Giant Impacts and the Thermal State of the Early Earth*
J. H. Jones and L. L. Hood: *Does the Moon Have the Same Composition as the Earth's Upper Mantle?*

Benz, Cameron, and Melosh all agree that the Giant Impact would have almost completely melted the Earth. In addition, Benz and Cameron calculate that the surface temperature of the Earth could reach 16,000 K. This temperature is believed to be sufficient to boil off the atmosphere, if it had not already been ejected during the Giant Impact. These effects are profound and should have had immense petrological and geochemical consequences. One of these consequences is the likely presence of a terrestrial magma ocean; this will be discussed in more detail in the next section.

If a giant impact did bring about the origin of the Moon, it is reasonable to suppose that the Earth and Moon may be similar in composition. Jones and Hood have used the bulk physical properties of the Moon (density and moment of inertia) to explore the likelihood that the Earth and Moon are compositionally similar. In general, the terrestrial upper mantle is not a good starting composition for the Moon. If the Giant Impact hypothesis is correct, a significant amount of the Moon may have been contributed by the impactor. This conclusion is consistent with the calculations of Mueller et al. (1988, *Journal of Geophysical Research, vol. 93*, 6338-6352).

There is little discussion in this section of competing hypotheses for the origin of the Moon (e.g., co-accretion or capture). This omission reflects, at least in part, the popularity of the Giant Impact hypothesis and the desire to explore the consequences of that model for the early Earth. Interested readers should consult *Origin of the Moon* (W. K. Hartmann, R. J. Phillips, and G. J. Taylor, eds., Lunar and Planetary Institute, 1986) to evaluate the status of other hypotheses such as co-accretion.

The papers in this section indicate the necessarily close intertwining of geophysical and geochemical investigations. The geophysical studies have geochemical implications and vice versa. To really decipher the events that formed the Earth, we will have to find solutions that satisfy both disciplines.

TERRESTRIAL EFFECTS OF THE GIANT IMPACT

W. Benz and A. G. W. Cameron

Harvard-Smithsonian Center for Astrophysics, 60 Garden Street, Cambridge, MA 02138

Although the central goal of our investigations of the Giant Impact has been to find the optimum parameters that may have led to the formation of the Moon, a secondary goal has been to find what effects the collision would have had on the early Earth. We have found that the effects are rather dramatic, and that they do not vary a great deal from one case to another. The iron from the core of the Impactor plunges through the mantle of the proto-Earth and settles on top of its core. This iron is heated to a temperature of typically several tens of thousands of degrees. This will become an important heat source to drive convection in the mantle. Most of the rock from the mantle of the impactor, together with a considerable amount of rock ejected from the mantle of the proto-Earth during the collision, falls back onto the proto-Earth. There is at least some heating of all parts of the interior of the proto-Earth, but the rock in the surface layers is vaporized; the rock vapor at the surface of the proto-Earth has a temperature of typically 16,000 K. This condition, combined with the presence of an orbiting disk of rock vapors and magmas, is expected to have caused the ejection of the early terrestrial atmosphere.

INTRODUCTION

In *Benz et al.* (1986, 1987, 1989) we described in some detail a three-dimensional smoothed particle hydrodynamics (SPH) code and gave some results of our initial investigations of the single-impact hypothesis for the origin of the Moon (we have been referring to this event as the Giant Impact). The SPH procedure used here divides the mass of the proto-Earth plus that of the impactor into 3008 particles, which are free to move around in response to the gravitational and pressure gradient forces exerted on them. Local values of density and pressure are determined using local averages over the neighboring particles. Both the proto-Earth and the impactor are represented as bodies with rock (i.e., dunite) mantles and iron cores, with equations of state calculated using the Chart D radiation-hydrodynamic code (*Thompson and Lauson,* 1984, generally called the ANEOS equation of state). We assumed the proto-Earth and the impactor to have been preheated by accumulative collisions to 4000 K, and we took the interior temperatures to be initially isothermal to facilitate interpretation of the effects of the Giant Impact. For some details of the code testing and the equation of state, see *Benz et al.* (1989).

For the last two years we have have been systematically investigating the parameter space of the Giant Impact, varying the mass ratio, the angular momentum, and the velocity of the collision. We have taken the angular momentum in the collision to vary from 1.13 to 1.9 times the present value for the Earth-Moon system. We have taken the mass ratio of impactor to proto-Earth to have values of 0.14, 0.16, and 0.25, with the sum of the masses always one Earth mass. We have taken velocities at infinity to have values of 0, 5, and 7 km/sec (i.e., approximately equal steps in initial kinetic energy). At the present time the rate of loss of angular momentum from the Earth-Moon system due to solar tides on the Earth is about 0.3 times the rate of transfer of angular momentum from the spin of the Earth to the lunar orbit by lunar tides, and this fractional value was obviously smaller in the past (*Munk and MacDonald,* 1960). Because of the nonlinearity of the ratio the initial value is uncertain, and it may have been significantly changed by major terrestrial impacts subsequent to the Giant Impact.

RESULTS OF OUR SIMULATIONS

In our complete series of runs, we have carried out 41 simulations of the Giant Impact. We have observed a considerable variety of behavior in these collisions. For lower values of the angular momentum, one collision takes place with the Impactor and typically one to two lunar masses of rock are placed into orbit, with this material typically divided about evenly between the region inside the Roche lobe and outside of it. For the 0.14 mass ratio, higher values of angular momentum lead to the injection of considerable amounts of iron into orbit, which would be inconsistent with a lunar formation scenario. For higher values of the mass ratio, increased angular momentum produces little or no injection of iron into orbit, and a wide range of angular momentum is consistent with the formation of the Moon, provided that subsequent collisions can produce suitable adjustments to the angular momentum in the system.

For the two higher mass ratios, as the angular momentum increases, much of the material placed into orbit beyond the Roche lobe sometimes clumps together to form a body that may range in mass from a quarter of that of the Moon to even a little more than the Moon. Such large clumps are usually in stable orbits, but occasionally they are in orbits that have perigees within the Roche lobe, and when making the first perigee passage these clumps become tidally disintegrated. They are too hot to have enough dissipation to resist this disintegration.

For still larger values of the angular momentum, the remnant of the Impactor draws itself together after passing out of the Roche lobe. After passing through an apogee, it plunges back and has a second collision with the proto-Earth. This time it is fully destroyed and is generally drawn out into a long arc of material, mostly rock. The material in this long arc usually draws itself together gravitationally into several relatively small clumps. These clumps generally either escape from the system or are on orbits with perigees within the Roche lobe, so that they will be tidally disintegrated.

Hence for both low and high values of the angular momentum the result of the Giant Impact is a disk of mostly or completely rock material in orbit about the proto-Earth, distributed between the regions inside and outside the Roche lobe. For intermediate values of the angular momentum, most of the material outside the Roche lobe may occasionally form a large clump, with the material inside the Roche lobe forming a disk. In both cases the disk inside the Roche lobe has a high enough surface density that the self-gravity of the disk is important.

It is not clear which of these two types of scenario should be preferred for formation of the Moon. We do not address this question further here. For the early history of the Earth all scenarios indicate the presence of an orbiting disk, the dissipation of which will have significant consequences.

FALL OUT ON THE PROTO-EARTH

In the initial stages of the Giant Impact, a large crater, extending down nearly to the core of the proto-Earth, is excavated, and much of the excavated material is temporarily thrown into space, only to fall out onto the surface of the proto-Earth shortly thereafter. The crater closes up within a few minutes. If the collision is the only one at low angular momentum or is the second one at higher angular momentum, the impactor is destroyed by being drawn out into a rather long arc. The impactor material in the hemisphere closest to the site of the collision mingles with the ejected proto-Earth material and falls out with it. That leaves the iron from the impactor core in the near half of the arc together with impactor mantle material, and the material in the far half of the arc is almost entirely impactor mantle material from the hemisphere away from the site of the collision.

The material from the arc starts falling onto the proto-Earth right away, but the point of fallout moves progressively across the limb of the planet. It is generally true that most or all of the iron in the arc falls onto the proto-Earth before the point of fallout moves off the limb. When that occurs, the fallout turns into passage through a perigee, and the remainder of the impactor mantle, together with a small amount of mantle from the proto-Earth, goes into orbit about the proto-Earth. The inner part of the orbiting arc predominantly contributes material of lower angular momentum likely to end up inside the Roche lobe when the orbits are circularized through collision, whereas the more remote material has more angular momentum and is more likely to end up in orbit outside the Roche lobe.

The impactor iron that falls out is denser in general than the proto-Earth dunite mantle, and within a few minutes it has plunged through the mantle and has settled on top of the proto-Earth core. The mantle rock which falls out has generally been highly shocked by the Giant Impact followed by the fallout collision, so this material is largely vaporized and forms a massive rock vapor atmosphere on the proto-Earth.

This situation is shown in Fig. 1, for the case DEF3, described in *Benz et al.* (1989). This is for a mass ratio of 0.14, an angular momentum of 1.13 present Earth-Moon units, and a velocity of 0.28 km/sec at infinity. This is one of the "gentler" cases that we have run; the damage shown is thus likely to be a minimal. The plot shows the particles that end up in the proto-Earth, plotted according to their radial distance and temperature.

There are two artifacts of the collision that show in this figure and tend to be confusing. One of these is an oversimplification associated with the use of the ANEOS equation of state. This was designed to give a proper treatment of the mixed phases at a given temperature and pressure, so that it will generally indicate, for example, that the material is composed of some fraction that is solid or liquid and some complementary fraction that is gaseous. However, a particle in the SPH treatment remains an indivisible particle. Thus it may be seen in Fig. 1 that the fallout iron particles are typically heated to a temperature of several tens of thousands of degrees, a surprisingly high value predicted by the shock Hugoniot using the ANEOS equation of state for iron. Under these conditions some of the material is predicted to be gaseous or near the critical point for iron. The mean density of the particle is high enough that the iron on top of the core is not bouyant in the presence of rock, but the iron vapor by itself

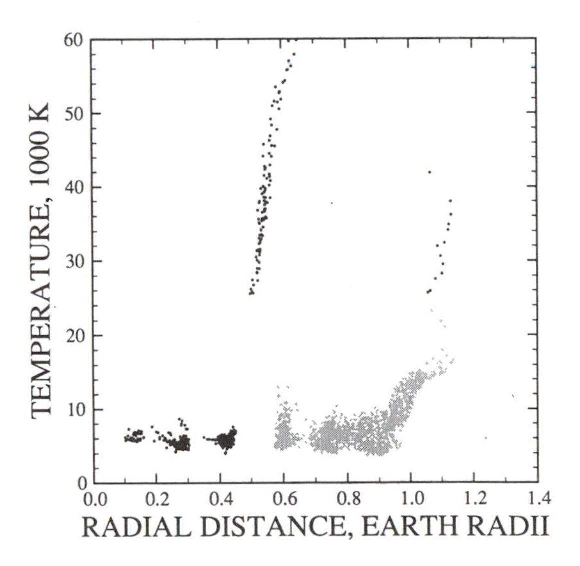

Fig. 1. *A plot showing all the particles that ended up in the proto-Earth in run DEF3. Iron particles are shown as filled black circles, and rock particles are shown as filled gray circles. See the text for the explanations of certain features as artifacts of the calculation.*

would be. Thus, our procedure obscures many important chemical consequences of the Giant Impact. A similar consideration concerning the hot rock at the surface of the proto-Earth indicates that, due to the indivisibility of the rock particles, the structure shown in Fig. 1 does not properly represent a structure that would consist of an underlying magma overlain by rock vapor decomposition products.

Since this application of SPH involves the division of the mass into 3008 particles, these particles tend to settle into a configuration of least energy, which means that they form a regular structure. This structure has the particles approximately evenly spaced on the surfaces of several concentric shells, which accounts for the radial clustering of the particles in the figures. This structure largely survives the collision in the deeper parts of the proto-Earth, although all particles deep in the proto-Earth receive at least some heating (at least 500 K). Thus the iron particles originally in the proto-Earth appear as clumps at particular values of the radius, rather than being smoothly distributed. In addition, since density is a locally averaged property of the particles in SPH, and since all the rock and iron particles have the same mass and effective smoothing length in these calculations, a very hot mixed-phase iron particle that becomes isolated in the process of the dynamical motion is not denser than the surrounding rock on which it lands. Thus, several such isolated iron particles show up in Fig. 1 at the surface of the proto-Earth; we know that in fact the condensed phases in this iron would also sink through the mantle to the core. There are also even hotter iron particles at the base of the mantle, but these are not quite bouyant enough to rise through the rock; the hot ANEOS iron is more compressible than the ANEOS dunite, so that there is no contradiction between these conditions.

We show in Fig. 2 one of our more energetic impacts in which the damage to the proto-Earth is somewhat greater than that shown in Fig. 1, but the general character of the results is the same. This is for run CO12, with a mass ratio of impactor to proto-Earth of 0.25, a velocity at infinity of 7 km/sec, and an angular momentum of 1.7 times the present angular momentum of the Earth-Moon system. Because of this relative insensitivity of the general character of the results to the details of the impact, our predictions for the terrestrial consequences of the Giant Impact are more secure than our predictions about the detailed distribution of the material left in orbit, which varies considerably.

One of the characteristic chemical features of the Earth's mantle is the departure from chemical equilibrium; nickel and other siderophile elements would be extensively scoured out from the mantle if it came into intimate contact with the infalling iron subsequent to the Giant Impact. However, there is relatively little contact between the two. Following the impact, the fall out of material traverses part of a great circle in the plane of the impact, and the impact itself sets the proto-Earth in rotation, so that the mantle

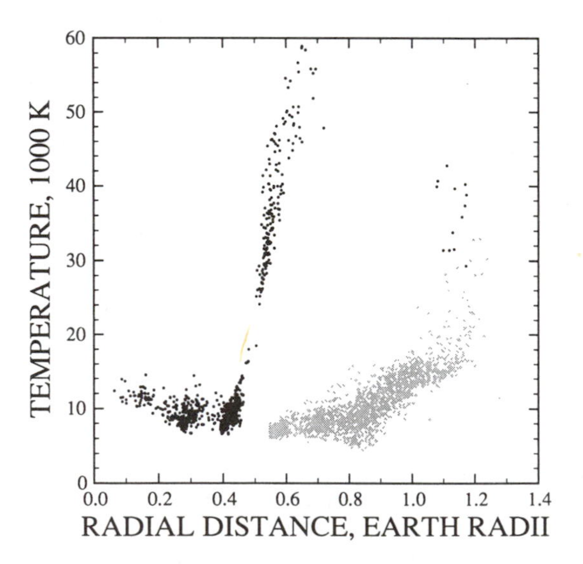

Fig. 2. *A plot showing all the particles that ended up in the proto-Earth in run CO12. Iron particles are shown as filled black circles, and rock particles are shown as filled gray circles. See the text for the explanations of certain features as artifacts of the calculation.*

moves in the same direction as the impact point of the infalling material. This confines the impact area of the iron fall out to within a small part of the great circle on the surface of the mantle. The iron can spread out once it reaches the bottom of the mantle, since it floats on top of the core. There will also be an addition of siderophile elements to the mantle from subsequent accretion of smaller planetesimals onto the surface of the proto-Earth.

LOSS OF THE EARLY TERRESTRIAL ATMOSPHERE

The Giant Impact itself thoroughly devastates nearly one hemisphere of the proto-Earth, and thus the atmosphere associated with that hemisphere would be at least temporarily ejected. However, most of the rock associated with that hemisphere falls back onto the proto-Earth, and we have no evidence from our calculations that would enable us to argue that the ejected atmosphere does not also fall back. The opposite hemisphere, at least initially, is not greatly heated by the Giant Impact, although this may change if there is a second collision (this depends on the site of the second impact). We do find that a layer of very hot vaporized rock flows around to this second hemisphere and covers it. The original terrestrial atmosphere would float on top of these rock vapors but could only be thermally ejected, if at all. Thus, the question of whether the Giant Impact could eject the early terrestrial atmosphere reduces to the question of whether that atmosphere could thermally escape due to the heat generated in the Giant Impact.

Consider first the initial structure of the rock vapor atmosphere. The rock is vaporized during either the initial collision or the fallout collision, so that the hottest vapors will rise to the top of the rock vapor atmosphere. This differs from the usual situation in an atmosphere where the heat source is at the base. Thus the initial atmosphere has the temperature increasing outward and convection is impossible. Remember that in all of our runs we have found a disk of rock to be placed in orbit, and any dissipation of this disk will lead to a continual addition of material from the inner part of the disk to the proto-Earth, thus maintaining a supply of hot rock vapors at the top of the rock-vapor atmosphere. However, dissipation of the disk is likely to be slow under conditions in which the disk is entirely vaporized.

The primitive atmosphere of the Earth is likely to have been dominated by CO_2 and H_2O. At 16,000 K, these gases will be mostly dissociated and ionized. Thermal escape from a planetary exosphere occurs fairly rapidly for particle velocities of a few $\times$ kT (say up to about 4kT for the case here). To be general, let us say that escape can occur for particle kinetic energies $mv^2/2 = \alpha kT$, where α is a factor of order 4. We may evaluate this as $v = 1.5\sqrt{\alpha/M}$ in units of the escape velocity from the Earth (11 km/sec). Here M is the mass of the escaping particle in atomic mass units, and T has been taken as 16,000 K. Thus it may be seen that hydrogen (from dissociation of water) would rapidly escape for any reasonable value of α. Carbon and oxygen atoms would also rapidly escape for $\alpha = 4$.

If this primitive atmosphere had a mass comparable to the present atmosphere of Venus, there would be a tendency for this atmosphere to form a blanket on top of the rock-vapor atmosphere, forming a convective upper atmosphere, and giving a reduced effective radiating temperature. However, at a temperature of 16,000 K at the base of this blanketing atmosphere, the scale height would be $H = 1.4 \times 10^4/M$ km, where M is the mean atomic weight in atomic mass numbers. At 16,000 K we would have complete dissociation and about one stage of ionization for the common constituents of the atmosphere, so $M \approx 8$. Thus, $H \approx 1800$ km, or about 30% of an Earth radius. Furthermore, the temperature would fall relatively slowly with distance because the ratio of specific heats becomes not much greater than 1 through the region of recombination and association, which thus would cover several scale heights, each getting larger because of the reduced gravity at the relevant distances. Thus, in this scenario it appears that the common constituents of the atmosphere would be rapidly lost in a hydrodynamic flow, resembling the hydrodynamic flow of the solar wind but with a much shorter timescale. The less common constituents of the atmosphere, including the heavier noble gases, would also mostly be lost through entrainment in this hydrodynamic flow.

Any atmospheric remnant after the hydrodynamic flow, including heavy noble gases, could escape, if at all, only in a different manner (see *Cameron,* 1983). Consider xenon by itself, a worst-case example. Its scale height would be about 240 km, which would extend to the vicinity of the orbiting rock disk, which would also be in the form of vapor in the radiation field of the proto-Earth. From our results it is reasonable to expect a large fraction of a lunar mass to exist in the disk within the Roche lobe, and to expect significant amounts of mass beyond the Roche lobe (too hot to undergo gravitational separation into separate bodies initially; see *Stevenson,* 1987). This means that the self-gravitation of the disk is likely to exceed the vertical component of gravity within the disk. The xenon atmosphere, coming within this disk gravitational field, would thus be free to spread out to the outer limits of the disk surfaces, which may extend beyond the Roche lobe. The rate at which this happens may be hindered by diffusion into the rock vapor, which, being heated from the surfaces, will not be convective.

At the Roche lobe (about 3 Earth radii) the escape velocity is reduced to about 6 km/sec, and the orbital velocity is reduced to about 4 km/sec. Thus the xenon atom or ion needs to gain about 2 km/sec to escape in the forward direction of the orbital motion. At 16,000 K the mean thermal motion of a xenon atom is about 1 km/sec, so escape at that temperature is possible with $\alpha = 4$. Allowing for geometrical dilution of the radiation energy density would require a value of α between 5 and 6. Thus, it may be possible for the residual xenon to slowly escape, but this is by no means assured.

It should not be overlooked that if xenon can escape by this mechanism, so can the rock vapor from the disk itself (in fact, it can escape more easily). This is a major effect that needs to be considered in working out the consequences of the Giant Impact.

In any case, it seems likely that the heavy noble gases will be mostly lost by entrainment in the hydrodynamic flow of the major constituents of the early atmosphere. Entrainment will be least efficient for xenon, and mass fractionation will be important. Xenon in the present terrestrial atmosphere is quite significantly mass fractionated, so this may indicate the presence of a remnant of the primitive terrestrial atmosphere (*Cameron,* 1983).

DISCUSSION

The iron from the impactor that settles on top of the proto-Earth core is extremely hot, and it will sort itself out so that the hottest iron is at the top of the core, at the core-mantle boundary. It may be seen in the figures that this has already happened on the timescales of our runs. The stratification of this iron protects it against any convective overturn, so that the heat can be transferred to the core only by the slow process of thermal conduction.

This hot iron thus serves as a heat source capable of driving vigorous convection in the mantle. However, a full system of convection cannot start right away. The very hot surface layers of the mantle are of higher entropy than the base of the mantle, and the surface layers must get rid of all their excess heat before the mantle as a whole can become slightly superadiabatic and thermally-driven convection can begin. Nevertheless, convection can start in the lower mantle as the rock at the core-mantle boundary is heated to the level of the underlying hot iron. In fact, what will happen at this initial stage is that much of the heat content of the hot iron will be transferred to the rock at the base of the mantle. The entropy of the base rock will be raised, and convection will establish an isentropic lower mantle extending outward to the level at which the entropy begins to increase toward the surface. Once this has been achieved, convection will cease for all practical purposes; only very slow convective motions will be needed to maintain the lower mantle isentropy. As the upper mantle cools, the isentropic condition can slowly spread to the surface. When that condition is achieved, the top layers of rock can actively radiate away their energy, thereby greatly increasing the heat flux from the interior. Only at that point can the convection within the whole mantle become vigorous.

Thus the Giant Impact makes predictions in a number of areas that have been controversial. The Earth as a whole has been totally molten (but later infalling material may have introduced new inhomogeneities). The upper layers of the mantle were heated so much that the Earth's primitive atmosphere thermally escaped, taking less abundant heavier volatiles with it. The removal of volatiles may have included a number of elements that are not normally part of an atmosphere but which are nevertheless significantly volatile. The Earth is greatly depleted in a number of such volatile elements (the Moon is even more depleted in them). The high temperature at the surface of the Earth plays an important role in the behavior of material in space around the Earth, and must be considered in the environment within which formation of the Moon is postulated to occur.

Acknowledgments. We thank J. Melosh for useful discussions. This work has been supported in part by NASA grants NGR 22-007-269 and NAGW-1598. W.B. thanks the Swiss National Science Foundation for partial support.

REFERENCES

Benz W., Slattery W. L., and Cameron A. G. W. (1986) The origin of the Moon and the single impact hypothesis I. *Icarus, 66,* 515-535.

Benz W., Slattery W. L., and Cameron A. G. W. (1987) The origin of the Moon and the single impact hypothesis II. *Icarus, 71,* 30-45.

Benz W., Cameron A. G. W., and Melosh H. J. (1989) The origin of the Moon and the single impact hypothesis III. *Icarus, 81,* 113-131.

Cameron A. G. W. (1983) Origin of the atmospheres of the terrestrial planets. *Icarus, 56,* 195-201.

Munk W. H. and MacDonald G. J. F. (1960) *The Rotation of the Earth.* Cambridge Univ., Cambridge.

Stevenson D. J. (1987) Origin of the Moon: The Collision Hypothesis. *Annu. Rev. Earth Planet. Sci., 15,* 271-315.

Thompson S. L. and Lauson H. S. (1984) *Improvement in the Chart D Radiation-Hydrodynamic Code III: Revised Analytic Equations of State.* Sandia National Laboratory Report SC-RR-71 0714.

GIANT IMPACTS AND THE THERMAL STATE OF THE EARLY EARTH

H. J. Melosh

Lunar and Planetary Laboratory, University of Arizona, Tucson, AZ 85721

The putative giant collision between the proto-Earth and a protoplanet half its diameter that created the Moon 4500 m.y. ago would have had a profound effect on the thermal state of the early Earth. The history of thermal models of the accreting Earth began with consideration of impacts by bodies small enough that most of their energy was radiated to space before it became incorporated in the growing Earth. Such models had difficulty explaining the onset of melting and differentiation in any planet. More recent models examined the physics of energy deposition by impacts of bodies larger than a few tens of kilometers and concluded that a large fraction of their energy is retained as heat. These models indicate that melting should have begun in the Earth when it had grown to about 10% of its present mass. In both types of model the Earth is assumed to grow gradually from bodies much smaller than the growing planet itself. However, recent work on the size-frequency distribution of the planetesimal swarm indicates that most of the mass and energy of the swarm resides in objects at the large end of the spectrum, so that most of the Earth's growth may have occurred in catastrophic impacts with bodies up to half its own diameter. In such collisions heat is deposited deep within the growing Earth, and at the latest stages of accretion the impacts may have melted the entire impacted hemisphere. The detailed effects of such a giant impact are explored using a 3-D numerical hydrocode to simulate the collision between the proto-Earth and a body half its diameter at two different impact velocities.

INTRODUCTION

The idea that the Moon was born in a gigantic collision between the proto-Earth and a Mars-sized protoplanet has continued to gain adherents since it was first proposed by *Hartmann and Davis* (1975) and by *Cameron and Ward* (1976). This theory received a great deal of attention at the 1984 Conference on the Origin of the Moon in Kona, Hawaii (*Hartmann et al.,* 1986), and has subsequently been the subject of a number of recent reviews (*Boss,* 1986; *Newsom and Taylor,* 1989; *Stevenson,* 1987). It is probably not an exaggeration to claim that it has become the current consensus theory of the Moon's origin. Although a great deal of work has been, and is being, done on the details of this scenario, most work up until now has concentrated on the Moon; little consideration has been given to the effects of such a large impact on the Earth. Nevertheless, it is clear that such an event would have profound effects on the Earth, most especially on its thermal state. Although the last word has by no means been written on this problem, in the following pages I present a summary of what is currently known about how impacts, particularly giant ones, may have affected the thermal regime of the early Earth. Although this paper is largely tutorial, near the end I will present some of the results of recent computations I have been doing in conjunction with Marlan Kipp on the early stages of a giant impact and its implications for the early thermal evolution of the Earth. Another view of this event can be found in the paper by *Benz and Cameron* (1990).

ENERGY DEPOSITION BY IMPACTS ON GROWING PLANETS

The history of accretion models of the Earth and planets mirrors a gradually increasing appreciation for the importance of large impacts. In one of the earliest models of the Earth's accretion (*Hanks and Anderson,* 1969; *Urey,* 1952, p. 110 ff) the gravitational energy of the infalling debris was exactly balanced by thermal radiation from the surface of the growing planet. Although these models did not include the effect of conduction into the planet's interior, this was included in later work (*Mizutani et al.,* 1972). I will distinguish models of this type as the "physicist's model" of accretion. Shown in Fig. 1, such models assume material falling onto the planet in time dt is added to the growing planet in thin uniform shells of thickness dR. The mass added in this time is thus $dM = 4\pi\rho R^2 dR$, where ρ is density. Each unit mass deposits a net energy consisting of gravitational binding energy and initial kinetic energy of the mass, $GM/R + \nu_\infty^2/2$, where G is Newton's gravitational constant, M is the planet's mass when it has grown to radius R, and ν_∞ is the velocity of encounter between the growing planet and the infalling material at great distance from the planet (I neglect here any initial thermal energy). The basic equation describing the equilibrium between infall energy, radiation and conduction is

$$\rho\left(\frac{GM(t)}{R(t)} + \frac{\nu_\infty^2}{2}\right)\frac{dR(t)}{dt} = \epsilon\sigma[T^4(R,t) - T_a^4] + \rho c_p T(R,t)\frac{dR(t)}{dt} + k\left(\frac{\partial T}{\partial r}\right)_{r=R} \quad (1)$$

where T(R,t) is the temperature at the surface of the growing planet, T_a is the effective temperature of the atmosphere, σ is the Stefan-Boltzman constant, ϵ is the emissivity (generally set equal to 1), c_p is heat capacity and k is the thermal conductivity. Radius r is the position *within* the growing planet of total radius R(t). The second term on the right side accounts for the heat content of the hot added material and should include latent heat if melting occurs; it is a standard term in the heat conduction equation when moving boundaries are present (*Turcotte and Schubert,* 1982, p. 174 ff). The thermal state of a growing planet is determined by equation (1) as soon as the growth rate, dR(t)/dt, is specified. Equation (1) implicitly assumes that the added layer of mass is both laterally uniform and infinitesimally thin, assumptions that, as we shall see shortly, lead to major errors.

The studies based on equation (1) had great difficulty explaining why any of the planets should be differentiated. Thermal radiation is so efficient at removing energy that the planets would have had to have grown extremely fast

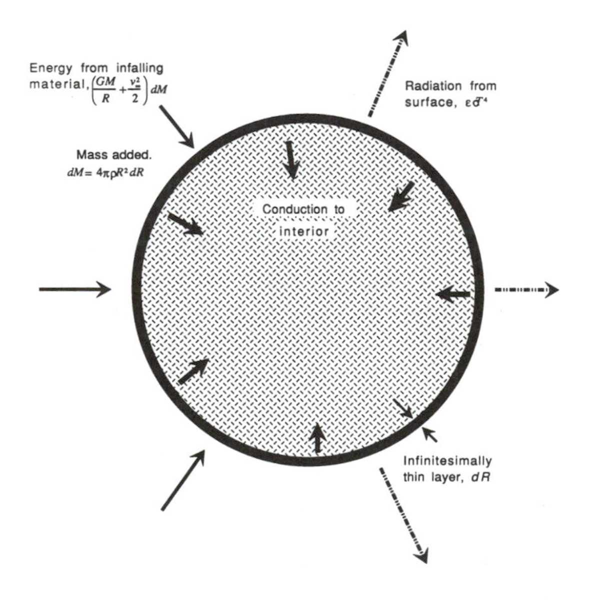

Fig. 1. *The "physicist's model" of planetary accretion, current between about 1950 and 1975, assumes that the gravitational energy of infalling material is balanced primarily by thermal radiation from the surface with a small additional contribution from conduction. This model is valid only if the planet grows from planetesimals smaller than about 10 km in diameter.*

for temperatures to reach the melting point anywhere in their interiors. Thus, *Hanks and Anderson* (1969) require the Earth to grow in 10^5 to 10^6 years for melting to occur, while *Mizutani et al.* (1972) find the Moon must accrete within 1000 years if melting is to occur in its outer portions. These times are far longer than the 10^7- to 10^8-yr timescale derived from standard planetesimal accretion models (*Wetherill,* 1980) and, taken at face value, suggest that the Earth and Moon might have accreted cold, after which the Earth differentiated when enough radiogenic heat had accumulated to cause internal melting. On the basis of this model the Moon never differentiated, a prediction that was quickly proved wrong when the Apollo missions returned Moon rocks to Earth for analysis.

The solution to this quandary was discovered by *Safronov* (1969, 1978) and more recently was elaborated by *Kaula* (1979). They realized that the physicist's idealization of the gradual addition of infinitesimally thin global layers is not a valid representation of the actual process in which individual planetesimals impact the planet's surface. Instead, each impact deposits an ejecta blanket of finite thickness and localized extent in addition to heating the target rocks directly beneath the impact site, as shown in Fig. 2. Under these conditions radiation is able to cool only a thin layer on the top of the ejecta blanket before yet another ejecta sheet is deposited on top of it. A large fraction of the initial energy of the infalling material is thus retained in the overlapping ejecta blankets.

It is easy to estimate the thickness δ of an ejecta sheet for which heat retention outweighs radiative loss. If the planet grows at a rate dR/dt the average time interval Δt between deposition of ejecta sheets of thickness δ at any given site is simply $\delta/(dR/dt)$. But the conductive cooling time of an ejecta sheet is of order δ^2/κ, where κ is the ejecta's thermal diffusivity. This cooling time equals the average time between deposition events when $\delta_0 = \kappa/(dR/dt)$. Layers thinner than δ_0 radiate all of their heat before the next ejecta deposition event, whereas thicker layers do not have time to cool between events. Making the conservative assumption that the Earth grew over a period of about 100 m.y. and that $\kappa \approx 10^{-6}$ m^2/sec yields a crossover thickness δ_0 of about 3 km. The maximum ejecta thickness

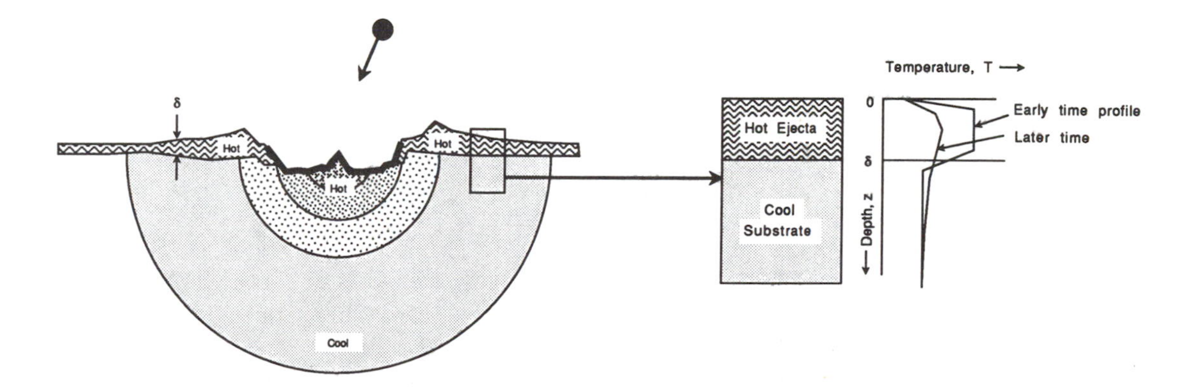

Fig. 2. *Heat deposition in the vicinity of a large impact. Shock waves deposit heat directly in the rocks beneath the crater, the amount of heating decreasing with increasing distance from the impact site, illustrated by the schematic temperature contours. The heavy black line inside the crater indicates a thick layer of melted target rocks. Such layers are observed inside large fresh lunar craters such as Copernicus and Tycho. The ejecta also contain a considerable amount of heat and thickly blanket the terrain to a distance of about 1 crater diameter from the rim. The temperature in the ejecta blanket (inset) declines as a result of thermal radiation from the surface and thermal conduction into the cooler substrate.*

of a fresh crater is about 0.02 of the crater diameter, which itself is about 10 times larger than the projectile diameter (*Melosh,* 1989), so the crossover condition is met by planetesimals about 15 km in diameter. Moreover, much of a projectile's energy is deposited more deeply beneath the crater floor than in the ejecta blanket, so that it is plausible that most of the heat added by 10 km or larger diameter planetesimals is retained by the growing planet rather than lost by radiation.

The details of the overall heat deposition process are complex and depend sensitively on impact mechanics. Thus, of the projectile's initial kinetic energy, about 30% is initially partitioned into the kinetic energy of the ejecta and most of the remainder is directly deposited as heat in the target rocks (*Melosh,* 1989, p. 66 ff). The kinetic energy of the ejecta is converted into heat when it comes to rest on the surface after mixing with a variable amount of preexisting surface material, resulting in a mixed sheet of hot rock debris thickly covering the surface within one or two crater diameters of the crater's original rim. The rocks beneath the crater (and the expelled ejecta) are directly heated by the shock from the impact which, depending upon the impact velocity, may melt or vaporize large quantities of material. The amount of material thus heated depends mainly upon the projectile's size and the square of its velocity (*Melosh,* 1989, p. 122 ff). On the other hand, a large impact cools the target planet by raising deeply-buried materials closer to the surface, where their heat may be more readily conducted to the surface and ultimately lost by radiation. Some of the initial heat of the ejecta may be lost by radiation during its ballistic flight from the impact crater to its site of deposition, while at high impact velocities a portion of the ejecta (especially that in the vapor plume) may travel at velocities greater than escape velocity and thus leave the planet entirely. The net thermal effect of an impact is thus a sum of gains and losses that tend to offset one another and hence make accurate estimation of the net heat deposition difficult. Crude consideration of these processes indicates that large impacts are more efficient at depositing energy than small impacts, and that for kilometer-sized planetesimals the radiation term in equation (1) is almost entirely negligible.

Kaula's (1980) solution to the uncertainty in estimating the net heat deposition of impacts is to lump all the poorly known processes into a single numerical factor h and write equation (1) in a form that neglects both radiation and conduction. Rearranging (1)

$$T(R,t) = \frac{h}{c_p}\left(\frac{GM(t)}{R(t)} + \frac{\nu_\infty^2}{2}\right) \qquad (2)$$

Note that with the neglect of radiation and conductive heat loss the accretion rate dR(t)/dt drops out, so that the interiors of planets growing over the 10^8-yr accretion timescale may easily reach temperatures high enough to initiate melting and differentiation. The unknown dimensionless factor h must lie somewhere between 0 (no net burial of heat) and 1 (all kinetic energy of infalling matter is retained as heat). In the absence of better information, h is generally assumed to be about 0.5 for kilometer-sized or larger planetesimals (*Stevenson,* 1981).

In *Safronov's* (1969) theory the random velocity component ν_∞ of approaching planetesimals at any time t is proportional to the escape velocity $\nu_{esc} = (2GM/R)^{1/2}$ of the planet growing in their neighborhood, so the entire right-hand side of equation (2) is proportional to GM/R times factors of the order of unity. Since M is proportional to $R^3(t)$, the surface temperature T of a growing planet at time t is proportional to $R^2(t)$. This temperature is "locked in" at radius $r = R(t)$ as more material accumulates on the former surface of the planet, establishing an internal temperature distribution of the form $T(r) \sim r^2$. This relation holds until T approaches the melting temperature in the outer portion of the planet, at which time convection begins and the mantle temperature remains near the melting point (*Kaula,* 1980). Melting in the Earth begins when it has reached about 10% of its final mass, or about half its final diameter. A core is presumed to form at about this time, further heating the mantle and stirring it so that the original $T \sim r^2$ thermal structure is wiped out in a manner similar to that described by *Stevenson* (1981).

Subsequent to core formation the mantle temperature must have been closer to the solidus than the liquidus, since convection in a completely liquid mantle is so vigorous that its cooling time is only about 10^4 years (*Tonks and Melosh,* 1989), far shorter than the timescale for Earth's growth from planetesimals. Once more than about 50% of the mantle material crystallizes, however, convection is regulated by the high viscosity of solid-state creep in the crystalline fraction, thus greatly lowering the cooling rates. Craters formed by impacts subsequent to core formation would therefore have excavated either the conductive boundary layer or, if sufficiently large, the semimolten convecting mantle. The extent to which the deposition of impact energy is altered in a planet with a hot mantle has not yet, to my knowledge, been investigated [although *Minear* (1980) studied the disrupting effect of impacts on the boundary layer of a cooling lunar magma ocean, concluding that impacts decrease the cooling time], nor has the related problem of convection in the presence of rapid accumulation of thick ejecta sheets been studied.

Although equation (2) is more realistic than equation (1), it is nevertheless limited by the implicit assumption that the impacting planetesimals are small in comparison to the growing planetary embryo. Impact energy is still assumed to be added in thin (although not infinitesimally thin) shells that are, on average, uniformly distributed over the planet's surface (Fig. 3). Thermal radiation losses can be neglected because most of the infalling planetesimals' energy is buried below the surface, but the overall pattern of energy deposition is much the same.

In recent years, however, it has seemed increasingly likely that the distribution of planetesimal sizes follows a rough power law extending from the smallest sizes up to objects half the size of the largest planetary embryo (*Hartmann and Davis,* 1975; *Wetherill,* 1985), at least during the later stages of accretion. This power distribution is expected to be of the form $N_{cum}(D) = CD^{-b}$, where N_{cum} is the number of planetesimals with diameters greater than or equal to D, C is a constant, and the power b is frequently observed to be close to 2 in numerical simulations of accretion processes (*Greenberg et al.,* 1978), impact fragmentation experiments (*Fujiwara* 1986), and in the size distribution of comet nuclei (*Delsemme,*

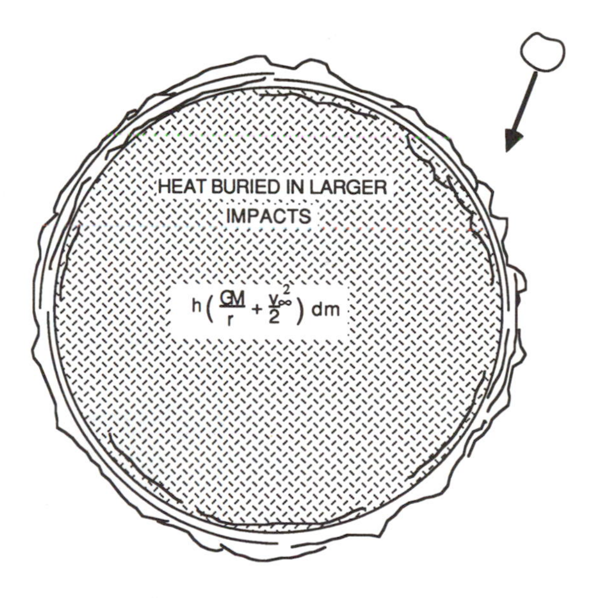

Fig. 3. *A more realistic model of planetary accretion in which a large fraction of the planetesimal's energy is trapped in the growing planet as heat. Melting temperatures are reached by the time the Earth has grown to about 10% of its present mass. In this model the planetesimals are assumed to be much smaller than the diameter of the planet so that growth is gradual; no provision is made for the effects of very large impacts that deposit their heat catastrophically through a substantial depth of the planetary embryo's mantle.*

1987). The b = 2 distribution has the special property that the total surface area of planetesimals in successive logarithmically increasing size intervals is constant. That is, the surface area of planetesimals with diameters between, say, D_1 and $2D_1$ is the same as the surface area of planetesimals with diameters between $2D_1$ and $4D_1$ (see Fig. 4). It seems natural that a distribution of this kind should arise from processes such as impact or coagulation that depend upon cross-sectional area (*Chapman and Morrison,* 1989, p. 57).

While the largest numbers of planetesimals in a b = 2 distribution are concentrated at the small sizes, most of the mass and energy resides in the largest objects (see Fig. 5). Thus, although a growing planetary embryo would be constantly battered by small planetesimals, most of the overall mass and heat transfer would occur in large, rare events that bury their heat deep within the embryo's interior. If this catastrophic mode of growth were indeed important, then the assumption that the infalling planetesimals are much smaller

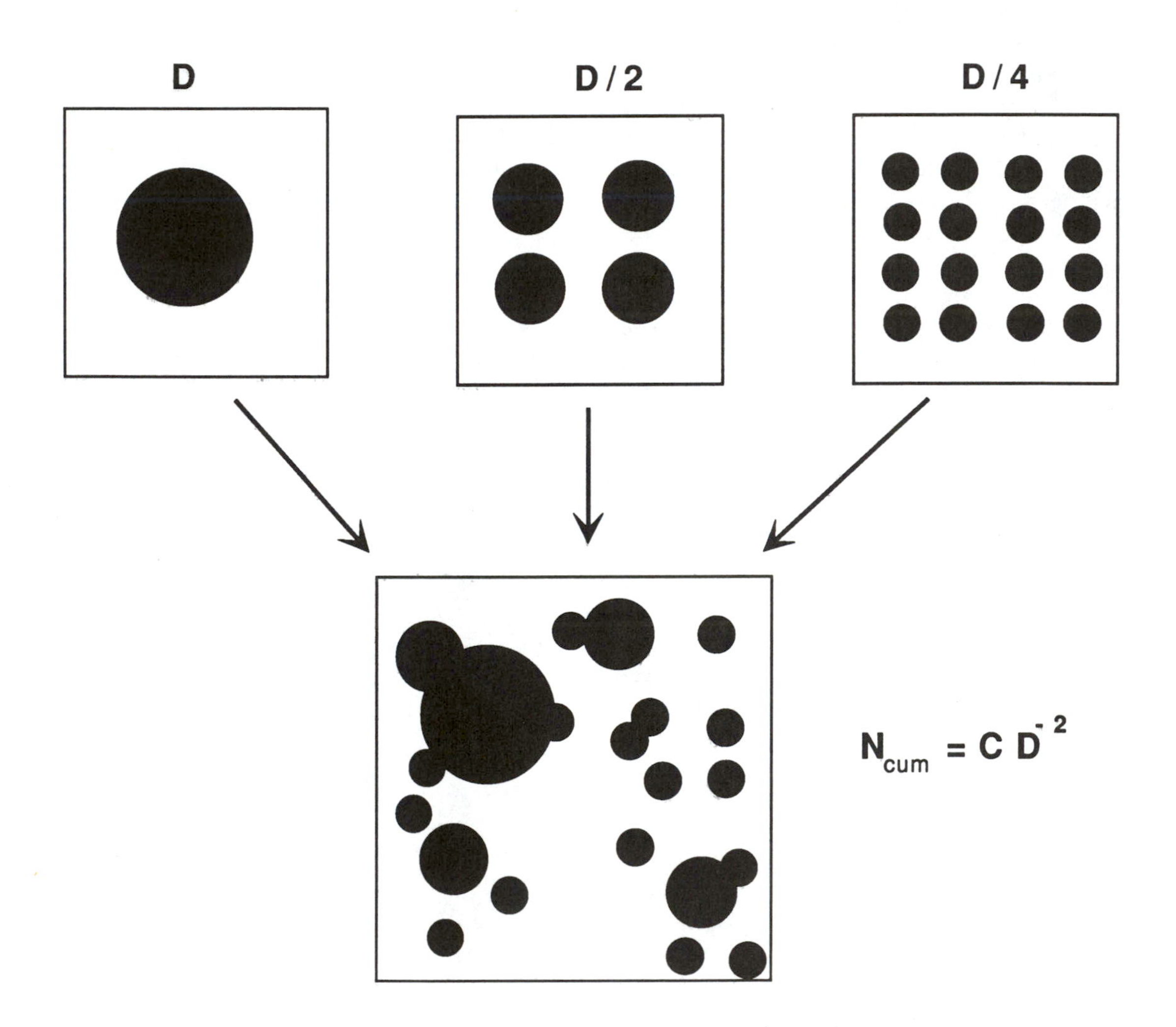

Fig. 4. *Schematic illustration of the distribution of planetesimal sizes in a population described by a b = 2 cumulative number distribution. In this distribution the projected area of planetesimals in logarithmically decreasing size intervals is constant.*

than the growing planet fails for the most significant events and a revised thermal analysis must take account of impacts between objects of comparable size. In the following pages I take the first step in such an analysis by examining the effects of a single such "giant impact" between the proto-Earth and an object half its size. Because the direct deposition of heat from a shock wave plays a major role in this process, I begin with a short review of the physics of shock heating.

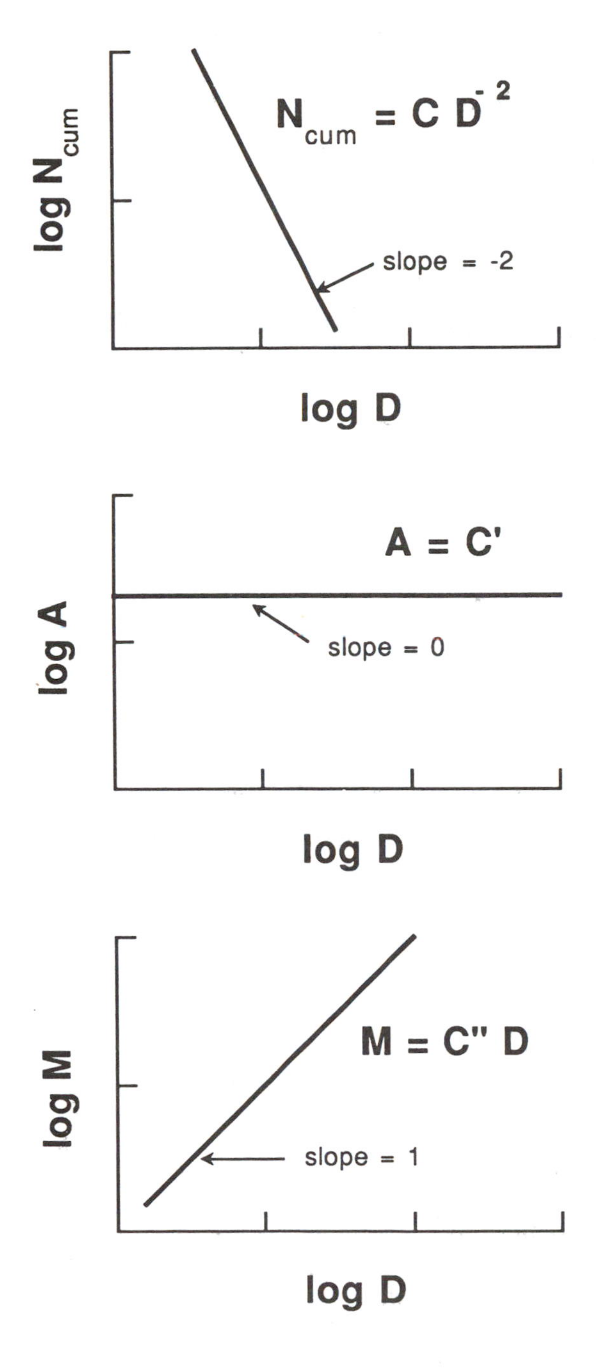

Fig. 5. *The dependence of various quantities on planetesimal diameter D in a population where the cumulative number N_{cum} is a function of D^{-2}. In such a population the number distribution is dominated by the smallest objects, the area (either total surface or cross-sectional) is uniformly distributed, and the mass distribution is dominated by the largest objects.*

PHYSICS OF SHOCK HEATING BY IMPACTS

Physics of Shock Compression and Release

During the earliest phases of an impact the incoming projectile meets the target and decelerates, transforming much of its initial kinetic energy into heat through sudden compression. Initially deposited in a region roughly comparable to the projectile in size, the impact energy spreads outward as shock waves advance into the target. Target material engulfed by these shock waves is at first suddenly compressed, then released more slowly to ambient pressure. The three equations describing this sudden compression were first derived by P. H. Hugoniot in 1887, and express the conservation of mass, momentum and energy (respectively) across the shock (*Melosh*, 1989)

$$\rho(U - u_p) = \rho_0 U \quad (3a)$$

$$P - P_0 = \rho_0 u_p U \quad (3b)$$

$$E - E_0 = \frac{1}{2}(P + P_0)\left(\frac{1}{\rho} - \frac{1}{\rho_0}\right) \quad (3c)$$

where ρ is material density behind the shock, ρ_0 is initial density, U is shock velocity, u_p is particle velocity behind the shock, P and P_0 are shock and initial pressures, respectively, and E and E_0 are specific internal energy (per unit mass) behind and before the shock. These equations are often written in terms of the specific volumes $V = 1/\rho$ and $V_0 = 1/\rho_0$ rather than density.

Although shock compression conserves mass, momentum, and energy, it does not conserve entropy; shock compression is an irreversible process that cannot be represented as a continuous path on a thermodynamic diagram, and in which heat is irreversibly deposited in the compressed material. In contrast, the slower release from high pressure is reversible and adiabatic, so that although the temperature declines as the pressure decreases, no net heat is transferred to the material during this release.

The three Hugoniot equations involve five unknowns and so are not sufficient to determine the thermal state of shocked material. One unknown such as P or u_p is established by the boundary conditions. In addition, a thermodynamic equation of state is needed to complete the system. This equation contains all of the material and chemical complexity of the shocked material, and is conventionally written $P = P(V,T)$, where the specific volume $V = 1/\rho$ and temperature T are thermodynamic state variables. For given initial conditions in an impact the equations (3a-3c) and the equation of state determine the thermodynamic state of material behind the shock wave. The release adiabat is computed from a simple relation derived from the first law of thermodynamics

$$\left.\frac{dE}{d\rho}\right|_S = \frac{P}{\rho^2} \tag{4}$$

where the derivative with respect to density ρ is taken at constant entropy S. Equation (4) can be integrated from the initial shock state to some desired final pressure or density. Most equations of state used in impact computations are written in the convenient form $P = P(\rho,E)$, even though E is not strictly a state variable (*Melosh,* 1989, Appendix II). This form makes integration of equation (4) straightforward using elementary numerical techniques.

The best equation of state currently available for impact computations is generated by an approximately 3000-line FORTRAN computer code called ANEOS (*Thompson and Lauson,* 1972). This code requires 24 input parameters to describe a given substance, from which it computes a numerical approximation to the Helmholtz free energy. Unlike other impact equations of state, thermodynamic quantities such as pressure, temperature, entropy, and internal energy are derived from the free energy by standard thermodynamic relations, so they are guaranteed to be consistent with one another, even through phase transitions. The ANEOS parameters for dunite were previously constructed and the resulting equation of state was found to agree well with the existing data (*Benz et al.,* 1989, Appendix I). Dunite is regarded as a reasonably good approximation to the Earth's mantle.

Three Hugoniot curves derived from this ANEOS equation of state are shown in Fig. 6. This plot shows the shock pressure as a function of particle velocity. This form is particularly useful here because, for the impact of two objects composed of the same material, the particle velocity is precisely half the normal component of the impact velocity. Thus, for an impact between the Earth and a large planetesimal the particle velocity is in the vicinity of 5 km/sec, and the corresponding shock pressure of material starting from surface conditions is a few hundred GPa. The dashed line on this figure shows the Hugoniot curve for material starting at conditions applicable deep in the Earth's mantle, $P = 100$ GPa and $T = 3000°K$. For such material the shock pressures are higher, nearly 500 GPa at the same particle velocity (note that the shock pressure plotted is the *increase* above the ambient pressure). The gray line is for material starting at $P = 100$ GPa and $T = 298°K$. It is clear from this curve that initial temperature has very little effect on the

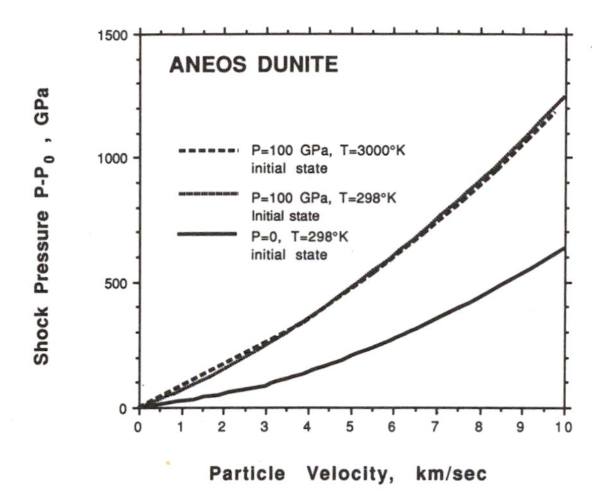

Fig. 6. *Hugoniot curves for dunite computed by the ANEOS equation of state package described in the text. The particle velocity is one half of the impact velocity for a head-on impact between objects composed of the same material. The solid curve is for material that starts at Earth-surface conditions, the dashed curve is the Hugoniot for material whose initial conditions are appropriate for the Earth's deep mantle, and the gray curve is for cold material at deep mantle pressures, illustrating that starting temperature has little effect on the Hugoniot.*

Hugoniot curve; the main variable affecting the curve is initial pressure.

The release curves are shown on a P-T plot in Fig. 7 for material starting at surface conditions, and the residual temperatures are plotted in Fig. 8 as a function of the shock pressure. It is clear that shock pressures in the vicinity of 200 GPa are sufficient to raise mantle material starting near the Earth's surface to the liquid-vapor phase boundary. A weaker shock of 100 GPa suffices to melt it. State changes in the deep mantle are less easy to gauge, as the liquidus is not known at pressures much above 25 GPa (*Ito and Takahashi,* 1987), although recent work on Perovskite, $MgSiO_3$, pushes the melting curve to nearly 100 GPa (*Knittle and Jeanloz,* 1989). Recent estimates of the current temperature at Earth's core-mantle boundary establish a lower bound of 3800°K (*Williams et al.,* 1987), at which temperature the mantle is still evidently in the solid phase. Nevertheless, shock pressures of a few hundred GPa result in temperature increases of 1000°K or more, which I here presume leads to melting of the deep mantle, even if it was not molten to begin with.

Shock Heating in Impacts

The pattern of pressure decline near an impact is a complex function of impact geometry and target material. This pattern is best understood for vertical impacts on a planar target, a geometry that is not particularly relevant to an oblique giant impact between the growing Earth and a large planetesimal of comparable size. Nevertheless, understanding the simple geometry may aid in interpreting the outcome of a more realistic impact geometry.

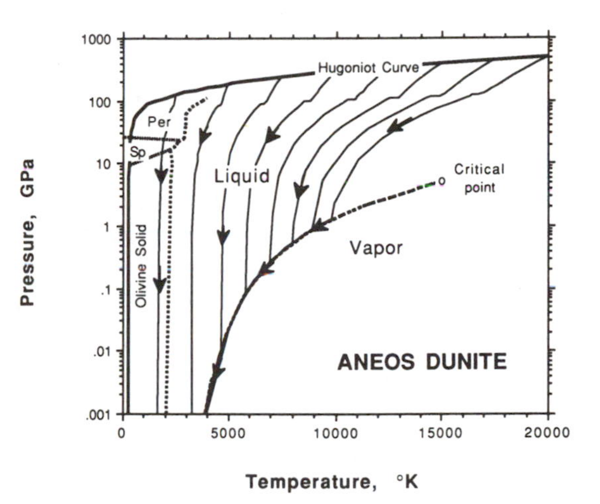

Fig. 7. *The Hugoniot curve and release adiabats for ANEOS dunite as a function of pressure and temperature. Most of the release adiabats intersect the liquid-vapor phase boundary and follow the curve to lower pressure as the vapor condenses. The liquid and vapor phase boundaries and critical point are idealized to approximate those of a simple material. In reality melting and vaporization of silicates is incongruent, although this subtlety is not easily shown on a plot covering such a broad range of P and T. Also shown is the computed critical point and the melting curve of candidate mantle materials up to 25 GPa as determined by Ito and Takahashi (1987). The Spinel-Perovskite phase boundary is from Ito and Takahashi (1989), and the melting curve from 25 to 100 GPa is from Knittle and Jeanloz (1989).*

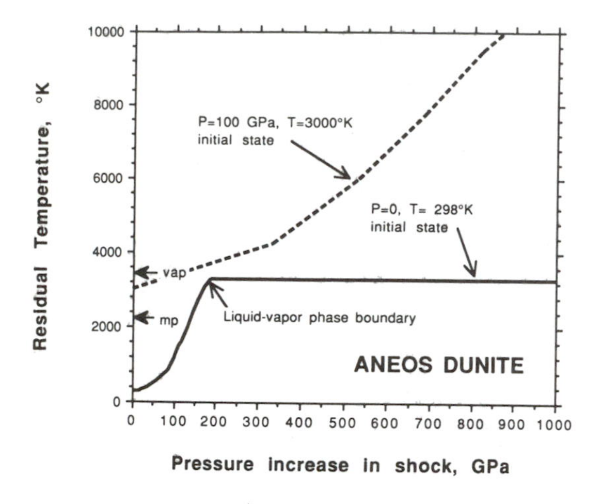

Fig. 8. *Residual temperatures after release from the shock pressure on the horizontal axis to a final pressure of 1 bar (0.1 MPa). The solid curve is for material that starts at Earth surface conditions and the dashed curve is for material that starts at conditions approximating those of the Earth's deep mantle. The plateau in the solid curve develops because both liquid and vapor are present at 1 bar over the indicated range of shock pressures. The temperature of 3380 K is the vaporization temperature of dunite at 1 bar. There is no corresponding plateau at 110 GPa, where melting begins, because the ANEOS is unable to treat both a melt transition and solid state transition simultaneously. In any event this step is small due to the relatively small enthalpy of melting compared to the enthalpy of vaporization.*

Both numerical computations and data from a limited number of experiments show that in a vertical impact a nearly isobaric core of high pressure develops around the site of the impact. This core is comparable in size to the projectile, and the mean pressure in this region is accurately given by the planar impact approximation (*Melosh,* 1989, p. 54 ff). Although islands of higher pressure develop during the earlier jetting phase, most of the projectile mass (and a roughly equal volume of the target) is shocked only to the mean pressure. As the shock waves from the impact spread out they weaken both because their energy is diluted as they engulf progressively more material and because of irreversible energy deposition in the target. The rate of this weakening is a function of maximum pressure and material, but is approximately given by

$$P(r) = P_{core}\left(\frac{a}{r}\right)^n \quad (5)$$

where P_{core} is the mean pressure in the isobaric core, a is the projectile radius, and r is now the distance from the impact site. The decay power n is typically in the range of 2 to 3, so that the pressure declines by a factor of 10 at a distance of 2 to 3 projectile radii from the impact site. This approximation is similar to the successful "gamma model" previously proposed by *Croft* (1982), except that in the gamma model r is the distance from a point located a distance equal to the projectile radius a below the surface.

Impacts at velocities on the order of 10 km/sec (particle velocity ~5 km/sec) in dunite generate pressures P_{core} on the order of 300 GPa and will thus melt and partially vaporize a few projectile masses of the target, but such melting will be confined to regions within about 2 projectile radii of the impact site. More distant regions of the planet will be warmed by impact heating, but will not necessarily melt unless they are close to melting already (see Fig. 9). Large impacts thus deposit their energy deeper than small impacts, and impacts with objects half the size of the proto-Earth can be expected to melt at least one hemisphere of the Earth's mantle right down to the core. The melt pool following a large impact will undergo further change in shape after it forms, since the melt will have a different (generally smaller) density than surrounding rocks at the same depth. Since the surrounding rocks were likely to be hot, and therefore relatively fluid, subsolidus viscous deformation subsequent to the melting event should close the initial melt-solid crater, producing a global magma ocean of nearly uniform depth overlying hot, more dense solid mantle material. Although the timescale of this relaxation is difficult to estimate, is was probably not shorter than the timescale of post-glacial relaxation in the present Earth; that is, a few thousand years.

Core formation in the context of this model is relatively quick and simple: Metallic iron trapped in the outer regions of a planet quickly sinks toward the bottom of any large melted region immediately following a large impact. The only requirement for core formation is thus the impact of a projectile sufficiently large and fast to melt a substantial portion of the planet. The precise time of core formation thus contains a stochastic element, although it cannot occur too early as encounter velocities must be large enough to generate melting without completely disrupting the planetary embryo.

EFFECT OF A GIANT IMPACT ON EARTH'S THERMAL STATE

The collision between a Mars-sized protoplanet and the proto-Earth adds a truly prodigious amount of energy to the Earth over a time interval measured in hours. The mass m, velocity v, and impact parameter b of the projectile are constrained only by the total angular momentum L of the Earth-Moon system, 3.49×10^{34} kg m^2/sec. Since the angular momentum is a product of all three terms, $L = mvb$, the value of each individual quantity is uncertain within broad limits. The impact parameter is bounded between zero and the sum of the Earth's and the projectile's radii, while the impact velocity is at least as large as the proto-Earth's escape velocity but, because of the overall geochemical similarity of the Earth and Moon, it was probably not as much as twice the escape velocity (i.e., the projectile had an initial orbit close to that of Earth's). These constraints point to an impact by a body with a mass about 10% of the Earth's mass, hence a diameter approximately half of the Earth's—about the size of Mars.

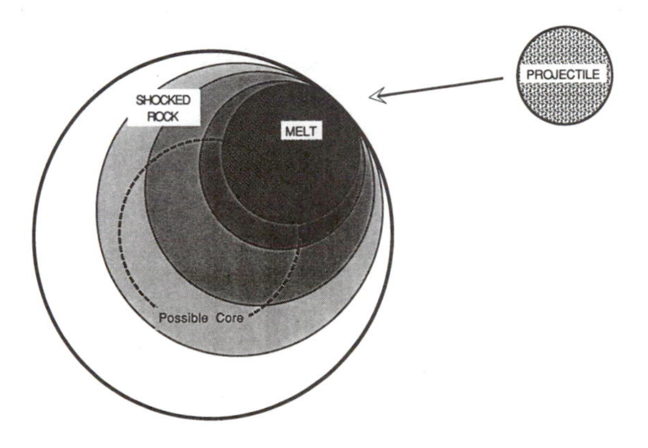

Fig. 9. *Schematic illustration of the pattern of heat deposition in a planet struck by a projectile of comparable size (1/4 its diameter in this figure). Adjacent to a melted region roughly twice the projectile's diameter, the shock level, and thus temperature, falls off steeply with increasing distance from the impact site.*

The energy released in such a collision is the sum of gravitational and kinetic energies, and was probably within the range 2×10^{31} to 5×10^{31} J. Averaged over the Earth's mass this is about 7.5×10^6 J/kg. Several authors (including myself) have used this energy in conjunction with a single value of silicate heat capacity to infer rather high average temperatures for the Earth. Thus, a silicate heat capacity of 10^3 J/kg K yields an average temperature of 7500 K for the Earth. However, such estimates neglect the relatively small latent heat of melting, $\sim 4 \times 10^5$ J/kg, the large latent heat of vaporization, $\sim 5 \times 10^6$ J/kg, and the rapid increase of heat capacity at temperatures above the Debye temperature (taken to be 676 K for dunite in ANEOS). According to the ANEOS equation of state, an internal energy of 7.5×10^6 J/kg corresponds to a temperature rise of about 3200 K for dunite near the Earth's surface and about 7000 K deep in the mantle where vaporization does not limit the temperature rise. Nevertheless, it is clear that melting, even vaporization, should be widespread in a giant collision.

Although the total energy available from the collision of a Mars-sized projectile with the proto-Earth is impressive, the ***distribution*** of the energy within the Earth is equally important. If, as has been suggested by ***Stevenson*** (1987), this energy is mainly expended in vaporizing the projectile, the Earth may acquire a transient silicate vapor atmosphere without strongly heating the deeper mantle. The simple considerations of shock wave generation and decay discussed above, however, indicate that only partial vaporization occurs at probable impact velocities, although jetting (*Melosh and Sonnett,* 1986) may enhance the local production of hot vapor, and that deep melting should be widespread in at least the hemisphere that the impact occurs.

An additional factor not previously considered here (but see also *Benz and Cameron,* 1990) is that the projectile might have an iron core that will sink through the Earth's mantle shortly after the collision and merge with the Earth's core, releasing its gravitational potential energy. Assuming a core equal to 30% of the projectile's mass, the energy released by sinking 3000 km through Earth's mantle is of order 3×10^{30} J, which itself will cause strong heating of the mantle through which it sinks and of the Earth's core when it arrives. After the cores have merged, this heat is applied to the bottom of the mantle so that any portion of the mantle that escaped melting by the direct shock wave will likely be melted by this means.

To address the temperature rise in the Earth more exactly, I performed a series of 3-dimensional numerical hydrocode computations in conjunction with M. E. Kipp of Sandia National Laboratory. These computations were designed to simulate the impact between the proto-Earth and a Mars-sized protoplanet. We used the code CTH, implemented on the Cray X/MP supercomputers of the Sandia National Laboratory. This computation uses the ANEOS equations of state for dunite in the mantles and iron in the cores of the two colliding planets. The Earth has a central gravitational field, and is adjusted so that its initial temperature profile is similar to that of the present-day Earth. These models thus start out relatively cold, with mantle temperatures well below the solidus of dunite. We have performed computations at a variety of initial velocities and impact parameters, including pairs that give the Earth-Moon system its present angular momentum.

At the lowest velocity ($v_\infty = 0$), for impact parameters b of 0.88 (Fig. 10a) and 1.25 times the Earth's radius R_e, the strongest heating upon impact is confined to the hemisphere on which the projectile strikes. Shock-induced temperature rises are typically 2000 to 3000 K between the site of the impact and the Earth's core. A crater forms that extends most of the way down to the core. The gravitational energy of this excavation itself is of order 10^{30} J, which appears as heat within an hour or two as mantle material flows inward to fill the crater cavity. Unfortunately, our computations do not extend to long enough times for the entire projectile to merge with the Earth in the high impact parameter runs. This limitation is a result of our finite grid size; by this time a substantial amount of material had left the grid and further computations could not take account of the fallback of this material. In these cases we had to stop the computation while some of the projectile was still falling on portions of the Earth more distant from the impact site. In these cases we believe that more than half the mantle will be strongly heated, so the quoted results must be seen as a lower limit. Figure 10b illustrates the temperature contours for the $b = 0.88R_e$ computation 1800 sec after the impact (this impact parameter and velocity correspond to the angular momentum of the present Earth-Moon system). Note that in this computation a very hot (>4200 K) low-velocity vapor plume is expelled backward from the impact site. This plume eventually spreads over the entire Earth, producing a transient silicate vapor atmosphere.

The results for the higher impact velocity $v_\infty = 7.8$ km/sec are more spectacular. For the impact parameters studied (0.59 and $1.25R_e$) the hemisphere near the impact was heated nearly uniformly by 1000 to 3000 K. The projectile's core was almost entirely vaporized and a much larger crater formed in the proto-

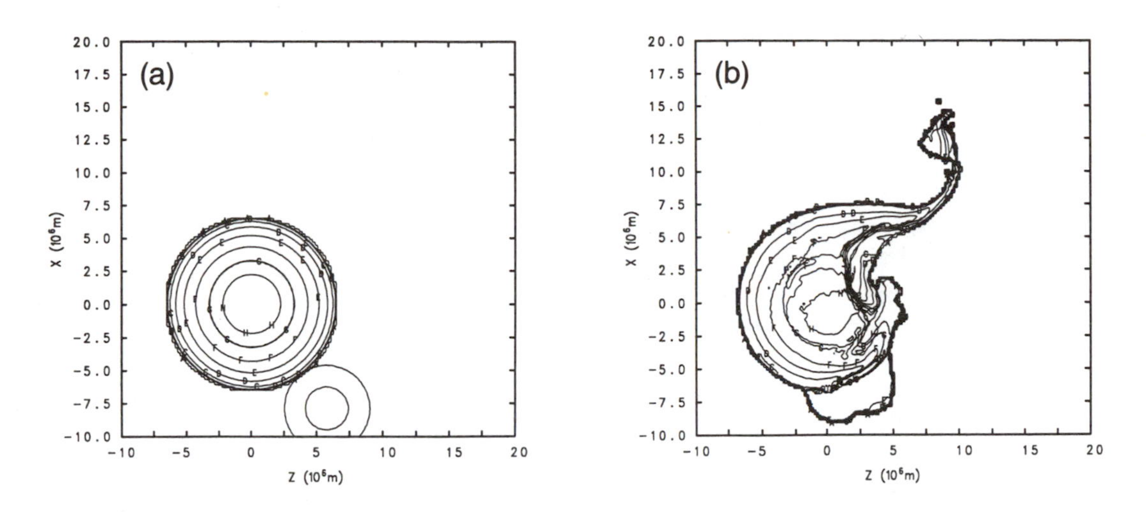

Fig. 10. *Temperature contours in the collision between the proto-Earth and a protoplanet half its diameter. This computation is for $v_\infty = 0$ km/sec (8 km/sec at contact) and an impact parameter of 0.88 R_e at contact.* **(a)** *The initial configuration before impact in which the projectile is traveling upward (positive x-direction) and the proto-Earth is at rest.* **(b)** *The configuration 1802 sec after contact. The contour values are A = 300 K, B = 600 K, C = 1200 K, D = 1800 K, E = 2400 K, F = 3000 K, G = 3600 K, and H = 4200 K. Figures 10 and 11 were computed by M. E. Kipp at Sandia National Laboratory, Albuquerque, NM, using the 3-D hydrocode CTH. These plots are in the symmetry plane of the two colliding spheres.*

Earth. A fast, hot vapor plume also carries several lunar masses of material out along trajectories that eventually take up elliptical orbits about the Earth. Figure 11b illustrates temperature contours for $b = 0.59R_e$, corresponding to the angular momentum of the present Earth-Moon system, at 1200 sec after the impact. Again, a hot low-velocity backward vapor plume is formed that will eventually cover the Earth's surface.

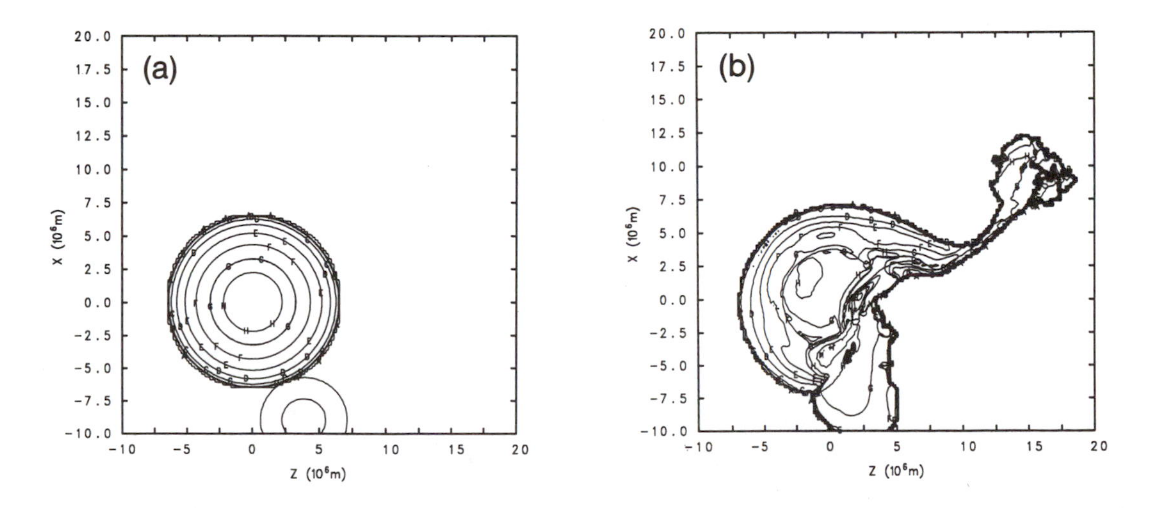

Fig. 11. **(a)** *Initial configuration and* **(b)** *temperature contours 1193 sec after the impact between the proto-Earth and a protoplanet half its diameter moving at $v_\infty = 7.8$ km/sec (12 km/sec at contact) and an impact parameter of $0.59R_e$ at contact. Contour values are the same as in Fig. 10.*

CONCLUSION

The 3-D hydrocode computations, in conjunction with the more general considerations described above, indicate that a Moon-forming impact would have had a profound effect on the Earth's thermal state. The shock produced by the impact would have heated the Earth to great depths, raising at least the hemisphere adjacent to the impact above the melting temperature. Later phenomena, such as the merger of the projectile's and proto-Earth's cores and the collapse of the mantle-deep crater created by the impact, would have added comparable amounts of energy to the Earth. There seems to be no way to avoid the conclusion that a large Moon-forming impact is inevitably accompanied by widespread melting of most or all of the Earth's mantle. This scenario has implications for the geochemistry of the Earth that are dealt with in an accompanying paper (*Tonks and Melosh,* 1990).

The effects of "giant" impacts (i.e., impacts by bodies roughly half the diameter of the primary) on the thermal state of a growing planetary embryo is, however, a more general problem than that of the hypothetical Moon-forming impact on the proto-Earth. If the cumulative spectrum of planetesimal sizes is close to a power law of slope -2, then most of the mass and energy added to a growing planet will be deposited by such "giant" impacts. In this case a simple pattern of temperature vs. radius may never develop, as the thermal state at any given era will depend upon the time, velocity, and obliquity of the last large collision. The process of core formation and differentiation of such a catastrophically growing body may be qualitatively different than that suggested by current gradualist models of planetesimal growth. The implications of such catastrophic growth have yet to be worked out, but it is clear that much more work needs to be done on the effects of large impacts on protoplanets of all sizes.

REFERENCES

Benz W., Cameron A. G. W., and Melosh H. J. (1989) The origin of the Moon and the single impact hypothesis III. *Icarus, 81,* 113-131.

Benz W. and Cameron A. G. W. (1990) Terrestrial effects of the giant impact. In *Origin of the Earth,* this volume.

Boss A. P. (1986) The origin of the Moon. *Science, 231,* 341-345.

Cameron A. G. W. and Ward W. R. (1976) The origin of the Moon (abstract). In *Lunar Science VII,* pp. 120-122. The Lunar Science Institute, Houston.

Chapman C. R. and Morrison D. (1989) *Cosmic Catastrophes.* Plenum, New York. 302 pp.

Croft S. K. (1982) A first-order estimate of shock heating and vaporization in oceanic impacts. In *Geological Implications of Impacts of Large Asteroids and Comets on the Earth* (L. T. Silver and P. H. Schultz, eds.), pp. 143-152. Geol. Soc. Amer. Spec. Pap. 190.

Delsemme A. H. (1987) Diversity and similarity of comets. In *Symposium on the Diversity and Similarity of Comets,* pp. 19-30. ESA SP-278.

Fujiwara A. (1986) Results obtained by laboratory simulations of catastrophic impact. *Mem. Soc. Astron. Ital., 57,* 47-64.

Greenberg R., Wacker J. F., Hartmann W. K., and Chapman C. R. (1978) Planetesimals to planets: Numerical simulation of collisional evolution. *Icarus, 35,* 1-26.

Hanks T. C. and Anderson D. L. (1969) The early thermal history of the Earth. *Phys. Earth Planet. Inter., 2,* 19-29.

Hartmann W. K. and Davis D. R. (1975) Satellite-sized planetesimals and lunar origin. *Icarus, 24,* 504-515.

Hartmann W. K., Phillips R. J., and Taylor G. J., eds. (1986) *Origin of the Moon.* Lunar and Planetary Institute, Houston. 781 pp.

Ito E. and Takahashi E. (1987) Melting of peridotite at uppermost lower-mantle conditions. *Nature, 328,* 514-517.

Ito E. and Takahashi E. (1989) Postspinel transformations in the system Mg_2SiO_4-Fe_2SiO_4 and some geophysical implications. *J. Geophys. Res., 94,* 10637-10646.

Kaula W. M. (1979) Thermal evolution of the Earth and Moon growing by planetesimal impacts. *J. Geophys. Res., 84,* 999-1008.

Kaula W. M. (1980) The beginning of the Earth's thermal evolution. In *The Continental Crust and its Mineral Deposits* (D. W. Strangway, ed.), pp. 25-34. Geol. Assoc. Can. Spec. Pap. 20, Waterloo.

Knittle E. and Jeanloz R. (1989) Melting curve of (Mg, Fe)SiO_3 perovskite to 96 GPa: Evidence for a structural transition in lower mantle melts. *Geophys. Res. Lett., 16,* 421-424.

Melosh H. J. (1989) *Impact Cratering: A Geologic Process.* Oxford Univ., New York. 245 pp.

Melosh H. J. and Sonnett C. P. (1986) When worlds collide: Jetted vapor plumes and the Moon's origin. In *Origin of the Moon* (W. K. Hartmann, R. J. Phillips, and G. J. Taylor, eds.), pp. 621-642. Lunar and Planetary Institute, Houston.

Minear J. W. (1980) The lunar magma ocean: A transient lunar phenomenon? *Proc. Lunar Planet Sci. Conf. 11th,* pp. 1941-1955.

Mizutani H., Matsui T., and Takeuchi H. (1972) Accretion process of the Moon. *Moon, 4,* 476-489.

Newsom H. E. and Taylor S. R. (1989) Geochemical implications of the formation of the Moon by a single giant impact. *Nature, 338,* 29-34.

Safronov V. S. (1969) *Evolution of the Protoplanetary Cloud and Formation of the Earth and Planets.* Nauka, Moscow. Translated by the Israel Program for Scientific Translation (1972).

Safronov V. S. (1978) The heating of the Earth during its formation. *Icarus, 33,* 1-12.

Stevenson D. J. (1981) Models of the Earth's core. *Science, 214,* 611-619.

Stevenson D. J. (1987) Origin of the Moon—The collision hypothesis. *Annu. Rev. Earth Planet. Sci., 15,* 271-315.

Thompson S. L. and Lauson H. S. (1972) *Improvements in the Chart D radiation-hydrodynamic CODE III: Revised analytic equations of state.* Sandia National Laboratory Report SC-RR-710714. 119 pp.

Tonks W. B. and Melosh H. J. (1989) Crystal settling in a vigorously convecting magma ocean (abstract). In *Lunar and Planetary Science XX,* pp. 1124-1125. Lunar and Planetary Institute, Houston.

Tonks W. B. and Melosh H. J. (1990) The physics of crystal settling and suspension in a turbulent magma ocean. In *Origin of the Earth,* this volume.

Turcotte D. L. and Schubert G. (1982) *Geodynamics: Applications of Continuum Physics to Geological Problems.* Wiley, New York. 450 pp.

Urey H. C. (1952) *The Planets: Their Origin and Development.* Yale Univ., New Haven. 245 pp.

Wetherill G. W. (1980) Formation of the terrestrial planets. *Annu. Rev. Astron. Astrophys., 18,* 77-113.

Wetherill G. W. (1985) Occurrence of giant impacts during the growth of the terrestrial planets. *Science, 228,* 877-879.

Williams Q., Jeanloz R., Bass J., Svendsen B., and Ahrens T. J. (1987) The melting curve of iron to 250 Gigapascals: A constraint on the temperature at the Earth's center. *Science, 236,* 181-182.

DOES THE MOON HAVE THE SAME CHEMICAL COMPOSITION AS THE EARTH'S UPPER MANTLE?

J. H. Jones

Mail Code SN2, NASA Johnson Space Center, Houston, TX 77058

L. L. Hood

Lunar and Planetary Laboratory, University of Arizona, Tucson, AZ 85721

If the Moon were derived from the Earth, then it is possible that the bulk silicate portions of the Earth and Moon have similar chemical compositions. We have used a combined geophysical and geochemical study to investigate this possibility. Models for the internal structure and internal temperature of the Moon have been combined with a terrestrial upper mantle composition and used to calculate the Moon's density and moment of inertia. These calculated values are then compared to measured values to evaluate whether the bulk silicate Moon could have the same composition as the Earth's upper mantle. Only a very restricted set of parameters allows the upper mantle composition to simultaneously meet the geophysical constraints of density and moment of inertia. In particular, if the Moon has the composition of the Earth's mantle, then the Moon must have a rather large core (~500 km radius). If the Moon had a core of such size and if the Moon's siderophile element abundances were initially those of the terrestrial upper mantle, then siderophile elements should be much more depleted in lunar materials than is observed.

INTRODUCTION

One of the most important pieces of information about a planet is its bulk chemical composition. Unfortunately in the case of the Moon, early differentiation events have frustrated efforts to arrive at bulk compositions that are well defined and universally accepted. However, in general, two extremes of compositions are currently in favor: (1) refractory-enriched compositions that contain $\sim 4 \times$ CI abundances of refractory elements such as Al and U (*Taylor,* 1982, 1987); and (2) "Earth-like" compositions that contain lower abundances of refractories (*Ringwood et al.,* 1986; *Jones and Delano,* 1989). Both of these general classes of models typically assume that the Moon is enriched in FeO compared to the Earth. However, there exists a variant of (2) that argues that the Moon is not enriched in FeO and that the Earth and the Moon have similar Mg#s [i.e., molar Mg/(Mg+Fe) * 100 = ~90; *Warren* (1986)].

Earlier, *Hood and Jones* (1987) showed that the compositional models of *Wänke et al.* (1977), *Morgan et al.* (1978), *Taylor* (1982), and *Delano* (1984) could all be made compatible with the measured geophysical parameters of bulk density and moment of inertia. In all cases, however, these compositional models required the presence of a significant metallic core—250 to ~500 km radius or 0.7 to

~5.5 wt.% Fe. However, all compositions that were explored in the Hood and Jones study were more enriched in FeO (>10 wt.%) than the Earth's upper mantle (~8 wt.%).

Similar conclusions were reached by *Mueller et al.* (1988), who used a different method of geophysical/petrological analysis. They preferred, on the basis of seismic velocities and density considerations, that the Moon be more magnesian and aluminous than the Earth. *Hood and Jones* (1987) showed that a wide range of bulk lunar compositions could be consistent with the geophysical data, although more aluminous models were in better accord with the seismic velocities inferred by *Nakamura et al.* (1982) for the mantle below 500 km depth. Both sets of authors demanded a significant lunar core to meet the bulk density requirement. *Mueller et al.* (1988) found that, for nearly all of the models that they investigated, *minimum* core sizes of 150 km radius were required and that the majority of the models required a core >250 km. Thus, even if disagreements concerning bulk lunar composition still exist, a general consensus is emerging that the Moon must have a small core of at least a few hundred kilometers radius.

Although the exact composition of the Moon will probably never be specified in detail, it is still of interest to determine if it is likely (or even possible) that the bulk composition of the Moon could be that of the terrestrial upper mantle. Some models of lunar origin that postulate a close relationship between the Earth and the Moon (e.g., *Ringwood,* 1986) also predict that the Earth and Moon should be compositionally similar. And regardless of the specific model, oxygen isotopic data require that the Earth and Moon be built from isotopically similar materials (*Grossman et al.,* 1974). Thus, in the present study, we have explicitly coupled a terrestrial upper mantle composition with reasonable models of the Moon's interior thermal structure and geological history to evaluate whether the Moon could have the same bulk composition as the Earth's upper mantle. As we will show below, only one thermal model allows the silicate portions of the Earth and Moon to be chemically equivalent. Even this model, however, requires metallic lunar cores that are prohibitively large. Thus, we find it difficult to relate the composition of the Moon to that of the Earth's upper mantle in any straightforward manner. If the Moon formed from the Earth, then either significant amounts of extraterrestrial materials were incorporated into the Moon or the composition of the Earth's mantle has changed over time.

METHOD

Geophysical Modeling

In order to calculate the bulk density and moment of inertia of the Moon, we follow the techniques of *Hood and Jones* (1987). Because of the many uncertainties in the composition, thermal state, and differentiation history of the Moon, we chose in our 1987 paper to permute thermal models, compositional models and models of the Moon's internal structure to investigate the sensitivity of our results to any particular assumption. For example permutation of 4 compositions, 3 thermal profiles and 4 differentiation scenarios led us to investigate a total of 48 models. Here, because we fix bulk lunar silicate composition to be that of the Earth's upper mantle, only 12 models are possible. One difference between this study and our earlier work is that here we have not attempted to model the seismic velocity profile of *Nakamura et al.* (1982). This is because our earlier study indicated that the errors associated with the seismic velocity profile permitted a wide range of bulk compositions. Thus, it is not clear to us that further modeling of the *Nakamura et al.* (1982) seismic data, without additional refinement of that data set, is particularly useful.

To delineate our model in more detail, a bulk silicate composition is chosen (in this case that of the Earth's upper mantle; Table 1) and from this a lunar highlands crust is

TABLE 1. *Summary of compositions.*

Oxide	Terrestrial Upper Mantle*	Lunar Crust†	Bulk Silicate Moon Minus Crust‡
SiO_2	45.16	45.	45.7
Al_2O_3	3.97	24.6	1.2
FeO	7.82	6.6	8.0
MgO	38.3	6.8	45.2
CaO	3.56	15.8	1.9

* *Jagoutz et al.* (1979).
† *Taylor* (1982).
‡ Crust taken to represent 11.6% of the Moon.

subtracted. The crustal composition that we remove is the same as that of *Taylor* (1982), but other estimates of the chemical composition of the lunar crust are similar (*Davis and Spudis,* 1985). The remaining material, bereft of its crustal component, is distributed within the interior of the Moon according to one of several simple differentiation models (Fig. 1). In this manner the radial chemical variation within the Moon is defined. Each model of radial chemical variation is then coupled with several models for the thermal state of the lunar interior (Fig. 2). The calculation proceeds from the base of the crust inward, and pressure is calculated by integrating over the mass of material above the depth of interest. Knowing temperature, pressure, and chemical composition as a function of depth, variations in mineralogy (and, by extension, density) can also be calculated. Compressibility and thermal expansion of individual mineral phases are taken into account in this calculation using data sources listed in Table 2 of *Hood and Jones* (1987). Radial variations in density can then be converted into the moment of inertia or integrated to yield the bulk density. If the calculated model Moons are not dense enough (as is always the case for this particular bulk composition), then a small core of metallic iron is added. The addition of a metallic core simultaneously increases the density of the model Moon and lowers the moment of inertia. This process is repeated until, by trial and error, an acceptable fit to both the bulk density and moment of inertia is achieved or it is found that there is no possible solution.

More details of the method can be found in *Hood and Jones* (1987). However, in terms of the mineralogical calculations, Al_2O_3 content is the key parameter. The Moon is divided into three regions of plagioclase, spinel, and garnet stability. Within the spinel and garnet stability zones, Al is also allowed to enter pyroxene as Ca- and Mg-tschermakite components. Thus, even if chemical composition remains constant, mineralogy is allowed to change in response to changes in temperature and pressure. Further, garnet and spinel are denser than other phases in the lunar interior,

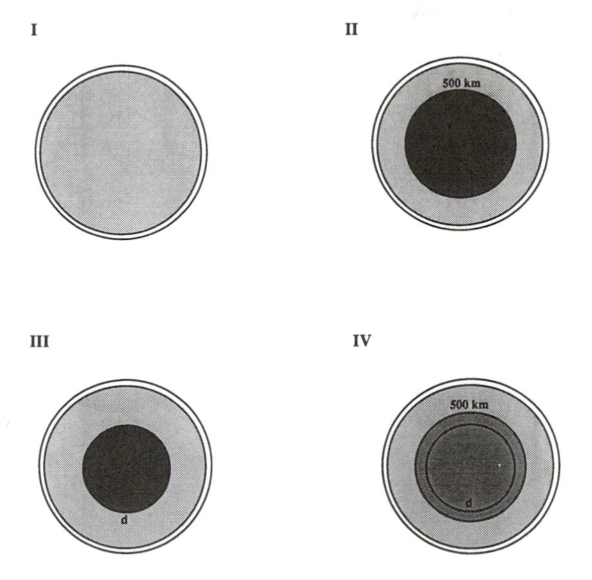

Fig. 1. *Schematic of the differentiation models used in this study (after Hood and Jones, 1987). I. Complete mixing of all mantle materials after removal of the crust. This model allows the most dilution of Al and, therefore, produces the minimum amount of garnet, a dense mineral phase. II. Removal of crust from only the upper 500 km, leaving the interior pristine. No solutions were found to this model because the terrestrial upper mantle composition that we chose did not have enough Al in the upper 500 km to make the crust. III. Differentiated to a depth d (just deep enough to supply enough Al to produce the lunar crust) with pristine material below. This model maximizes the amount of garnet in the Moon. IV. Same as III except that mixing is allowed beneath 500 km.*

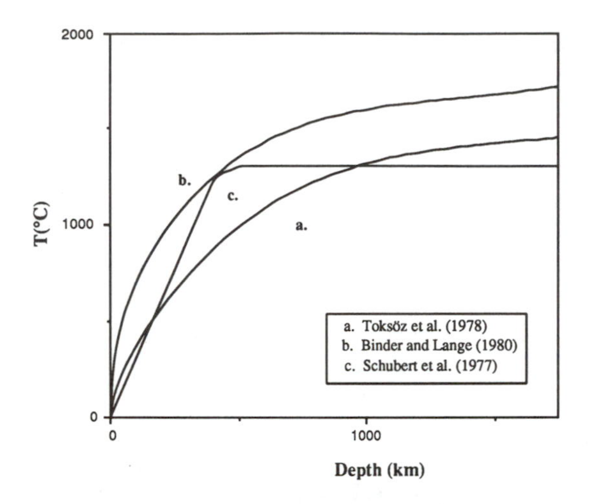

Fig. 2. *The lunar temperature profiles used in this study originate in the thermal models of: (1) Toksöz et al. (1978); (2) Binder and Lange (1980); and (3) Schubert et al. (1977). Only the Toksöz et al. thermal model allowed a terrestrial upper mantle composition to simultaneously meet bulk density and moment of inertia data for the Moon.*

so the modal abundance of these minerals is important in determining the bulk density of the Moon. If Al concentrations are low, then all the Al may reside in pyroxene. If Al abundances are high, then spinel and garnet are stabilized, producing a denser Moon.

In our earlier study (*Hood and Jones,* 1987), as a check on our parameterization of Al solubility in pyroxene, we compared the Al content of a pyroxene just above the spinel-garnet transition (in the spinel field) to that of a pyroxene just below that boundary (in the garnet field). In principle, when garnet, spinel, and pyroxene are all in equilibrium, the garnet-pyroxene parameterization should yield the same alumina content for pyroxene at the spinel-garnet transition as the spinel-pyroxene parameterization. The two pyroxenes from above and below the spinel-garnet transition agreed to within ~30%, which we consider satisfactory. The thermal profile for this particular model was that of *Toksöz et al.* (1978).

Geochemical Modeling

As a further restriction on our geophysical models, we require that any acceptable geophysical model for the Moon also approximately reproduce the correct lunar siderophile element pattern (Fig. 3), using the terrestrial upper mantle as a starting composition. The model that we use is that of *Newsom* (1985), who considered equilibrium of siderophile elements between three reservoirs: a solid metal core, a silicate liquid, and a solid silicate residuum. This model should be appropriate for bodies that are depleted in volatiles, such as S and P, which lower the melting point of iron and stabilize liquid metal, which would constitute a fourth reservoir. Under volatile-depleted conditions the fraction (X) of metal in a planetary body that is required to produce the observed siderophile element depletions is given by

$$X = (\alpha - 1)/(<D> + \alpha - 1) \qquad (1)$$

where α is the observed depletion relative to the initial conditions and $<D>$ is the bulk partition coefficient between metal and silicate given by

$$<D> = D_{sm/ls} / [F + D_{ss/ls}(1 - F)] \qquad (2)$$

where $D_{sm/sl}$ is the partition coefficient between solid metal and silicate liquid, $D_{ss/ls}$ is the partition coefficient between solid silicate and liquid silicate, and F is the degree of partial melting.

In order to evaluate a particular model of core formation, the variation in X as a function of F is calculated for a single siderophile element and plotted on an X vs. F diagram. The process is repeated for other elements, and if there is a viable solution, then there will be an intersection of the X vs. F trends at some specific value of X and F. In general, $<D>$ of equation (2) is a function of temperature, pressure and f_{O_2}. The effects of pressure on D are unknown but are thought to be small, since pressures in the Moon can never exceed

50 kbar. The effects of temperature and f_{O_2} are quite strong, however, and the exact temperature of lunar core formation is unknown.

These difficulties may be at least partially circumvented by defining core formation in terms of conditions relative to a buffer curve such as iron-wüstite (IW) or quartz-fayalite-iron (QFI) rather than demanding that the T and f_{O_2} of core formation be known exactly. For example, *Rammensee* (1978) has shown that metal-silicate partition coefficients change only weakly as long as the system follows a (T, f_{O_2}) path that is either along a buffer curve such as IW or is parallel to a buffer curve. Thus, if the approximate redox state of the Moon is known, then partition coefficients can be corrected from the conditions of the laboratory to those of the Moon. This is accomplished by assuming that nonidealities in the solution properties of the siderophile elements are small compared to the energy required for oxidation/reduction (*Rammensee,* 1978), so that changes in D between different redox conditions occur essentially ideally. The set of solid metal/liquid silicate partition coefficients that will be used here are those of *Jones and Drake* (1986), which are internally consistent for an f_{O_2} of ~1 log unit below QFI. Solid silicate/liquid silicate partition coefficients are also taken from Jones and Drake and represent either typical experimental results or inferences from natural basalts. A summary of partition coefficients used in this study is given in Table 2.

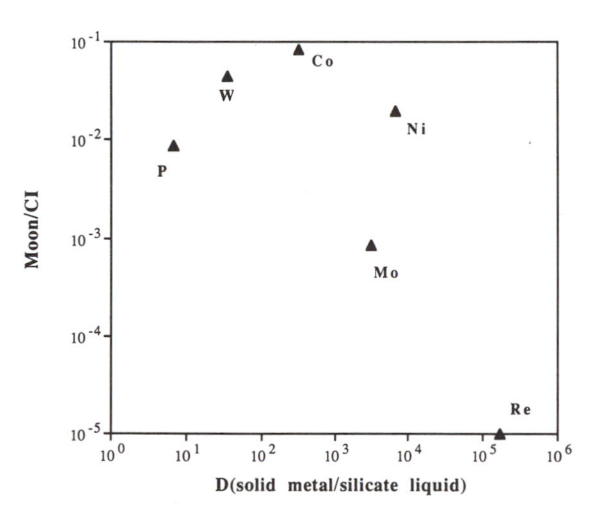

Fig. 3. *Siderophile element depletions in the Moon. Values are taken from Newsom (1986) and Delano (1984).*

TABLE 2. *Summary of partition coefficients.*

Element	D(solid metal/ silicate liquid)*	D(solid silicate/ liquid silicate)
W	112	.01
Re	3.9×10^5	0.5-0.05
Mo	1×10^4	0.01
Ni	1.3×10^4	10
Co	620	3
P	30	0.02

*For 2 log units below IW (*Newsom and Drake,* 1983).

The elements to be modeled are Re, W, Mo, Ni, Co, and P, a suite that covers a wide range of geochemical behaviors. In our models we have calculated depletions assuming either that (1) siderophile element abundances in the bulk Moon are the same as the present-day terrestrial upper mantle or (2) the Moon contains chondritic abundances of siderophiles relative to a refractory element. Listed in order of decreasing siderophility, Re, Mo, and W are refractory and are expected to be undepleted by volatility-related processes. Nickel and Co are somewhat less refractory but should still be present in chondritic proportions if no core formation event has fractionated them; of these two, Co is the least siderophile. Phosphorus is moderately siderophile and is also volatile. If no metal is present, Ni and Co behave compatibly during silicate partial melting, while Re, W, Mo, and P behave incompatibly. Thus, our suite of elements covers a wide range of siderophility and exhibits a variety of partitioning behaviors when metal is absent.

In the models explored here, the P abundance is taken to be either (1) that of the Earth's upper mantle, (2) chondritic, relative to a refractory element, or (3) 0.1 × chondritic refractories. The third case assumes that P has been depleted because of its volatility, as well as by core formation. The tenfold depletion is calculated by assuming that the volatility of P is intermediate to that of Li and Na (Fig. 4) as predicted by *Wasson* (1985).

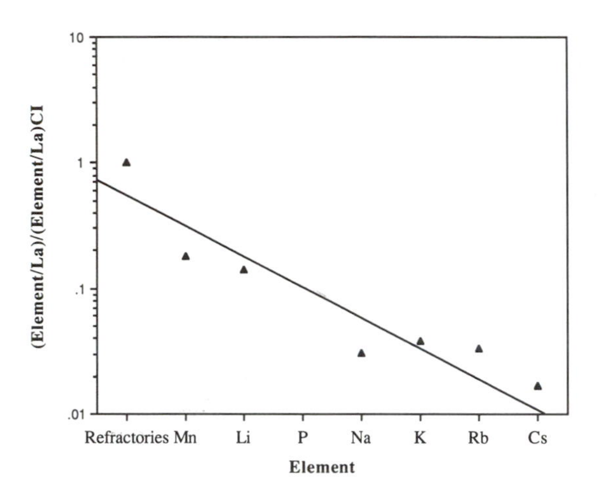

Fig. 4. *Estimation of the depletion of P in the Moon based on the depletion of the alkali metals. Phosphorus is assumed to have a volatility intermediate to that of Li and Na (Wasson, 1985).*

RESULTS

Geophysics

Table 3 gives a summary of the results. A dash denotes no possible solution; numbers refer to the core radius necessary to fit the lunar density of 3.344 ± 0.003 g/cm^3 and the lunar moment of inertia (C/MR^2) of 0.3905 ± 0.0023 (*Hood,* 1986). In all, 4 different differentiation models and 3 different thermal models were explored—a total of 12 different permutations. No solutions to differentiation model II were possible because there was not enough Al_2O_3 in the upper 500 km of the Moon to make 70 km of lunar crust. Of the three thermal models, only the model of *Toksöz et al.* (1978) produced acceptable solutions. The "hotter" thermal models of *Binder and Lange* (1980) and *Schubert et al.* (1977) produced moons that were not dense enough. If metallic cores of sufficient sizes were added to make the Binder and Lange and Schubert et al. models have the correct density, then the moment of inertia became too small.

It should be noted that the *Toksöz et al.* (1978) model moons, while acceptable in terms of the density and moment of inertia criteria, are only marginally so. To produce acceptable fits, we have used cores that approach the limits of acceptability (~500 km; *Hood,* 1986) and crusts of minimum thickness (60 km) and maximum density (3.05 g/cm^3). Thus, several input parameters have been forced to their limits in order to simultaneously match the measured density and moment of inertia.

Geochemistry

The core sizes that we have calculated here for the Toksöz et al. thermal model are rather large (~500 km in radius) but are within the range of maximum core sizes that are geophysically permissible (450-500 km; *Hood,* 1986). This range of upper limits to the size of the lunar core is at least approximately consistent with the geochemically-derived maximum size of *Newsom* (1984). In Newsom's model the maximum lunar core size is 5.6 wt.%, which corresponds to a Fe-metal core with a 500-km radius.

However, the maximum core size derived by *Newsom* (1984) to explain siderophile element depletions was for chondritic starting compositions. If terrestrial upper mantle

TABLE 3. *Summary of results for core sizes of successful models.*

Thermal model	Differentiation Model			
	I	II	III	IV
Toksöz et al.	500 km	—	500 km	510 km
Schubert et al.	—	—	—	—
Binder and Lange	—	—	—	—

siderophile element abundances are used instead, then much smaller cores are required. This is shown graphically in Figs. 5-7, which summarize the study of *Jones et al.* (1988). Using the model of *Newsom* (1985), as outlined in the preceding section, geochemically permissible cores can be calculated for a variety of siderophile elements. The model assumes that the core is solid metallic iron and that, at the time of core formation, the Moon could have been partially molten. Conditions of core formation are parameterized in terms of fraction of the mass of the planet that is core and degree of partial melting of silicate.

Figure 5 shows the allowed elemental paths for core formation using a chondritic starting composition. In this model, Re is taken to be moderately incompatible ($^{Re}D_{ss/ls} = 0.5$). No intersection of the paths of the various elements is seen. Figure 6 explores a variant of the chondritic model, where P is depleted relative to the other siderophiles by a factor of 10 and where Re behaves more incompatibly ($^{Re}D_{ss/ls} = 0.05$). Here there is a relatively small region of intersection between 10-30% silicate partial melting and 3-5 wt.% of metallic core. This type of core size is consistent with the geophysical data and the degree of partial melting is sufficient to have produced a lunar magma ocean several hundreds of kilometers deep. Thus, with only moderate adjustment, the chondritic model can be made consistent with what we know or infer about the structure and early differentiation of the Moon. On the other hand, if the proto-Moon began with terrestrial upper mantle abundances of siderophile elements, then only a very small core is required to deplete lunar siderophiles to their present concentrations (Fig. 7). For most of the elements in Fig. 7, a core of 0.2 wt.% would suffice to produce the observed depletions. This corresponds to a core of 165-km radius, which although within the range of core sizes expected by *Mueller et al.* (1988), is insufficient to eliminate the discrepancy between the density of the Moon and that of the Earth's upper mantle. No real intersection of the various elemental paths is observed (note change of scale on the abscissa to logarithmic). Also note that W is not modeled here, because the nominal concentration of W in the Moon is actually higher than in the Earth, even though it could be the same within error (but see *Sims et al.*, 1990).

Thus, our calculations are at odds with the original hypothesis that the Moon could have the same composition as the terrestrial upper mantle. Geophysical constraints dictate that if the Moon is made entirely of upper mantle material then there must be a relatively large (~500 km) lunar core to meet the density requirement. If, however, such a core existed, it could not have equilibrated with the lunar mantle and crust, otherwise the silicate

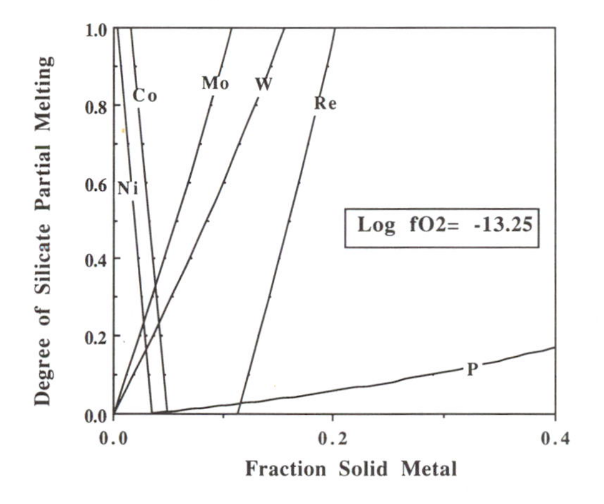

Fig. 5. *Degree of silicate partial melting vs. fraction of solid metal that is required to produced the observed depletions of various siderophile elements, starting with CI chondritic abundances of these elements. The method of calculation is that of Newsom (1985). Partition coefficients are from Jones and Drake (1986), extrapolated to an oxygen fugacity appropriate for the Moon (Newsom and Drake, 1983). No real agreement among the elements evaluated (Co, Mo, Ni, P, Re, and W) is observed. For example, Co and Ni require only very small amounts of metal segregation, regardless of the degree of partial melting. Phosphorus, however, demands small degrees of partial melting and is rather insensitive to the amount of metal. In this model, Re is assumed to behave moderately compatibly during silicate partial melting events ($^{Re}D = 0.5$).*

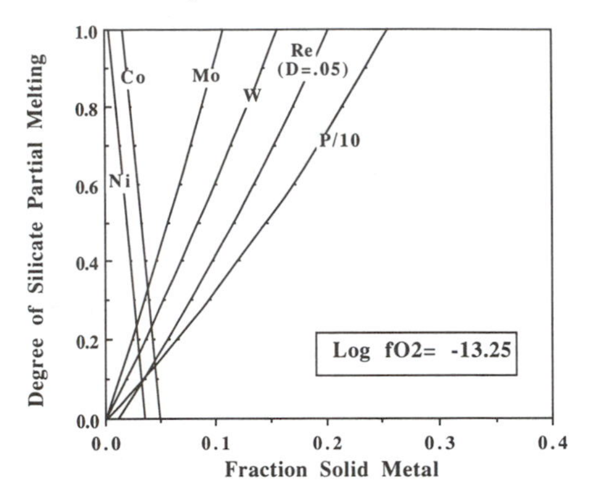

Fig. 6. *Same as Fig. 5 except that P is assumed to be depleted by 10× in the bulk Moon and Re is assumed to be a very incompatible element. The P depletion was estimated by interpolation of alkali metal depletions (Wänke, 1981) and by assuming that P is intermediate in volatility to Li and Na. It was also assumed that refractory elements such as La are undepleted in the bulk Moon. Here a modest overlap of the different analytical solutions is observed. Metal fractions of 3-5 wt.% at 10-30% silicate partial melting can account for the depletions of a variety of siderophile elements. These values are consistent with upper limits on the size of the lunar core (~5 wt.%) and with the amount of partial melting that must have been required to produce a magma ocean several hundreds of kilometers deep.*

portion of the Moon would be much more depleted in siderophiles than is presently observed.

DISCUSSION

Evaluation of Results

Our calculations indicate that it is not feasible to make a Moon from the terrestrial upper mantle alone. If the composition of the upper mantle is indicative of the bulk silicate Earth, then there is no terrestrial reservoir that can produce a Moon of the correct composition. If so, this is a significant result, and thus it is proper to question how our model could be incorrect. We feel that the conclusions of our model are rather firm, not because the details of lunar core formation are well known, but because of the magnitude of error that would be required to make the model suspect. The difference in calculated masses between the "geochemical core" and the "geophysical core" is approximately a factor of 25 (0.2 vs. 5 wt.%) and thus our error must be correspondingly large to invalidate our calculations.

We note that the lunar geophysical data do not allow us to judge between these two extreme core sizes. Seismic data indicate that, within error, P wave transmission in the lower lunar mantle appears to remain constant down to a depth of ~1300-1400 km (*Nakamura et al.,* 1982), suggesting a maximum size of ~450 km for the lunar core. We have already

Lunar Siderophile Model: Terrestrial Source

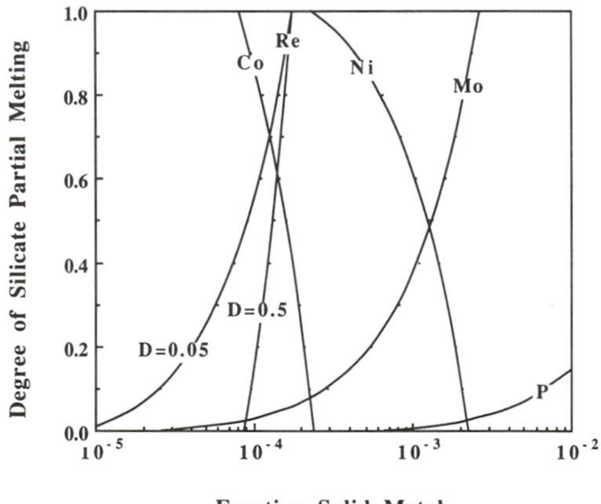

Fig. 7. *Same as Figs. 5 and 6, except that the composition of the starting material is assumed to be that of the terrestrial upper mantle. Note the change of scale on the abscissa. As in Fig. 3, no real intersection is observed. However, if P is excluded, an upper limit of 0.2 wt.% may be placed on the size of the lunar core. Phosphorus can be made to be consistent with this upper limit if a small amount of liquid metal is present during core formation. This amount of metal is about a factor of 4 smaller than the smallest core calculated by Hood and Jones (1987), a factor of 5-10 smaller than typical minimum core sizes of Hood and Jones (1-2 wt.%), and a factor of 25 smaller than the core size necessary to allow a Moon comprised of terrestrial upper mantle materials to have the correct density (~5 wt.%) (see text for discussion). There are no solutions for W, which is not depleted in the Moon compared to the Earth.*

seen that variation in thermal and compositional parameters within their present uncertainties also allows the bulk density and moment of inertia to be consistent with a fairly large range of core sizes (*Hood and Jones,* 1987). In general, therefore, most of the geophysical evidence that we have could be made consistent with any core smaller than ~500 km radius (e.g., *Hood,* 1986).

Geochemical modeling. While we would not dispute that our choice of partition coefficients could be in error by a factor of 2 to 3—typically the largest discrepancy between the experiments of different research groups—order of magnitude errors seem prohibitively large. Additionally, although we have not specifically investigated the change that would occur if the Moon's core were primarily FeS liquid instead of solid (Fe,Ni) metal, this too seems improbable as the cause of the discrepancy in core sizes—firstly, because the Moon appears to be depleted in volatiles such as S (e.g., Fig. 3), and secondly because both the geophysical and geochemical calculations presume that the core is solid. If the core is liquid FeS, then both models will require larger cores.

It is possible that our choice of redox conditions is incorrect. *Ringwood* (1990) argues that the redox conditions assumed by *Newsom and Drake* (1983) are too reducing and, if the Moon is more oxidizing than we have assumed, then larger cores are required to produce the same levels of siderophile element depletion. The f_{O_2} estimate of Newsom and Drake was derived by assuming that the bulk silicate Moon contains 13 wt.% FeO and that this material was once in equilibrium with pure iron metal. The *Newsom and Drake* (1983) FeO content is slightly higher than recent estimates by *Ringwood et al.* (1987) and *Jones and Delano* (1989), but not significantly so. More importantly perhaps, this amount of FeO is approximately the correct amount to produce the observed density of the Moon (*Taylor,* 1982). If the lunar core is Ni-rich and contains ~40 wt.% Ni (*Seifert et al.,* 1988; *Ringwood,* 1990), then the Newsom and Drake f_{O_2} estimate is somewhat too reducing, but not seriously so. If the lunar core contains even as much as 50 mol.% Ni, log f_{O_2} increases by only 0.6 units. Thus, it appears most probable that the f_{O_2} at the time of lunar core formation was 1.5-2.0 log units below IW. For most of the elements in our geochemical model, this amount of uncertainty in f_{O_2} corresponds to about a factor of two uncertainty (or less) in our partition coefficients, which, as discussed above, we deem acceptable. Uncertainties in the lunar f_{O_2} do not appear able to reconcile the difference between the geophysical and geochemical calculations.

Geophysical modeling. In our geophysical models the uncertainty in chemical composition is small and our results are not tied to a particular model of igneous differentiation (Table 3). Thus, in our view, our results mainly depend on the correctness of the range of thermal models that we have utilized. For example, if the interior of the Moon is much colder than we have assumed, then our calculated densities will be too low, both because of thermal contraction and because at lower temperatures more Al_2O_3 will will form garnet and spinel rather than tschermakite components of pyroxene. Therefore, it is reasonable that the "coldest" lunar thermal model, that of *Toksöz et al.* (1978), gives the best agreement with the measured geophysical parameters. How much confidence, then, may we place in the Toksöz et al. model?

It is difficult to evaluate the correctness of the *Toksöz et al.* (1978) model from first principles, although this model was preferred by *Hood* (1986) in his review of lunar geophysics. Factors such as the initial lunar temperature, the spatial variation of temperature within the early Moon, and exact mechanisms of heat transport are not well known. Bulk concentrations and distributions of heat-producing radioactive elements are also uncertain. However, it seems probable that, below 1000 km depth (40 kbar), temperatures in the Moon approach or exceed the peridotite solidus. Otherwise, there is no obvious means of accounting for the observed attenuation of S-waves below that depth (*Nakamura et al.,* 1982). In the absence of water, the solidus of terrestrial peridotite at 40 kbar is approximately 1600° C (*Takahashi and Scarfe,* 1985). If terrestrial materials may be used as a guide, then the deep lunar interior should have a temperature of 1400°-1600° C, with the lower temperature estimate attempting to take into account possible differences in Mg# between the Earth and Moon. This estimate is not very different from those of the thermal models that we have utilized, all of which predict that the deep lunar interior has a temperature of ~1300°-1700° C. Thus, we see no evidence that the interior of the Moon is significantly colder than we have assumed. If the Moon is hotter than the *Toksöz et al.* (1978) model, then the difficulties involved in making a terrestrial mantle composition compatible with the lunar geophysical measurements are exacerbated.

Therefore, on the basis of this analysis, we see no realistic escape from the inconsistency of the geochemical and geophysical calculations. The simplest solution to this problem is that the Moon was not solely derived from terrestrial materials.

Application to Theories of Lunar Origin

Fission origin of the Moon and its variants (including giant impact). Historically, the driving impetus for making the Moon from terrestrial materials has come from the low siderophile element abundances in the Moon—coupled with the low density of the Moon and the supposed absence of a lunar core (e.g., *Ringwood,* 1979). A simple means of explaining these diverse observations is for the metal and siderophile elements of the Moon to be in the Earth's core and for the Moon to have formed from terrestrial mantle materials subsequent to core formation on the Earth (the "fission" model). However, in recent years the contention that highly siderophile elements are more depleted in the lunar mantle than in the terrestrial mantle has gained general acceptance (*Newsom,* 1984, 1986; *Ringwood,* 1987, 1990), implying that the Moon does have a small core $\lesssim 5$ wt.%. Also, as discussed above, the models of *Hood and Jones* (1987) and *Mueller et al.* (1988) required the presence of a small core to meet the geophysical constraints of density and moment of inertia. *Thus, with the general acceptance of the existence of a lunar core, the original motivation for demanding that the Moon be comprised of only terrestrial materials, already depleted in siderophiles, is obviated.*

It seems most probable that the Moon did not totally form from terrestrial materials. If, however, the Moon formed from *predominantly* terrestrial materials, then, of course, some other component is necessary. One possibility is that the Moon was formed by a giant impact onto the Earth and that some of the impactor material was incorporated into the Moon (*Benz et al.,* 1986; *Kipp and Melosh,* 1986). *Ringwood* (1987) concluded that the siderophile element signatures of the Earth and Moon are so similar that only ~20% of the Moon could have been derived from the impactor. The terrestrial upper mantle has 8 wt.% FeO and, if the silicate portion of the Moon is taken to have ~12 wt.% FeO (*Jones and Delano,* 1989) and 20% is taken to be an upper limit to the contribution of impactor materials, then a minimum FeO concentration for the impactor of 28 wt.% is required. This is quite a lot of FeO. The only meteorites in our collections that contain such large quantities of FeO are carbonaceous chondrites, all of which have different oxygen isotopic signatures than the Earth. Whether an impactor can be found that is both rich in FeO and has the same oxygen isotopic composition as the Earth is uncertain. Therefore, we feel that we may confidently take 20% as a lower limit to the amount of material contributed by the impactor. A more detailed analysis of impactor-Moon compositional relations has been given by *McFarlane* (1989). Using known meteorite types as impactors, McFarlane found that, in general, the *Jones and Delano* (1989) bulk lunar silicate compositions were easier to achieve if the Earth contributed less than 50% of the protolunar materials, in agreement with our conclusions here.

Another possibility is that, at the time of the impact, the upper mantle of the Earth contained more FeO than at present (*Delano and Stone,* 1985; *Ringwood,* 1986). In this scenario FeO removal from the upper mantle over geologic time is required. It is unlikely that this FeO now resides as Fe or FeO in the Earth's core, because Ni and Co have remained in approximately chondritic relative abundances in the Earth's upper mantle. If some version of the fission or giant impact model is correct, then Ni and Co abundances in the Earth's upper mantle were established prior to the formation of the Moon and FeO must be removed from the Earth's mantle while leaving Co and Ni relatively unfractionated. This is difficult, in that both Ni and Co have quite different siderophile characters and, more importantly, are both more siderophile than Fe. Additionally, FeO is somewhat difficult to fractionate during typical igneous processes (e.g., *Jones,* 1984). Changes in Mg# during igneous events are mainly due to MgO fractionation—not to selective partitioning of FeO. Thus, if the FeO content of the upper mantle has changed over time, the mechanism for this change is not immediately clear.

Disequilibrium between the lunar core and mantle. One possibility is that the present lunar core was never truly in equilibrium with the silicate portion of the Moon (e.g., *Newsom and Taylor,* 1989). Partial equilibration occurred, depleting siderophiles from their initial abundances, but not to the extent that would have occurred at equilibrium. If so, our geochemical models, which assume equilibrium, would be invalid and the lunar core could be much larger than our geochemical calculations indicate. This is a difficult model to evaluate, as nonequilibrium events are, by their very nature, intractable and untestable. However, some general statements may be made.

Firstly, if there were inhomogeneities in siderophile element abundances in the silicate portion of the Moon immediately following core formation, then these have been largely erased. The Ni contents of the primitive lunar glasses of *Delano* (1986) correlate well with MgO, as they should if there were no gross heterogeneities in Ni. *Dickinson and Newsom* (1985) have argued that Ge in the Moon is indeed heterogeneous and that, because Ge is difficult to fractionate in processes that only involve silicates (*Malvin and Drake,* 1987), core formation and/or accretion resulted in heterogeneities that were never erased. However, *Dickinson et al.* (1989) have observed that the only lunar landing site with truly anomalous Ge is Apollo 14 and here the strong association with KREEP argues that some type of fluid phase may have been important in fractionating Ge. Thus, it is not clear that the Ge enrichment at the Apollo 14 site is primordial, but may instead reflect late-stage processes. In any event, we see no strong evidence that Ge distributions constrain the early Moon to have been heterogeneous.

Secondly, the smooth variation and continual decrease in abundance of siderophile elements as a function of siderophility (e.g., *Newsom,* 1986) argues that there was at one time a close approach to metal-silicate equilibrium. If this signature were inherited from another body, such as the putative giant impactor, then the Moon's inherited fraction of terrestrial mantle material, which does not show such a pattern, must have been small. Therefore, within our ability to evaluate an inherently untestable model, we see no strong evidence that the Moon should be dominated by terrestrial materials or that disequilibrium effects frustrate our ability to evaluate the Earth's contribution to the present Moon.

Independent or coaccretion origin of the Moon. Our results effectively rule out one endmember model for the origin of the Moon ("fission" from the Earth, with no change in bulk chemistry). What then do our calculations imply for the other extreme case, an origin totally independent of the Earth? The basic, *qualified* answer appears to be, "Yes, it is possible." Figure 6 shows that, with minor modification, a source possessing "chondritic" relative abundances of siderophile elements can be processed to produce a Moon similar to that we infer. This does not imply that the Moon should be made of material totally foreign to the 1-A.U. region of the solar system. We feel that the oxygen isotopic evidence indicates a clear relationship between prototerrestrial and protolunar materials. However, if the geochemical model is correct, the Moon's siderophile element abundances could be explained by 3-5% of core formation on a body whose bulk abundances of refractory siderophiles were chondritic.

The qualification is that, regardless of how the Moon formed, a major depletion of Fe must have occurred either prior to or during the Moon's accretion. How major must this depletion be? Optimistically, if the silicate portion of the Moon contains 12 wt.% FeO and there is a 5 wt.% core that is 100% Fe, then the total Fe content of the Moon could be as large as 14 wt.%. In this case the Fe content of the Moon is about 25% lower than CI chondrites or, more importantly, is depleted relative to H, L, and LL chondrites by factors of 50%, 35%, and 25%, respectively. For comparison, LL chondrites are depleted in total iron relative to H chondrites by ~35% and EL chondrites are depleted relative to EH chondrites by ~25%. Thus, it seems to us that it is possible that the Moon is no more depleted in Fe, relative to L or LL chondrites, than the LL chondrites are to H chondrites. Further, the ordinary chondrites, with their significant range of Fe contents, all have CI-chondritic abundances of the refractory siderophile elements to within ~20% (*Wasson and Kallemeyn,* 1988). The mechanism for fractionating Fe from lithophile elements such as Mg and Si and (to a degree) from other siderophile elements is unknown. However, it seems no more implausible to us that the materials that formed the Moon were involved in an extreme version of the process that formed the various chondrites than to require that the Moon be formed by a giant impact or by some version of terrestrial fission. Thus, we believe that the independent or co-accretion model is still chemically viable and is not necessarily ruled out either by our knowledge of the geochemistry of the Moon or by the chemical variation observed within the ordinary chondrite suite. The primary objections to co-accretion or capture are physical not chemical (e.g., *Weidenschilling et al.,* 1986).

SUMMARY AND CONCLUSIONS

Within the confines of the present model, we conclude that it is unlikely that the Moon can be wholly derived from a terrestrial upper mantle composition. In general the density of terrestrial mantle is too low to meet the bulk density requirement of the Moon. Supplementation of terrestrial silicate materials with a large metallic core can overcome the difficulty of bulk density, but runs the risk of unacceptably lowering the lunar moment of inertia. Consequently, very few of the models that we have investigated meet both density and moment of inertia requirements simultaneously. Additionally, cores that are large enough to meet the density requirement would deplete the Moon so severely in siderophiles that the terrestrial upper mantle by itself is not a viable starting material. Comparison of lunar, terrestrial, and meteoritic FeO contents imply that the terrestrial contribution to the Moon is less than 80%. Thus, at least 20% of "exotic" material is necessary to make the Moon. If some of the Fe budget in this exotic material were in the metallic state, then the percentage of nonterrestrial material in the Moon must proportionally increase.

It appears to us, therefore, that the Earth and the Moon are compositionally similar but compositionally distinct. This conclusion is in agreement with some recent estimates of the Moon's composition (*Ringwood et al.,* 1987; *Jones and Delano,* 1989). In terms of physical models for the origin of the Moon, we cannot rule out either an origin by giant impact or by some form of coaccretion. Further investigation will be necessary to decide between these two endmember models that have such profound implications for the origin and early evolution of the Earth and Moon.

REFERENCES

Benz W., Slattery W. L., and Cameron A. G. W. (1986) Short note: Snapshots from a three-dimensional modeling of a giant impact. In *Origin of the Moon* (W. K. Hartmann, R J. Phillips, and G. J. Taylor, eds.), pp. 617-620. Lunar and Planetary Institute, Houston.

Binder A. and Lange M. A. (1980) On the thermal history, thermal state, and related tectonism of a moon of fission origin. *J. Geophys. Res., 85,* 3194-3202.

Davis P. A. and Spudis P. D. (1985) Petrologic province maps of the lunar highlands derived from orbital geochemical data. *Proc. Lunar Planet. Sci. Conf. 16th,* in *J. Geophys. Res., 90,* D61-D71.

Delano J. W. (1984) Abundances of Ni, Cr, Co, and major elements in the silicate portion of the Moon: Constraints from primary lunar magmas (abstract). In *Papers Presented to the Conference on the Origin of the Moon,* p. 15. Lunar and Planetary Institute, Houston.

Delano J. W. (1986) Pristine lunar glasses: Criteria, data and implications. *Proc. Lunar Planet. Sci. Conf. 16th,* in *J. Geophys. Res., 91,* D201-D213.

Delano J. W. and Stone K. (1985) Siderophile elements in the Earth's upper mantle: Secular variations and possible cause for their overabundances (abstract). In *Lunar and Planetary Science XVI,* pp. 181-182. Lunar and Planetary Institute, Houston.

Dickinson T. and Newsom H. E. (1985) A possible test of the impact theory for the origin of the Moon (abstract). In *Lunar and Planetary Science XVI,* pp. 183-184. Lunar and Planetary Institute, Houston.

Dickinson T., Taylor G. J., Keil K., and Bild R. W. (1989) Germanium abundances in lunar basalts: Evidence of mantle metasomatism? *Proc. Lunar Planet. Sci. Conf. 19th,* pp. 189-198.

Grossman L., Clayton R. N., and Mayeda T. K. (1974) Oxygen isotopic constraints on the composition of the Moon. *Proc. Lunar Sci. Conf. 5th,* pp. 1207-1212.

Hood L. L. (1986) Geophysical constraints on the lunar interior. In *Origin of the Moon* (W. K. Hartmann, R. J. Phillips, and G. J. Taylor, eds.), pp. 361-410. Lunar and Planetary Institute, Houston.

Hood L. L. and Jones J. H. (1987) Geophysical constraints on lunar bulk composition and structure: A reassessment. *Proc. Lunar Planet. Sci. Conf. 17th,* in *J. Geophys. Res., 92* E396-E410.

Jagoutz E., Palme H., Baddenhausen H., Blum K., Cedales M., Dreibus G., Spettel B., Lorenz V., and Wänke H. (1979) The abundances of major, minor and trace elements in the earth's mantle as derived from primitive ultramafic nodules. *Proc. Lunar Planet. Sci. Conf. 10th,* pp. 2031-2050.

Jones J. H. (1984) Temperature- and pressure-independent correlations of olivine/liquid partition coefficients and their application to trace element partitioning. *Contrib. Mineral. Petrol., 88,* 126-132.

Jones J. H. and Delano J. W. (1989) A three-component model for the bulk composition of the Moon. *Geochim. Cosmochim. Acta, 53,* 513-527.

Jones J. H. and Drake M. J. (1986) Geochemical constraints on core formation in the Earth. *Nature, 322,* 221-228.

Jones J. H., Treiman A. H., Janssens M.-J., Wolf R., and Ebihara M. (1988) Core formation on the Eucrite Parent Body, the Moon and the AdoR Parent Body (abstract). *Meteoritics, 23,* 276-277.

Kipp M. E. and Melosh H. J. (1986) Short note: A preliminary numerical study of colliding planets. In *Origin of the Moon* (W. K. Hartmann, R. J. Phillips, and G. J. Taylor, eds.), pp. 643-647. Lunar and Planetary Institute, Houston.

Malvin D. J. and Drake M. J. (1987) Experimental determination of crystal/melt partitioning of Ga and Ge in the system forsterite-anorthite-diopside. *Geochim. Cosmochim. Acta, 51,* 2117-2128.

McFarlane E. A. (1989) Formation of the Moon in a giant impact: Composition of the impactor. *Proc. Lunar Planet Sci. Conf. 19th,* pp. 593-605.

Morgan J. W., Hertogen J., and Anders E. (1978) The Moon: Composition determined by nebular processes. *Moon and Planets, 18,* 465-478.

Mueller S., Taylor G. J., and Phillips R. (1988) Lunar composition: A geophysical and petrological synthesis. *J. Geophys. Res., 93,* 6338-6352.

Nakamura Y., Latham G. V., and Dorman J. (1982) Apollo lunar seismic experiment—final summary. *Proc. Lunar Planet. Sci. Conf. 13th,* in *J. Geophys. Res., 87,* A117-A123.

Newsom H. E. (1984) The lunar core and the origin of the Moon. *Eos Trans. AGU, 65,* 369-370.

Newsom H. E. (1985) Molybdenum in eucrites: Evidence for a metal core in the eucrite parent body. *Proc. Lunar Planet. Sci. Conf. 15th,* in *J. Geophys. Res., 90,* C613-C617.

Newsom H. E. (1986) Constraints on the origin of the Moon from the abundance of molybdenum and other siderophile elements. In *Origin of the Moon* (W. K. Hartmann, R. J. Phillips, and G. J. Taylor, eds.), pp. 203-230. Lunar and Planetary Institute, Houston.

Newsom H. E. and Drake M. J. (1983) Experimental investigation of the partitioning of phosphorus between metal and silicate phases: Implications for the Earth, Moon and Eucrite Parent Body. *Geochim. Cosmochim. Acta, 47,* 93-100.

Newsom H. E. and Taylor S. R. (1989) The single impact origin for the Moon. *Nature, 338,* 29-34.

Rammensee W. (1978) Verteilungsgleichgewichte von spurenelementen zwischen metallen und silikaten. Ph.D. thesis, Universität Mainz. 160 pp.

Ringwood A. E. (1979) *Origin of the Earth and Moon.* Springer-Verlag, New York. 295 pp.

Ringwood A. E. (1986) Composition and origin of the Moon. In *Origin of the Moon* (W. K. Hartmann, R. J. Phillips, and G. J. Taylor, eds.), pp. 673-698. Lunar and Planetary Institute, Houston.

Ringwood A. E. (1987) Terrestrial origin of the Moon. *Nature, 322,* 323-328.

Ringwood A. E. (1990) Earliest history of the Earth-Moon system. In *Origin of the Earth,* this volume.

Ringwood A. E., Seifert S., and Wänke H. (1987) A komatiite component in Apollo 16 highland breccias: Implications for the nickel-cobalt systematics and bulk composition of the Moon. *Earth Planet. Sci. Lett., 81,* 105-117.

Schubert G., Young R. E., and Cassen P. (1977) Subsolidus convection models of the lunar internal temperature. *Philos. Trans. R. Soc. London, A285,* 523-536.

Seifert S., O'Neill H. St. C., and Brey G. (1988) The partitioning of Fe, Ni and Co between olivine, metal, and basaltic liquid: An experimental and thermodynamic investigation, with application to the composition of the lunar core. *Geochim. Cosmochim. Acta, 52,* 603-616.

Sims K. W. W., Newsom H. E., and Gladney E. S. (1990) Chemical fractionation during formation of the Earth's core and continental crust: Clues from As, Sb, W, and Mo. In *Origin of the Earth,* this volume.

Takahashi E. and Scarfe C. M. (1985) Melting of peridotite to 14 GPa and the genesis of komatiite. *Nature, 315,* 566-568.

Taylor S. R. (1982) *Planetary Science: A Lunar Perspective.* Lunar and Planetary Institute, Houston. 481 pp.

Taylor S. R. (1987) The origin of the Moon. *Am. Sci., 75,* 468-477.

Toksöz M. N., Hsui A. T., and Johnson D. H. (1978) Thermal evolutions of the terrestrial planets. *Moon and Planets, 18,* 281-320.

Wänke H. (1981) Constitution of the terrestrial planets. *Philos. Trans. R. Soc. London, A303,* 287-302.

Wänke H., Baddensausen H., Blum K., Cendales M., Dreibus G., Hofmeister H., Krause H., Jagoutz E., Palme C., Spettel B., Thacker R., and Vilcsek E. (1977) On the chemistry of lunar samples and achondrites. Primary matter in the lunar highlands: A reevaluation. *Proc. Lunar Sci. Conf. 8th,* pp. 2191-2213.

Warren P. H. (1986) The bulk-Moon MgO/FeO ratio: A highlands perspective. In *Origin of the Moon* (W. K. Hartmann, R. J. Phillips, and G. J. Taylor, eds.), pp. 279-310. Lunar and Planetary Institute, Houston.

Wasson J. T. (1985) *Meteorites.* Freeman, New York. 267 pp.

Wasson J. T. and Kallemeyn G. W. (1988) Compositions of chondrites. *Philos. Trans. R. Soc. London, A325,* 535-544.

Weidenschilling S. J., Greenberg R., Chapman C. R., Herbert F., Davis D. R., Drake M. J., Jones J. H., and Hartmann W. K. (1986) Origin of the Moon from a circumterrestrial disk. In *Origin of the Moon* (W. K. Hartmann, R. J. Phillips, and G. J. Taylor, eds.), pp. 731-762. Lunar and Planetary Institute, Houston.

WAS THERE A TERRESTRIAL MAGMA OCEAN?

It has become a tenet in lunar studies that the Moon once had a magma ocean several hundred kilometers deep. While this, like any model, deserves scrutiny, it has nevertheless become fairly accepted. One may wonder then if the Moon's nearest neighbor also passed through a magma ocean stage. The six papers in this section cover various aspects of this question:

Ringwood covers many topics in his paper, but one of the key constraints on theories of the early Earth are the ultrahigh pressure experiments of his group in Canberra. These experiments imply that certain elements such as Sc and La should fractionate during the solidification of a magma ocean. Interestingly, there are fertile mantle xenoliths (lherzolites) where such fractionations are not observed, perhaps implying that they have never undergone the types of melting envisioned in a magma ocean scenario. McFarlane and Drake echo this sentiment. Their high-pressure experiments also imply that fractionations that should have occurred during the magma ocean stage did not.

Tonks and Melosh counter the views of Ringwood, McFarlane, and Drake by noting that a magma ocean would have been extremely turbulent and that crystals would not have had a chance to settle. Tonks and Melosh believe that there was no large-scale fractionation during the crystallization of the magma ocean. Crystals stay with the packet of liquid that they crystallized from until the system "locks up" at high degrees of crystallization and rapid convection is no longer possible.

The other authors in this section do not address the magma ocean hypothesis directly, but still do so obliquely. A major conclusion of Davies' essay on heat and mass transport in the Earth is that there is no absolute necessity to postulate a magma ocean during accretion in the absence of a giant impact or a blanketing, water-rich atmosphere (the greenhouse effect). Sasaki points out that such an atmosphere is quite probable and that high surface temperatures should be expected, along with a magma ocean that dissolved significant amounts of the atmospheric water.

The final paper in this section is transitional. Ahrens gives us his grand overview of accretion and leads us into the next section on core formation. For the same reasons as Sasaki, Ahrens too favors the Earth having a magma ocean stage. Ahrens also calculates the size of impact necessary to eject the Earth's protoatmosphere. In some ways his paper is a summary (albeit a personal one) of the processes that might be expected to have been operative during the earliest history of the Earth. Ahrens' model of core formation is not traditional. Most of us have assumed that the early Earth was reducing and became more oxidized—a consequence of the redox state of the accreting material. In Ahrens' view, the oxidation state of the Earth is highly dependent on the mechanism of accretion and the temperature dependence of redox reactions. Ahrens argues that, if accretion processes are taken into account, the Earth became more reduced with time.

The subject of the terrestrial magma ocean is covered rather well and could be called the central theme of the book. However, there are differing views on the interpretation of the experiments of Ringwood that are not represented [e.g., see Walker and Agee (1989) *Earth and Planetary Science Letters, vol. 96,* 49-60].

The crux of the issue is simple. There are many good reasons to believe that the early Earth was very hot. Some would go so far as to suggest that a terrestrial magma ocean was unavoidable. The recent advances in our understanding of the accretion of the Earth appear to further promote the idea that the early Earth must have undergone large degrees of partial melting. On the other hand, there are rocks from the mantle that appear to have experienced minimal processing. How do we reconcile these two opposing observations?

Perhaps the Tonks and Melosh model is correct and the molten mantle never truly differentiated, although after the magma ocean "locks up" it is not obvious why the remaining melt should stay in place. How is this situation different from that of modern basalt genesis (i.e., how do you tell the difference between large degrees of crystallization and small degrees of partial melting?)? Perhaps solid state convective mixing in the post-magma-ocean phase was rapid and efficient and removed the evidence of magma ocean fractionation. Perhaps the standard accretion models are incorrect and the Earth somehow accreted slowly and quiescently from small planetesimals incapable of heating the Earth to great depths. Clearly, to sufficiently evaluate these diverse possibilities we have much more work ahead of us.

EARLIEST HISTORY OF THE EARTH-MOON SYSTEM

A. E. Ringwood

Research School of Earth Sciences, Australian National University, Canberra, ACT 2601, Australia

The thermal and chemical development of the Earth during its accretion is considered, particularly in the context of the single giant impact hypothesis for the origin of the Moon. Impact of a martian-sized planetesimal during accretion of the Earth would have caused complete or extensive melting of the mantle. Differentiation of a molten mantle would then have produced strong chemical and mineralogical stratification, causing it to become gravitationally stable. The resulting composition and mineralogy of the upper mantle would have been quite different from those that have existed during the past 3.8 Ga. The absence of these fractionations implies that the upper and lower mantle possess similar bulk compositions. The formation of metallic cores within planetary bodies is accompanied by drastic fractionations of siderophile elements. Because of the complexity of the core-formation process within a given body and the multiplicity of chemical and physical processes involved, the resultant abundance pattern of siderophile elements that remains in the silicate mantle is expected to be uniquely characteristic of the particular body. Thus, the siderophile signature of Mars and of the eucrite parent body are quite distinct from that of the Earth's mantle. Because of the tiny size of the lunar core and the vastly different conditions under which it formed, as compared to the Earth's core, it would be expected that the siderophile signature of the lunar mantle would be quite different from that of the terrestrial mantle. However, a detailed review of lunar siderophile geochemistry demonstrates that many moderately siderophile elements possess similar abundances in the terrestrial and lunar mantles. This similarity implies that a major proportion of the material now in the Moon was derived from the Earth's mantle after core formation. Chromium, vanadium, and manganese are of particular significance since these elements are depleted in the terrestrial mantle and Moon to similar degrees, but are undepleted in the mantles of Mars and the eucrite parent body. High-pressure experimental studies show that Cr, V, and Mn are lithophile under conditions of core formation in a martian-sized planetesimal. However, at the much higher pressures accompanying core formation in the Earth, Cr and V (and possibly Mn) become siderophile. Thus, the depletions of these elements in the Moon are consistent with derivation of protolunar material from the Earth's mantle, but not with its derivation from the mantle of a martian-sized planetesimal. An assessment is made of the single giant impact hypothesis, which maintains that the Moon was formed as a byproduct of the impact of a martian-sized planetesimal upon the Earth, and that this event was also responsible for the high angular momentum of the Earth-Moon system. This giant impact must have occurred *very* late during Earth's accretion, otherwise the Moon would be contaminated with siderophiles to a far greater extent than is observed. However, numerical simulations of the Earth's accretion indicate that most giant impacts occur relatively early, and also indicate that the probability that a giant impact occurred both very late *and* with an impact parameter capable of producing the angular momentum of the Earth-Moon system is less than 1 in 300. The single giant impact model is also incapable of explaining the siderophile geochemistry of the Moon and the lack of evidence for total melting of the Earth's mantle at a late stage of accretion. It may be possible to avoid these difficulties by relinquishing the requirement that a single giant impact was responsible for *both* the angular momentum density of the Earth-Moon system and the origin of the Moon. A giant impact by a planetesimal (e.g., 0.2 M_E) occurring when the Earth had accreted to about 60% of its present size could provide the required angular momentum. It would also cause melting and differentiation of the mantle. However, subsequent accretion of the remaining ~0.2 M_E in the form of small, dense, Fe-rich planetesimals could cause convective rehomogenization of the mantle, without extensive melting. Protolunar material may have been ejected into orbit from the Earth's mantle at a very late stage of accretion by collisions of relatively small high-velocity (20-30 km/sec) icy planetesimals scattered by Jupiter from the outer solar system. The Moon may have accreted from the family of moonlets thereby placed in Earth orbit.

INTRODUCTION

Whether the Earth began as a hot, molten body, as a cool, unmelted and undifferentiated planet, or in some intermediate state, was strongly influenced by the manner of its accretion within the primordial solar nebula. Accretion is believed to have been a hierarchical process in which dust particles coagulated to form small planetesimals that in turn accreted to form larger, intermediate-sized bodies that were ultimately assembled to form the Earth. The process involved a complex equilibrium between planetesimal growth, collisions, and fragmentation.

Modern cosmogonic models display widely divergent views concerning the size distribution of planetesimals and intermediate-sized bodies. According to the scenario developed by *Wetherill* (1986), planetesimals grew to form very large intermediate-sized bodies before finally combining to form the Earth. His model proposes that 1 or 2 martian-sized bodies, 2-4 mercurian-sized bodies, and numerous lunar-sized bodies may have participated in the formation of the Earth. This model leads to relatively efficient retention of the gravitational energy of accretion within the Earth, causing strong heating (e.g., *Kaula,* 1979). The resultant temperatures are believed to have been sufficiently high that core formation was synchronous with accretion (e.g., *Stevenson,* 1981). Moreover, it is argued that the larger intermediate-sized bodies were themselves sufficiently hot to have experienced extensive melting, differentiation, and core formation (e.g., *Cameron,* 1986; *Benz et al.,* 1987). If the Earth had formed via accretion of such massive hot bodies, it would have been completely melted because of effective conversion of the gravitational and kinetic energy of accreting bodies to heat that is retained within the Earth. *Benz et al.* (1987), *Cameron and Benz* (1988), and *Melosh and Kipp* (1988) have analyzed the consequences of an impact by a single martian-sized body and concluded that this alone would have caused complete melting of the Earth. These authors also consider that the Moon was formed as a by product of this same giant impact.

Other cosmogonic models such as those of *Safronov* (1969), *Greenberg et al.* (1978), *Hayashi et al.* (1985), *Weidenschilling and Davis* (1985), *Weidenschilling et al.* (1986), and *Patterson and Spaute* (1988) lead to very different size distributions of planetesimals that would have had very different early geothermal implications. For example, Safronov's model, although closely related to that of Wetherill, nevertheless postulates a much smaller population of planetesimals. According to his model, the largest intermediate-sized body to impact the Earth was about one thousandth of an Earth-mass. This accretion scenario can lead to a hot initial state sufficient to permit core formation during accretion, but is unlikely to have resulted in total melting of the entire Earth. Other scenarios, in which the Earth formed by accretion from small bodies, can lead to a relatively cool initial state, in which core formation occurred long after accretion had been completed (e.g., *Urey,* 1952).

The above considerations suggest that if we could establish some constraints upon the Earth's initial thermal regime and degree of differentiation, these in turn might provide insights into the nature of the accretion process and could also help to evaluate the giant impact model for the origin of the Moon.

CONSTRAINTS ON EARLY MANTLE DIFFERENTIATION

If the Earth had been born in a largely molten state, as implied by current versions of the giant impact hypothesis, extensive petrological and geochemical differentiation of the mantle would be expected to occur during its crystallization. Thus, according to *Kumazawa* (1981), *Ohtani* (1985), and *Agee and Walker* (1988), the entire mantle (of chondritic bulk composition) became thoroughly differentiated, yielding a perovskitic lower mantle enriched in SiO_2 and a peridotitic upper mantle relatively depleted in SiO_2.

The model can be tested experimentally. $MgSiO_3$ perovskite is found to be the liquidus phase in chondritic and pyrolite compositions below about 700 km (*Ito and Takahashi,* 1987) and because of its high relative abundance (~80% of a chondritic lower mantle) it is expected to display a broad crystallization field before being joined by magnesiowüstite (~10%). $CaSiO_3$ perovskite is not expected to crystallize as a liquidus phase in the lower mantle because of its low abundance (5%). The partition coefficients of many minor elements between silicate perovskites, garnets, and a range of ultrabasic liquids were determined by *Kato et al.* (1988a). Some of their results are summarized in Table 1. They demonstrate that extensive separation of $MgSiO_3$ perovskite into the lower mantle would have caused certain characteristic geochemical signatures in the upper mantle. Some of these are illustrated in Fig. 1. It is seen that a large degree of $MgSiO_3$ perovskite fractionation is necessary in order to cause substantial changes in the Si/Mg ratio of a chondritic mantle. However, the separation of only 5-10% of $MgSiO_3$ perovskite would have

TABLE 1. *Partition coefficients for selected elements between liquidus $MgSiO_3$ perovskite, pyrope-rich garnets and ultrabasic melts (from Kato et al., 1988a).*

Element	$MgSiO_3$ (pv) D_{mpv}/liq*	Garnet D_{gnt}/liq
Ca	0.2	0.6
Al	0.5-0.8	2.5
Na	0.02	0.1
Sc	5	1.7
Y	3	1.3
Yb	2	1.4
Sm	0.3	0.2
La	<0.1	<0.1
Ti	3	0.4
Hf	14	0.8
Zr	9	0.6
Nb	~1	<0.1
[K, Rb, Cs, Ba, Sr, Th, U]	<0.1	<0.1

*This is the preferred set of partition coefficients obtained in the investigations of *Kato et al.* (1988a,b). It is believed that they are likely to be accurate to ±50% and in many cases, to ±30%. The set of partition coefficients obtained by *Kato et al.* (1988b) represent limits only, owing to constraints imposed by the experimental method employed. Nevertheless, they demonstrate that D_{mpv}/liq (Sc,Zr,Hf) > 3 and D_{mpv}/liq (Ca,Sm) < 0.2.

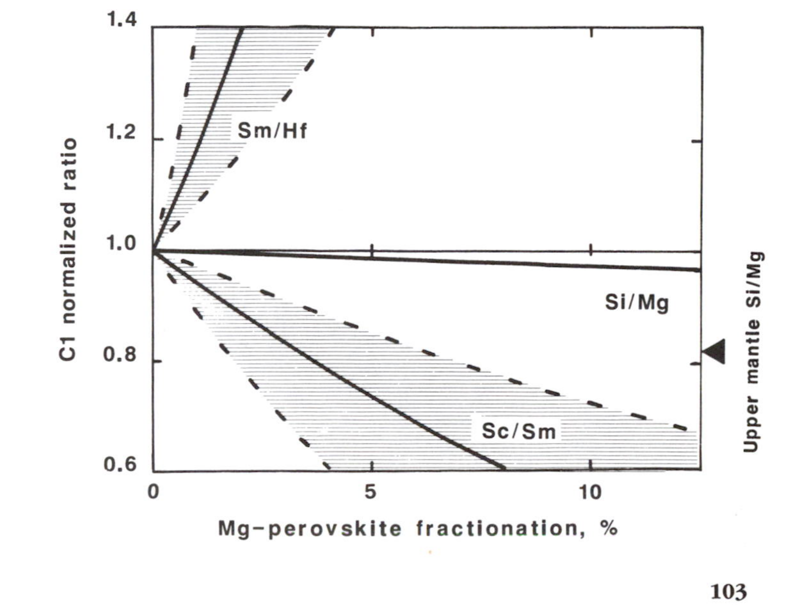

Fig. 1. *Variation of Sm/Hf, Sc/Sm, and Si/Mg ratios as a function of Mg-perovskite fractionation from a chondritic mantle composition, calculated from partition coefficients given in Table 1. The Si/Mg ratio of the present upper mantle is also indicated. Shaded areas indicate uncertainties that would be caused by errors of ±2 in the Sm/Hf and Sc/Sm ratios used in the calculation. From Kato et al. (1988a).*

driven the Sm/Hf and Sc/Sm ratios of the residual melts well outside the near-chondritic ratios that are now observed in the extensive upper mantle source region of midocean ridge basalts. The results of *Kato et al.* (1988a) imply rather strongly that the present mantle does not reflect the operation of the comprehensive perovskite-controlled differentiation processes postulated by *Agee and Walker* (1988). They suggest, moreover, that the Si/Mg ratio of the lower mantle is similar to that of the present upper mantle.

Kato et al. (1988a) also tested an alternative scenario. The Earth was postulated to have been born in a totally molten state and the mantle then differentiated as postulated by *Agee and Walker* (1988); however, it was assumed that all traces of this early differentiation were subsequently removed via homogenization caused by pervasive subsolidus mantle convection. The recent discovery of 4.2 Ga zircons in Western Australia (*Froude et al.,* 1983; *Compston and Pidgeon,* 1986) provides an important constraint on this alternative model. *Kinny* (1987) demonstrated that these zircons have "primitive" hafnium isotope ratios, i.e., they crystallized in a geochemical environment characterised by an approximately chondritic Hf/Lu ratio.

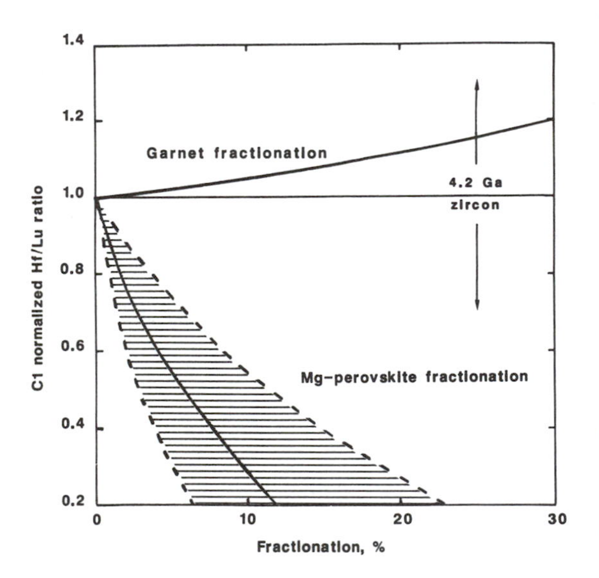

Fig. 2. *Effect of majorite garnet and Mg-perovskite fractionation on Hf/Lu ratios in a chondritic model mantle composition, calculated from partition coefficients given in Table 1. The shaded region indicates uncertainties that would be caused by errors of ±2 in the Hf/Lu ratio used in the calculation. Also shown are the permitted bounds in which 4.2 Ga zircons from Mt. Narryer, Western Australia, evolved between 4.55 and 4.2 Ga. From Kato et al. (1988a).*

If the reasonable assumption is made that the partition coefficient of Lu is similar to that of Yb, the results of *Kato et al.* (1988a), depicted in Fig. 2, show that only a few percent of fractionation of $MgSiO_3$ perovskite into the lower mantle would have caused the Hf/Lu ratio of the upper mantle to change beyond the limits implied by the ancient zircons. Thus, if the mantle had experienced extensive melting and differentiation around 4.5 Ga, accompanied by extensive separation of $MgSiO_3$ perovskite into the lower mantle, it must have become convectively rehomogenized by 4.2 Ga. This requirement encounters certain problems that are discussed below.

STRATIFICATION OF AN INITIALLY MOLTEN MANTLE

Mineralogical Zoning and Buoyancy

The course of differentiation of an extensively molten pyrolite [this composition contains ~0.4% Na_2O and 0.03% K_2O (e.g., *Sun,* 1982)] mantle normatively equivalent to ~73% $MgSiO_3$ perovskite, 19% (Mg,Fe)O magnesiowüstite, and 7% $CaSiO_3$ perovskite, was explored by *Ringwood* (1989a) in the light of recent data on relevant melting equilibria at very high pressures and temperatures (e.g., *Ito and Takahashi,* 1987; *Kato et al.,* 1988a). The melting point gradient of the mantle is greater than the adiabatic gradient and hence crystallization of the liquidus phase, $MgSiO_3$ perovskite, proceeds from the base upward until it is joined by (Mg,Fe)O magnesiowüstite. Crystallization-differentiation causes the lower mantle to be composed mainly of $MgSiO_3$ perovskite and lesser magnesiowüstite, thereby producing strong enrichment of Na and Ca (which are excluded from the crystal lattice of $MgSiO_3$ pv) in the residual liquid and modest enrichment of Al. By the time crystallization has progressed upward to 650 km, the overlying liquid would contain 10-12% CaO, ~1.5% Na_2O, and 5-8% Al_2O_3. At this depth, crystallization of $MgSiO_3$(pv) would cease, being replaced by pyrope-rich garnet $\pm\gamma,\beta(Mg,Fe)_2SiO_4$ (*Irifune and Ringwood,* 1987). Partition coefficient data (Table 1) show that crystallization of these phases would have caused further enrichment of Na and Ca in the residual magma but Al becomes depleted because of its preferential entry into garnet ($D_{ga/liq} \sim 2.5$). At the stage that crystallization had progressed upward to 400 km, the residual liquid would be expected to be ultrabasic, enriched in CaO (~12%) and Na_2O (~2.5%) and low in Al_2O_3 (~4%). If it crystallized without further fractionation, the magma ocean above 400 km would yield an assemblage consisting mainly of olivine + omphacite (Ca,Na)(Mg,Fe,Al)Si_2O_6. Orthopyroxene would be absent because of the high CaO content of the liquid while garnet would be suppressed or absent because of high (Na+K)/Al.

Continued differentiation of the upper mantle (above 400 km) would be dominated by the separation of Mg-olivine and omphacite. The results of *Takahashi* (1986) and *Scarfe and Takahashi* (1986) indicate that little fractionation of the jadeite component ($NaAlSi_2O_6$) of clinopyroxene would occur between depths of 400 and 200 km. However, above 200 km, sodium would become enriched in residual liquids (this effect becoming stronger at shallower depths). Potassium would be strongly fractionated into residual liquids throughout the entire differentiation sequence. At the stage by which crystallization had extended upward to a depth of 30 km or so, the residual liquid would contain 2-3% K_2O and 5-8% Na_2O together with appreciable contents of H_2O. Since jadeite is unstable at low pressure and elevated temperature, the residual liquid would be expected to crystallize to an assemblage containing alkali felspar and nepheline as major phases. Amphibole, biotite, and Fe-rich olivine would probably be the principal subsidiary phases. Because of the low density of this assemblage, it would form a highly buoyant primitive crust. This crust would have contained most of the Earth's inventory of highly incompatible and heat-producing elements (e.g., K, Rb, Ba, Sr, Pb, lREE, LREE, U, and Th).

The stratified structure proposed for the differentiated mantle is illustrated in Fig. 3. This structure would be gravitationally stable and highly resistant to rehomogenization via thermally driven convection. The crust would be intrinsically buoyant relative to the upper mantle. Likewise, the upper mantle would be intrinsically stable relative to the underlying region because the omphacite component

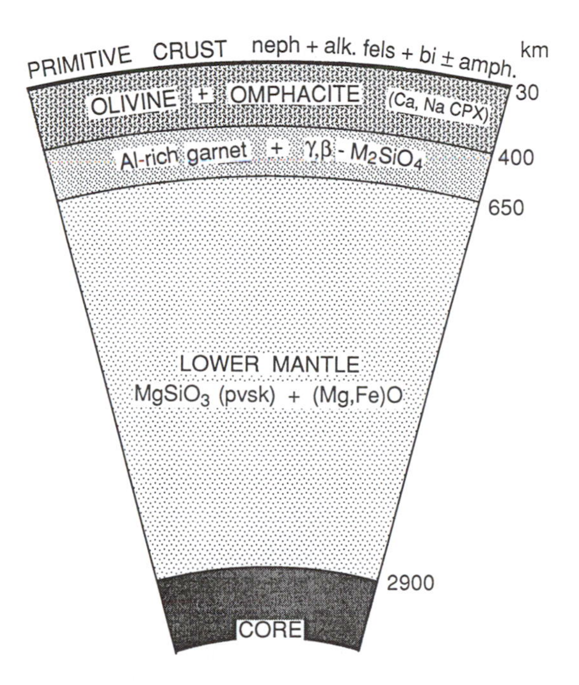

Fig. 3. *Section through Earth's mantle illustrating the gravitationally stable, chemically and mineralogically zoned structure that would ensue during its crystallization and differentiation from a molten state according to the processes described in the text.*

does not transform to denser phases until depths of 500-600 km (*Liu,* 1980; *Irifune et al.,* 1989). Thus, any upper mantle material somehow transported below 400 km would be buoyant because of untransformed omphacite, which possesses a much lower density than the garnet, and $\beta,\gamma(Mg,Fe)_2SiO_4$ comprising the region between 400-650 km. Likewise, material of the transition zone between 400-650 km is intrinsically gravitationally stable relative to the perovskitic lower mantle. Aluminum-rich garnet from this region requires a higher pressure to transform to perovskite than does the relatively Al-poor composition that is characteristic of the lower mantle. [See *Irifune and Ringwood* (1987) and *Ringwood and Irifune* (1987) for detailed discussion of this point.]

The above analysis has ignored the effects of Fe-Mg fractionation. Because early phases would be Mg-rich, and would be overlain by increasingly Fe-rich and therefore denser phases, fractional crystallization may have led to gravitational instability within the cumulates. However, for a wide range of initial conditions, the Fe-rich high-level cumulates would develop instabilities and would sink through the less-dense Mg-rich cumulates to collect near the base of the mantle. Continuation of this process would also be likely to lead, ultimately, to a gravitationally stable density distribution within the mantle cumulate column that would inhibit convective rehomogenization. In some circumstances, the effects of Fe/Mg fractionation may counterbalance the buoyancy effects caused by the phase stability relationships discussed earlier. *Ringwood* (1989a) evaluated these effects and concluded that Fe/Mg fractionation may overcome the buoyancy of the transition zone (400-650 km) relative to the lower mantle, caused by the presence of Al-rich garnet in the transition zone. However, Fe/Mg fractionation would not overcome the intrinsic buoyancy of the upper mantle as compared to underlying regions, nor of the crust as compared to the upper mantle.

The Primitive Crust

We concluded above that crystallization of the entire mantle from a molten state would ultimately produce a buoyant primitive crust with alkali felspar and nepheline as major phases. This crust may have been about 30 km thick and would have contained most of the Earth's inventory of highly incompatible and heat-producing elements. The composition of this primitive crust would have been dramatically different from that of the continental crust that is known to have existed at 3.8 Ga. All traces of the existence of this primitive crust must therefore have been removed prior to this time. It is difficult to understand how this could have occurred. Even if plate tectonics were operative at this very early stage, subduction of highly buoyant primitive crust appears unlikely. If indeed it occurred, subducted primitive crust would fuse at shallow depths because of its low melting point and the high temperature of surrounding mantle. Most of the melted material would have been promptly returned to the surface. Extraction of most of the radioactive heat sources from the mantle into the crust would diminish the driving forces for thermal convection in the mantle. Once a 30-km-thick, buoyant, low-melting point crust containing most of the Earth's K, U, and Th had been formed, it would be unlikely ever to be destroyed by subsequent convection.

Mantle Homogeneity and the Mg/Si Ratio

The previous discussion showed that the upper mantle does not display any evidence of the geochemical signatures that would be caused by a process of mantle differentiation from an extensively molten state that involved a significant degree of segregation of perovskite into the lower mantle. This implies that either the mantle was formed in a relatively cool and largely unmelted state or, if extensive melting and differentiation indeed occurred, all geochemical traces of this process were removed by subsolidus convection at a very early stage, probably prior to 4.2 Ga. In the previous section we noted that extensive melting followed by crystallization-differentiation would be likely to produce a gravitationally-stable stratification of the mantle that would be extraordinarily difficult to obliterate by subsequent thermally driven convection, especially in the brief interval between 4.5 and 4.2 Ga. These conclusions provide a powerful constraint on the giant impact hypothesis for the formation of the Moon, which invokes a process that would be expected to have melted the entire mantle of the Earth. The implications of this constraint are considered further in the section on the origin of the Moon.

An important conclusion arising from the above discussion is that the Mg/Si ratios of the upper and lower mantle are likely to be similar. The Mg/Si (atomic) ratio of the upper mantle is close to 1.27 (e.g., *Sun,* 1982). This is significantly larger than the corresponding ratio of 1.05 displayed by CI chondrites. It is usually assumed that this latter value represents that of the primordial solar nebula and hence the upper mantle appears to be relatively depleted in silicon. The results of *Kato et al.* (1988a) imply that this depletion is characteristic of the entire Earth's mantle.

Ringwood (1989b) has recently described evidence that suggests that the terrestrial mantle Mg/Si ratio may actually be characteristic of the inner solar system. He showed that the similar Al_2O_3 contents of terrestrial and venusian basalts imply that the ratio of normative pyroxene to olivine in the venusian mantle resembles that of the terrestrial mantle and that the venusian Mg/Si ratio is therefore terrestrial rather than chondritic. He also showed that geochemical relationships between pallasites, howardites, diogenites, and eucrites implied that the Mg/Si ratio of the eucrite parent body is closer to the terrestrial than to the chondritic value. Moreover, he noted that the spectra of S-type asteroids that predominate in the inner asteroid belt possess high olivine/pyroxene ratios and terrestrial rather than chondritic Mg/Si ratios.

The Mg/Si ratios of the Earth's mantle and CI chondrites are both consistent with the Mg/Si ratio in the solar photosphere within the errors of measurement of the latter (*Anders and Grevesse,* 1989). We should therefore consider the possibility that rather than being depleted in Si, the inner solar system (including the Earth's mantle) may actually possess the primordial solar Mg/Si ratio and that CI chondrites could have been enriched in Si via a cosmochemical process. *Rietmeijer* (1987, 1988) showed that both chondritic interplanetary dust particles and Comet Halley dust particles from the outer parts of the solar system are substantially enriched in relatively volatile elements (Mn, Cu, K, Na, S, Zn, Bi) and also in silicon relative to CI chondrites (asteroid belt). He suggested that there had been a radial chemical differentiation of volatile elements (including Si) between the two source regions in the inner and outer solar system and that the existence of this chemical zonation challenges the assumption that the CI chondrites necessarily possess primordial compositions.

Ringwood (1989b) attempted to interpret these results and proposed that during formation of the terrestrial planets, varying proportions of elements more volatile than silicon may have been transported outwards to recondense in the lower temperature environment outwards from the asteroid belt. Some Si may also have been lost in this manner, although not enough to significantly alter planetary Mg/Si ratios. However, recondensation of some of this Si on the relatively small mass of dust particles in the outer asteroid belt may have caused a substantial enrichment of Si relative to Mg, which would in turn have been inherited by CI chondrites that subsequently formed in this region. This suggests that the Mg/Si ratio of the inner planets (~1.27) may be more representative of the solar nebula value than is the CI ratio.

CORE FORMATION IN PLANETARY BODIES

Formation of the core also provides important constraints on the early thermal history of the Earth. During the 1950s and 1960s, it was almost universally believed that the Earth had accreted in a relatively cool, unmelted state, and that it was then heated by radioactive decay of U, Th, and K over a period of a billion years or so until temperatures were high enough to permit core segregation, which occurred over an extended period (e.g., *Urey,* 1952, 1962; *Elsasser,* 1963; *Birch,* 1965). It was pointed out by *Ringwood* (1960, 1966a,b) that core segregation would have substantially altered the U/Pb ratio of the mantle-crust system. The "age of the Earth" of 4.5 Ga that had been obtained by *Patterson* (1956) from the lead isotopic systematics of crustal rocks must therefore date the time of core formation. Ringwood concluded that the Earth had accreted in a relatively hot condition, which would enable core formation to occur during or very soon after accretion. This conclusion was further supported by *Oversby and Ringwood* (1971), who measured the partition coefficients of lead between iron alloys and silicates and showed that compared to uranium, lead would have been preferentially partitioned into the core.

Early formation of the core must be reconciled with the evidence for lack of gross differentiation within the mantle, as discussed in the section on constraints on early mantle differentiation. The core is known to be about 10% less dense than pure metallic iron under equivalent P,T conditions, implying the presence of substantial quantities of one or more light elements. Recent experiments have demonstrated that the solubility of FeO in molten iron increases rapidly at high pressures and temperatures and that, essentially, complete miscibility of Fe and FeO is attained around 25 GPa, 2000°C (*Ohtani et al.,* 1984; *Ringwood,* 1984; *Kato and Ringwood,* 1989a). Equilibrium conditions dictate that metallic iron segregating from the lower mantle into the core would necessarily dissolve large amounts of FeO and strongly suggest that oxygen is the principal light element in the core. The density of the outer core would be explained if it contained 35-40 mol.% FeO, equivalent to 8-10 wt.% of oxygen. Geochemical arguments suggest that about 2-3 wt.% of sulphur may also be present (*Ringwood,* 1984).

Kato and Ringwood (1989a) demonstrated that the melting point of iron is greatly reduced by the solution of FeO and that at pressures above 16 GPa, the Fe-FeO eutectic temperature lies below the solidus of mantle pyrolite; moreover, this situation is likely to prevail through most of the lower mantle. *Urakawa et al.* (1987) showed that a further large reduction in melting temperature of the Fe-O system is caused by the presence of modest amounts of sulphur. Thus, if the core contained, say, 7% O and 3% S, as seems plausible on geochemical and geophysical grounds, it could have segregated from the mantle during or immediately after the Earth's accretion without being accompanied by significant partial melting of the mantle (*Stevenson,* 1981).

When a metallic core forms within a planetary body, varying proportions of siderophile elements are left behind in the silicate mantle. The resultant siderophile signatures in these mantle silicates reflect the nature of the core-forming process in that particular planet. This may be illustrated by reference to core-forming processes in the Earth, Mars, and the eucrite parent body. The abundance patterns displayed by siderophile elements in the Earth's mantle are extraordinarily complex. A simplified account of the major trends may be summarized as follows (*Wänke and Dreibus,* 1982; *Wänke et al.,* 1984; *Ringwood,* 1984, 1986a,b; *Wänke,* 1987):

1. A major group of siderophiles including Ni, Co, Cu, Fe, Ga, W, Mo, P, As, Sb, and Ge is present in abundances ranging from 1-15% of their abundances in Cl meteorites (Mg-normalized).

2. Several of these elements (e.g., Ni, Co, Mo) are much more abundant in the mantle than would be expected if they had been partitioned into the Earth's core on the basis of their iron/silicate partition coefficients as determined under low-pressure conditions.

3. The relative abundances of these elements in the mantle are not significantly correlated with the above metal/silicate partition coefficients.

4. The platinum-group metals and also gold and rhenium display quite different abundance characteristics as compared to this first group. In contrast to the highly variable abundances (normalized to Cl chondrites) of the first group, the "noble" metals are present in chondritic relative abundances, but at levels corresponding to about 0.3% of Cl abundances (Mg-normalized). These levels are also far higher than would be expected on the basis of low-pressure iron/silicate partition coefficients.

5. Vanadium, chromium, and manganese are depleted in the mantle compared to Cl abundances (Mg-normalized) by factors of 0.6-0.2. Although these elements are not normally regarded as being "siderophile," increasing evidence (discussed in the section on chromium, vanadium, and manganese in the lunar siderophile signature) suggests that the depletions have been caused mainly by their incorporation in the Earth's core.

This highly complex signature of the siderophile abundance patterns in the mantle is the end product of several physico-chemical processes that were involved in the formation of the Earth's metallic core. The following factors were among those involved:

1. Partitioning of siderophiles according to their chemical affinities between a predominantly metallic iron phase amounting to ~30% of the mass of the Earth and mantle silicate phases during the formation of the Earth's core.

2. The circumstance that this process occurred within a body of planetary dimensions under conditions of very high pressure and temperature. Metal/silicate partition behavior for siderophile elements is likely to have been substantially influenced by these high P,T conditions.

3. The processes that formed the Earth's core also led to the incorporation of about 10 (wt.)% of light elements. As discussed above, it is now believed that oxygen is the principal member of this group. The presence of light element(s) would necessarily influence the partitioning of siderophile elements between core and mantle.

4. It is unlikely that complete chemical equilibrium between metal and silicates was achieved during the formation of the Earth's core (*Ringwood,* 1960, 1966a; *Stevenson,* 1981; *Jones and Drake,* 1986). Moreover, several widely supported cosmogonic hypotheses have proposed that the Earth accreted inhomogeneously (e.g., *Turekian and Clark,* 1969; *Wänke,* 1981) in a manner that would have prevented chemical equilibrium being achieved between core and mantle.

The detailed operation and relative importance of the several physico-chemical processes involved in formation of the Earth's core are poorly understood. However, it is readily recognized that this particular combination of complex processes is most unlikely to have operated similarly in all planetary bodies. It seems reasonable to assume, therefore, that the siderophile signature of the Earth's mantle should be unique to the Earth or, at the very least, to a planet of similar size that had experienced similar geochemical evolution. Venus is the only candidate that could be in this class.

The relative abundances of siderophile elements in the mantles of Mars and of the EPB have been estimated by *Dreibus and Wänke* (1980, 1984) and are shown in Fig. 4. It is seen that they are quite different from the Earth's mantle and also differ markedly from each other. These differences add credence to the view expressed above that the mantles of differentiated planetary bodies are characterized by unique siderophile signatures.

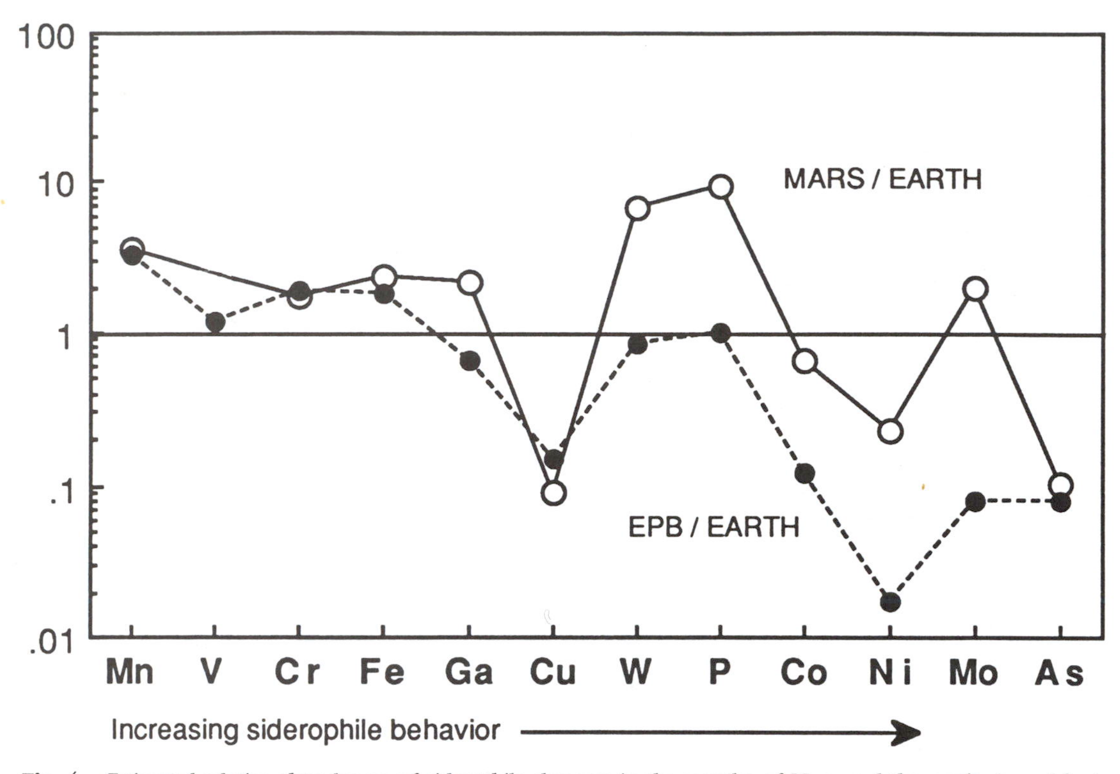

Fig. 4. *Estimated relative abundances of siderophile elements in the mantles of Mars and the eucrite parent body normalized to siderophile abundances in the terrestrial mantle. Volatility corrections were applied to Ga and Cu in the EPB as described in the text. After Ringwood and Seifert (1988).*

The mantle of Mars is depleted in Cu, Ni, Co, and As (Fig. 4) and also in Tl and In (*Wänke et al.,* 1984) as compared to the Earth's mantle. In contrast, the martian mantle is enriched in P, W, Mn, and Cr, which appear to be present in near-chondritic abundances (Mg-normalized). *Dreibus and Wänke* (1984) and *Treiman et al.* (1986) have pointed out that these characteristics could be explained if the martian core were rich in sulphur. The mantle of the EPB is depleted in Ni, Co, Cu, Mo, Ga, Ge, and Ir in comparison to the Earth's mantle and enriched in Mn, V, and Cr. Phosphorus and tungsten are present at levels similar to those in the Earth's mantle but are depleted in comparison to the martian mantle. These characteristics are readily explained if the EPB core had been composed dominantly of metallic iron and was depleted in sulphur relative to the martian core (*Delano and Ringwood,* 1978a,b; *Newsom,* 1984).

DIFFERENTIATION OF THE MOON

An extensive region of the Moon extending to a depth of 400 km or more experienced melting and differentiation during or soon after accretion (e.g., *Taylor,* 1982). One of the products of crystallization from the resultant magma ocean was the plagioclase-rich lunar crust with a mean thickness of about 70 km. The upper mantle of the Moon is believed to be geochemically complementary to the crust, and to be composed of pyroxene and olivine cumulates from the magma ocean together with refractory peridotite, depleted in fusible components by the partial melting process that produced the magma ocean (e.g., *Ringwood*, 1979).

Mare basalts comprise only a small proportion of the volume of the lunar crust. Nevertheless, they are of profound petrological and geochemical significance because they are believed to have been formed by subsequent partial melting of the ferromagnesian cumulates and peridotitic residua beneath the crust at depths of 100-400 km. Mare basalts are therefore capable of providing key information on the geochemistry of the lunar upper mantle, analogous to corresponding information on the Earth's upper mantle that is provided by terrestrial basalts. *Ringwood and Kesson* (1976a) showed that the low-titanium mare basalts have had a simpler petrogenetic evolution than the high-titanium basalts and retain a more direct "memory" of their source regions. Accordingly, it is preferable to use geochemical data derived from low-titanium basalts in order to constrain the composition of their source regions in the lunar mantle. Most mare basalts have experienced varying degrees of fractionation (mainly of olivine and pyroxene) after leaving their source regions. It is therefore desirable to select the most "primitive" basalts, i.e., those that can be shown by standard criteria such as Mg/(Mg+Fe) ratios to have experienced the least fractionation after leaving their source regions. The recognition by *Delano* (1986) of some 25 discrete families of volcanic glasses in lunar soils from widely separated locations is of major importance to lunar basalt petrogenesis. These glassy spherules apparently formed by volcanic fire-fountaining. Their high liquidus temperatures combined with melting relationships at elevated temperatures imply an origin at considerable depths, probably 300-400 km. Delano and colleagues demonstrated that the least fractionated members of each of these families closely approximate primary magmas that ascended from their source regions without appreciable fractionation of olivine (or pyroxene).

Delano and Livi (1981), *Delano and Lindsley* (1983), and *Jones and Delano* (1989) showed that the source regions from which these primary magmas were derived appear to consist of mixtures of two distinct components, Ti-rich and Ti-poor, corresponding to different kinds of cumulates. Their data provide strong support for a petrogenetic model proposed by *Ringwood and Kesson* (1976a), who pointed out that during crystallization of the magma ocean, intermediate and late ferromagnesian cumulates formed at shallow depths would have become gravitationally unstable because of their high iron (and titanium) contents. Consequently they would have been subducted into the lunar interior where hybridization with less dense layers of early olivine and orthopyroxene cumulates and refractory residua occurred. *Ringwood and Kesson* (1976a) showed that many characteristics of mare basalts could be explained if they had formed by subsequent partial melting of these hybrid source regions at considerable depth.

The green volcanic glasses possess the highest Mg numbers and nickel contents among primary lunar magmas and have been found at four of the Apollo sites (*Delano*, 1986). They were evidently derived from source regions in which Mg-rich olivine residua and/or early cumulates played a major role. From a detailed investigation of major and trace element compositions of individual spherules, *Ma et al.* (1981) showed that they had been produced by rather small degrees ($\leq 10\%$) of partial melting of olivine-rich cumulate source regions. Accordingly, the compositions of liquidus olivines crystallized from these melts closely approach the compositions of olivines in their respective source regions. Experimental measurements of these liquidus compositions can therefore provide the abundances of key siderophile elements such as nickel and cobalt in the source regions.

In view of the limited sample of lunar volcanic glasses available for study, it seems doubtful whether the green glasses so far recovered represent the most primitive liquids erupted at the lunar surface. The occurrence of negative Eu and Sr anomalies in green glasses also provides clear evidence that this material contains a more evolved component. Thus, experimentally derived compositions of olivine on the liquidus of green glasses may tend to underestimate the contents of nickel and cobalt in olivine from the primordial lunar interior (*Ringwood and Seifert,* 1986).

Geophysical and geochemical data strongly suggest that the Moon possesses a small metallic core, probably amounting to about 1-5% of the lunar mass (e.g., *Newsom,* 1984; *Hood and Jones,* 1987). This small core may have segregated during the early differentiation process that formed the lunar crust and upper mantle. Experimental investigations show that if the lunar core had formed by segregation of metal initially dispersed throughout the lunar mantle, it would have contained about 40% of nickel (*Ringwood and Seifert,* 1986; *Seifert et al.,* 1988). This tiny core formed under vastly different pressures, temperatures, and other physico-chemical conditions to those prevailing during formation of the Earth's core. As discussed in the section on core formation of planetary bodies, it would be expected that the lunar siderophile signature would differ drastically from that of the Earth's mantle. In fact, as shown in the following section, there are some remarkable similarities in the siderophile abundance patterns of the Earth's mantle and Moon. There are also some important differences, but it will be seen that these are readily explained by the operation of well-known processes.

More than a decade ago, Wänke and his colleagues in Mainz developed a powerful set of geochemical arguments pointing toward a genetic relationship between the Moon and the Earth's mantle (*Dreibus et al.,* 1976; *Rammensee and Wänke,* 1977; *Wänke et al.,* 1978, 1979). Independently, in Canberra, my colleagues and I reached similar conclusions (*Ringwood and Kesson,* 1976b, 1977; *Delano and Ringwood,* 1978a,b). Both groups recognized certain key resemblances between the siderophile element abundances in the Earth's mantle and in the Moon, and concluded that most of the material now in the Moon had been derived ultimately from the Earth's mantle. This conclusion generated intense controversy at the time. Most lunar scientists were reluctant to accept the proposition that the vexed question of the Moon's origin could be settled by such a "simple" argument. Nevertheless, the controversy had beneficial results because it stimulated the acquisition of additional analytical data on the abundances of siderophile elements in lunar and terrestrial rocks and on the partition coefficients of siderophile elements between silicate and

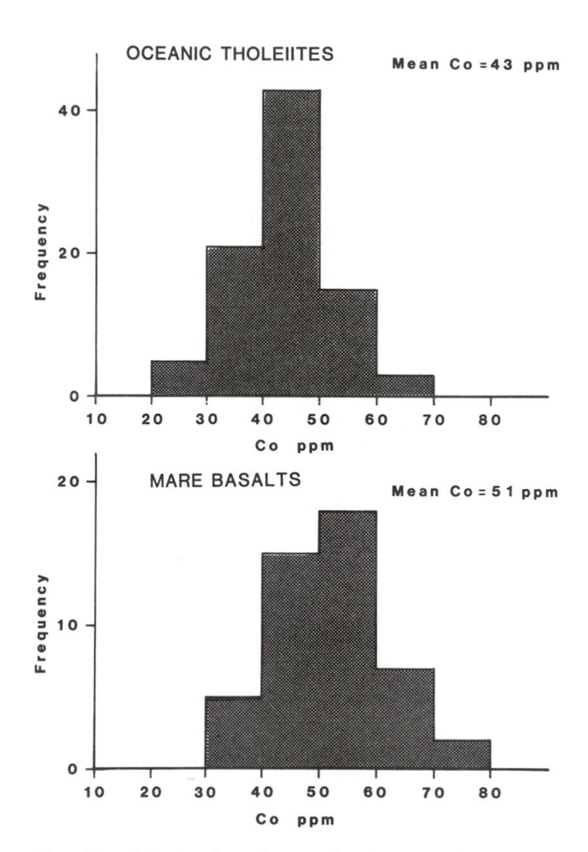

Fig. 5. *Cobalt abundances in low-titanium mare basalts and MORB oceanic tholeiites. After Ringwood and Seifert (1986).*

metal phases. Because of the fundamental importance of this topic to the origin of the Moon, a detailed discussion of some key aspects of lunar siderophile geochemistry follows.

THE LUNAR SIDEROPHILE SIGNATURE

Cobalt

The abundances of cobalt in low-Ti mare basalts and terrestrial oceanic tholeiites are shown in Fig. 5. Mean abundances and dispersions are very similar in both rock-types. The dispersion among both mare and terrestrial basalts is caused by low-pressure fractionation controlled mainly by the separation of olivine. *Ringwood and Kesson* (1977) concluded from the above relationships that the CoO content of the lunar mantle is very similar to that of the terrestrial mantle. This is confirmed by the experimental studies of *Seifert et al.* (1988), which show that olivine crystallizing near the liquidus of a relatively primitive lunar green glass (82 ppm Co, *Ma et al.,* 1981) contains 135 ppm Co (as oxide), implying that residual olivine in its source region in the lunar mantle contained a similar amount of cobalt. This may be compared with the mean value of 136 ppm Co found in a collection of olivines from terrestrial peridotite xenoliths (*Stosch,* 1981).

Cobalt present in lunar highland breccias is a mixture of indigenous lunar Co with a component provided by meteoritic contamination. Indigenous Co abundances can be obtained by removing the effects of meteoritic contamination according to the procedures of *Delano and Ringwood* (1978a,b) and *Ringwood et al.* (1987). A plot of indigenous Co vs. total (Mg+Fe) in Apollo 16 highland breccias and primitive lunar volcanic glasses is shown in Fig. 6. This figure reveals a good correlation between indigenous Co and (Mg+Fe) among the highland breccias. Moreover, the line of best fit extrapolates directly toward the field of the least fractionated volcanic glasses, the cobalt contents of which are unquestionably indigenous. This demonstrates the strong similarity of Co-Mg-Fe systematics both in the lunar highlands and in primitive lunar basalts and that these are indeed of global extent.

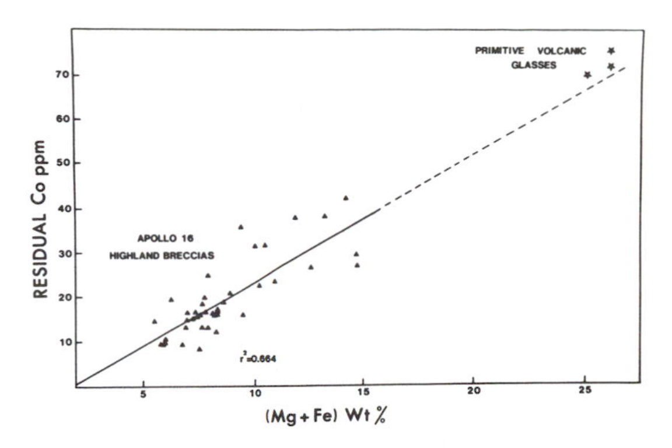

Fig. 6. *Cobalt in primitive lunar volcanic glasses and Apollo 16 breccias and soils (corrected for meteoritic contamination) vs. (Mg+Fe). Note that the solid line representing the best fit to lunar highland data projects directly into the field of primitive volcanic glasses showing that the correlation is of global significance. After Ringwood et al. (1987).*

The close similarity in cobalt abundances of the lunar and terrestrial mantles is of considerable genetic significance. The lunar mantle shares with the terrestrial mantle the characteristic that its cobalt abundance is much too high ever to have been in equilibrium with the metallic iron phase that would have been present in the solar nebula during accretion of the Earth. The composition of this "cosmic" metallic phase can be estimated from the relative abundances of Fe, Ni, and Co in the Earth's core (*Hart and Zindler,* 1986) or from the primordial solar system abundances (*Anders and Grevesse,* 1989). Both approaches yield almost identical relative abundances for Fe, Ni, and Co and imply that the cosmic metallic phase contained 0.26% Co. The results of *Seifert et al.* (1988) show that olivine (Fo_{83}) in equilibrium with metal of cosmic composition at 1400°C would contain only ~21 ppm Co (as oxide), compared to ~135 ppm Co present in olivines from both the terrestrial and lunar mantles. Thus, the latter are enriched in CoO by a factor of 6.4 above the levels that would be produced in equilibrium with metal of cosmic composition at this temperature. (The enrichment factor increases as the metal-silicate equilibration temperatures decreases.)

This considerable "overabundance" of CoO in lunar olivine places strict limits upon the amount of metallic iron that might have been accreted by the Moon and incorporated into its mantle during and after the major early differentiation process that formed the lunar crust and upper mantle. Under equilibrium conditions, the addition of only 1% of iron of cosmic composition would have extracted nearly all of this excess cobalt from lunar mantle silicates and sequestered it into the metallic phase. Therefore, the amount of "cosmic" iron that might have been introduced into the lunar mantle from accreting planetesimals of terrestrial composition was much smaller than 1%. The above relationships also imply that any iron-rich metal phase that might have been produced during the early lunar differentiation event would have contained 1.7% Co, since it necessarily equilibrated with olivine containing ~135 ppm Co. Thus, it would be much richer in Co and distinct from any infalling metal from the solar nebula (0.26% Co). The lunar geochemistry of cobalt clearly provides some important boundary conditions for the early development of the Moon that will be considered later in this paper.

Nickel

The relationships between the nickel contents of terrestrial ocean-floor tholeiites and low-Ti mare basalts and their corresponding Mg-numbers is shown in Fig. 7. It is seen that the two fields almost completely overlap, indicating that the lunar and terrestrial magmatic systems possessed similar amounts of nickel at corresponding stages of differentiation, as indicated by Mg numbers. The strong decrease of nickel contents in terrestrial basalts with falling Mg number is known to be caused by high-level fractionation, controlled principally by olivine crystallization. The corresponding lunar trend is caused by the same process (*Delano and Ringwood,* 1978a). In both the terrestrial and lunar systems, it is therefore necessary to select magmas that have experienced minimal degrees of high-level fractionation in order to constrain the composition of the source region. Noncumulus magmas possessing the highest Mg numbers, such as the green volcanic glasses, are of particular importance in this context (e.g., *Delano,* 1986; *Jones and Delano,* 1989).

Although Fig. 7 indicates the existence of a close relationship between lunar and terrestrial nickel geochemistry, it does not imply that the NiO contents of the mantles of both bodies are identical. The situation is complicated by evidence (e.g., *Ringwood,* 1979) that the lunar mantle is substantially enriched in FeO compared to the terrestrial mantle. *Seifert*

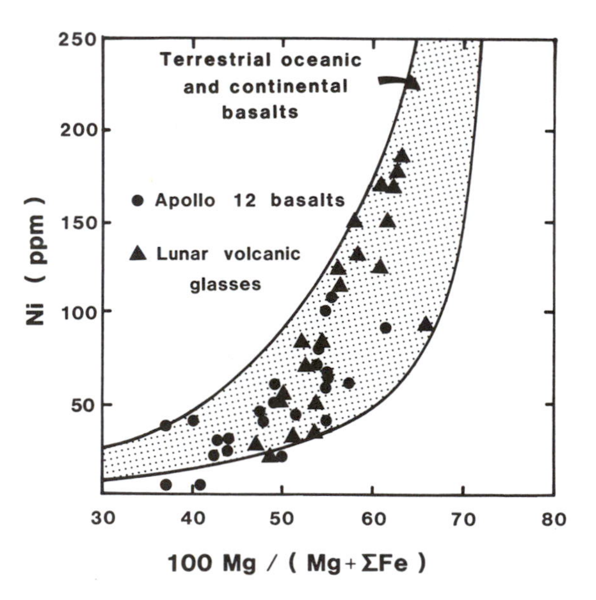

Fig. 7. *Nickel contents vs. MgO/(MgO+FeO) ratios in primitive lunar volcanic glasses (Delano, 1986) and Apollo 12 basalts compared with corresponding field observed for terrestrial basalts. The boundaries of the latter are shown by the stippled region inside the solid lines. Based on Delano and Ringwood (1978b) and Ringwood and Seifert (1986).*

et al. (1988) demonstrated experimentally that olivine crystallizing near the liquidus (1420°C) of a relatively primitive green glass (188 ppm Ni) contained 830 ppm Ni (as oxide). In view of the evidence (discussed previously) that the green glass magmas were themselves derived from a geochemically evolved source region, this estimate probably represents a lower limit to the nickel content of olivine from the most primitive regions of the lunar mantle. [*Newsom* (1989) has given a hypothetical "example" that purports to show that green glass (188 ppm Ni) could have been derived from a lunar mantle that itself possessed a bulk composition containing 188 ppm Ni. His argument is essentially circular and ignores most of what is known about the petrogenesis of lunar volcanic glasses in general and of green glass in particular. It makes unsubstantiated propositions about degrees of melting involved in forming the lunar magma ocean; moreover, its calculations are invalid because they ignore the relative proportions and types of minerals present in the lunar mantle prior to and after the melting event. The inadequacy of Newsom's logic has recently been demonstrated by *O'Neill* (1989).] The lunar mantle olivine composition derived by Seifert et al. is depleted in nickel by a factor of 3.5 as compared with olivine from the terrestrial mantle, which typically contains ~2900 ppm Ni (as oxide) (*Archibald,* 1979). *Delano* (1986) and *Wänke and Dreibus* (1986) estimated that the lunar mantle was depleted in nickel by a factor of 4.0 ± 0.5 compared to the terrestrial mantle on the basis of Ni vs. Mg correlations for a broad family of lunar basalts, volcanic glasses, and terrestrial basalts and komatiites. This value is consistent with the depletion factor obtained by *Seifert et al.* (1988); however, the methodology employed by Seifert et al. is more rigorous and their result is therefore preferred.

The conservative lower limit of ~830 ppm Ni in olivine from the lunar mantle nevertheless implies that the lunar mantle resembles the Earth's mantle (in terms of its nickel inventory) far more closely than it does the mantles of the EPB or Mars, which are much more depleted in nickel. The lunar mantle shares with the terrestrial mantle the characteristic that it is far too rich in NiO ever to have been in equilibrium with an iron-rich metallic phase of cosmic composition containing 5.7% Ni. The data of *Seifert et al.* (1988) show that olivine in equilibrium with such a metal phase would contain only 121 ppm Ni (as NiO) at 1400°C and that the NiO level would decrease with falling temperature. The level of ~830 ppm Ni in lunar olivine thus represents an "overabundance" of NiO in the lunar mantle amounting to a factor of 6.8. A similar overabundance of cobalt in the lunar mantle has previously been noted.

The geochemical implications of the high NiO abundance of the lunar mantle are similar to those discussed previously in relation to its high CoO content. Firstly, the mass fraction of metallic iron incorporated by the Moon from the solar nebula during and after the differentiation of the lunar upper mantle-crust system was much smaller than 1%. Secondly, if a metallic phase had segregated to form a core during the major differentiation process that produced the anorthositic crust from the lunar mantle, the resultant core would have contained about 40% Ni.

Chromium, Vanadium, and Manganese

These elements are usually regarded as being lithophile, rather than siderophile, in geochemical processes. However, recent evidence (discussed below) suggests that in the context of geochemical relationships between the Earth and Moon, they have behaved as siderophile elements.

The abundances of Cr, V, and Mn (normalized to Mg) are very similar in the Moon's and Earth's mantle (e.g., *Wänke et al.,* 1977, 1978; *Ringwood et al.,* 1987; *Seifert and Ringwood,* 1988; *Drake et al.,* 1989). However, they are substantially depleted in the mantles of both bodies as compared to CI chondrites. The depletion factors obtained from the above references are V: 0.6-0.7; Cr: 0.4-0.5; and Mn: 0.25-0.35. *Dreibus and Wänke* (1979, 1980) showed that Mn, Cr, and V are present in near-chondritic abundances in the mantle of the EPB while *Dreibus et al.* (1982) showed that Mn and Cr were likewise undepleted in the martian mantle. (The martian vanadium abundance is poorly constrained.)

Ringwood (1966a) suggested that the depletions of Mn, Cr, and V in the Earth's mantle were caused by their entry into the core. *Wänke* (1981) proposed that the Earth accreted inhomogeneously and that V, Cr, and Mn became siderophile under the highly reducing conditions that prevailed during the early and intermediate stages of accretion of the Earth. *Dreibus and Wänke* (1979) also proposed an alternative explanation, suggesting that the extensive solubility of FeO in molten iron at high pressures and temperatures inferred by *Ringwood* (1977) and later confirmed by *Kato and Ringwood* (1989a) might enhance the partitions of Mn, Cr, and V into the core.

Drake et al. (1989) proposed that the depletions of Mn, Cr, and V in the Earth and Moon were caused by volatility-controlled fractionations in the solar nebula prior to accretion. This explanation is untenable for V, which condenses at a temperature about 110 K higher than Mg (as Mg_2SiO_4). It is also untenable in the case of Cr, which condenses in a temperature interval that overlaps that of Mg_2SiO_4 and is substantially higher than the temperature interval over which $MgSiO_3$ condenses (*Grossman and Olsen,* 1974; *Grossman and Larimer,* 1974). As pointed out by *Dreibus and Wänke* (1979) and *Wänke and Dreibus* (1986), the Mg/Cr ratio is near constant in different chondrite groups (CI = 36, CM = 38, CO = 41, CV = 41, L = 40, H = 42, E(av) = 41) and is uncorrelated with Si/Al ratios in these groups, which range from 9 to 20 and have probably been fractionated by processes involving selective volatility.

The nebula condensation temperature of Mn is ~150 K below that of Mg and it is possible that the Mn depletions in the Earth and Moon could be due to higher volatility. However, this interpretation is not supported by the observation that Mn is undepleted in the mantle of the EPB, which experienced much stronger depletions of many other volatile elements than did the Earth's mantle (*Dreibus and Wänke,* 1979). Moreover, Mn is slightly more abundant in the lunar mantle than in the terrestrial mantle (*Ringwood and Kesson,* 1977) despite much greater depletion of most other volatile elements in the Moon.

Kato and Ringwood (1989b) measured the partition coefficients of Cr, V, and Mn between metallic iron and silicate phases over the range of pressures that would occur in the mantle of a martian-sized planetesimal. They show that in the P,T regime in which separation of an iron core from the mantle of such a planetesimal would occur, these elements are lithophile and are preferentially retained in the mantle. Thus, separation of an iron core in a martian-sized planetesimal is incapable of causing depletions of Cr, V, and Mn in its silicate mantle. [*Drake et al.* (1989) obtained a similar result for the case of separation of an Fe-S core.] These experiments are entirely consistent with the prior conclusion by *Dreibus et al.* (1982) and *Wänke and Dreibus* (1988) that Cr and Mn are not depleted in the martian mantle. More importantly, they also show that *the depletions of Mn, Cr, and V in the lunar mantle cannot have been inherited from the mantle of a martian-sized impactor.*

Ringwood et al. (1989) have recently obtained evidence that shows that the siderophile nature of Cr, V, and Mn only becomes apparent at very high pressures when significant quantities of oxygen become soluble in the molten metallic phase. In a series of experiments at 16 GPa and at 1700°C, they showed that the presence of ~2 wt.% oxygen in the liquid metal phase caused the partition coefficients for these elements between liquid metal and silicate phases to be increased by factors of 4-7 in favor of the molten metal phase. At higher pressures, much larger amounts of oxygen would dissolve (*Kato and Ringwood,* 1989) and corresponding increases in the metal-silicate partition coefficients for these elements are expected to occur. The results of *Ringwood et al.* (1989) strongly suggest that the depletions in the Earth's mantle of Cr and V, and perhaps also of Mn, could be caused by their enhanced siderophile tendencies under the conditions of core formation in the very high-pressure and high-temperature regime provided by a terrestrial-sized planet.

These depletions cannot be achieved by core formation in a martian-sized planetesimal. Nor can they be explained by volatility-controlled fractionations in the solar nebula prior to accretion. The similarity of abundances of V and Cr (and Mn) in the lunar and terrestrial mantles assumes great significance in the light of these circumstances, as was first pointed out by *Dreibus and Wänke* (1979). The only plausible explanation for the lunar depletions is that they were inherited from material derived originally from the Earth's mantle.

Additional Siderophile Elements (W, P, S, Se, Ga, Sn, Cu, As, Ge, Mo, Re, and Au)

The lunar and terrestrial geochemistry of this group of siderophile elements has recently been reviewed by *Ringwood* (1989a) and is summarized below. Tungsten and phosphorus are both moderately siderophile and their abundances in the Earth's mantle have been depleted by a factor of about 25 compared to their primordial abundances, owing to core formation. The geochemistry of oxidized species of these elements in the terrestrial and lunar crust-mantle systems have been studied in great detail. The absolute abundances of tungsten in both these lunar and terrestrial systems are almost identical (*Rammensee and Wänke,* 1977; *Wänke and Dreibus,* 1986; *Newsom and Palme,* 1984). Phosphorus is only modestly depleted in the overall lunar mantle-crust system, by factors of 0.8-0.5, as compared to the corresponding terrestrial system (*Wänke and Dreibus,* 1986).

Earth-Moon abundance comparisons of Ga, Sn, Cu, As, and Ge are more complex because these siderophile elements have been subjected to an additional depletion process in the Moon because of their high volatilities. Their abundances in protolunar material prior to the volatilization event were estimated by *Seifert and Ringwood* (1987) by normalizing them to lithophile elements possessing similar volatilities. They found that the abundances of the moderately siderophile elements Ga, Sn, and Cu in protolunar material were similar (within a factor of 2) to their abundances in the Earth's mantle. On the other hand, the highly siderophile elements Ge and As remained strongly depleted in protolunar material, even after this correction had been made.

In contrast to the moderately siderophile elements discussed above, several studies have shown that some strongly siderophile, involatile elements, including Re, Au, and Mo are also greatly depleted in the lunar mantle as compared to the terrestrial mantle (*Ringwood and Kesson,* 1977; *Wolf and Anders,* 1980; *Newsom and Palme,* 1984).

Implications of the Lunar Siderophile Signature

Ratios of the abundances of siderophile elements in the lunar and terrestrial mantles are plotted against their (iron) metal-silicate partition coefficients in Fig. 8. The database for involatile siderophiles is from *Ringwood* (1986b). In the cases of the volatile siderophiles (Cu*, Sn*, Ga*, Ge*, and As*), abundances have been corrected to correspond to the levels in protolunar material prior to partial loss by volatilization, as discussed above. (The asterisks denote incorporation of this correction.) Diagrams analogous to, but less comprehensive than Fig. 8 have been con-

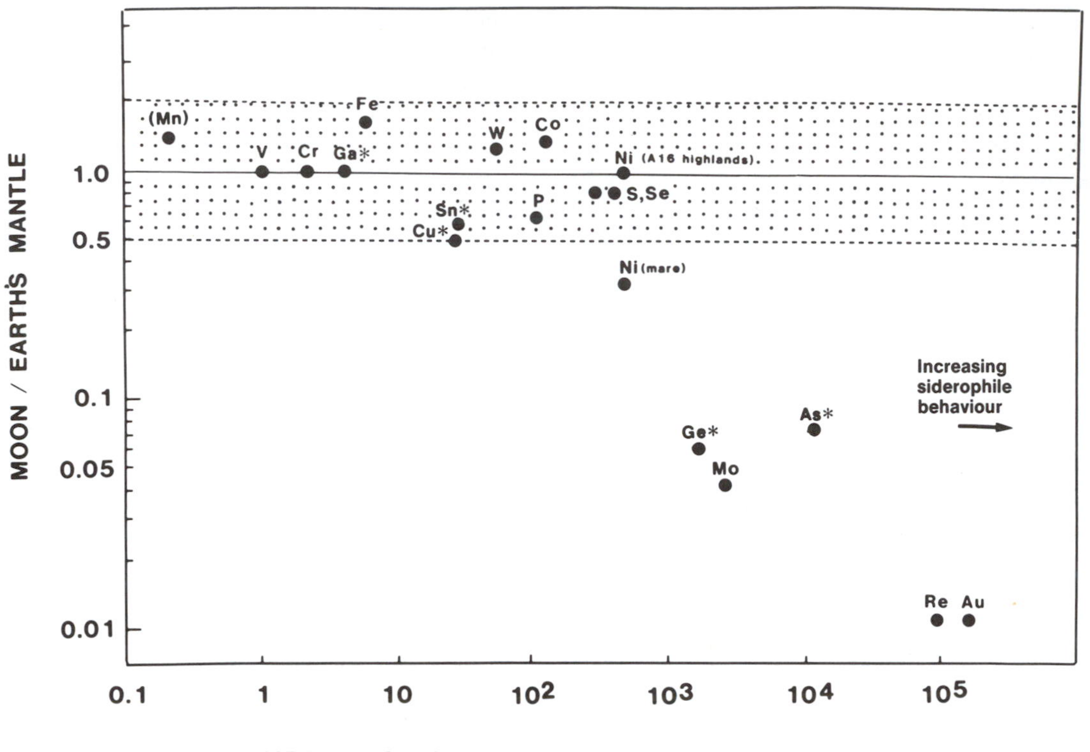

Fig. 8. *Abundance ratios of siderophile elements in the Moon and Earth's mantle vs. their metal/silicate partition coefficients. Asterisks denote elements whose abundances have been corrected to allow for losses as volatile species during the formation of the Moon as described in the text. Based on Ringwood (1986b) and Ringwood and Seifert (1988).*

structed by *Wänke and Dreibus* (1986) and by *Newsom* (1984). It is seen that the abundances of elements less siderophile than nickel (Mn, V, Cr, Ga*, Fe, Sn*, Cu*, W, P, Co, S, and Se) are similar (within a factor of 2) in both bodies. Nickel itself is depleted by a factor of 3.5 in the source regions of the most primitive lunar basalts but is present in terrestrial abundances (relative to Mg) in breccias at the Apollo 16 lunar highlands site (*Ringwood et al.,* 1987).

The similarities in abundances of a large group of siderophiles in both the terrestrial and lunar mantles is quite remarkable. It was pointed out previously that because of the complexity and multiplicity of processes connected with core formation, which ultimately established the siderophile pattern of the terrestrial mantle, this pattern was likely to be unique to the Earth. As seen in Fig. 4, it is quite different from those in the EPB and Mars. No way is known whereby such a pattern can be established in the solar nebula prior to planetary accretion, nor by internal differentiation of a core within a lunar-sized body, nor even by internal differentiation of a core within a martian-sized body (see section on chromium, vanadium, and manganese above). The similarities in terrestrial and lunar abundances of this group of siderophiles thus provide powerful support for the hypothesis that a large proportion of the material now in the Moon was derived from the Earth's mantle subsequent to core formation (*Ringwood and Kesson,* 1977; *Rammensee and Wänke,* 1977).

It is seen from Fig. 8 that elements more siderophile than nickel such as Ge*, Mo, As*, Re, and Au are depleted in the lunar mantle as compared to the terrestrial mantle. The degrees of depletion of this group of elements correlate roughly with their metal-silicate partition coefficients. *Wänke and Dreibus* (1986) and *Newsom* (1984) have shown that these depletions can readily be explained by separation of a small metallic core amounting to <1% of the lunar mass. This would not markedly affect the abundances of moderately siderophile elements but would nevertheless have strongly depleted the highly siderophile elements. The presence of a small (<2% by mass) lunar core is also implied by the lunar moment of inertia coefficient (*Blackshear and Gapcynski,* 1977) and by the observation of a phase shift in the forced precession of the lunar figure (*Yoder,* 1984).

HETEROGENEOUS ACCRETION AND ITS BEARING ON THE LUNAR GEOCHEMISTRY OF SIDEROPHILE AND VOLATILE ELEMENTS

It has been suggested that the high levels of siderophile elements such as Ni, Co, and Cu that are present in the Earth's mantle were caused by a process of heterogeneous accretion, whereby a low-temperature, volatile-rich condensate from the solar nebula was mixed into the terrestrial mantle after segregation of the core (e.g., *Wänke,* 1981). The possibility that this same process may have influenced siderophile abundances in the lunar mantle should also be considered. The abundance of Co in the lunar mantle would require that about 10% (Mg-normalized) of a low-temperature condensate had been incorporated into the Moon. However, this would necessarily have introduced chondritic relative abundances of other moderately siderophile elements such as P and W. In fact the Co/P and Co/W ratios of the lunar mantle are respectively 3 and 2 times higher than the chondritic ratios, showing that only a minor proportion of the Moon's endowment of Co could have been derived in this manner.

The observation that Co and Ni are present in the lunar mantle as oxidized species (as in the terrestrial mantle) also provides a significant constraint. In the solar nebula, Co and Ni become oxidized only at low temperatures, below 273 K (*Grossman and Larimer,* 1974). Under these conditions, most elements are fully condensed and would therefore be present in cosmic (CI) relative abundances in the accreting material. However, volatile elements such as Zn, Cd, In, Bi, and Tl are depleted, relative to Co (CI-normalized) in the Moon by factors of 20 to 50 (*Wolf and Anders,* 1980). It is difficult to envisage how these elements, once condensed, could be separated so efficiently from oxidized Co and Ni. This would require the operation of an exceptionally efficient devolatilization process after the Moon had been formed. Although escape of volatiles from the primordial lunar magma ocean may well have been an important auxiliary process, it seems doubtful whether it would have been as efficient as this.

The geochemistry of volatile elements in the Moon has been discussed by *Wolf and Anders* (1980). They propose that the volatiles were introduced into protolunar material by mixing with about 0.2-1% of a volatile-rich carrier similar to CI-chondrites. The bulk of the material that formed the Moon was thus believed to have been extremely depleted in volatiles—well below the levels now present. This initial lunar volatile element signature would have been modified further by processes such as differential volatilization from the lunar magma ocean and core formation. *Wolf and Anders'* (1980) interpretation of lunar volatile-element geochemistry is consistent with *Ringwood's* (1966b) model of the formation of the Moon. Protolunar material derived from the Earth's mantle was believed to have been extremely depleted in volatiles and was emplaced as a ring of planetesimals orbiting the Earth. This ring captured a small proportion of oxidized CI-like material from the nebula at a very late stage of the Earth's accretion. The sediment ring then coagulated to form the Moon, thereby incorporating most of its small volatile inventory from the CI-like component.

Ringwood's model contains a simple explanation for two geochemical characteristics that were later invoked in a criticism of the hypothesis that the Moon was derived mainly from terrestrial material. *Kreutzberger et al.* (1986) noted that the Cs/Rb ratio of the Moon seems to be higher than that of the Earth. Since Cs is more volatile than Rb, and the Moon is more depleted overall in volatiles than the Earth, they argued that the Moon should possess a lower Cs/Rb ratio than the Earth if it had been derived from terrestrial material. However, this would not be expected if the lunar volatile inventory had been derived mainly from a low-temperature CI-like component. *Hinton et al.* (1988) have produced evidence suggesting that the $^{39}K/^{41}K$ isotopic composition of the Moon is similar to the chondritic ratio and slightly different from the terrestrial ratio. *Newsom and Taylor* (1989) argued that this was inconsistent with a terrestrial origin for protolunar material. However, it is readily explicable by the hypothesis that the volatile inventory of the Moon was introduced by a CI-like carrier.

Previously we noted that, when normalized to terrestrial mantle abundances, highly siderophile elements such as Mo, Re, and Au are substantially depleted in the lunar mantle compared to moderately siderophile elements such as Co, W, and P (Fig. 8). These observations implied that a small amount ($<1\%$) of metal phase had been present in the region of the Moon that differentiated to form the anorthositic crust and the underlying olivine-pyroxene cumulates. This metallic phase may have remained as a residual phase at depths of 200-400 km, or alternatively, it may have segregated to form a small lunar core. The latter interpretation is preferred on the basis of geophysical evidence discussed previously. The modest ($\times 3.5$) depletion of NiO relative to CoO that is evident in the lunar mantle was probably caused by this latter process since the metal phase in equilibrium with the present NiO content of the lunar mantle would contain about 40% Ni (*Seifert et al.*, 1988).

BULK COMPOSITION OF THE MOON

If a large proportion of the material in the Moon had indeed been derived from the Earth's mantle, one might hope to find supporting evidence in a comparison of the chemical compositions of both bodies. Unfortunately, there are divergent views among geochemists regarding the major element composition of the Moon, estimates of which are necessarily model dependent.

Jones and Delano (1989) derived a range of lunar bulk compositions from the remarkable set of systematic geochemical correlations that they recognized among the 25 defined groups of lunar volcanic glasses. Except for enrichment in FeO and depletion of alkalis, the lunar bulk compositions that they obtained were very similar to that of the Earth's mantle. *Ringwood et al.* (1987) employed an entirely independent methodology based upon systematic mixing relationships displayed by lunar highland breccias in order to deduce the bulk composition of the lunar interior. The composition that they obtained was very similar to those derived by Jones and Delano.

A third estimate of lunar bulk composition has been obtained primarily from Al_2O_3 mass-balance considerations between the lunar crust and mantle (*Taylor*, 1982). This composition depends sensitively upon estimates of the total Al_2O_3 content of the lunar crust, the proportion of the Moon that differentiated to form the crust, and the amount of Al_2O_3 that remained behind in the refractory, residual region. These estimates possess substantial uncertainties. The lunar bulk composition thereby derived is about 50% richer in Al_2O_3,

CaO, and other refractory oxides than compositions based directly upon the petrogenesis of lunar volcanic and highland rocks. However, these Al_2O_3 mass balance constraints are quite permissive and cannot be used to exclude the less refractory bulk compositions.

It was once thought that observations of lunar heat flow favored refractory-enriched compositions. However, further geophysical analyses showed the data could also be reconciled readily with the less refractory bulk compositions (*Ringwood,* 1979; *Warren and Rasmussen,* 1987). Recently, it has been suggested that refractory-enriched bulk compositions can be reconciled more readily with the distributions of seismic velocities in the lunar mantle and, in particular, with the inferred presence of a seismic discontinuity near 500 km (*Hood and Jones,* 1987; *Mueller et al.,* 1988). However, the seismic velocity gradients and existence of a discontinuity near 500 km are sensitively dependent upon the assumption that the physical properties of the lunar upper mantle are radially symmetrical. Since this region is believed to be olivine rich and is likely to have experienced convection at some stage of its history, it would be surprising if a significant degree of velocity anisotropy were not present. This, in turn, would eliminate evidence for a seismic discontinuity near 500 km. Accordingly, it should be recognized that the presently preferred seismic depth-velocity profiles (*Nakamura,* 1983) provide only a weak constraint on the refractory content of the lunar mantle.

The hypothesis that the Moon is enriched in refractories (*Taylor,* 1982, 1986) encounters some serious difficulties:

1. Oxygen isotope ratios are essentially identical in the Earth and Moon (*Clayton and Mayeda,* 1975). If the Moon were indeed enriched (by ~50%) in refractory elements and this enrichment had been caused by high-temperature vapor-phase fractionation processes, either in the nebula prior to accretion, or during the impact of a giant planetesimal upon the Earth, significant thermal and chemical fractionations of oxygen isotopes along the chemical fractionation trend ($^{17/16}O/^{18/16}O = 0.5$) might be expected (*Grossman et al.,* 1974; *Clayton and Mayeda,* 1975). These fractionations are not observed.

2. Vanadium is a member of the refractory element group. If the Moon were indeed enriched in refractories (relative to Mg) its abundance of vanadium should be correspondingly enhanced. However, it was noted in the section on chromium, vanadium, and manganese that vanadium is actually *depleted* in the Moon by a factor of 0.6 as compared to Mg. [See *Seifert and Ringwood* (1988) for further discussion of this topic.]

3. According to *Taylor* (1986), the enrichment of refractories in the Moon is caused by the circumstance that protolunar material was derived predominantly from the mantle of a martian-sized planetesimal that collided with the Earth. It is arbitrarily assumed that the mantle of the impactor was itself enriched in refractory elements, but no explanation is offered concerning the processes that caused this refractory enrichment. Any such process would inevitably be confronted by the difficulties mentioned in (1) and (2) above.

4. The Safronov-Wetherill scenario of planetary accretion is specifically adopted by *Taylor* (1986). According to this scenario, accretion of planetesimals to form larger bodies was a hierarchical process involving repeated episodes of growth, disruption and further growth, in turn resulting in a size distribution in which the second-largest bodies were 10–30% of the masses of their primaries. This process was accompanied by considerable radial mixing of planetesimals and their debris, and efficient mixing and homogenization of material within the feeding zones of individual planets. Even if the original population of asteroid-sized precursors had possessed vastly different compositions (which is debatable), it is difficult to conceive how the Safronov-Wetherill planetary accretion scenario could lead to a situation where the second largest planetesimal in the Earth's feeding zone was

substantially different in composition (e.g., in its refractory and volatile element abundances) as compared to the primary planet. Indeed, the bulk composition of the impactor would be expected to be similar to that of the proto-Earth.

Much has been made in the literature concerning the significance of differences between the bulk compositions of the Moon and Earth's mantle for hypotheses of lunar origin. It has often been asserted that the existence of significant chemical differences between both bodies demonstrates that the Moon could not have been derived primarily from the Earth's mantle. This is a simplistic argument. It is only the most extreme (and implausible) versions of the "fission" hypothesis that imply identity in bulk compositions. The models for a terrestrial origin of the Moon that I have developed over many years (e.g., *Ringwood,* 1966a,b, 1970, 1972, 1979, 1986a,b) all lead to significant differences between the bulk compositions of the Earth's mantle and Moon. The same applies to recent models of impact-induced Earth fission (e.g., *Wänke and Dreibus,* 1986).

There is indeed widespread consensus that the Moon is depleted in volatiles and enriched in FeO (~13 wt.%) as compared to the Earth's mantle (~8 wt.% FeO). *Jones and Hood* (1990) have cogently demonstrated that the net iron content of the Moon is substantially higher than that which is now present in the Earth's mantle. Depletion of the Moon in volatiles could arise from their loss by evaporation during the high-temperature processing that occurred when protolunar material was removed from the Earth to form the Moon (e.g., *Ringwood,* 1986a). Enrichment of FeO in the Moon may have been consequential to the process of core formation within the Earth. The extraction of FeO from the Earth's mantle and its transfer into the core as discussed by *Ringwood* (1977) may have occurred over an extended period (e.g., $1\text{-}2 \times 10^8$ yr). It is possible that ejection of terrestrial material to form the Moon occurred before this FeO-extraction process had been completed (*Jagoutz and Wänke,* 1982; *McCammon et al.,* 1982). Thus, protolunar material could have been intrinsically richer in FeO than is the Earth's mantle today.

Other factors that may have contributed to differences between the compositions of the terrestrial mantle and Moon are discussed by *Ringwood* (1989a). These include the capture of a small proportion of iron-rich, sun-orbiting planetesimals by a circumterrestrial sediment ring that was parental to the Moon. The same factors would also influence comparisons of siderophile signatures in both bodies. I have never maintained that these patterns are identical. It is the ***similarity*** of abundances of a large group of siderophile elements in both bodies, within a factor of about two, that is of genetic significance. This unique compositional signature has survived, despite the effects of other factors, mentioned above, that would tend to degrade it.

ORIGIN OF THE MOON

Background

The densities of terrestrial planets vary substantially, even after corrections for self-compression. Between 1950 and 1970, the density variations were usually explained by assuming that terrestrial planets are composed of varying proportions of silicate and metallic iron phases. It was assumed that these materials had somehow become fractionated in the different regions of the solar nebula prior to planetary accretion (e.g., *Urey,* 1952). The considerable depletion of metallic iron in the Moon, evidenced by its low density, was usually interpreted as representing an extreme case of this fractionation process. In this sense, the Moon was regarded as an "independent planet" that, owing to special circumstances, had accreted in orbit around the Earth or had been captured into Earth orbit.

Ringwood (1960, 1966b) rejected this interpretation because the physical mechanisms invoked as causes of the iron-silicate fractionation seemed to be implausible. He proposed instead that most of the material now in the Moon had been derived by a special process from the Earth's mantle after the core had segregated. In one sense, this represented a return to an earlier hypothesis of "Earth fission" proposed by *Darwin* (1880). However the physical mechanism proposed for extracting protolunar material from the Earth's mantle was quite different from Darwin's hypothesis.

According to the models developed by *Ringwood* (1966b, 1970, 1979), accretion of the Earth occurred on a short timescale ($\sim 10^6$ yr) and was completed before the primordial gases of the solar nebula had been dissipated. Accretion was accompanied by the formation of a primitive terrestrial atmosphere composed mainly of hydrogen gravitationally captured from the solar nebula and mixed with higher molecular weight gases produced by impact degassing of accreting planetesimals. The primitive atmosphere was coupled to the Earth's rotation through turbulent viscosity and hydromagnetic torques and was thereby spun out into a co-rotating disk (period ~5 hr). During the later stages of accretion of the Earth, high temperatures were produced by a combination of rapid accretion and thermal insulation by the primitive atmosphere. In these conditions, material from the mantle was evaporated into the primitive atmosphere and spun out into the disk. As the primitive atmosphere cooled and was dissipated, mainly by strong particle and UV radiation during the solar T-Tauri phase, the silicate components were precipitated to form a ring of Earth-orbiting planetesimals. Further fractionation due to volatility occurred during the precipitation process, since the more volatile components would be precipitated at relatively low temperatures, forming

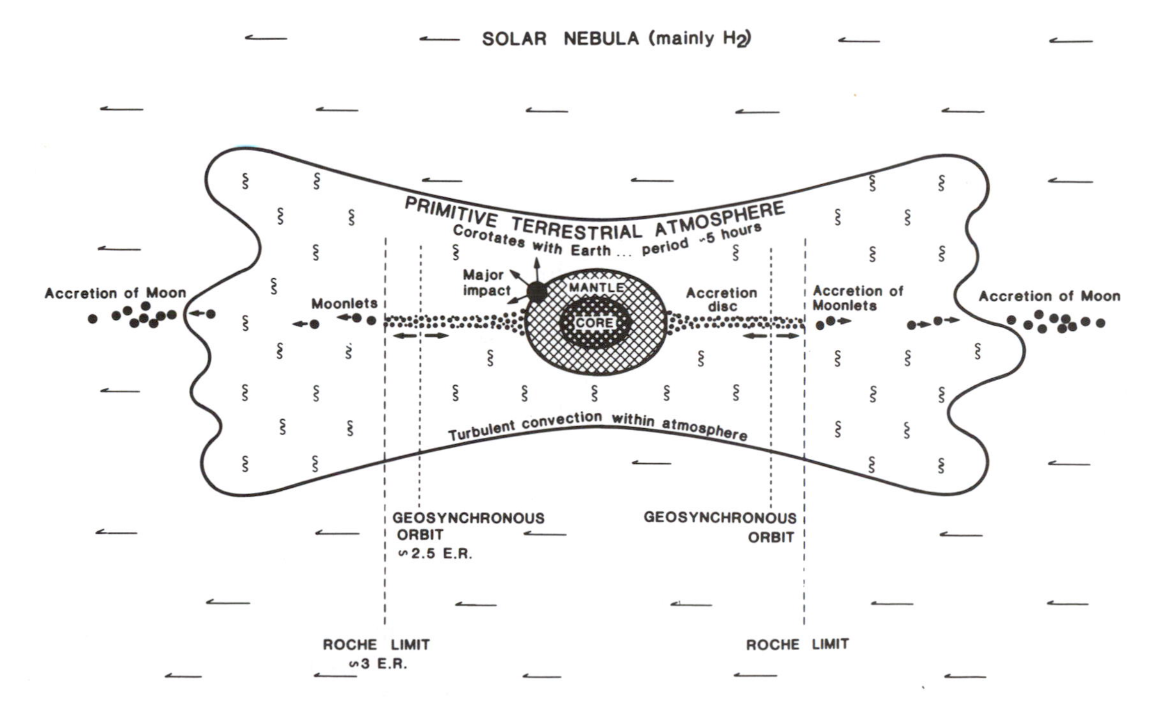

Fig. 9. *Model showing the growth of the Earth via an accretion disk and the formation of the Moon via the ejection of material from the Earth's mantle by impacts from late-accreting planetesimals in the presence of a co-rotating primitive terrestrial atmosphere. From Ringwood (1986b).*

micrometer-sized smoke particles. These remained coupled to the escaping gases and hence were removed from the system. The Moon accreted from the ring of devolatilized Earth-orbiting planetesimals. The model is illustrated in Fig. 9.

The principal problem with this scenario was the short timescale ($\sim 10^6$ yr) over which the Earth must have formed if the gravitational potential energy of impacting planetesimals was to have generated the sustained high temperatures necessary to evaporate part of the upper mantle during the later stages of accretion. An attractive alternative was provided by the models of *Hartmann and Davis* (1975) and *Cameron and Ward* (1976). These authors invoked one or a few impacts by very large, late-accreting planetesimals to evaporate material from the Earth's mantle and place it into orbit. The planetesimals are believed to range in mass from sublunar ($<0.01\ M_E$) to martian ($\sim 0.1\ M_E$). As noted by *Hartmann and Davis* (1975) and *Stevenson* (1987), the physical processes involved in the impact scenarios for the formation of the Moon were closely related to Ringwood's earlier models. Rather than achieving high mantle-evaporation temperatures via the "steady-state" liberation of gravitational potential energy over a short accretion timescale ($\sim 10^6$ yr), large impact models achieved these conditions in one or in a few transient ultrahigh energy events, and thus permitted a much longer accretion timescale for the Earth (e.g., 10^7-10^8 yr).

These advantages led *Ringwood* (1979) to support a modified version of the impact-ejection model. However, apart from Wänke's group at Mainz, the lunar and planetary science community remained indifferent to these proposals until the Conference on the Origin of the Moon held at Kona in November, 1984. The stage had been set by some elegant modeling of the accretion of planets by *Wetherill* (1985, 1986). This work was described at the conference and provided strong support for *Hartmann and Davis'* (1975) inference that accretion of the Earth had been a hierarchical process proceeding via the formation of intermediate-sized bodies that might have been as large as Mars. *Cameron and Ward* (1976) had proposed an impact by a martian-sized planetesimal as a means not only of ejecting the Moon, but also of providing the high angular-momentum density of the Earth-Moon system. At the Kona conference, *Cameron* (1984) presented numerical simulations of a giant impact of this type, which supported his proposal that the impact could have resulted in the emplacement of a lunar mass into orbit around the Earth.

The confluence of these factors, fueled by the seductive appeal of a single process that promised to explain both the origin of the Moon and the high angular-momentum density of the Earth-Moon system proved irresistable to planetary scientists, and a bandwagon of support for the single giant impact hypothesis soon developed. This was further stimulated by the widening recognition of seemingly insuperable difficulties faced by the "traditional" hypotheses of lunar origin—capture, rotational fission, and binary accretion (e.g., *Boss and Peale,* 1986; *Stevenson,* 1987).

This enthusiasm may be premature, however. It is suggested below that the currently popular version of the single giant impact hypothesis of lunar origin is faced by certain problems—both dynamical and geochemical—that have not been adequately recognized. Before discussing these problems, a diversion is needed to consider an important general aspect of the model, namely, the stage in the accretion of the Earth at which the giant impact is required to have occurred.

Timing of the Giant Impact

The process of ejecting material during a giant impact upon the Earth to form a circumterrestrial ring of planetesimals, followed by coagulation to form the Moon, was necessarily very short, and probably on the order of 10^2-10^3 years (*Stevenson,* 1987). The Moon was thereby formed in an extensively or fully molten state (*Stevenson,* 1987). Crystallization of the Moon from this initial state has been investigated experimentally by *Ringwood* (1976) and *Kesson and Ringwood* (1977). They showed that the melting-point gradient is much higher than the adiabatic gradient of the convecting magma ocean, which therefore crystallizes from the base upward, forming a mantle of olivine and orthopyroxene cumulates. Rising convection currents arrive at the lunar surface with temperatures well above the liquidus and hence heat is lost by radiation very efficiently. It is not until crystallization has proceeded from the base of the molten zone to a depth of less than 150 km that differentiation would have caused sufficient enrichments of Al_2O_3 and CaO in the residual magma to result in the crystallization of plagioclase. After this stage has been reached, plagioclase may float in the magma because of its low density, leading to the formation of a buoyant anorthositic crust. Prior to the appearance of plagioclase, crystallization of the magma ocean from its base (which may have extended to the center of the Moon) upward to a depth of ~150 km would have been completed in an interval of only 10^3-10^4 years (e.g., *Jeffreys,* 1959). However, after the onset of plagioclase-precipitation from the crystallizing magma ocean and the development of a buoyant anorthositic crust, the rate of heat loss and further crystallization would have become considerably slower.

Constraints imposed by experimental petrology and by the Moon's thermal history imply that the most primitive lunar volcanic glasses represent magmas that were formed within the early cumulate system (olivine ± orthopyroxene) at depths of 300-500 km around 3.2-3.8 Ga (*Delano,* 1986; *Ringwood and Kesson,* 1976a; *Ma et al.,* 1981). As discussed in the section on the lunar siderophile signature, the high CoO and NiO contents and the nonchondritic siderophile element ratios that are characteristic of this region demonstrate that, at most, only minor amounts (<1%) of either "cosmic" metal or of oxidized solar nebula material were incorporated into the lunar mantle from planetesimals that had yet to accrete on the Earth or Moon. It follows that the giant impact that is hypothesized to have led to the formation of the Moon must have occurred at a *very* late stage of accretion of the Earth when only a small proportion of the original planetesimal population remained to be accreted.

Siderophile abundances in the anorthositic lunar crust substantiate this conclusion. As noted above, a buoyant crust probably began to form within 10^4 years after the giant impact. Thereafter, the growing crust would have provided an efficient trap for all but the largest infalling planetesimals. Iridium contents of lunar highland breccias show that the uppermost layer of the lunar crust contains only about 1-2% of a "meteoritic" (=planetesimal) component. The concentration of this component doubtless decreases with depth. If we make the generous assumption that the entire crust contains 0.5% of a planetesimal component, it follows that the amount of this material that fell on the Moon after development of the lunar crust was only about 0.05% of a lunar mass. The low Re abundances of mare basalts likewise indicate that the lunar mantle contains only a minute amount (less than 0.005%) of a primary planetesimal component. Moreover, much of the observed siderophile contamination of the lunar crust probably occurred after accretion had been completed, over the period from 4.4 Ga onward. Pristine lunar ferroan anorthosites that are believed to have crystallized from the magma ocean that formed the crust, contain much lower abundances of highly siderophile elements than the highland breccias. Differentiation of the lunar crust seems to have been largely completed by about 4.45 Ga (*Carlson and Lugmair,* 1988).

Flaws in the Single Giant Impact Hypothesis

In the course of 13 numerical simulations of the accretion of the Earth, *Wetherill* (1986) found that the growing planet would be struck by 17 planetesimals of martian (0.11 M_E) or greater size (an average of 1.3 such encounters per simulation). *Benz et al.* (1989) found that the preferred size for an impactor to produce both the angular momentum of the Earth-Moon system and also to produce a lunar mass of iron-free (= silicate) material in Earth orbit, was 0.14 M_E. According to *Wetherill* (1986) the probability of the Earth being struck by a planetesimal of this size during its accretion is 0.8. In the light of these results, and to facilitate a discussion of the giant impact hypothesis, we will assume that during accretion, the Earth was hit by a single planetesimal of martian or larger size.

Wetherill's numerical experiments show that this giant impact is much more likely to occur at an early or intermediate stage of the Earth's accretion than towards the end. However, it was noted in the previous section that if the Moon were formed by a single giant impact, the episode must have occurred *very* late in the accretion of the Earth in order to avoid excessive contamination of the Moon by material that had yet to be accreted. It seems quite likely that, prior to the giant impact, the Earth had reached 95-99% of its final mass (apart from mass added by the impactor itself). The probability of such an impact occurring as a function of the growth of the Earth can be obtained from *Wetherill* (1986, Fig. 8) and is shown in Fig. 10. It is seen that only one giant impact out of a total of 17 such events occurred sufficiently late to be compatible with the compositional constraints discussed earlier.

In order to produce the angular momentum required for the Earth-Moon system, a martian-sized impactor must strike the proto-Earth in a region that is constrained to be toward the outer circumference of the planet and within about 40° of the equatorial (ecliptic) plane. *Boss and Peale* (1986) showed that there is only a 1 in 20 chance that the impact would actually occur within the region needed to produce the observed angular momentum of the Earth-Moon system. Combining this with

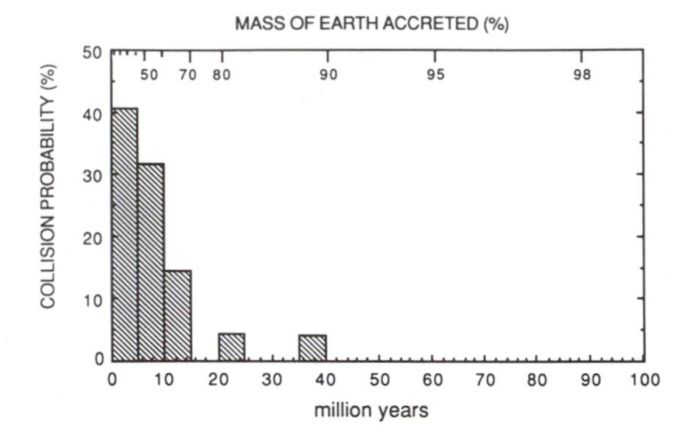

Fig. 10. *The probability of impact on Earth by a planetesimal of martian or greater size is shown as a function of time and the prortional mass of the Earth accreted. Based on the results of 13 calculations of terrestrial accretion by Wetherill (1986). Figure from Ringwood (1989c).*

the low probability for a very late impact, as discussed above, the overall probability of forming the Earth-Moon system by a single giant impact falls to about 1 in 340. These odds are not encouraging.

A second major problem for the single giant impact hypothesis arises because the catastrophe would have caused essentially complete melting of the Earth's mantle followed by differentiation, leading to a gravitationally stable compositional stratification, as discussed in the sections on constraints on early mantle differentiation and on stratification of an initially molten mantle. However, the evidence reviewed in those sections demonstrated rather clearly that the mantle did not possess this structure immediately after accretion of the Earth had been completed.

Thirdly, it seems that the giant impact hypothesis fails to provide a satisfactory source of protolunar material. Numerical simulations of the impact of a martian-sized planetesimal on the Earth indicate that most of the material that would be emplaced in Earth orbit would be derived from the mantle of the impactor and not from the Earth's mantle (*Cameron,* 1986; *Benz et al.,* 1987, 1989). This is contradicted by the detailed geochemical evidence reviewed in the section on the lunar siderophile signature, which demonstrates that most of the material in the Moon was derived not from the mantle of the impactor, *but from the Earth's mantle.*

A Modified Impact Hypothesis of Lunar Origin

Cumulatively, the evidence above is highly unfavorable to the single giant impact hypothesis in its present form. Is there any way in which the hypothesis could be modified so as to avoid the difficulties, while still remaining consistent with the general model of planetary accretion advocated by *Wetherill* (1986)? This might be achieved if we relinquish the requirement that a single giant impact was responsible *both* for the high angular momentum density of the Earth-Moon system and for the origin of the Moon.

Wetherill's (1986) model implies that the largest planetesimal to impact the Earth was between 10-30% of the final mass of the Earth. The larger the mass ratio between the impacting planetesimal and the proto-Earth, the greater is the probability that a random impact could provide the required angular-momentum density. Thus, if a planetesimal amounting to, say, 20% of the Earth's mass were accreted after the proto-Earth had grown to about 60% of its final mass, there could be a reasonable probability of spinning the proto-Earth up to a period of 3-4 hr. Providing we employ the giant impact *only* to provide the angular-momentum requirements, but not to form the Moon, the unfavorable dynamical probabilities discussed above might be avoided. The collision may have placed material temporarily in Earth orbit, but this material is likely to have been dispersed and returned to Earth by the flux of planetesimals ($\sim$20% M_E) yet to accrete.

A giant impact of this kind would totally melt and differentiate the Earth's mantle as discussed above. It would also be likely to temporarily remove any primitive atmosphere that may then have been present (*Cameron,* 1986), thereby permitting rapid cooling of the Earth by thermal radiation from its molten surface. Because the melting point gradient within the Earth is greater than the adiabatic gradient, crystallization would proceed from the base of the mantle outwards, and most of the mantle could have solidified in a period as short as 10^4 yr (*Jeffreys,* 1959). The structure would initially have been highly stratified, as described above. However, 20% of the Earth had yet to accrete. Most of this material may have been in the form of relatively small planetesimals ($<$0.01 M_E) that caused only localized melting on impact. The addition of this dense, iron-rich material to the surface of the solidified, differentiated mantle would provide a powerful force for compositionally driven subsolidus convection, that would have been maintained by much larger density gradients than those which drive thermal convection. This process might well have been strong enough to rapidly rehomogenize the mantle and to remove all traces of its previous differentiated condition. The second problem mentioned earlier, relating to the lack of evidence for gross differentiation of the Earth shortly after its formation, might thereby be accommodated.

It was pointed out above that formation of the Moon must have occurred at a very late stage of the Earth's accretion, otherwise the Moon would have become excessively contaminated by iron and volatiles intercepted from Earth-accreting material. Moreover, while the Moon is believed to have formed mainly from material derived from the Earth's mantle, the process responsible did not lead to extensive melting of the terrestrial mantle. These conditions might be satisfied if protolunar material had been ejected from the Earth's mantle via impacts from a number of large (e.g., 0.001-0.01 M_E) but not giant ($>$0.1 M_E) high-energy planetesimals (*Ringwood,* 1979, 1986a,b). These late-accreting bodies may have belonged to the terrestrial planetesimal swarm or, alternatively, they may have been derived from the outer solar system via scattering by Jupiter (*Kaula and Bigeleisen,* 1975, *Hartmann and Vail,* 1986). Planetesimals in this latter class, consisting mainly of ices and possessing velocities (relative to Earth) of 20-30 km/sec, would have been highly effective in placing terrestrial material in orbit. Experimental and theoretical studies (*Boslough and Ahrens,* 1983) have shown that planetesimals impacting the Earth at only 15 km/sec would vaporize about 5 times their own mass of terrestrial mantle material and should shock melt 100 times their own mass. The shock-melted material from a large, high-energy impact would probably have formed a spray of droplets that would be largely devolatilized at the high temperatures prevailing. Rapid expansion of the gases of the impact cloud would cause acceleration to high velocities. The impact cloud is likely to become coupled to the Earth via turbulent viscosity and spun out into an equatorial disc. Condensed material in the outer region of the disk could then aggregate to form Earth-orbiting moonlets. Large numbers of moonlets so formed in successive large impacts may have subsequently coagulated to form the Moon (*Ringwood,* 1986a,b; *Stevenson,* 1987). The model outlined above marks a return to the spirit of the earlier proposal of *Hartmann and Davis* (1975) to form the Moon by impact-induced fission of the Earth's mantle. These authors did not seek to solve the angular-momentum problem simultaneously with the formation of the Moon. It has been the attempts by most subsequent workers to achieve this combined objective in a single event that has led to the problems outlined earlier.

Accretion in the Presence of Nebula Gas

The model developed in the previous section sought to utilize the conceptual framework provided by the Safronov-Wetherill hypothesis of planetary accretion. These authors assume that accretion of the Earth occurred after the gases of the solar nebula had been dissipated. Since accretion of the massive "nuclei" of Jupiter and Saturn must have occurred before the gases of the nebula were dissipated, it seems plausible that accretion of the terrestrial planets should also have proceeded in the presence of nebula gasses, as proposed, for example, in the "Kyoto" cosmogony (*Hayashi et al.,* 1985). The dynamics of accretion may be greatly modified under these conditions.

The models of *Weidenschilling and Davis* (1985) and of *Patterson and Spaute* (1988) indicate that a relatively large proportion of the mass of the planetesimal population may have consisted of small bodies. Aerodynamic friction would tend to circularize the orbits of small (e.g., 10^{10}g) planetesimals, causing them to contract radially inward. *Ward* (1988) has suggested that the orbits of much larger planetesimals (e.g., 10^{25}g) may also be circularized and would therefore contract inward because of dissipation caused by density waves generated by the planetesimals in the gases of the nebula. Under these conditions, the growth of planets may be dominated by accretion of planetesimals from orbits possessing low eccentricities. Planetesimals accreting in this manner preferentially deliver prograde angular momentum to the growing planet (*Giuli,* 1968; *Nakazawa et al.,* 1983). This may provide an alternative mechanism to the giant impact for producing a primitive Earth spinning with a period of about 4 hr (*Ringwood,* 1989a). Further detailed investigations of accretion in the presence of nebula gases will be necessary in order to clarify the processes that govern the spins of planets.

Acknowledgments. The author is indebted to Dr. S. E. Kesson for helpful comments on the manuscript.

REFERENCES

Anders E. and Grevesse N. (1989) Abundance of the elements: Meteoritic and solar. *Geochim. Cosmochim. Acta, 53,* 197-214.

Agee C. B. and Walker D. (1988) Mass balance and phase density constraints on early differentiation of chondritic mantle. *Earth Planet. Sci. Lett., 90,* 144-156.

Archibald P. N. (1979) Abundance and dispersions of some compatible and volatile siderophile elements in the mantle. M.Sc. thesis, Australian National University, Canberra.

Benz W., Slattery W., and Cameron A. G. W. (1987) The origin of the Moon and the single impact hypothesis II. *Icarus, 66,* 515-535.

Benz W., Cameron A. G. W., and Melosh H. J. (1989) The origin of the Moon and the single impact hypothesis III. *Icarus, 81,* 113-131.

Birch F. (1965) Speculations on the earth's thermal history. *Bull. Geol. Soc. Am., 76,* 133-154.

Blackshear W. and Gapcynski J. (1977) An improved value for the lunar moment of inertia. *J. Geophys. Res., 82,* 1699-1701.

Boslough M. and Ahrens T. (1983) Shock melting and vaporization of anorthosite and implications for an impact-origin of the Moon (abstract). In *Lunar and Planetary Science XIV,* pp. 63-64. Lunar and Planetary Institute, Houston.

Boss A. P. and Peale S. J. (1986) Dynamical constraints on the origin of the Moon. In *Origin of the Moon* (W. K. Hartmann, R. J. Phillips, and G. J. Taylor, eds.), pp. 59-101. Lunar and Planetary Institute, Houston.

Cameron A. G. W. (1984) Formation of the prelunar accretion disk (abstract). In *Papers Presented to the Conference on Origin of the Moon,* p. 58. Lunar and Planetary Institute, Houston.

Cameron A. G. W. (1986) The impact theory for origin of the Moon. In *Origin of the Moon* (W. K. Hartmann, R. J. Phillips, and G. J. Taylor, eds.), pp. 609-616. Lunar and Planetary Institute, Houston.

Cameron A. G. W. and Benz W. (1988) Effects of the giant impact on the Earth (abstract). In *Papers Presented to the Conference on Origin of the Earth,* pp. 11-12. Lunar and Planetary Institute, Houston.

Cameron A. G. W. and Ward W. (1976) The origin of the Moon (abstract). In *Lunar Science VII,* pp. 120-122. The Lunar Science Institute, Houston.

Carlson R. W. and Lugmair G. W. (1988) The age of ferroan anorthosite 60025: Oldest crust on a young Moon? *Earth Planet. Sci. Lett., 90,* 119-130.

Clayton R. and Mayeda T. (1975) Genetic relationships between the Moon and meteorites. *Proc. Lunar Sci. Conf. 6th,* pp. 1761-1769.

Compston W. and Pidgeon R. T. (1986) Jack Hills, evidence of more very old detrital zircons from Western Australia. *Nature, 321,* 766-769.

Darwin G. H. (1880) On the secular changes in the orbit of a satellite revolving around a tidally disturbed planet. *Philos. Trans. R. Soc. London, 171,* 713-891.

Delano J. W. (1986) Abundances of cobalt, nickel and volatiles in the silicate portion of the Moon. In *Origin of the Moon* (W. K. Hartmann, R. J. Phillips, and G. J. Taylor, eds.), pp. 231-247. Lunar and Planetary Institute, Houston.

Delano J. W. and Lindsley D. (1983) Mare glasses from Apollo 17: Constraints on the Moon's bulk composition. *Proc. Lunar Planet. Sci. Conf. 14th,* in *J. Geophys. Res., 88,* B3-B16.

Delano J. W. and Livi K. (1981) Lunar volcanic glasses and their constraints on mare petrogenesis. *Geochim. Cosmochim. Acta, 45,* 2137-2149.

Delano J. W. and Ringwood A. E. (1978a) Indigenous abundances of siderophile elements in the lunar highlands: Implications for the origin of the Moon. *Moon and Planets, 18,* 385-425.

Delano J. W. and Ringwood A. E. (1978b) Siderophile elements in the lunar highlands: Nature of the indigenous component and implications for origin of the Moon. *Proc. Lunar Planet. Sci. Conf. 9th,* pp. 111-159.

Drake M. J., Newsom H., and Capobianco C. (1989) V, Cr and Mn in the Earth, Moon, EPB and SPB and the origin of the Moon. *Geochim. Cosmochim. Acta, 53,* 2101-2111.

Dreibus G. and Wänke H. (1979) On the chemical composition of the Moon and eucrite parent body and comparison with composition of the Earth: The case of Mn, V and Cr (abstract). In *Lunar and Planetary Science X,* pp. 315-317. Lunar and Planetary Institute, Houston.

Dreibus G. and Wänke H. (1980) The bulk composition of the eucrite parent asteroid and its bearing on planetary evolution. *Z. Naturforsch., 359,* 204-216.

Dreibus G. and Wänke H. (1984) Accretion of the Earth and the inner planets. *Proc. 27th Int. Geol. Congress, 11,* pp. 1-20. VNU Science, Utrecht.

Dreibus G., Spettel B., and Wänke H. (1976) Lithium as a correlated element, its condensation behaviour and its use to estimate the bulk composition of the Moon and of the eucrite parent body. *Proc. Lunar Sci. Conf. 7th,* pp. 3383-3396.

Dreibus G., Palme H., Rammensee W., Spettel B., Weckwerth G., and Wänke H. (1982) Composition of Shergotty parent body: Further evidence of a two component model of planet formation (abstract). In *Lunar and Planetary Science XIII,* pp. 186-187. Lunar and Planetary Institute, Houston.

Elsasser W. M. (1963) Early history of the Earth. In *Earth Science and Meteorites* (J. Geiss and E. Goldberg, eds.), pp. 1-30. North Holland, Amsterdam.

Froude F. O., Ireland T. R., Kinny P., Williams I. L., and Compston W. (1983) Ion microprobe identification of 4100-4200 Myr-old terrestrial zircons. *Nature, 304,* 616-618.

Giuli R. (1986) On the rotation of the Earth produced by gravitational accretion of particles. *Icarus, 8,* 301-323.

Greenberg R., Wacker J., Hartmann W., and Chapman C. (1978) Planetesimals to planets: Numerical simulation of collisional evolution. *Icarus, 35,* 1-26.

Grossman L. and Larimer J. (1974) Early history of the solar system. *Rev. Geophys. Space Phys., 12,* 71-101.

Grossman L. and Olson E. (1974) Origin of the high temperature fraction of C2 chondrites. *Geochim. Cosmochim. Acta, 38,* 173-187.

Grossman L., Clayton R. N., and Mayeda T. (1974) Oxygen isotopic constraints on the composition of the Moon. *Proc. Lunar Sci. Conf. 5th,* pp. 1207-1212.

Hart S. R. and Zindler A. (1986) In search of a bulk-earth composition. *Chem. Geol., 57,* 247-267.

Hartmann W. K. and Davis D. (1975) Satellite-sized planetesimals and lunar origin. *Icarus, 24,* 504-515.

Hartmann W. K. and Vail S. (1986) Giant impactors: Plausible sizes and populations. In *Origin of the Moon* (W. K. Hartmann, R. J. Phillips, and G. J. Taylor, eds.), pp. 551-566. Lunar and Planetary Institute, Houston.

Hayashi C., Nakazawa K., and Nakagawa Y. (1985) Formation of the solar system. In *Protostars and Planets II* (D. C. Black and M. S. Matthews, eds.), pp. 110-153. Univ. of Arizona, Tucson.

Hinton R. W., Clayton R. N., Davis A., and Olsen E. (1988) Isotope mass fractionation of potassium in the Moon (abstract). In *Lunar and Planetary Science XIX,* pp. 497-498. Lunar and Planetary Institute, Houston.

Hood L. and Jones J. (1987) Geophysical constraints on lunar bulk composition and structure: A reassessment. *Proc. Lunar Planet. Sci. Conf. 17th,* in *J. Geophys. Res., 92,* E396-E410.

Irifune T. and Ringwood A. E. (1987) Phase transformations in a harzburgite composition to 26 GPa: Implications for dynamical behaviour of the subducting slab. *Earth Planet. Sci. Lett., 86,* 365-376.

Irifune T., Hibberson W. A., and Ringwood A. E. (1989) Eclogite-garnetite transformation at high pressure and its bearing on the occurrence of garnet inclusions in diamond. In *Kimberlites and Related Rocks, vol. 2: Their Mantle/Crust Setting, Diamonds and Diamond Exploration* (J. Ross et al., eds.), pp. 877-882. Geol. Soc. Aust. Spec. Publ. No. 14, Blackwell.

Ito E. and Takahashi E. (1987) Melting of peridotite under the lower mantle condition. *Nature, 258,* 514-517.

Jagoutz E. and Wänke H. (1982) Has the earth's core grown over geological times? (abstract). In *Lunar and Planetary Science XIII,* pp. 358-359. Lunar and Planetary Institute, Houston.

Jeffreys H. (1959) *The Earth,* 4th edition. Cambridge Univ., New York. 420 pp.

Jones J. H. and Delano J. W. (1989) A three-component model for the bulk composition of the Moon. *Geochim. Cosmochim. Acta, 53,* 513-527.

Jones J. H. and Drake M. J. (1986) Geochemical constraints on core formation in the Earth. *Nature, 322,* 221-228.

Jones J. H. and Hood L. (1990) Does the Moon have the same chemical composition as the Earth's upper mantle? In *Origin of the Earth,* this volume.

Kato T. and Ringwood A. E. (1989a) Melting relationships in the system Fe-FeO at high pressures: Implications for the composition and formation of the Earth's core. *Phys. Chem. Minerals, 16,* 524-538.

Kato T. and Ringwood A. E. (1989b) Was the Moon formed from the mantle of a martian-sized planetesimal? (abstract). In *Lunar and Planetary Science XX,* pp. 510-511. Lunar and Planetary Institute, Houston.

Kato T., Ringwood A. E., and Irifune T. (1988a) Experimental determination of element partitioning between silicate perovskites, garnets and liquids: Constraints on early differentiation of the mantle. *Earth Planet. Sci. Lett., 89,* 123-145.

Kato T., Ringwood A. E., and Irifune T. (1988b) Constraints on element partition coefficients between $MgSiO_3$ perovskite and liquid determined by direct measurements. *Earth Planet. Sci. Lett., 90,* 65-68.

Kaula W. M. (1979) Thermal evolution of Earth and Moon growing by planetesimal impacts. *J. Geophys. Res., 84,* 999-1008.

Kaula W. M. and Bigeleisen P. (1975) Early scattering by Jupiter and its collision effects in the terrestrial zone. *Icarus, 25,* 18-33.

Kesson S. E. and Ringwood A. E. (1977) Further limits of the bulk composition of the Moon. *Proc. Lunar Sci. Conf. 8th,* pp. 411-431.

Kinny P. D. (1987) An ion-microprobe study of U-Pb and Hf isotopes in natural zircons. Ph.D. thesis, Australian National University, Canberra. 128 pp.

Kreutzberger M., Drake M., and Jones J. (1986) Origin of the Earth's Moon: Constraints from alkali volatile trace elements. *Geochim. Cosmochim. Acta, 50,* 91-98.

Kumazawa M. (1981) Origin of materials in the earth's interior and their layered distribution. *Jpn. J. Petrol. Mineral Econ. Geol., Spec. Issue, 3,* 239-247.

Liu L. G. (1980) Phase relationships in the system diopside-jadeite at high pressures and temperature. *Earth Planet. Sci. Lett.,* 47, 398-402.

McCammon E., Jackson I., and Ringwood A. E. (1982) Thermodynamics of the system Fe-FeO at high pressure and temperature and a model for formation of the earth's core. *J. Geophys. Res., 88,* 501-506.

Ma M. S., Liu Y., and Schmitt R. (1981) A chemical study of individual green glasses and brown glasses from 15426: Implications for their petrogenesis. *Proc. Lunar Planet. Sci. 12B,* pp. 915-933.

Melosh H. J. and Kipp M. (1988) Giant impact theory of the Moon's origin: Implications for the thermal state of the Earth (abstract). In *Papers Presented to the Conference on Origin of the Earth,* pp. 57-58. Lunar and Planetary Institute, Houston.

Mueller S., Taylor G., and Phillips R. (1988) Lunar composition: A geophysical and petrological synthesis. *J. Geophys. Res, 93,* 6338-6352.

Nakamura Y. (1983) Seismic velocity structure of the lunar mantle. *J. Geophys. Res., 88,* 677-686.

Nakazawa K., Komuro T., and Hayashi C. (1983) Origin of the Moon: Capture by gas drag of the Earth's primordial atmosphere. *Moon and Planets, 28,* 311-327.

Newsom H. E. (1984) The lunar core and the origin of the Moon. *Eos Trans. AGU, 65,* 369-370.

Newsom H. E. (1989) The nickel content of the lunar core (abstract). In *Lunar and Planetary Science XX,* pp. 784-785. Lunar and Planetary Institute, Houston.

Newsom H. E. and Palme H. (1984) The depletion of siderophile evidence from molybdenum and tungsten. *Earth Planet Sci. Lett., 69,* 354-364.

Newsom H. E. and Taylor S. R. (1989) Geochemical implications of the formation of the Moon by a single great impact. *Nature, 338,* 29-34.

Ohtani E. (1985) The primordial terrestrial magma ocean and its implications for stratification of the mantle. *Earth Planet. Sci. Lett., 78,* 70-80.

Ohtani E., Ringwood A. E., and Hibberson W. (1984) Composition of the core, II. Effect of high pressure on solubility of FeO in the molten iron. *Earth Planet Sci. Lett., 71,* 85-93.

O'Neill H. (1989) Reply to Newsom. *Geochim. Cosmochim. Acta,* in press.

Oversby V. M. and Ringwood A. E. (1971) Time of formation of the Earth's core. *Nature, 234,* 463-465.

Patterson C. (1956) Age of meteorites and the Earth. *Geochim. Cosmochim. Acta, 10,* 230-237.

Patterson C. and Spaute D. (1988) Planetary accretion by runaway growth: Formation of the Earth (abstract). In *Papers Presented to the Conference on Origin of the Earth,* p. 67. Lunar and Planetary Institute, Houston.

Rammensee W. and Wänke H. (1977) On the partition coefficient of tungsten between metal and silicate and its bearing on the origin of the Moon. *Proc. Lunar Sci. Conf. 8th,* pp. 399-409.

Rietmeijer F. (1987) Chondritic interplanetary dust and primitive chondrite matrices: The search for chemically pristine solids in the solar system (abstract). In *Lunar and Planetary Science XVIII,* pp. 832-834. Lunar and Planetary Institute, Houston.

Rietmeijer F. (1988) On a chemical continuum in early solar system dust at >1.8 A.U. (abstract). *Chem. Geol., 70,* 33.

Ringwood A. E. (1960) Some aspects of the thermal evolution of the Earth. *Geochim. Cosmochim. Acta, 20,* 241-259.

Ringwood A. E. (1966a) The chemical composition and origin of the Earth. In *Advances in Earth*

Sciences (P. M. Hurley, ed.), pp. 287-356. MIT, Cambridge.

Ringwood A. E. (1966b) Chemical evolution of the terrestrial planets. *Geochim. Cosmochim. Acta, 30,* 41-104.

Ringwood A. E. (1970) Origin of the Moon: The precipitation hypothesis. *Earth Planet. Sci. Lett., 8,* 131-140.

Ringwood A. E. (1972) Some comparative aspects of lunar origin. *Phys. Earth Planet. Inter., 6,* 366-376.

Ringwood A. E. (1976) Limits on the bulk composition of the Moon. *Icarus, 28,* 325-349.

Ringwood A. E. (1977) Composition of the core and implications for origin of the Earth. *Geochem. J., 11,* 111-135.

Ringwood A. E. (1979) *Origin of the Earth and Moon.* Springer-Verlag, New York. 295 pp.

Ringwood A. E. (1984) The Earth's core: Its composition, formation and bearing upon the origin of the Earth. *Proc. R. Soc. London, A395,* 1-46.

Ringwood A. E. (1986a) Composition and origin of the Moon. In *Origin of the Moon* (W. K. Hartmann, R. J. Phillips, and G. J. Taylor, eds.), pp. 673-698. Lunar and Planetary Institute, Houston.

Ringwood A. E. (1986b) Terrestrial origin of the Moon. *Nature, 322,* 323-328.

Ringwood A. E. (1989a) The Earth-Moon connection. *Z. Naturforsch, 44a,* 891-923.

Ringwood A. E. (1989b) Significance of the terrestrial Mg/Si ratio. *Earth Planet Sci. Lett., 95,* 1-7.

Ringwood A. E. (1989c) Flaws in the giant impact hypothesis of lunar origin. *Earth Planet. Sci. Lett., 95,* 208-214.

Ringwood A. E. and Irifune T. (1987) Nature of the 650 km discontinuity: Implications for mantle dynamics. *Nature, 331,* 131-136.

Ringwood A. E. and Kesson S. E. (1976a) A dynamic model for mare basalt petrogenesis. *Proc. Lunar Sci. Conf. 7th,* pp. 1697-1722.

Ringwood A. E. and Kesson S. E. (1976b) *Basaltic Magmatism and the Composition of the Moon, Part II: Siderophile and Volatile Elements in Moon, Earth and Chondrites. Implications for Lunar Origin.* Publication 1221, Research School of Earth Sciences, Australian National University, Canberra. 46 pp.

Ringwood A. E. and Kesson S. E. (1977) Basaltic magmatism and the bulk composition of the Moon II. Siderophile and volatile elements in Moon, Earth and chondrites: Implications for lunar origin. *The Moon, 16,* 425-464.

Ringwood A. E. and Seifert S. (1986) Nickel-cobalt abundance systematics and their bearing on lunar origin. In *Origin of the Moon* (W. K. Hartmann, R. J. Phillips, and G. J. Taylor, eds.), pp. 249-278. Lunar and Planetary Institute, Houston.

Ringwood A. E. and Seifert S. (1988) Lunar siderophile signature and its genetic significance (abstract). In *Lunar and Planetary Science IX,* pp. 984-985. Lunar and Planetary Institute, Houston.

Ringwood A. E., Seifert S., and Wänke H. (1987) A komatiite component in Apollo 16 highland breccias: Implications for the nickel-cobalt systematics and bulk composition of the Moon. *Earth Planet. Sci. Lett., 81,* 105-117.

Ringwood A. E., Hibberson W., Ware N., and Kato T. (1989) High pressure geochemistry of Cr, V and Mn: Implications for the formation of the Earth's core and origin of the Moon. *Eos Trans. AGU, 70,* 1419.

Safronov V. S. (1969) *Evolution of the Protoplanetary Cloud and Formation of the Earth and Planets.* Nauka, Moscow. Translated by the Israel Program for Scientific Translation (1972).

Scarfe C. M. and Takahashi E. (1986) Melting of garnet peridotite to 13 GPa and the early history of the upper mantle. *Nature, 322,* 354-356.

Seifert S. and Ringwood A. E. (1987) Metal-silicate partition coefficients for some volatile siderophile elements and implications for lunar origin (abstract). In *Lunar and Planetary Science XVIII,* pp. 904-905. Lunar and Planetary Institute, Houston.

Seifert S. and Ringwood A. E. (1988) The lunar geochemistry of chromium and vanadium. *Earth, Moon and Planets, 40,* 45-70.

Seifert S., O'Neill H., and Brey G. (1988) The partitioning of Fe, Ni and Co between olivine, metal and basaltic liquid: An experimental and thermodynamic investigation with application to the composition of the lunar core. *Geochim. Cosmochim. Acta, 52,* 603-616.

Stevenson D. J. (1981) Models of the Earth's core. *Science, 214,* 611-619.

Stevenson D. J. (1987) Origin of the Moon: The collision hypothesis. *Annu. Rev. Earth Planet. Sci., 15,* 271-315.

Stosch H. (1981) Sc, Cr, Co and Ni partitioning between minerals from spinel peridotite xenoliths. *Contrib. Mineral. Petrol., 78,* 166-174.

Sun S. (1982) Chemical composition and origin of the Earth's primitive mantle. *Geochim. Cosmochim. Acta, 46,* 179-192.

Takahashi E. (1986) Melting of a dry peridotite KLB-1 up to 14 GPa: Implications on the origin of peridotitic upper mantle. *J. Geophys. Res., 91,* 9367-9382.

Taylor S. R. (1982) *Planetary Science: A Lunar Perspective.* Lunar and Planetary Institute, Houston. 481 pp.

Taylor S. R. (1986) The origin of the Moon: Geochemical considerations. In *Origin of the Moon* (W. K. Hartmann, R. J. Phillips, and G. J. Taylor, eds.), pp. 125-143. Lunar and Planetary Institute, Houston.

Treiman A., Drake M., Janssens M., Wolf R., and Ebihara M. (1986) Core formation in the Earth and shergottite parent body (SPB): Chemical evidence from basalts. *Geochim. Cosmochim. Acta, 50,* 1071-1091.

Turekian K. and Clark S. P. (1969) Inhomogeneous accumulation of the Earth from the primitive solar nebula. *Earth Planet. Sci. Lett., 6,* 346-348.

Urakawa S., Kato M., and Kumazawa M. (1987) Experimental study on the phase relations in the system Fe-Ni-O-S up to 15 GPa. In *High Pressure Research in Mineral Physics* (M. H. Manghnani and Y. Syono, eds.), pp. 95-111. AGU, Washington, D.C.

Urey H. C. (1952) *The Planets.* Yale Univ., New Haven. 283 pp.

Urey H. C. (1962) Evidence regarding the origin of the Earth. *Geochim. Cosmochim. Acta., 26,* 1-13.

Wänke H. (1981) Constitution of terrestrial planets. *Philos. Trans. R. Soc. London, A303,* 287-302.

Wänke H. (1987) Chemistry and accretion of the Earth and Mars. *Bull. Soc. Geol. France, (8)t.III,* 13-19.

Wänke H. and Dreibus G. (1982) Chemical and isotopic evidence for the early history of the Earth-Moon system. In *Tidal Friction and the Earth's Rotation II* (P. Brosche and J. Sündermann, eds.), pp. 322-677. Springer-Verlag, Berlin.

Wänke H. and Dreibus G. (1986) Geochemical evidence for the formation of the Moon by impact-induced fission of the proto-Earth. In *Origin of the Moon* (W. K. Hartmann, R. J. Phillips, and G. J. Taylor, eds.), pp. 649-672. Lunar and Planetary Institute, Houston.

Wänke H. and Dreibus G. (1988) Chemical composition and accretion history of terrestrial planets. *Philos. Trans. R. Soc. London, A325,* 545-557.

Wänke H., Baddenhausen H., Blum K., Cendales M., Dreibus G., Hofmeister H., Kruse H., Jagoutz E., Palme C., Spettel B., Thacker R., and Vilsek E. (1977) On the chemistry of lunar samples and achondrites. Primary matter in the lunar highlands: A re-evaluation. *Proc. Lunar Sci. Conf. 8th,* pp. 2191-2213.

Wänke H., Dreibus G., and Palme H. (1978) Primary matter in the lunar highlands: The case of the siderophile elements. *Proc. Lunar Planet. Sci. Conf. 9th,* pp. 83-110.

Wänke H., Dreibus G., and Palme H. (1979) Nonmeteoritic siderophile elements in the lunar highland rocks: Evidence from pristine rocks. *Proc. Lunar Planet. Sci. Conf. 10th,* pp. 611-626.

Wänke H., Dreibus G., and Jagoutz E. (1984) Mantle chemistry and accretion history of the Earth. In *Archaean Geochemistry* (A. Kröner, G. Hanson, and A. Goodwin, eds.), pp. 1-24. Springer-Verlag, Berlin.

Ward W. (1988) Density waves and planetesimal dispersion velocities (abstract). In *Papers Presented to the Conference on Origin of the Earth,* pp. 96-97. Lunar and Planetary Institute, Houston.

Warren P. H. and Rasmussen K. (1987) Megaregolith insulation, internal temperatures and bulk uranium content of the Moon. *J. Geophys. Res., 92,* 3453-3465.

Weidenschilling S. and Davis D. (1985) Orbital resources in the solar nebula: Implications for planetary accretion. *Icarus, 62,* 16-29.

Weidenschilling S., Greenberg R., Chapman C., Herbert F., Davis D., Drake M., Jones J., and Hartmann W. (1986) Origin of the Moon from a circumterrestrial disk. In *Origin of the Moon* (W. K. Hartmann, R. J. Phillips, and G. J. Taylor, eds.), pp. 731-762. Lunar and Planetary Institute, Houston.

Wetherill G. W. (1985) Occurrences of giant impacts during the growth of the terrestrial planets. *Science, 228,* 877-879.

Wetherill G. W. (1986) Accumulation of the terrestrial planets and implications concerning lunar origin. In *Origin of the Moon* (W. K. Hartmann, R. J. Phillips, and G. J. Taylor, eds.), pp. 519-550. Lunar and Planetary Institute, Houston.

Wolf R. and Anders E. (1980) Moon and Earth: Compositional differences inferred from siderophiles, volatiles and alkalis in basalts. *Geochim. Cosmochim. Acta, 44,* 2111-2124.

Yoder C. (1984) The size of the lunar core (abstract). In *Papers Presented to the Conference on Origin of the Moon,* p. 6. Lunar and Planetary Institute, Houston.

ELEMENT PARTITIONING AND THE EARLY THERMAL HISTORY OF THE EARTH

E. A. McFarlane and M. J. Drake

Lunar and Planetary Laboratory, University of Arizona, Tucson, AZ 85721

The partitioning of Ni, Co, Sc, and La between olivine and natural basaltic melt and between various subsolidus phases has been determined at 1800°C and 75 kbar. Aliquots of the mantle composition material KLB-1 were doped with 1-2 wt.% each of Ni, Co, and Sc, were compressed to high pressures, and heated in a uniaxial split-sphere anvil apparatus for approximately 1 hr. Successful run products typically consist of a subsolidus assemblage of olivine, orthopyroxene, clinopyroxene, spinel, and probably garnet at the cold end, and silicate melt containing quench crystals of olivine at the hot end. The liquidus boundary within the charge is defined by the appearance of sizable equant olivine crystals (instead of quench-textured olivine crystals, which are smaller and more elongate). Olivine/melt partition coefficients (D) at 75 kbar and 1800°C, rounded to one significant figure, are D(Ni) = 2, D(Co) = 1, D(Sc) =0.1, and D(La) < 0.007. These partition coefficients may be used to test the hypothesis that the high Mg/Si ratio in the upper mantle of the Earth relative to most chondritic meteorites results from the floating of olivine in a magma ocean, with subsequent mixing of that olivine into the upper mantle of the Earth. For example, the Ni/Co ratio inferred for the upper mantle is approximately chondritic. The experimentally determined partition coefficients imply that the addition of 30% olivine into the upper mantle to raise the Mg/Si ratio from CI chondritic to its present value yields a Ni/Co ratio 20-25% higher than its initial value. This result is inconsistent with the olivine flotation hypothesis as a means of explaining the elevated Mg/Si ratio of the upper mantle. The implication of these experiments and those of Kato et al. (1987, 1988a,b) is that minor and trace element abundances and ratios in the upper mantle of the Earth do not presently show the effects of extensive olivine, majorite garnet, or perovskite fractionation. One possibility is that the Earth was never substantially molten. If so, the accretional process must have delivered gravitational potential energy more slowly than current theory predicts, and an origin of the Moon in a giant impact would be unlikely. Alternatively, if the Earth were indeed substantially molten, then it is possible that minerals remained entrained in magma and were unable to segregate. In either case, the high Mg/Si ratio in the Earth relative to most classes of chondrites would be intrinsic to the Earth, implying that the accretional process did not mix material efficiently between 1 A.U. and 2-4 A.U. where most chondritic meteorites are presumed to originate.

INTRODUCTION

Theories of the formation of the Earth strongly suggest that the Earth should have been substantially molten at some point in the past. There are many potential heat sources for melting some or all of the Earth, including extinct radioactive elements, gravitational energy from core formation, and gravitational energy from accretion. Some models of accretional history strongly suggest that the Earth should have been substantially molten during and immediately after accumulation from planetesimals (e.g., *Wetherill,* 1985). Monte Carlo simulations of planetesimal accretion in the absence of gas drag imply that toward the end of the accretional process, most of the accreting mass may reside in larger bodies (approximately three times the mass of Mars) that have relatively large relative velocities (approximately 9 km/sec). The kinetic energy resulting from accretion is several times greater than the energy required to melt the entire Earth (*Wetherill,* 1985). The fact that the Earth's equatorial plane and the Moon's orbital plane are offset by nearly 28° may also imply that the Earth was struck by fairly large planetesimals.

Experimental determinations of phase equilibria may also imply that the Earth was at least partially molten at some point in its history, producing a magma ocean. When a multicomponent system is melted there is one composition, either a peritectic or a eutectic composition, at which melting will begin. On a phase diagram, the peritectic and eutectic compositions are points at which the solidus and liquidus converge. *Takahashi* (1986) has shown that the liquidus and solidus for a composition representative of the upper mantle of the Earth (KLB-1) converge at 170 kbar and 2000°C. Partial melting of a primitive chondritic Earth at depth could possibly yield a peridotitic upper mantle composition. The convergence of the liquidus and solidus for KLB-1 is consistent with, although does not require, partial melting of a more primitive composition.

The composition of the primitive upper mantle presumably holds clues as to whether or not the Earth was ever substantially molten. There are potentially two sources of information for the composition of the primitive upper mantle: partial melts from the mantle (as represented by basalts) and xenoliths, which are assumed to directly represent mantle material. Attempts to determine the composition of the primitive upper mantle have included the use of mantle nodules (e.g., *Jagoutz et al.,* 1979), basaltic magmas (e.g., *Chou,* 1978; *Chou et al.,* 1983), and an approach combining evidence including mantle nodules and elemental ratios (e.g., *Palme and Nickel,* 1985).

An assumption inherent to the mantle nodule approach is that mantle nodules can be identified that closely represent the composition of the primitive upper mantle, i.e., that have not had basalt extracted from them and have not been significantly altered by metasomatism or during their ascent toward the surface. For instance, *Jagoutz et al.* (1979) used spinel lherzolite xenoliths, which are similar world-wide in bulk and mineral chemistry and which were identified as "primitive" on the basis of their complements of highly incompatible lithophile elements, to determine the composition of the upper mantle.

The use of basalts to determine the composition of the upper mantle of the Earth involves the model assumption that magmas, which are derived from the mantle via partial melting, yield accurate compositional information about their source. *Palme and Nickel* (1985) assumed chondritic ratios among refractory elements in the bulk Earth, meaning that these elements have abundance ratios characteristic of primitive chondritic meteorites that are thought to have undergone almost no chemical processing. They also used spinel lherzolite data and several other lines of evidence to calculate an inferred composition for the primitive upper mantle. Regardless of the method used, estimates of the composition of the primitive upper mantle indicate that many elements are present at approximately chondritic ratios. For example, refractory lithophile elements such as Sc and the REE are present at approximately CI chondritic ratios in approximately CI chondritic concentrations. The moderately siderophile elements Ni and Co are present at approximately the CI chondritic ratio, but are depleted relative to chondritic abundances. Highly siderophile elements such as the noble metals are also present at approximately CI chondritic ratios, but at less than 1% of chondritic abundances. These and other elemental abundances are illustrated in Fig. 1.

The origin of chondritic ratios, particularly of the siderophile elements, in the primitive mantle following core formation is a subject of debate (see *Jones and Drake,* 1986; *Newsom,* 1990, for a discussion). Nevertheless, their existence provides strict constraints upon substantial early Earth melting events.

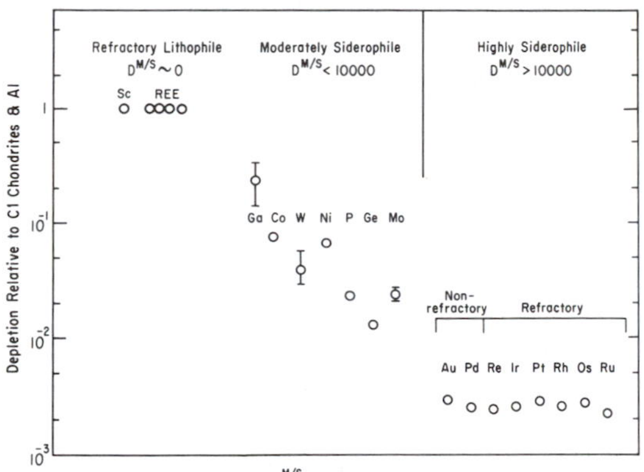

Fig. 1. *Abundances of various elements in the upper mantle of the Earth, normalized to CI chondrites and Al. Elements are arranged in order of increasingly siderophile behavior. Both Ir and Au lie on a horizontal line, and therefore the Ir/Au ratio is chondritic. Likewise, the Ni/Co ratio and the Sc/REE ratio are also chondritic. After Newsom and Palme, 1984.*

For example, *Kato et al.* (1987, 1988a,b) address fractional crystallization of the high-pressure minerals Mg-perovskite and majorite garnet, minerals that are expected to be on the liquidus as the result of substantial amounts of melting. Kato et al. examine in particular the effects that such mineral fractionations will have upon elemental ratios.

Mineral/melt partition coefficient values (a partition coefficient is the ratio of the concentration of an element in one phase vs. the concentration of the same element in another phase) for these phases at elevated temperatures and pressures seem to contradict the notion that the Earth was at one time substantially molten. Despite accretional and experimental petrological evidence in support of such melting, experimental measurements of element partition coefficients show that segregation of Mg-perovskite and majorite garnet would fractionate the ratios of certain elements away from the chondritic values that are currently observed (*Kato et al.,* 1987, 1988a,b). It should be noted that the work of Kato et al. has engendered controversy (*Agee and Walker,* 1989; *Kato et al.,* 1989), and not all experimental ratios (e.g., Lu/Hf) claimed by them in the earlier papers to be inconsistent with Mg-perovskite fractionation are actually inconsistent. However, we believe the general conclusions of Kato et al. to be robust.

While there is some debate about the value of certain element ratios in the upper mantle of the Earth, for instance the Ca/Al ratio (*Palme and Nickel,* 1985; *Kato et al.,* 1987, 1988a,b), there is consensus that the Mg/Si ratio inferred for the upper mantle of the Earth is higher than in most chondritic meteorites (Fig. 2). The class of primitive meteorites with a Mg/Si ratio most similar to the upper mantle is the C3V carbonaceous chondrites. These meteorites are not good candidates for proto-Earth matter, however, as their Ir/Au ratio is almost twice as high as both the CI ratio (Fig. 3) and estimates for the upper mantle of the Earth (Fig. 1) (*Kallemeyn and Wasson,* 1981). Oxygen isotopes also imply that C3V chondrites are not good candidates for proto-Earth material since C3V chondrites lie off the

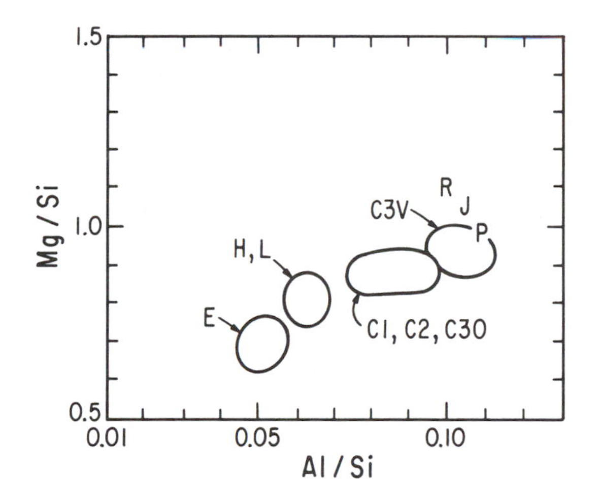

Fig. 2. *The Mg/Si ratio vs. Al/Si ratio for different types of undifferentiated meteorites. Also included are three estimates of the upper mantle of the Earth: R refers to pyrolite (Ringwood, 1979); J refers to the preferred composition of Jagoutz et al. (1979); P refers to the preferred composition of Palme and Nickel (1985). After Jagoutz et al. (1979).*

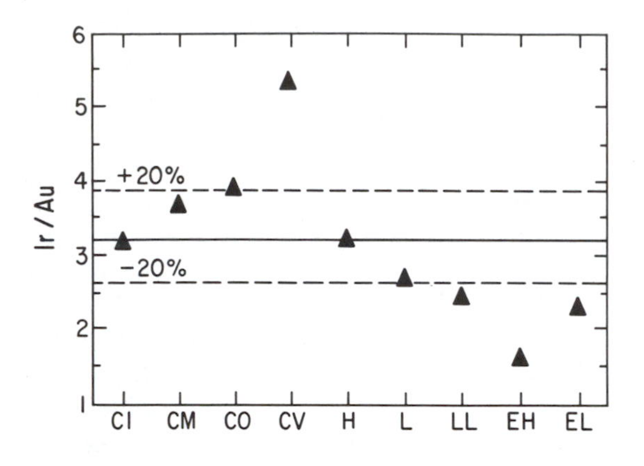

Fig. 3. *The Ir/Au ratio in various types of chondritic meteorite. While the Mg/Si ratio of CV chondrites is similar to the Mg/Si ratio of the Earth, the Ir/Au ratio in CV chondrites is substantially higher than the CI ratio. The Ir/Au ratio in the upper mantle of the Earth is CI chondritic (Fig. 1), making CV chondrites a poor candidate for proto-Earth material.*

terrestrial fractionation line and are, therefore, from a material reservoir distinct from proto-Earth material (*Clayton et al.,* 1976).

Agee and Walker (1988) have offered an innovative proposal to produce a high Mg/Si ratio from material that is initially CI chondritic. The basis of their proposal is a careful experimental study that shows that olivine becomes neutrally buoyant in a melt representative of upper mantle composition at about 80 kbar. This experimental observation has led them to propose that the high Mg/Si ratio in the upper mantle results from mixing into the upper mantle of up to 30% olivine after solidification of a terrestrial magma ocean. This olivine crystallized from a magma ocean that was initially continuous in depth and formed a neutrally buoyant septum dividing the molten outer part of the Earth into two separated oceans (Fig. 4).

This septum is less dense than both the compressed liquid below and the subsequently crystallized Fe enriched garnet and pyroxene solids above. This density inversion causes Rayleigh-Taylor instabilities to develop, resulting in convection that mixes the olivine septum into the upper layer, leaving the material below the septum largely unaffected. The result of this convective mixing is a homogeneous upper mantle with a Mg/Si ratio greater than found in any class of chondritic meteorite except, possibly, the C3V chondrites. *Agee and Walker* (1988) also call for segregation of Mg-perovskite at depth into the lower mantle, noting that estimates of upper mantle bulk major element composition may be related to chondritic composition by addition of olivine and subtraction of Mg-perovskite.

The possibility of olivine flotation into the upper mantle may be tested against olivine/melt partition coefficients for elements present in the upper mantle in chondritic ratios (*Drake,* 1989). For example, Sc and La are present at CI chondritic ratios and CI chondritic abundance levels, while Ni and Co are present at approximately CI chondritic ratios at about 20% of CI chondritic abundance levels (Fig. 1). If the elements in each pair have sufficiently different partition coefficient values, in principle their approximately chondritic ratios may be used to limit possible olivine addition into the upper mantle. This contribution focuses on the partitioning of Ni, Co, Sc, La, and other elements between olivine and natural basaltic melt at 75 kbar and 1800°C, a temperature and pressure relevant to neutral buoyancy of olivine in a melt of upper mantle composition.

EXPERIMENTAL AND ANALYTICAL PROCEDURES

All experiments were run in a uniaxial split-sphere anvil high-pressure apparatus (USSA-2000) (see *Gasparik,* 1989) at SUNY Stony Brook. The initial starting material used for all experimental runs is KLB-1 (*Takahashi,* 1986), which represents undepleted mantle composition (kindly provided by E. Takahashi). Aliquots of KLB-1 were ground and doped with 1-2 wt.% each of Ni, Co, and Sc. The charge was then packed into a graphite container (approximately 3 mm in length, with a 1.5 mm inner diameter) and capped with a graphite lid. The container has an approximate volume of 5 mm^3.

The graphite sample container is then placed atop an alumina ceramic cap within a lanthanum chromate ($LaCrO_3$) sleeve that serves as the furnace. The graphite sample container also serves as an auxiliary heater, in addition to the $LaCrO_3$, until the graphite transforms to diamond at high pressure. We note that *Takahashi* (1986) found that a simple $LaCrO_3$ heater is more stable than a composite heater at temperatures between 700°C and 2000°C, but our composite furnace assembly did not exhibit instability. The $LaCrO_3$ sleeve is contained within a zirconia sleeve, which provides thermal insulation. Both sleeves reside within a standard 10-mm MgO octahedron (see Fig. 5 for a cross-sectional view).

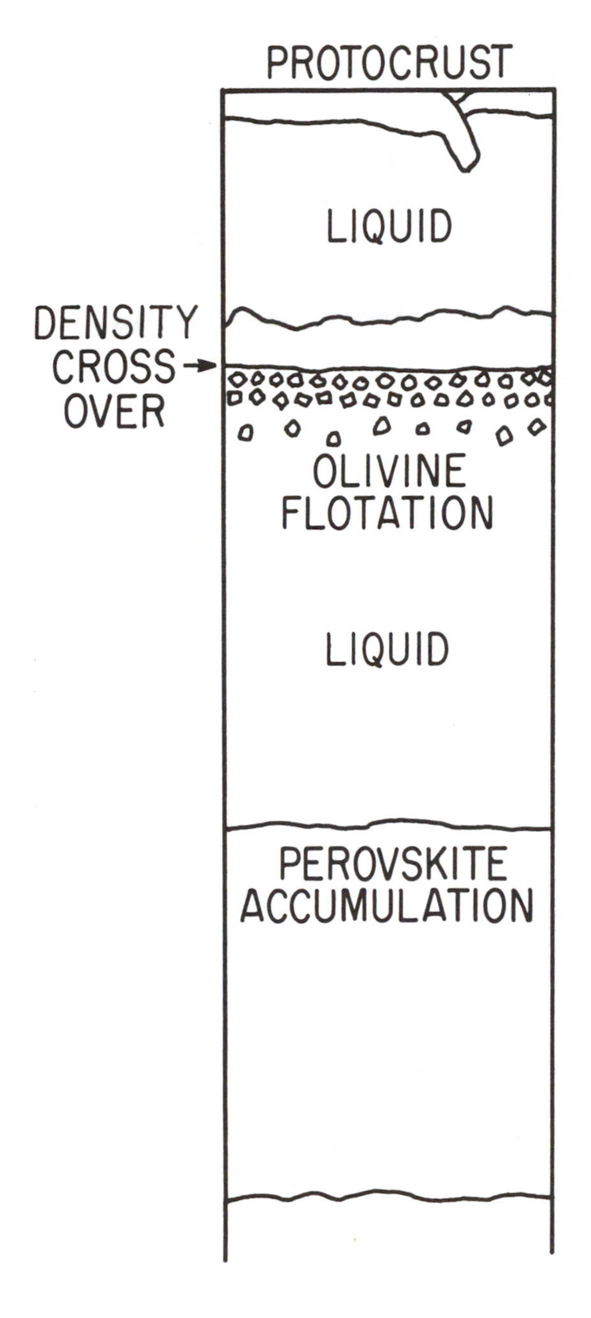

Above and below the $LaCrO_3$ furnace sit two Mo rings, which comprise part of the heating circuit, and help to improve electrical contact. Current flows from one ring through the $LaCrO_3$ sleeve and out the other ring to heat the sample. Temperatures are monitored with W-Re thermocouples, composed of W_3Re and $W_{25}Re$. The wires are held in place with a two-hole alumina plug that sits atop the graphite sample container. The wires are electrically insulated from the heating circuit with alumina ceramic tubes placed in grooves carved in one face of the MgO octahedron. The assembly is cemented closed, and any excess space is filled with zirconia cement.

Initially, a Mo disc was placed between the thermocouple wires and the graphite bottom of the sample container in order to maintain an electric current through the thermocouple wires in the event of graphite transformation to diamond. A possible reaction between the W-Re thermocouple, the Mo lid, and the graphite capsule may have been responsible for thermocouple failure during early runs. For later experimental runs, the assembly design was modified so that the thermocouple wires were welded together and the Mo disc was no longer necessary.

Fig. 4. *Schematic cross-section of the mantle of the Earth after accretion is largely complete, and heat loss from the Earth by radiation to space exceeds heat input (modified from Agee and Walker, 1988).*

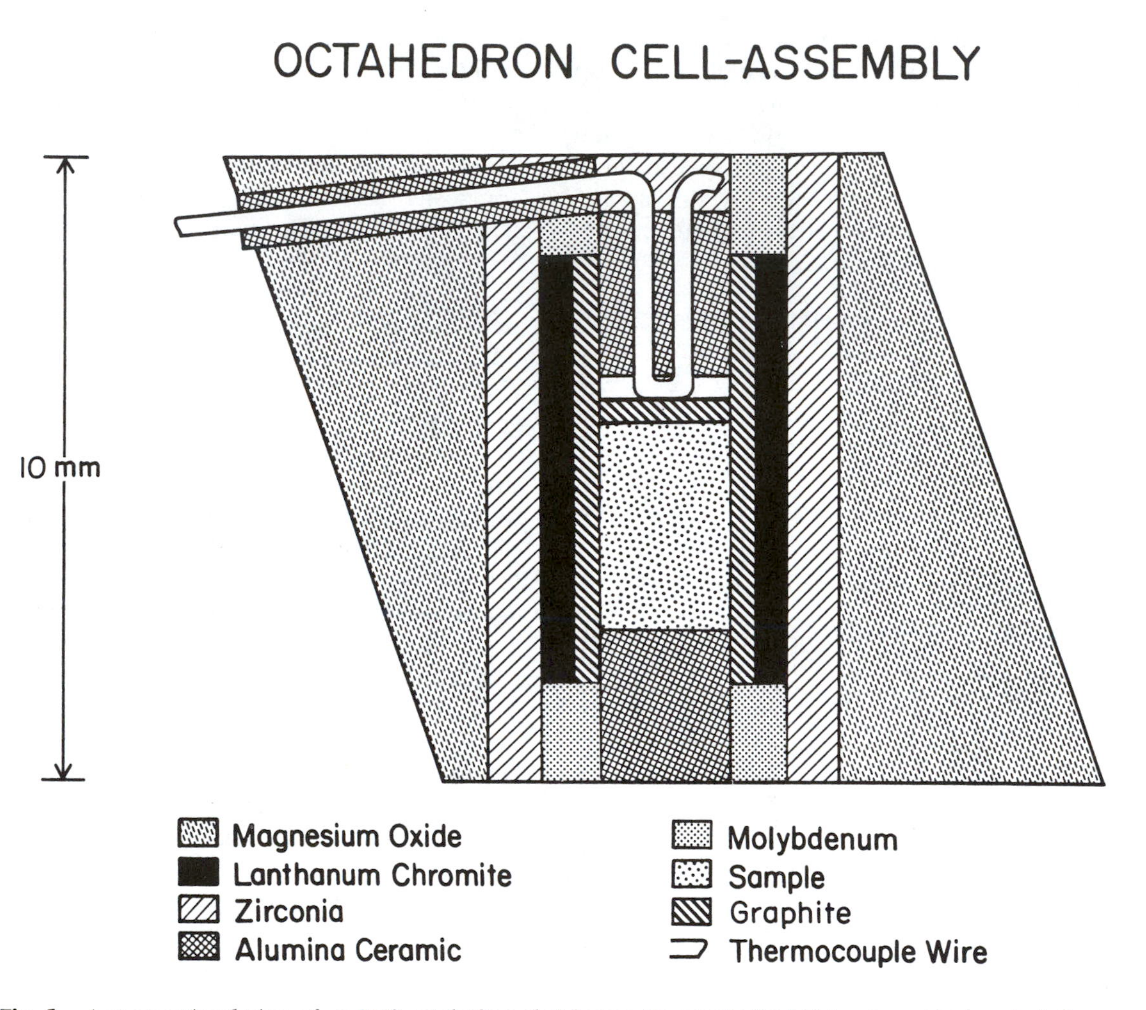

Fig. 5. *A cross-sectional view of an MgO octahedron that houses the sample for high-pressure experiments (after Gasparik, 1989; see text for details). Compositions of the components are noted within the figure.*

The sample assembly is heated in a furnace flushed continuously with Ar for 1 hr at ~1100°C to remove water immediately prior to being subjected to elevated pressures and temperatures. The MgO octahedron is then surrounded by eight tungsten-carbide cubes. Each cube is 32 mm on edge and is truncated at one corner. The cubes are separated from one another by pyrophyllite gaskets, teflon back-up gaskets, and balsa wood spacers, and are electrically insulated from one another by teflon tape. The cubes housing the octahedron are inserted into a uniaxial split-sphere cubic anvil apparatus (USSA-2000) and submitted to high pressure [see *Gasparik* (1989) for a detailed description of the pressure calibration].

Full pressure is applied cold, and the temperature is subsequently raised. All experimental runs were compressed to 75 kbar and heated to approximately 1800°C. A temperature gradient of about 200°C exists from the cold to the hot end of each charge (*Gasparik,* 1989). During the experiments, La and Cr from the furnace contaminated the charges. The longest run time was approximately 1 hr. Charges were quenched by turning off power to the $LaCrO_3$ furnace. Pressure was released gradually over a period of several hours to avoid gasket blowout. After the sample assembly is removed from the high pressure apparatus, the MgO octahedron is split along the axis of the heater. Samples can be fragile and easily crumbled, but it is not necessary to separate the sample container from the octahedron. The sample, with or without any part of the octahedron attached, is then mounted in epoxy, ground, and polished.

Diamond was often present in the charges, making grinding and polishing a formidable task. Protruding diamonds were manually ground down using diamond dental drills. Once the diamond was recessed, normal grinding procedures on the lapidary were used. The process was rather painstaking and slow, and oxidizing the diamond by heating the sample in a furnace may be a preferable approach. Each polished thick section was analyzed by electron microprobe at an accelerating voltage of 15 KeV and a beam current of approximately 40 namps. A square raster 100 μm on a side was typically used to analyze the quench melt region. A finely focused beam was used for the crystalline phases.

RESULTS

Nine experimental runs were attempted, but only one yielded olivine/melt partition coefficient values. Experiments #1, 2, 3, 4, 6, and 8 were doped with the trace elements Ni, Co, and Ga. Experiments #5 and 9 were doped with Ni, Co, and Sc. Experiment #7 was doped with P, Mo, and Ge. In experiments #1, 2, and 3 the thermocouple failed (possibly due to a postulated reaction between the W-Re, Mo, and graphite). Experiment #4 also had no thermocouple reading. Experiment #5 was run at a nominal temperature of 1800°C (32.62 mV set point), and was successful. A photomicrograph of experiment #5 is shown in Fig. 6. The experiment was extremely stable, and could have lasted longer than the run time of 1 hr were it not for other time constraints.

Experiment #6, run at a nominal temperature of 1850°C (33.52 mV set point), yielded only melt, and experiment #7, run at a nominal temperature of 1800°C, has not yet been successfully polished. Experiment #8 was also held at 1800°C for 1 hr. No quenched

Fig. 6. *Results for experiment #5 (sample R5), which ran at 75 kbar and 1800°C for approximately 1 hr. Regions of melt are at the far right and a subsolidus assemblage is to the left. There is a gap between the subsolidus and the melt. Olivine/melt partition coefficient values (Table 1) are from a region in one corner of the melt where several olivine grains reside immediately adjacent to the melt. Partition coefficient values for solid phases are in Table 2. The sample is approximately 3 mm long.*

Fig. 7. *Experiment #8 (sample R8), which ran at 75 kbar and 1800° C for approximately 1 hr. There is no melt in this cross-section of the charge. Partition coefficient values for solid phases are in Table 2. Results for Run #8 correspond fairly well to those of Run #5. The sample is approximately 3 mm long.*

melt was recovered from experiment #8, possibly because of loss during polishing, and only subsolidus phase partition coefficient values could be obtained. A photomicrograph of experiment #8 is shown in Fig. 7. Experiment #9 was run at a nominal temperature of 1825°C (33.07 mV set point), and yielded only melt.

Successful run products typically consist of a subsolidus assemblage of olivine, orthopyroxene, clinopyroxene, spinel, and presumably garnet at the cold end (although the presence of garnet has not been confirmed), and in experiment #5 silicate melt containing quench crystals of olivine at the hot end. The liquidus is marked by the appearance of sizable equant olivine crystals (instead of smaller more elongate olivine crystals that typically form from a previously molten region upon rapid cooling). Olivine grains were generally ~15-50 μm in diameter. A metallic phase also nucleated.

All the experiments run thus far are synthesis experiments. Until reversals have been conducted, formal demonstration of how closely these experiments approached equilibrium is not possible. Homogeneity of phase compositions in the same region of the charge may indicate an approach to local equilibrium. Analyses of olivine and immediately adjacent melt for experiment #5 are given in Table 1, together with olivine/melt partition coefficient values. Lanthanum and some of the Cr are the result of contamination from the furnace. Whether this contamination occurred prior to nucleation of the olivine crystals or throughout the run is unknown. Thus, olivine/melt partition coefficients for La and Cr should be considered lower limits as it has not been demonstrated that the run duration was sufficient to achieve diffusional equilibrium for these elements in the crystalline phases. The melt also presumably contains CO_2 derived from the graphite container, accounting for the low totals.

Olivine/melt partition coefficients (D) at 75 kbar and 1800°C, rounded to one significant figure are D(Ni) = 2, D(Co) = 1, D(Sc) = 0.1, and D(La) < 0.007. The partition coefficient for Ni is identical to that estimated by *Drake* (1989) on the basis of extrapolation of trends from experiments conducted at lower temperatures and pressures. The partition coefficient of unity for Co is higher than the value of 0.6 estimated by *Drake* (1989). The value for Sc of 0.1 is lower than the value of about 0.3 used by *Drake* (1989) and derived from experiments conducted at one bar in the 1100°C to 1250°C temperature range. Lanthanum is highly incompatible as expected.

Average analyses of subsolidus phases are given in Table 2 for experiment #5 (sample R5) and experiment #8 (sample R8). Partition coefficients between olivine and orthopyroxene, orthopyroxene and spinel, orthopyroxene and clinopyroxene, clinopyroxene and spinel, olivine and clinopyroxene, and olivine and spinel are given in Table 3. Because of the temperature gradient across the charge, representative analyses at three locations from the cold end of the charge to close to the liquidus (termed hot) are given in Tables 2 and 3. Thus far, none of the partition coefficient values for any of the elements or phase pairs included in Table 3 has exhibited significant systematic variations from one end of the charge to the other, although variations in phase composition do occur.

DISCUSSION

Abundance Ratio of Ni/Co in the Upper Mantle

Mantle nodules exhibit a variety of Ni/Co ratios, as shown in Fig. 8. Figure 9 shows Ni/Co ratios plotted against La/Sm ratios. *Jagoutz et al.* (1979) argued that those nodules with the most chondritic light rare Earth element patterns are most likely to be primitive. These nodules are plotted in Fig. 8 at 28% olivine addition for purposes of clarity. All show a sub-CI chondritic Ni/Co ratio of ~0.9. While some nodules do have a Ni/Co ratio that plots closer to or above the CI chondrite line, Fig. 9 shows that all such nodules have an elevated La/Sm ratio, and are probably not as primitive as those nodules that exhibit flat REE patterns (*Jagoutz et al.,* 1979).

Note that other classes of chondritic meteorites, plotted arbitrarily at 20% addition of olivine for purposes of clarity, have Ni/Co ratios of 0.95× CI or greater, with the exception of the EL chondrites. We conclude in concurrence with *Jagoutz et al.* (1979) that the Ni/Co ratio of the upper mantle of the Earth is less than CI-chondritic, and is probably close to 0.9 × CI.

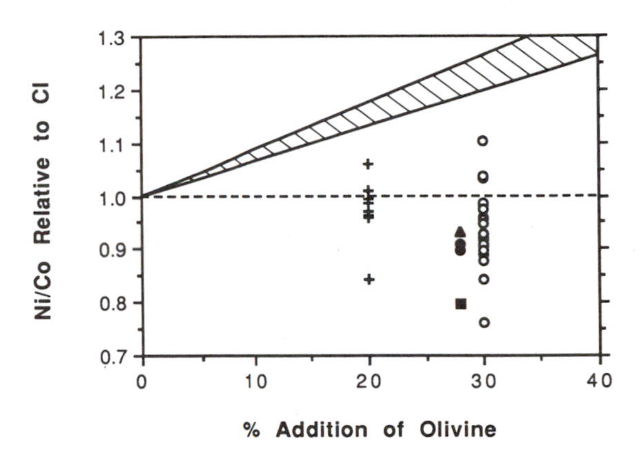

Fig. 8. *The variation of the Ni/Co ratio that would be observed in the upper mantle of the Earth with the addition of olivine into the upper mantle is shown in the shaded region. Approximately 30% olivine addition to the upper mantle is necessary to explain the Mg/Si ratio currently observed, assuming an initially chondritic Mg/Si ratio. The upper line corresponds to 90% fraction of liquid remaining, and the lower line to 80% fraction of liquid remaining. The olivine/melt partition coefficient values used in the computations are $D_{Ni} = 2$ and $D_{Co} = 1$ (Table 1). The currently observed Ni/Co ratio for the upper mantle of the Earth is approximately 0.9× the CI chondritic value. Open circles at 30% olivine addition show the Ni/Co ratio of 26 mantle nodules (20 from Basaltic Volcanism Study Project, 1981, pp. 282–310; 6 from Jagoutz et al., 1979). Closed circles, arbitrarily plotted at 28% olivine addition for clarity, represent the most primitive of the mantle nodules of Jagoutz et al. (1979); the closed triangle is an average of the six mantle nodule Ni/Co values (Jagoutz et al., 1979); the closed square is the Ni/Co ratio of the theoretical upper mantle composition pyrolite (from Jagoutz et al., 1979). The eight crosses are arbitrarily plotted at 20% olivine addition to illustrate the variation of the Ni/Co ratio among the undifferentiated meteorite types CI, CM, CO, CV, H, L, LL, EH, and EL (D. Malvin, personal communication). All Ni/Co ratios are normalized to CI chondrites (Anders and Ebihara, 1982).*

TABLE 1. *Analyses of four olivine grains and adjacent melt for experiment #5, and calculated olivine/melt partition coefficient values (D).*

	MgO	SiO_2	FeO	Al_2O_3	CaO	CoO	Cr_2O_3	NiO	Sc_2O_3	MnO	Na_2O	La_2O_3	Total
Olivine	52.17	41.03	5.00	0.19	0.12	0.63	0.42	0.25	0.11	0.07	0.04	<0.02	100.05
Melt	29.34	42.93	7.89	4.41	3.24	0.60	1.28	0.13	1.26	0.14	0.43	3.89	95.54
D ol/m	**1.8**	**0.96**	**0.6**	**0.04**	**0.04**	**1.1**	**0.3**	**1.9**	**0.09**	**0.5**	**0.09**	**<0.005**	
Olivine	52.86	40.99	4.98	0.20	0.11	0.62	0.36	0.29	0.12	0.07	0.04	<0.03	100.67
Melt	29.51	42.87	7.82	4.52	3.34	0.63	1.31	0.13	1.24	0.15	0.46	4.10	96.08
D ol/m	**1.8**	**0.96**	**0.6**	**0.04**	**0.03**	**1.0**	**0.3**	**2.2**	**0.10**	**0.5**	**0.09**	**<0.007**	
Olivine	53.05	41.82	4.84	0.21	0.14	0.63	0.40	0.22	0.11	0.06	0.05	<0.03	101.56
Melt	30.75	43.95	7.84	4.44	3.21	0.66	1.21	0.15	1.18	0.12	0.45	3.51	97.47
D ol/m	**1.7**	**0.95**	**0.6**	**0.05**	**0.04**	**1.0**	**0.3**	**1.5**	**0.09**	**0.5**	**0.11**	**<0.009**	
Olivine	52.88	41.10	4.81	0.16	0.13	0.59	0.43	0.26	0.11	0.07	0.04	<0.03	100.61
Melt	30.27	43.43	7.81	4.49	3.23	0.63	1.25	0.16	1.23	0.12	0.46	4.10	97.18
D ol/m	**1.7**	**0.95**	**0.6**	**0.04**	**0.04**	**0.9**	**0.3**	**1.6**	**0.09**	**0.6**	**0.09**	**<0.007**	

Experiment #5 was run at a pressure of 75 kbar and a temperature of 1800°C.

TABLE 2. *Average subsolidus analyses from the cold end, middle, and hot end of the charge.*

Smpl	Phse	Rgn	MgO	FeO	SiO_2	Al_2O_3	CaO	NiO	CoO	TiO_2	Na_2O	MnO	Cr_2O_3	Total
R5	ol	cold	47.9	10.8	40.4	0.04	0.09	0.45	0.1	x	0.01	0.14	0.03	100.0
R5	ol	hot	51.9	5.4	40.9	0.2	0.1	0.4	0.7	x	0.04	0.07	0.4	100.1
R5	opx	cold	31.5	6.9	54.5	5.2	0.86	0.16	0.07	0.15	0.12	0.15	0.3	99.3
R5	sp	cold	18.6	11.2	0.1	59.9	0.04	0.46	0.4	0.14	0.01	0.09	7.8	99.1
R5	cpx	cold	14.2	3.5	50.9	7.6	19.2	0.13	0.06	0.62	1.7	0.10	0.7	98.6
R8	ol	cold	49.5	9.7	38.8	0.1	0.08	0.4	0.1	x	0.02	0.16	0.14	99.0
R8	ol	mid	48.8	9.9	39.8	0.1	0.08	0.5	0.5	0.01	0.02	0.14	0.06	99.8
R8	ol	hot	46.6	8.9	38.6	0.9	0.25	1.3	3.0	0.04	0.06	0.14	0.08	99.8
R8	opx	mid	32.4	6.5	53.6	5.2	0.8	0.2	0.1	0.15	0.1	0.16	0.4	99.7
R8	opx	hot	32.9	6.8	49.7	6.6	1.4	0.6	1.4	0.19	0.3	0.16	0.5	100.6
R8	sp	cold	20.6	10.2	0.1	60.5	0.01	0.42	0.11	0.13	x	0.08	8.5	100.7
R8	sp	mid	20.6	10.6	0.1	59.7	0.03	0.5	0.27	0.13	x	0.10	8.7	100.7
R8	cpx	mid	15.1	3.3	50.3	7.44	19.4	0.13	0.18	0.62	1.67	0.09	0.8	99.0

An x means that the element was not detected. Both samples R5 and R8 were run at a pressure of 75 kbar and a temperature of 1800°C. La, Zr, and Mo were not detected at statistically significant levels in these phases in either sample. Likewise, Sc was not detected in sample R5, and Ga was not detected in sample R8. The number of significant digits reported for the elements varies according to levels of uncertainty.

TABLE 3. *Subsolidus partition coefficient values for different phase pairs.*

Sample	Phase	Region	Mg	Fe	Si	Al	Ca	Ni	Co	Sc	Ti	Na	Mn	Cr
R5	ol/opx	cold	1.51	1.58	0.75	0.009	0.1	3	1	1.8	0.1	0.1	0.9	0.04
R5	opx/sp	cold	1.7	0.6	890	0.08	20	0.4	x	0.6	0.8	x	1.9	0.04
R5	opx/cpx	cold	2.2	2.1	1.1	0.70	0.04	1.4	x	x	0.2	0.1	1.6	0.5
R5	cpx/sp	cold	0.7	0.3	630	0.12	630	0.3	x	x	4	160	1.1	0.1
R5	ol/cpx	cold	3.3	3.1	0.8	x	x	4	1	x	x	0.01	1.3	0.1
R5	ol/sp	cold	2.5	0.9	330	0.001	2.8	0.9	x	0.6	0.1	1	2	0.002
R8	ol/opx	cold	1.53	1.55	0.75	0.01	0.1	3	1	x	x	x	1.0	0.3
R8	ol/opx	middle	1.52	1.55	0.75	0.01	0.1	3	4	x	x	0.1	0.9	0.1
R8	ol/opx	hot	1.44	1.45	0.76	0.1	0.4	5	8	x	0.2	0.4	0.9	0.2
R8	opx/sp	cold	1.5	0.6	460	0.09	200	0.4	2		1.1	26	2.1	x
R8	opx/sp	middle	1.6	0.6	280	0.09	20	0.4	1		1.2	x	1.7	x
R8	opx/cpx	cold	2.2	1.9	1.1	0.69	0.04	1.1	1		0.2	0.06	2	0.5
R8	opx/cpx	middle	2.1	2.0	1.1	0.70	0.04	1.4	1	x	0.2	0.08	2	0.5
R8	opx/cpx	hot	1.5	1.4	1.1	0.62	0.09	0.1	0		0.3	0.15	1	0.5
R8	cpx/sp	cold	0.7	0.3	440	0.12	4800	0.4	2		4	430	1	0.1
R8	cpx/sp	middle	0.8	0.3	260	0.13	450	0.3	x		5	x	1	0.1
R8	ol/cpx	cold	3.4	3.0	0.8	0.007	x	3	1		x	0.01	1.5	0.2
R8	ol/cpx	middle	3.2	3.0	0.8	0.008	0.01	4	3	x	x	0.01	1.5	0.1
R8	ol/cpx	hot	2.2	2.1	0.8	0.1	0.03	1	1		0.11	0.04	1.2	0.1
R8	ol/sp	cold	2.3	1.0	420	0.001	x	1.0	1	x	x	x	2	x
R8	ol/sp	middle	2.4	1.0	270	0.001	2	1.1	3	x	x	3	2	x

A blank means that the element was not analyzed. An x means that the element was not detected in one or both of the phases. Both samples R5 and R8 were run at a pressure of 75 kbar and a temperature of 1800°C. La, Zr, and Mo were not dectected at significant levels. The number of significant digits reported for the elements varies according to levels of uncertainty.

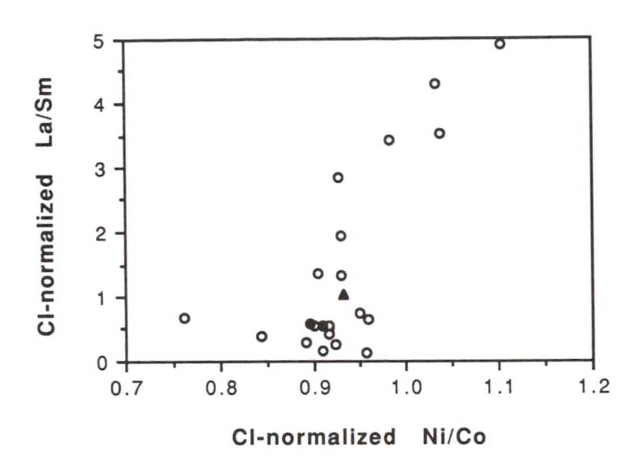

Fig. 9. *Open circles show the CI-normalized La/Sm ratio vs. CI-normalized Ni/Co ratio of 26 mantle nodules (20 from Basaltic Volcanism Study Project, 1981, pp. 282–310; 4 from Jagoutz et al., 1979); closed circles represent the most primitive of the mantle nodules (Jagoutz et al., 1979); the closed triangle is an average of six mantle nodule Ni/Co values, while the La/Sm ratio is calculated from element/Sc ratios in CI chondrites (Jagoutz et al., 1979). Those nodules that have a Ni/Co ratio of >1 also tend to have a La/Sm ratio of >1, or a LREE enrichment. Generally, samples showing flat REE patterns are thought to be more primitive and to better represent the composition of the upper mantle of the Earth (Jagoutz et al., 1979).*

Modeling Addition of Olivine to the Upper Mantle

The low value of the olivine/melt partition coefficient for Sc of 0.1 indicates that chondritic Sc/REE ratios cannot be used to detect addition of olivine to the upper mantle. The concentration of Sc in olivine is so low that even substantial accumulation of up to 30% olivine will not significantly affect Sc/REE ratios in the upper mantle of the Earth. The Ni/Co ratio is a potential indicator of olivine accumulation, however, because the olivine/melt partition coefficient values are neither identical to one another nor significantly less than unity. The partition coefficients for Ni and Co are large enough that the concentrations of these two elements in olivine are nontrivial, and substantial olivine accumulation will have a significant effect upon the Ni/Co ratio in the upper mantle of the Earth.

There are many uncertainties concerning the physical state of the Earth toward the end of accretion. In order to model the addition of olivine into the upper mantle, certain assumptions must be made. Unknown are the concentrations of elements in the upper mantle during the postulated magma ocean stage. Efficient metal segregation during core formation will readily fractionate Ni from Co, yielding a substantially subchondritic Ni/Co ratio. For example, the metal/silicate partition coefficients of *Jones and Drake* (1986) may be used to calculate the abundances of Ni and Co in the mantle of the Earth following core formation under the assumption that metal achieved perfect equilibrium with silicate. Table 4 contains these abundances and the resultant Ni/Co ratios for the four limiting cases arising from the possibility that both metal and silicate may vary between being entirely molten and completely solid.

In practice, both metal and silicate were probably at least partially molten, and the completely solid case has no relevance to the problem discussed here. Nevertheless, even in the completely solid case, which results in the smallest deviation of the Ni/Co ratio from the CI chondritic value, the Ni/Co ratio is less than

TABLE 4. *Abundances of Ni and Co in ppm and the resultant Ni/Co ratio in the upper mantle of the Earth following equilibrium segregation of metal from silicate during core formation.*

	Ni	Co	Ni/Co
LM/LS	7.4	12	0.03
SM/LS	5.5	5.1	0.05
LM/SS	73	34	0.10
SM/SS	55	15	0.17

The symbols LM, LS, SM, and SS represent sulfur-bearing liquid metal, liquid silicate, solid metal, and solid silicate respectively. The metal/silicate partition coefficients of *Jones and Drake* (1986) were used, and initial CI abundances of 11,000 ppm and 509 ppm were assumed for Ni and Co respectively (*Anders and Ebihara*, 1982).

one fifth of CI chondritic. We shall see from the calculations below that olivine accumulation alone is insufficient to raise the Ni/Co ratio back up to the slightly subchondritic value of 0.9 that is currently observed.

Instead, inefficient core formation may have occurred. *Jones and Drake* (1986) have considered the case where the Ni/Co ratio of the upper mantle of the Earth would remain approximately chondritic. They point out that in addition to a number of other problems, some of which are shared by the heterogeneous accretion hypothesis discussed below, the chondritic Ni/Co ratio is preserved by sequestering Ni in metal and Co in olivine, a somewhat artificial construct. Their model required a low fraction of partial melting of the silicates (approximately 10%) to be viable, however, and is not pertinent to the case of substantial melting of the outer part of the Earth, required to yield a substantial magma ocean as envisioned in Fig. 4.

Finally, heterogeneous accretion may have occurred. If matter continued to accumulate to the Earth after separation of an upper from a lower magma ocean, it seems plausible that the upper magma ocean would have a chondritic ratio resupplied by the accreting material (see *Jones and Drake,* 1986, for a discussion). Therefore, we will assume that the Ni/Co ratio of the magma ocean was chondritic prior to olivine addition. Although olivine will crystallize over a range of pressures and temperatures in a real planet, we will assume for the purposes of calculation that olivine equilibrates with melt only at 1800°C and 75 kbar. This assumption is justified by the relatively rapid diffusion rates of Ni and Co at 1800°C, and the relatively long timescales for the solidification of a planetary-scale magma ocean.

With these assumptions, the effect of addition of olivine to the upper mantle of the Earth may be calculated. Results are given in Fig. 8. It is seen that the deviation of the Ni/Co ratio (represented by the shaded region) of the upper mantle from the CI chondritic ratio depends on the fraction of liquid remaining in the magma ocean. Nevertheless, for olivine crystallizing at 90% and 80% liquid remaining, the Ni/Co ratio is 20-25% higher than its initial value for a 30% addition of olivine to the upper mantle. It seems difficult to avoid the conclusion that addition of enough olivine to the upper mantle to yield a Mg/Si ratio significantly higher than in CI chondrites will yield a super-CI chondritic ratio of Ni/Co, contrary to the sub-CI chondritic Ni/Co ratio observed.

CONCLUSIONS

The partition coefficient values derived from this work enable direct testing of the *Agee and Walker* (1988) hypothesis that olivine flotation into the upper mantle can explain the elevated Mg/Si ratio of the upper mantle of the Earth relative to CI chondrites. This work is inconsistent with the olivine flotation hypothesis, which would cause the Ni/Co ratio of the upper mantle to become super-CI chondritic, contrary to the sub-CI chondritic ratio currently observed. The implication of our experiments and those of *Kato et al.* (1987, 1988a,b) is that the Earth does not presently show the geochemical effects of extensive mineral fractionation.

One conclusion consistent with this implication is that the Earth never fractionated minerals from melt because the Earth was never substantially molten. If the Earth were never substantially molten, then the accretional process must have delivered gravitational potential energy more slowly than current theory predicts. This would favor a more quiescent solar system, and an origin of the Moon as the result of a giant impact with the Earth would be unlikely (*Drake,* 1989).

An alternative possibility is that there was substantial terrestrial melting but, nevertheless, mineral fractionation did not occur. *Tonks and Melosh* (1990) have proposed that a terrestrial magma ocean had fluid dynamical properties more akin to a planetary atmosphere than a highly viscous fluid, and that minerals remained entrained in magma and were unable to segregate. This proposal would eliminate the apparent conflict between accretional theory and geochemical observation. Regardless of whether the Earth was substantially molten or not, the lack of geochemical evidence for planet-wide mineral fractionation implies that the high Mg/Si ratio in the Earth relative to most classes of chondritic meteorites is intrinsic to the Earth. If the chondritic meteorites are from the Main Belt of asteroids as is generally assumed (e.g., *Kerridge and Matthews,* 1989), then the accretional process did not efficiently mix material between the Earth at 1 A.U. and the asteroid belt at 2-4 A.U. where most chondritic meteorites are presumed to originate.

Acknowledgments. The high-pressure experiments reported above were conducted in the Stony Brook High Pressure Laboratory, which is jointly supported by NSF grant EAR 86-07105 and SUNY Stony Brook. Many thanks to T. Gasparik for technical assistance with all experiments. Thanks also to J. Delano and R. Brett for helpful reviews. This work was supported by NSF grant EAR 86-18266.

REFERENCES

Agee C. B. and Walker D. (1988) Mass balance and phase density constraints on early differentiation of chondritic mantle. *Earth. Planet. Sci. Lett., 90,* 144-156.

Agee C. B. and Walker D. (1989) Comments on "Constraints on element partition coefficients between $MgSiO_3$ perovskite and liquid determined by direct measurements" by T. Kato, A. E Ringwood, and T. Irifune. *Earth. Planet. Sci. Lett., 94,* 160-161.

Anders E. and Ebihara M. (1982) Solar-system abundances of the elements. *Geochim. Cosmochim. Acta, 46,* 2363-2380.

Basaltic Volcanism Study Project (1981) *Basaltic Volcanism on the Terrestrial Planets.* Pergamon, New York. 1286 pp.

Chou C.-L. (1978) Fractionation of siderophile elements in the Earth's upper mantle. *Proc. Lunar Planet. Sci. Conf. 9th,* pp. 219-230.

Chou C.-L., Shaw D. M., and Crocket J. H. (1983) Siderophile trace elements in the Earth's oceanic crust and upper mantle. *Proc. Lunar Planet. Sci. Conf. 13th,* in *J. Geophys. Res., 88,* A507-A518.

Clayton R. N., Onuma N., and Mayeda T. K. (1976) A classification of meteorites based on oxygen isotopes. *Earth. Planet. Sci. Lett., 30,* 10-18.

Drake M. J. (1989) Geochemical constraints on the early thermal history of the Earth. *Z. Naturforsch., 44a,* 883-890.

Gasparik T. (1989) Transformation of enstatite-diopside-jadeite pyroxenes to garnet. *Contrib. Mineral. Petrol., 102,* 389-405.

Jagoutz E., Palme H., Baddenhausen H., Blum K., Cendales M., Dreibus G., Spettel B., Lorenz V., and Wänke H. (1979) The abundances of major, minor and trace elements in the Earth's mantle as derived from primitive ultramafic nodules. *Proc. Lunar Planet. Sci. Conf. 10th,* pp. 2031-2050.

Jones J. H. and Drake M. J. (1986) Geochemical constraints on core formation in the Earth. *Nature, 322,* 221-228.

Kallemeyn G. W. and Wasson J. T. (1981) The compositional classification of chondrites—I. The carbonaceous chondrite groups. *Geochim. Cosmochim. Acta, 45,* 1217-1230.

Kato T., Irifune T., and Ringwood A. E. (1987) Majorite partition behavior and petrogenesis of the Earth's upper mantle. *Geophys. Res. Lett., 14,* 546-549.

Kato T., Ringwood A. E., and Irifune T. (1988a) Experimental determination of element partitioning between silicate perovskites, garnets and liquids: Constraints on early differentiation of the mantle. *Earth. Planet. Sci. Lett., 89,* 123-145.

Kato T., Ringwood A. E., and Irifune T. (1988b) Constraints on element partition coefficients between $MgSiO_3$ perovskite and liquid determined by direct measurements. *Earth. Planet. Sci. Lett., 90,* 65-68.

Kato T., Ringwood A. E., and Irifune T. (1989) Constraints on element partition coefficients between $MgSiO_3$ perovskite and liquid determined by direct measurements—reply to C. B. Agee and D. Walker. *Earth. Planet. Sci. Lett., 94,* 162-164.

Kerridge J. F. and Matthews M. S., eds. (1988) *Meteorites and the Early Solar System.* Univ. of Arizona, Tucson. 1269 pp.

Newsom H. E. and Palme H. (1984) The depletion of siderophile elements in the Earth's mantle: New evidence from molybdenum and tungsten. *Earth. Planet. Sci. Lett., 69,* 354-364.

Newsom H. E. (1990) Accretion and core formation in the Earth: Evidence from siderophile elements. In *Origin of the Earth,* this volume.

Palme H. and Nickel K. G. (1985) Ca/Al ratio and composition of the Earth's upper mantle. *Geochim. Cosmochim. Acta, 49,* 2123-2132.

Ringwood A. E. (1979) *Origin of the Earth and Moon.* Springer-Verlag, New York. 295 pp.

Takahashi E. (1986) Melting of a dry peridotite KLB-1 up to 14 GPa: Implications on the origin of peridotitic upper mantle. *J. Geophys. Res., 91,* 9367-9382.

Tonks W. B. and Melosh H. J. (1990) The physics of crystal settling and suspension in a turbulent magma ocean. In *Origin of the Earth,* this volume.

Wetherill G. W. (1985) Occurrence of giant impacts during the growth of the terrestrial planets. *Science, 228,* 877-879.

THE PHYSICS OF CRYSTAL SETTLING AND SUSPENSION IN A TURBULENT MAGMA OCEAN

W. B. Tonks and H. J. Melosh

Lunar and Planetary Laboratory, University of Arizona, Tucson, AZ 85721

It is widely assumed that a magma ocean on either the Earth or the Moon would necessarily differentiate by fractional crystallization. Geochemists and petrologists making this assumption, however, have so far failed to take into account the physics of crystal suspension in a planetary-scale convecting system. In the first part of this paper we show that convective velocities in a deep ultramafic magma ocean on either the Earth or the Moon are sufficiently rapid to suspend large crystals. This demonstration involves aspects of convective theory and the mechanics of suspension that may be unfamiliar to most planetary scientists. Indeed, the problem has more affinities to aeolian transport in a deep planetary atmosphere than to traditional studies of magma convection. A vigorously convecting system can be divided into three distinct regions: an upper conductive boundary layer, a sublayer dominated by viscous forces, and a volume of turbulent fluid whose flow is controlled by inertial forces. This inertial flow zone, which does not exist in the more familiar subsolidus convection, supports large-scale turbulent eddies that are analogous to gusty winds in a planetary atmospheric convective system. The criterion for suspension is derived from empirical studies of sediment transport and is given by the Rouse number, which is the ratio between the terminal velocity of the crystals and the turbulent friction velocity. Settling (or flotation) only takes place for Rouse numbers greater than about 1. We present several studies of magma ocean evolution with realistic rheological properties and cooling histories. The results show that on both the Earth and Moon crystals up to 1-2 cm in diameter can be suspended from the onset of cooling. In the later stages of cooling, as viscosity increases, crystals of 10 cm to more than 1 m in diameter may be suspended. A surprising feature of our analysis is that the ability of a magma ocean to suspend crystals is only weakly dependent on gravity and depth: Viscosity is the major factor determining suspension. This result raises a serious problem for the lunar magma ocean hypothesis, because the differentiation of an anorthositic crust strongly suggests that crystal fractionation did occur. We offer an explanation for this apparent anomaly by appealing to the low lunar gravity. Since pressure in the Moon increases only slowly with depth, the solidus and liquidus temperatures are nearly independent of depth. In a convecting magma ocean on the Moon adiabats lie between the solidus and liquidus at all depths, so that growing crystals circulate through the ocean for long periods of time without dissolving. Under these circumstances they may grow large enough to settle out, resulting in the observed differentiation. In contrast, on the Earth the liquidus and solidus profiles are steeper than an adiabat so crystallization only takes place over a restricted depth range near the bottom of the ocean. Convective velocities are so high that crystals nucleating in this zone do not have time to grow larger than a few microns and are thus incapable of settling out. In this way even a totally molten initial Earth may have failed to differentiate by fractional crystallization.

INTRODUCTION

Ideas on the thermal state of the early Earth and planets are currently undergoing a revolution. Modern accretion theory (*Safronov,* 1969, 1978; *Wetherill,* 1976, 1986; *Taylor,* 1988) recognizes that the Earth and other terrestrial planets may have accumulated from a hierarchy of protoplanetary bodies rather than from the uniform distribution of ~ 10-km-diameter planetesimals proposed in older theories. Toward the end of accretion some of these protoplanetary bodies ranged up to the size of Mars, with important consequences for the initial thermal state of the Earth. Collisions between the proto-Earth and protoplanets of this size would have caused nearly complete melting of the Earth (e.g., *Kaula,* 1979; *Stevenson,* 1981; *Melosh,* 1990). Even in the absence of such collisions, segregation of the core could have released enough energy to raise the temperature of the whole Earth by 1500°K (*Verhoogen,* 1980). Additionally, the Earth and possibly the Moon may have had an early steam atmosphere that interfered with the planet's ability to radiate accretional heat (*Abe and Matsui,* 1985; *Matsui and Abe,* 1986). If the Moon was formed by the giant impact of a Mars-sized body, not only did a large portion of the Earth melt, but much of it was vaporized (*Cameron and Benz,* 1988; *Benz and Cameron,* 1990; *Melosh and Kipp,* 1988; *Melosh,* 1990). These lines of argument suggest that the early Earth underwent widespread melting events, after which a magma ocean would have covered its surface to substantial depths.

On the other hand, a growing body of geochemical evidence suggests that the Earth could *not* have had a deep magma ocean. *Kato et al.* (1988) argue that fractionation of a small amount of perovskite and/or majorite garnet from a terrestrial magma ocean would drive the nearly chondritic ratios of most refractory lithophile elements substantially away from their observed chondritic values. Fractionation would also result in a primitive crust rich in Ba, K, Rb, and Cs and relatively depleted in U, Th, Pb, Sr, Zr, and Hf. *Drake et al.* (1988) and *McFarlane and Drake* (1990) argue that fractionation of olivine by flotation at high pressure would yield ratios of Sc/Sm, Ni/Co, and Ir/Au more than 10% above their observed chondritic values, although *Agee* (1988) suggests that the interpretation of these results should wait until experimental measurements are performed under the same temperature and pressure conditions as existed in the magma ocean.

Ringwood (1988) argues that *Kato et al.'s* (1988) results can be used as a test of the giant impact hypothesis and Safronov accretion theory. Because the giant impact leads to global melting of the Earth, he infers that such melting could not have occurred or else the oldest (~4.2 Ga) zircons and most ancient (~3.9 Ga) crustal rocks would show some signature of the element fractionations noted above. Ringwood thus proposes that the Earth accreted without widescale melting and suggests several conditions under which this might occur. This proposal makes the implicit assumption that a magma ocean necessarily undergoes fractional crystallization.

Hofmeister (1983) modeled the solidification of a 120-km-deep terrestrial magma ocean in detail, assuming that the ocean undergoes a batch fractional crystallization process. The resulting hypothetical planet had a chemical structure quite unlike that of the present Earth. *Ohtani* (1985) also examined the solidification of a magma ocean and concluded that it could result in either a stratified or a homogeneous structure depending on the stirring action of convection. Thus, the reconciliation of the theoretical arguments for widespread melting in the Earth and the geochemical arguments prohibiting it may be rooted in the settling behavior of crystals in a convecting magma ocean.

Although the case for substantial melting in the Earth is confused at present, it is widely believed that the Moon was largely molten early in its history. Fractional crystallization of a cooling magma ocean is generally accepted as the explanation of the observed dichotomy between the Eu-enriched anorthositic crust and the Eu-depleted basaltic source regions of the Moon (i.e., *Wood,* 1970). Although the fact of lunar melting is not greatly disputed, the precise amount of melting is not well constrained. Thermal models of the Moon reviewed by *Solomon* (1986) indicate that the magma ocean cannot have been deeper than about 300 km, based on the absence of compressional features on the lunar highland surface. Recent studies that incorporate crustal differentiation, however, now stretch this limit to 640 km (*Kirk and Stevenson,* 1988). Trace element analyses of the lunar samples indicate that the magma system from which the lunar crust and mare basalt source region differentiated comprised about 20% of the lunar mass and had concentrations of refractory lithophile elements (e.g., Sr, U, Th, and the REEs) 10-20 times chondritic abundances (*Palme et al.,* 1984). Two possible interpretations of these data are (1) the bulk Moon has a high abundance of these elements, or (2) the crust and mare basalt source regions themselves represent a residual magma from a totally molten Moon. In this latter case, the bulk Moon began with 3-4 times chondritic abundances of these elements (*Binder,* 1986). In support of this hypothesis, the Apollo heat flow data are consistent with a bulk lunar abundance of U and Th of 4-5 times the chondritic value (*Langseth et al.,* 1976). This estimate thus appears to support an initially wholly molten Moon. Whether the Moon's magma ocean was deep or shallow, it is difficult to understand how the Moon could have undergone such widespread melting while the Earth did not melt at all.

We propose here that the key to understanding the apparent discrepancy between the theoretical prediction of widespread melting in the Earth and the geochemical data that seem to imply little net differentiation lies in the behavior of crystals in a convecting magma. Convection in a magma ocean is expected to be vigorous throughout cooling so long as the liquid can loose its heat near the free outer surface. Although a thin conductive skin may form at the immediate surface, differential convective velocities are generally high enough to overcome its strength and keep the heat loss near the theoretical value. This situation is quite different than that of a magma chamber, where a thick roof may form that strongly inhibits convection by lowering the differential temperature across the liquid body (*Brandeis and Marsh,* 1989). Convection may be similarly inhibited in a hypothetical deep mantle magma ocean (*Agee and Walker,* 1988) if it is overlain by a crystallized upper mantle. In spite of its importance, the effect of vigorous convection in keeping a magma ocean well stirred has not previously been well studied. It seems obvious that if convection is sufficiently vigorous, it will cause crystals to remain in intimate contact with the magma throughout the entire cooling process. This would result in the ocean cooling via equilibrium crystallization instead of fractionating as is normally assumed. If so, the Earth's upper mantle could have been extensively melted, yet not show the geochemical fractionations described above. Since the much smaller Moon does show signs of geochemical fractionation, any successful theory of magma ocean evolution must allow separation of crystals from the magma on a small body such as the Moon but not on the Earth. It seems that to first order the effect of the acceleration of gravity on convective vigor and crystal settling must be responsible for the divergent geochemical histories of the two bodies.

The main purpose of this paper is thus to critically examine the implicit assumption that a magma ocean differentiates by fractional crystallization in the context of a realistic model. First, the physics of convection applicable to the regime of a planetary-scale magma ocean is described. The mechanics of crystal settling and suspension is then reviewed in the context of modern sedimentation theory. Next, we estimate the minimum crystal size required for separation from a turbulently convecting magma ocean and the maximum crystal fraction that allows the ocean to remain turbulent enough to suspend crystals. This crystal fraction is then compared to the crystal fraction (48-52%) at which the crystal-magma mush locks up and is no longer free to flow as a liquid. The effect of the acceleration of gravity on crystal suspension, convective vigor, and the liquidus and solidus temperature profiles is examined in some detail to test the assertion made at the end of the last paragraph.

Finally, we crudely estimate the ability of the magma to percolate through the crystals during the remaining cooling history, from about 50% to complete crystallization.

CONVECTION AND CRYSTAL SETTLING

Convection in a Magma Ocean

The convection of a body of liquid several hundred to a thousand kilometers deep occurs in a flow regime quite different from that of the present subsolidus convection familiar to most solid earth geoscientists. The viscosity of liquid magma is typically 10 to 15 orders of magnitude lower than that of the present mantle of the Earth and convection is correspondingly more vigorous. Studies of subsolidus convection traditionally (and correctly) neglect acceleration and inertial forces in the flow. However, as we show below, such factors actually dominate convection in a planetary magma ocean, where buoyancy forces can only be balanced by inertial forces. The convective flow is thus not only unsteady, but is turbulent in the strict sense that the ratio of inertial forces to viscous forces is much greater than 1 (i.e., the Reynolds number is large). The convective regime of a planetary magma ocean thus has more in common with convection in a planetary atmosphere than with subsolidus convection: Indeed, the Rayleigh number for atmospheric convection is not very different from that of a magma ocean.

The convective regime of a system is largely defined by two dimensionless numbers, the Rayleigh and Prandtl numbers. (In the Boussinesq approximation these two numbers define the flow uniquely. In the real world of substances with depth- and temperature-dependent viscosities these other factors may play roles as well, although the success of parameterized convection models suggests that the Rayleigh and Prandtl numbers dominate.) The Rayleigh number characterizes the vigor of the convective flow. It is essentially the ratio of the forces that drive convective flow (buoyancy) to the forces that tend to damp it (viscosity and heat diffusion). It is given by the following equation for a fluid layer of constant viscosity heated from below

$$\mathrm{Ra} = \frac{\alpha g \Delta T D^3}{\kappa \nu} \tag{1}$$

where α is the thermal expansion coefficient, γ is the acceleration of gravity, D is the depth of the convecting fluid, κ is the fluid's thermal diffusivity, ν is its kinematic viscosity, and ΔT is the superadiabatic temperature difference (potential temperature in meteorological parlance), given by

$$\Delta T = \Delta T_{\text{fluid}} - \Delta T_{\text{adiabatic}} \tag{2}$$

ΔT_{fluid} is the temperature difference across the convecting fluid, given practically by the temperature at the bottom of the magma ocean minus the surface temperature. It includes the temperature drop across the conductive boundary layer. $\Delta T_{\text{adiabatic}}$ is found by using the adiabatic temperature gradient, $dT_{\text{adiabatic}}/dz$, and is approximately equal to

$$\Delta T_{\text{adiabatic}} \cong \frac{dT_{\text{adiabatic}}}{dz} \cdot D \tag{3}$$

The adiabatic temperature gradient gives the rate at which the temperature of an isolated parcel of fluid changes as it rises or sinks. In a vigorously convecting system conductive heat transfer is negligible in the interior of the flow, so the fluid's internal temperature is controlled mainly by adiabatic compression or expansion. The mean temperature of the convecting fluid thus changes according to the adiabatic temperature gradient. The largest changes in temperature occur across a thin conductive boundary layer at the top or bottom of the convecting layer. The adiabatic temperature gradient is given by the following equation

$$\frac{dT_{\text{adiabatic}}}{dz} = \frac{\alpha g T}{c_p} \tag{4}$$

where c_p is the specific heat capacity at constant pressure and the other terms are the same as in the definition of the Rayleigh number (*Turcotte and Schubert,* 1982, pp. 190-192). The adiabatic gradient in small-scale convective flows is normally quite small and in many applications the fluid can be considered as isothermal. However, in a large convecting body such as a magma ocean or the Earth's mantle, the adiabatic temperature gradient is often large enough to be important. Values of 0.3 K/km for the Earth and 0.05 K/km for the Moon were used in the calculations that follow.

An equivalent form of the Rayleigh number appropriate for a system heated uniformly from within (applicable to a mass of magma in which latent heat production is important) is

$$Ra = \frac{\alpha \rho_0 g H D^5}{K \kappa \nu} \tag{5}$$

where ρ_0 is the fluid's density, H is the rate of internal heat generation per unit mass, K is the thermal conductivity of the fluid, and the other terms are the same as defined above (*Turcotte and Schubert,* 1982, p. 279). The larger the Rayleigh number, the more vigorous the convection (e.g., a larger Rayleigh number usually implies a higher convective velocity). Before a convective system heated from below and cooled from above can begin to flow, however, the Rayleigh number must exceed a certain minimum value that depends on the exact boundary conditions (*Turcotte and Schubert,* 1982, p. 278). Linear stability theory (*Chandrasekhar,* 1961) shows that this value is of the order of 10^3 for most boundary conditions. This is much lower than that characterizing any of the magma oceans considered in this paper. The subsolidus mantle of the Earth is estimated to have a Rayleigh number of the order of 10^6-10^7 at the present time (*Schubert,* 1979), a typical 1-km magma chamber has an initial Rayleigh number of about 10^{14}-10^{15}, but a planetary magma ocean of reasonable depth on either the Earth or Moon has a Rayleigh number in the range 10^{24}-10^{29}, which is also comparable to that of the Earth's atmosphere.

The Prandtl number (σ) is the ratio of viscosity to the thermal diffusivity and measures the importance of thermal diffusion in regulating the flow

$$\sigma = \frac{\nu}{\kappa} \tag{6}$$

where the terms are the same as defined above.

In a large planet such as the Earth, the pressure differences between the top and bottom of the mantle may involve variations in the physical quantities that go into the Rayleigh and Prandtl numbers. Thus, the viscosity ν may be depth (pressure) dependent. Moreover α, κ, and especially ν are temperature dependent. Because of these dependencies the convective pattern is not rigorously determined by the Rayleigh and Prandtl numbers alone. Nevertheless, we use these dimensionless numbers to get a first-order estimate of the convective regime, although it is probable that resolution of some details may require a higher degree of accuracy than we maintain here.

Both experiments and numerical calculations show that for systems with Rayleigh numbers less than about 10^7 the flow takes on the form of steady cells or rolls (*Elder,* 1976). However, when the Rayleigh number exceeds about 10^7 this steady pattern breaks down and the flow becomes unsteady. Even in this unsteady flow regime, however, the dominant factor resisting flow is viscosity and the flow, while unsteady, is not turbulent. The mean flow velocity in this regime is given by a number of equations now familiar to geoscientists working on subsolidus convection in the Earth's mantle (*Schubert,* 1979). However, the inadequacy of the standard subsolidus convection theory quickly becomes apparent if these equations are blindly applied to the case of a magma ocean, where the mean convective velocities predicted are of the order of 1 km/sec, clearly too high to be reasonable. The reason for this discrepancy is that subsolidus convection theory considers only viscous forces and neglects inertial forces, which can be shown to dominate magma ocean convection.

Any study of magma ocean convection must thus consider the central role of inertial forces. Fortunately, a reasonable theory of convection under these conditions has been worked out by *Kraichnan* (1962), who based his analysis on mixing length theory. He found that a vigorously convecting system can be logically divided into three distinct regions in which different physical processes control the heat transfer and fluid flow. Figure 1 illustrates such a system. It is interesting to note that Kraichnan used an approximation for the mechanical response of the upper conductive boundary layer that is actually exact for a magma ocean with an upper boundary of negligible strength.

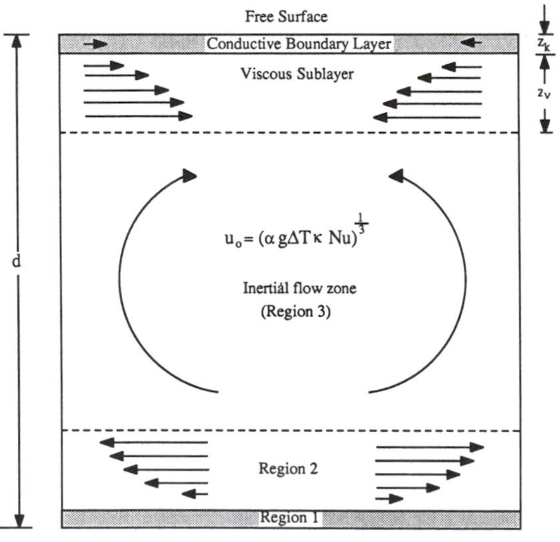

Fig. 1. *Convection breaks into three distinct regions of behavior at very high Rayleigh number (Kraichnan, 1962). Region 1 is a conductive boundary layer, region 2 is a viscous sublayer, and region 3 is the turbulent region where flow is dominated by inertial forces that produce large-scale eddies (the inertial flow zone).*

Region 1 is the conductive boundary layer, present in all convective systems. This region is extremely thin in high Rayleigh number systems and its thickness is greatly exaggerated in Fig. 1. If the magma cools from above and is insulated below, region 1 is absent from the bottom. In region 2, the viscous sublayer, flow is dominated by viscous forces but may be unsteady. Heat in this region is transferred by mainly by convection. In region 3, the inertial flow zone, flow is dominated by inertial forces that permit the formation of large-scale turbulent eddies with spatial dimensions of the order of the depth of the ocean. These eddies sweep into both the conductive boundary layer and the viscous sublayer, giving rise to shear stresses that transport momentum upward from the bottom boundary of the ocean (more on this later). According to *Kraichnan* (1962), this inertial flow region exists if

$$Ra^{1/3} \geq 35\sigma^{1/2} \tag{7}$$

Figure 2 illustrates the different convective regimes as a function of the Rayleigh and Prandtl numbers using equation (7) for the existence of the inertial flow zone. The Earth's mantle lies far below this boundary and is well described by the equations of *Schubert* (1979) because the flow is dominated by viscous forces throughout the convecting region. A planetary magma ocean at a few hundred kilometers depth lies high above this boundary, however, and the inertial flow zone is well developed, occupying the largest fraction of the ocean's volume. A 1-km magma chamber straddles the boundary, implying that an inertial flow zone may exist shortly after the formation of the chamber but that it probably disappears well before the chamber solidifies, especially if the Rayleigh number is lowered by the formation of a thick rigid roof. As we shall demonstrate below, the existence of this inertial flow zone bears directly on the ability of a magma ocean to keep crystals in suspension.

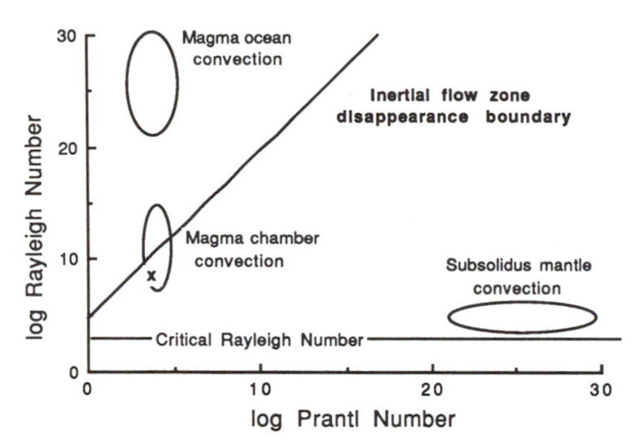

Fig. 2. *Rayleigh number vs. Prandtl number, showing the convective regimes of common geological processes and of planetary magma oceans. The inertial flow zone exists above and to the left of the inertial flow zone disappearance boundary and disappears to the right of the boundary. See the text and Fig. 1 for further explanation. The x marks where the highest Rayleigh-Prandtl number experiment of Martin and Nokes (1988) plots on this diagram.*

Another useful dimensionless number is the Nusselt number (Nu), which is the ratio of the actual heat flux, h, transferred by a convecting system to the heat flux, h_c, that would exist if the system were cooling by conduction alone. Thus

$$h = Nu \cdot h_c \tag{8}$$

where h_c is the heat that would be conducted through a slab of thickness D with the same temperature difference that exists over the convective region (including the conductive boundary layer). If there is no heat generation in the convecting region and if the conductivity is constant, the conductive heat flux is given by

$$h_c = \frac{K \Delta T}{D} \quad , \; K = \kappa \rho c_p \tag{9}$$

K is the thermal conductivity of the fluid in the convecting region and the other terms were defined previously. The Nusselt number itself is related to the Rayleigh number through the following semi-empirical relationship (*Schubert,* 1979)

$$Nu = A_\sigma Ra^n \tag{10}$$

where A_σ may be a function of Prandtl number σ and n is a constant. *Kraichnan* (1962) proposes that $A_\sigma = 0.089$, independent of σ, for $\sigma > 0.1$ and $n = 1/3$.

Equation (10) determines the overall cooling rate of a convecting fluid and forms the foundation of all parameterized convection models, so the precise values of the constants in this equation have been the object of several intensive studies. *Priestley* (1954) argued that if a fluid layer is heated from below at high Rayleigh number, heat transfer is controlled by conduction through a thin boundary layer and should be independent of the thickness D of the convective system. In this case the definitions of the Rayleigh and Nusselt numbers require that n must be 1/3, an argument supported by *Kraichnan* (1962). *Schubert* (1979) reviewed experiments that show that the power n depends weakly on the Prandtl and Rayleigh numbers and on the boundary conditions. He concludes that a power law can be used to model the heat transport by a vigorously convecting system in general. Experiments by *Chu and Goldstein* (1973), *Garon and Goldstein* (1973), and *Threlfall* (1975) show that at Rayleigh numbers up to 10^9 this Nusselt-Rayleigh number relationship holds with n between 0.278 and 0.293. *Long* (1975), however, argued that as the Rayleigh number increases, n should approach 1/3. Further experiments of *Goldstein and Tokuda* (1980) at Rayleigh numbers up to 10^{11} show that n does indeed approach 1/3. Finally, *Hartke et al.* (1988) showed theoretically that $Nu = A_\sigma Ra^{1/3}$ at large Rayleigh number. The coefficient A_σ is a function of the Prandtl number, in agreement with earlier theoretical arguments. Hartke et al. further derived a value of A_σ that agrees with the experimental value of Goldstein and Tokuda. This power law can be used with confidence to model the heat transfer properties of the magma ocean. Unfortunately, Hartke et al. did not determine A_σ as a general function of σ, so that it is difficult to generalize from the experimental data for water ($\sigma \cong 6.6$) to magma ($\sigma \approx 10^4$). On the positive side, our results for crystal settling do not depend strongly upon A_σ: Equation (19) shows that the Rouse number depends on A_σ to the 1/9 power. Following these arguments, we chose $n = 1/3$ with a coefficient of $A_\sigma = 0.089$ in the models that follow.

Crystal Settling

Daily experience indicates that small particles can be suspended in a liquid by sufficiently vigorous stirring. This everyday observation, however, seems to defy explanation when considered logically. All particles denser than the liquid, no matter how small, fall with a finite velocity (called the terminal velocity) with respect to the liquid. Even though the liquid may be stirred with random velocity components that exceed the terminal velocity, if the average velocity is zero (as it must be in a closed container with no liquid flux in or out) it seems inevitable that the particles' steady downward motion will eventually cause them to settle out. This argument has, in fact, been put forward by a number of students of crystal settling in magma chambers (e.g., *Martin and Nokes,* 1988, and references therein). On the other hand, the commonly observed suspension of dust in the atmosphere or silt in a stream seems to contradict this simple argument. This conundrum was first addressed by *Bagnold* (1965), who inferred that there must be a net momentum transport away from the boundary of a turbulent streamflow that would tend to keep small particles in suspension. Bagnold's speculation has since been confirmed by the discovery of the "bursting" phenomenon near the boundary of turbulent flows. This phenomenon, although still not fully understood, is now known to explain how sediments are suspended (*Sumer,* 1985). The essence of this phenomenon appears to be turbulence, so that particles in the unsteady but not turbulent convective flows typical of magma chambers may indeed settle out according to the above argument. On the other hand, when turbulence develops in a convective flow sufficiently small particles may be suspended indefinitely. The development of turbulence in a convecting magma ocean thus plays a crucial role in the ability of crystals to separate from the liquid. In the remainder of this section we examine the physics of crystal settling quantatively and in more detail.

As crystals nucleate they tend to settle out of the magma under the influence of gravity. The terminal velocity of a small spherical object in a viscous fluid is given by Stokes' law (*Turcotte and Schubert,* 1982, p. 266)

$$v_{fall} = \frac{2\Delta\rho g a^2}{9\rho\nu} \tag{11}$$

where a is the crystal radius, ν is the kinematic viscosity of the fluid through which the object is falling, ρ is the fluid's density, and $\Delta\rho$ is the density difference between the crystals and the fluid. Stokes' law is strictly valid for conditions in which the fluid flow around the object is laminar. Laminar flow around the crystal corresponds to the local Reynolds number being less than 1. The local Reynolds number is given by

$$Re = \frac{2 v_{fall} a}{\nu} \tag{12}$$

where a, the characteristic length scale, is equal to the radius of the crystal, v_{fall} is the fall speed of a crystal, and ν is the kinematic viscosity. For small crystals (<1 mm), this condition holds even in the extremely turbulent convection of a magma ocean. For larger crystals, this condition breaks down and turbulence increases the drag above that calculated by Stokes' law as the fluid flows past the crystal. Thus, a large crystal will fall less rapidly than its Stokes' velocity. Stokes' law thus serves as an *upper limit* to the settling velocity. This law is used in the theoretical arguments that follow (the numerical models described later use a more general expression for terminal velocity).

The criterion for particle suspension in a turbulent fluid has been addressed by a large number of workers in the sedimentological literature. In all these studies it is found that solids remain in suspension when their terminal velocity is about equal to the effective friction velocity associated with the turbulence. The suspension criterion for a turbulent magma ocean should be similar. In particular, if the effective friction velocity generated by the large eddies of the inertial flow zone (the equivalent of turbulence in stream flow) is greater than or equal to the terminal velocity, crystals will stay suspended in the magma. This condition is best expressed by a dimensionless ratio called the Rouse number (denoted by S). The Rouse number is used extensively in sedimentology and works well in describing the suspension of particles in a stream or river (e.g., *Vanoni,* 1975, p. 76). For the turbulent convection of a magma ocean, the Rouse number is simply

$$S = v_{fall}/v_{*} \qquad (13)$$

where v_* is the effective friction velocity resulting from the eddies of the inertial flow zone. In terms of the Rouse number, if $S \leq 1$, crystals should remain suspended, but if $S > 1$, the turbulence of convection will not be able to keep the crystals in suspension. If the inertial flow zone disappears we presume that the crystalls will eventually fall out.

The importance of the inertial flow zone for the suspension of crystals has received independent (albiet inadvertent) experimental support. *Martin and Nokes* (1988) proposed that crystals in a convecting magma chamber may settle out even when the mean convective speed is much larger than the settling velocities of particles. They argued that the convective motion vanishes in a boundary layer at the base of the convective region and, by continuity, is small just above it. Thus, particles entrained in the main body of the convecting system can settle in this quiet boundary. Martin and Nokes also performed experiments with solid polystyrene spheres in water that demonstrate the essential correctness of the theory. The water was mixed with varying amounts of NaCl to increase its density and sodium carboxymethyl cellulose to increase its viscosity. However, they noted that in their highest Rayleigh-Prandtl number experiments (Rayleigh number of about 5×10^7 and viscosity of 570 cSt $= 5.7 \times 10^{-4} m^2/sec$) substantial reentrainment of the solids at the bottom occurred. Using the viscosity and density values reported by these authors and assuming that the thermal conductivity (0.006 W/cm·K) and heat capacity (1 cal/g·K) of the fluid were the same as for pure water, the Prandtl number of the fluid was about 4000. Using Kraichnan's criterion expressed in equation (7), the Rayleigh number must be about 10^{10} for the existence of the inertial flow zone. This experiment plots somewhat below the boundary where the inertial flow zone disappears and is approximately marked by the x on Fig. 2. If the Rayleigh number were increased in this experiment, the solids would be even more efficiently entrained as flow develops toward full turbulence. By the time the Rayleigh number reaches the value given by Kraichnan's criterion, suspension should be fully established. Martin and Nokes noted the need for development of a reentrainment criterion. It appears that the existence of the inertial flow zone, in conjunction with the Rouse number, is a highly *conservative* criterion, as we argued above.

If the terminal velocity is given by Stokes' law (equation (11)) and if the inertial flow zone exists, the dependence of the Rouse number on parameters such as the acceleration of gravity, depth of the magma ocean, crystal radius, and viscosity of the liquid-crystal suspension can be derived theoretically. The effective friction velocity is given by (*Kraichnan,* 1962, equation 6.3)

$$v_* = u_o/b \qquad (14)$$

where u_o is the mean convective velocity and b is the solution to the following transcendental equation

$$b + A \ln b = A \ln Re_o + B' \qquad (15)$$

A and B′ are constants and Re_o is the characteristic Reynold's number for the top of the viscous sublayer. It is given by

$$Re_o = Du_o/2\nu \tag{16}$$

Using *Laufer's* (1951) measurements of flow in channels, A and B′ are equal to 3.0 and 4.8, yielding solutions to equation (15) with b between 19.1 and 61.2 for reasonable values of Re_o. We assumed a nominal value of 42 to simplify the expressions since the factor of 3 variation in b is smaller than the probable error in our estimate of the viscosity of the suspension.

The mean convective velocity u_o is derived from the theory of high-Rayleigh number convection (*Kraichnan,* 1962, equation 6.7)

$$u_o = \{\alpha g \kappa \Delta T\}^{1/3} \left\{\frac{D}{2z_k}\right\}^{1/3} \tag{17}$$

where z_k is the thickness of the conductive boundary layer. From Kraichnan's equation 8.4 and the definition of the Nusselt number, the expression in the second parentheses of equation (17) can be recognized as the Nusselt number. (Note that Kraichnan's equation 6.7 has z_k in the numerator. This is evidently a typo, since it must be in the denominator for the expression to be dimensionally correct.) Using the power law relation in equation (10) between the Nusselt and Rayleigh numbers, equation (17) becomes

$$u_o = \{\alpha g \Delta T\}^{\frac{1+n}{3}} \kappa^{\frac{1-n}{3}} D^n \nu^{-\frac{n}{3}} A_\sigma^{\frac{1}{3}} \tag{18}$$

where the terms are defined above. Note that the acceleration of gravity is raised to a positive power in this equation, implying that a magma ocean on a larger planet convects more vigorously than one on a smaller planet, other conditions being the same. Increasing viscosity lessens the convective vigor. Combining equations (13), (14), and (11), the Rouse number S is given by

$$S = \frac{28}{3} \frac{a^2 \Delta\rho g^{\frac{2-n}{3}}}{A_\sigma^{\frac{1}{3}} \rho \{\alpha \Delta T\}^{\frac{1+n}{3}} \kappa^{\frac{1-n}{3}} D^n \nu^{1-\frac{n}{3}}} \tag{19}$$

Taking $n = 1/3$ and $A_\sigma = 0.089$ as noted above, this equation becomes

$$S = \frac{20.9\, a^2\, \Delta\, \rho\, g^{\frac{5}{9}}}{\rho \{\alpha \Delta T\}^{\frac{4}{9}} \kappa^{\frac{2}{9}} D^{\frac{1}{3}} \nu^{\frac{8}{9}}} \tag{20}$$

Equation (20) illustrates the qualitative behavior of settling crystals. First, even though a higher acceleration of gravity increases convective vigor, it increases the terminal velocity of crystals even more, with the net effect that a higher acceleration of gravity *aids* crystal separation. Increasing the depth of a magma ocean decreases the Rouse number, thus enhancing the suspension of crystals. However, since D is raised to the 1/3 power, a factor of 1000 increase in depth (e.g., from 1 km to 1000 km) only decreases S by a factor of 10. By far the strongest dependences are on the crystal radius, the fractional density contrast between liquid and crystals and, most importantly, the magma viscosity. Since the effective viscosity of the mixture of crystals and liquid increases exponentially with the concentration of crystals (*Murase and McBirney,* 1973; *McBirney and Murase,* 1984), the Rouse number decreases as crystallization proceeds, enhancing the magma's ability to suspend crystals as time passes.

Combining our analyses of convection and crystal settling, there are two factors of importance to the question of crystal settling in magma oceans: (1) the minimum crystal size necessary for fractionation and (2) the longevity of the inertial flow zone. If the inertial flow zone disappears before the crystal-magma mush locks up at about 50% crystallization, the theory of Martin and Nokes should apply and crystals of all sizes should settle out from the magma. Even if crystals were suspended by inertial flow early in the history of the magma ocean, the later evolution in this case should be marked by geochemical fractionation. To estimate the relative importance of crystal settling and the evolution of the inertial flow zone, we have constructed the realistic model of magma ocean cooling reported below.

MODELING THE SUSPENSION OF CRYSTALS IN A MAGMA OCEAN

We modeled the cooling of a very simple magma ocean to address the two problems noted above: (1) to crudely estimate the radii of crystals that might be suspended by a magma ocean and (2) to estimate the longevity of the inertial flow zone. Additionally, we want to understand the effect of the acceleration of gravity on the ability of a magma ocean to suspend crystals, because gravity is the first-order difference between the Earth and Moon. For simplicity, we assume that the fraction of crystals is constant with depth at any stage of the cooling process. Crystals nucleate preferentially toward the bottom of a deep magma ocean because silicate melting curves have steeper slopes than an adiabat. The ratio of the adiabatic gradient to the melting curve gradient derived from the Clausius-Clapeyron equation is independent of the acceleration of gravity. However, the smaller the planet, the smaller both gradients become and in a magma ocean on a small planet the adiabat can lie between the liquidus and solidus temperatures (see Fig. 7). In the limit of a very small planet, the fraction of the magma that has crystallized is independent of depth. Recent experiments on the melting curves of peridotite at very high pressures (*Ito and Takahashi,* 1987) indicate that this simplification may not be unrealistic for the Earth between about 200 km to 400 km deep (see Fig. 9).

We assume that the planet was initially melted uniformly to depth D and has already cooled to the point that temperatures have just reached the liquidus. The temperature difference between the surface and the magma sets up free thermal convection. We assume that convection is vigorous enough to destroy any rigid crust that might tend to form because of the strong temperature dependence of viscosity, so the upper boundary is treated as free. The heat in the model is simply from the cooling and solidification of the magma; heating of the ocean from the bottom is not considered. If core formation on the Earth was concurrent with the formation of the magma ocean, additional heat would be added at the bottom of the ocean. However, the additional heat flux would only lengthen the cooling time: The behavior of the magma ocean as a function of the degree of crystallization would remain unchanged. Figure 3 illustrates the initial conditions of this calculation. The slopes of the adiabat, liquidus, and solidus are so

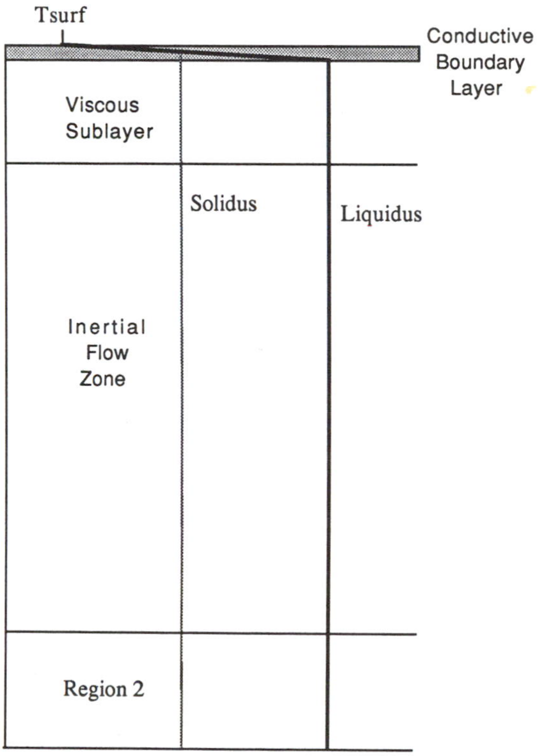

Fig. 3. *The functional initial conditions of the magma ocean at the beginning of the computation. The three distinct regions of behavior defined by Kraichnan (1962) are shown.*

small that they can be treated as vertical. As pointed out above, this is only an approximation, but it allows the crystal fraction to be readily calculated throughout the magma ocean. To begin the computation of magma ocean evolution, we first determine the initial Rayleigh, Nusselt, and Prandtl numbers from the initial conditions assumed (discussed below). After a time Δt, a unit volume of the ocean (which equals a unit area of the surface times the depth) has lost an amount of energy equal to $h\Delta t$, where h is the heat flux. Equations (9) and (10) are used to determine this heat flux. From the heat loss, a new magma temperature is calculated by assuming the heat loss is related to an effective specific heat capacity. This effective heat capacity is taken to be

$$c_{eff} = \frac{L + c_p\,(T_{liq} - T_{sol})}{T_{liq} - T_{sol}} \qquad (21)$$

where T_{liq} and T_{sol} are the liquidus and solidus temperatures, L is the latent heat of fusion, and c_p is the specific heat capacity of the liquid fraction. The crystal fraction is then calculated by linear interpolation (the "lever rule") between the solidus and liquidus. In the next time step a new effective magma viscosity is determined from the temperature, from which we compute new values of the Rayleigh, Nusselt, and Prandtl numbers. A test is performed using equation (7) to see if the inertial flow zone still exists at the new crystal fraction. If it does, the program then calculates the terminal velocity and effective friction velocity and takes their ratio to find the Rouse number. This process continues until the magma ocean is about 50% crystallized or until the inertial flow zone disappears.

The effective viscosity of a magma mixed with crystals is complex. If the crystal concentration is fairly low, it can be adequately treated as a Newtonian viscous liquid (*McBirney and Murase,* 1984). Its behavior undoubtedly becomes more complex as the crystal fraction increases. This does not pose a major difficulty in estimating the largest crystal radius that the ocean can suspend because the effective viscosity increases with increasing crystal fraction regardless of the suspension's detailed rheological behavior. Consequently, the Rouse number decreases, making crystal separation progressively more difficult as crystallization continues and increasing the radius of crystals that the ocean can suspend.

The rheology of a crystal-magma mixture can be roughly approximated as a viscous fluid with an effective viscosity determined by both temperature and crystal content. Experimental data of *Murase and McBirney* (1973) taken up to about 50% crystallization indicates that the effective viscosity increases nearly exponentially with increasing crystal fraction. Above about 48-52% crystallization, there is a major transition in the rheology. At this point, the crystal density approaches that of hexagonal closest packing so the mixture "locks up," the crystals and liquid are unable to separate easily, and any further convection occurs by the mechanisms of subsolidus convection (*McBirney and Murase,* 1984). If the inertial flow zone still exists at the time the lockup point is reached and if crystals are less than the minimum radius for separation, the crystals and magma can not separate by crystal settling. The crystals and magma will have remained in intimate contact up to this point. If any subsequent separation processes (e.g., percolation of the melt through the crystal matrix) are sufficiently slow, the crystals and residual magma continue to remain in intimate contact for the rest of the cooling history and the ocean cools without fractionation.

We estimated the effective viscosity of the crystal-magma mixture in the following manner. The viscosity of the liquid fraction was first calculated using the computer program published by *McBirney and Murase* (1984), which is itself based on the method of *Shaw* (1972). The program requires input of the major element composition of the magma and the fraction of each liquidus phase crystallizing from the melt. For simplicity, it was presumed that olivine is the liquidus phase throughout the cooling of the ocean to 50% crystals. This assumption is probably valid for the Moon, but a magma ocean deeper than ~400 km on the Earth would have perovskite or majorite garnet as the liquidus phase. With olivine as the liquidus phase, the silica content of the residual magma increases as crystallization proceeds. This results in a substantial increase in the viscosity of the liquid fraction. Thus the viscosity of the liquid fraction is likely somewhat overestimated for the Earth when perovskite or majorite garnet is the liquidus phase. Once the liquid fraction's viscosity is estimated, the effective viscosity of the mechanical suspension of the solid crystals and liquid is calculated using the following equation, derived from the empirical data of *Murase and McBirney* (1973)

$$\nu_{\text{eff}} = \nu_{\text{liquid}} 10^{10\Phi} = \nu_{\text{liquid}}\, e^{23.03\Phi} \qquad (22)$$

where ν_{eff} is the effective viscosity of the suspension, ν_{liquid} is the viscosity of the liquid fraction, and Φ is the crystal fraction. The viscosities calculated by these methods are shown in Fig. 4. Bulk Earth and Moon compositions input into the program of McBirney and

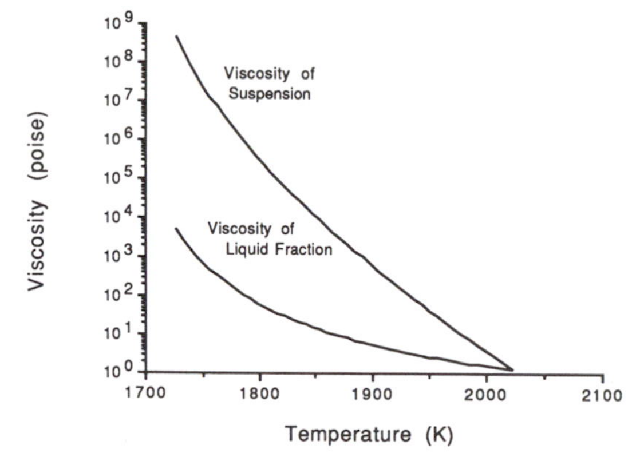

Fig. 4. *The viscosity of the suspension and the liquid fraction calculated according to Shaw's (1972) method (McBirney and Murase, 1984) using bulk Earth composition and assuming that olivine is the liquidus phase throughout solidification.*

Murase were taken from *Taylor* (1986). However, there is virtually no difference in the resulting viscosities of the two planets because the compositions of their silicate portions are similar and olivine is assumed to be the liquidus phase for both bodies.

Equation (20) shows that an overestimate in the viscosity overestimates the maximum crystal radius for suspension for a given Rouse number. However, overestimating the viscosity by a factor of 10 only causes an overestimate in the crystal radius of a factor of about 2.8, which will not change our results qualitatively.

The settling velocity is calculated using Stokes' law if the local Reynolds number was less than 1. If the Reynolds number was greater than 1, a numerical routine that calculates the settling velocity using the measured drag coefficient for spheres is invoked (*Rouse,* 1946, section 40).

We examined the evolution of magma oceans of 400 km and 1000 km depth on the Earth and a 400-km lunar ocean. Additionally, a 1-km lava lake on the Earth was studied under the same assumptions for comparison (note that, unlike magma chambers or real lava lakes such as the well-studied Alae Lava Lake in Hawaii, this computation assumes that no rigid roof forms, so that cooling takes place by free convection). The crystal radius was left as a free parameter because neither the size of crystals at nucleation nor the growth rate are well known. The overall rate of cooling is quite sensitive to the assumed surface temperature. The surface temperature will be the higher of either the equilibrium temperature of the Earth (about 250 K) or the blackbody temperature necessary to radiate the heat flux of the magma ocean to space. We found that the radiation constraint dominated the models presented here, so we determined the temperature difference $\Delta T = T_{liq} - T_{surf}$ between the magma ocean's surface and the liquidus temperature T_{liq} at 1 bar (1973 K; *Ito and Takahashi,* 1987) by requiring that the surface temperature T_{surf} be high enough to radiate the heat flux predicted via equations (1) and

(10) to space, $h = \epsilon\sigma T_{surf}^4$ (ϵ is emissivity, taken here to equal 1, and σ is Boltzman's constant). Although we assumed that the magma ocean's surface could radiate its heat unimpeded to space, we recognize that the temperature difference may be decreased by atmospheric opacity of the proto-Earth and proto-Moon. If the steam atmosphere scenario of *Abe and Matsui* (1985) is correct, the surface temperature of the Earth may in fact be quite high. However, some experimentation with different assumptions about the Earth's and Moon's surface temperatures convinced us that the major effect of such variations is to change the cooling time of the ocean: The settling behavior of the crystals as a function of degree of crystallization (Fig. 5) is nearly independent of assumptions about surface temperature, within broad limits.

Figures 5 and 6 show the behavior of the models described in the previous paragraph. As noted, the Rouse number decreases as the ocean crystallizes due to the rapid increase in the viscosity of the suspension. Thus, as the ocean crystallizes, the minimum crystal radius needed for crystal separation grows, as argued above. In the early stages of both the 1000-km Earth run and the 400-km lunar run, crystals smaller than about 0.5-1 cm initially remain suspended and crystals larger than ~1 cm can initally overcome the convective stirring and settle out on the average. These values must be regarded as order of magnitude estimates only due to significant uncertainty in a number of the parameters, although most of these enter rather weakly into the calculation.

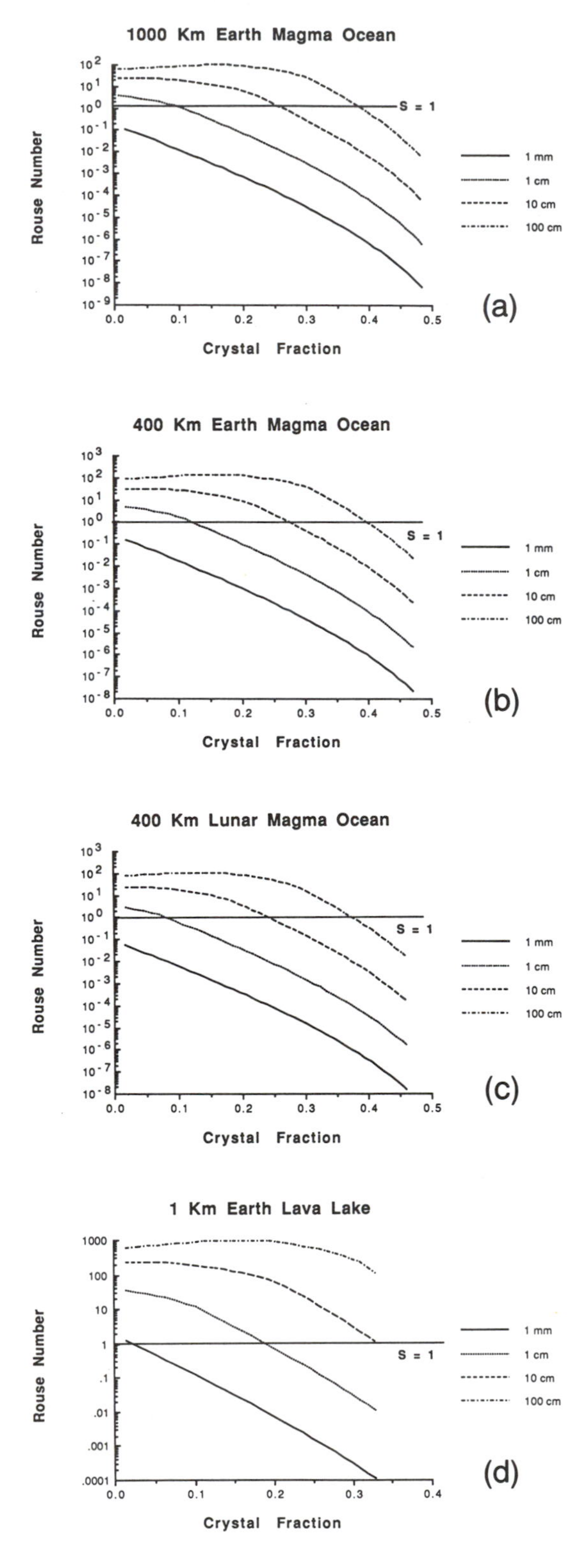

Fig. 5. *Comparison of four model runs showing the suspension criterion, the Rouse number, vs. the crystal fraction, with the crystal radius being a free parameter. If the Rouse number is less than 1, the crystals should remain suspended in the fluid. Note that the 1000-km Earth magma ocean and the 400-km lunar magma ocean are very similar to each other in their ability to suspend crystals. The 400-km Earth magma ocean cannot suspend crystals as well. Also note the strong effect of the crystal radius on the ability of the magma ocean to suspend crystals.*

Note that even if small crystals do initially settle through the low-viscosity magma, the effect of a small amount of fractional crystallization (less than about 10%) could probably not be detectable in the current MORB incompatible/compatible element ratios, especially since the "chondritic" ratios themselves vary by roughly a factor 3 from chondrite to chondrite (M. Drake, personal communication, 1989). As time passes much larger crystals can be suspended in the flow. Thus, if the inertial flow zone exists and the radii of crystals remain smaller than the minimum up to the lockup point, the crystals and magma should remain in intimate contact during this stage of the cooling process.

Comparison of the 1000- and 400-km Earth runs shows that deepening the ocean increases its ability to suspend crystals but that this effect is not strong, as expected from equation (20). Comparison of the 400-km Earth and 400-km lunar runs shows that the effect of the acceleration of gravity is also not great. The 400-km lunar magma ocean is more likely to keep crystals in suspension than is a 400-km terrestrial ocean, consistent with the behavior expected from equation (20), but the difference is very small and could be reversed if initial compositions were allowed to differ. The point remains that the effect of the higher acceleration of gravity on the Earth actually increases the ability of crystals to settle. A 1-km lava lake suspends crystals that are initially 10 times smaller.

The timescale for cooling is very short, being limited by the thickness of the conductive boundary layer and the rate of heat

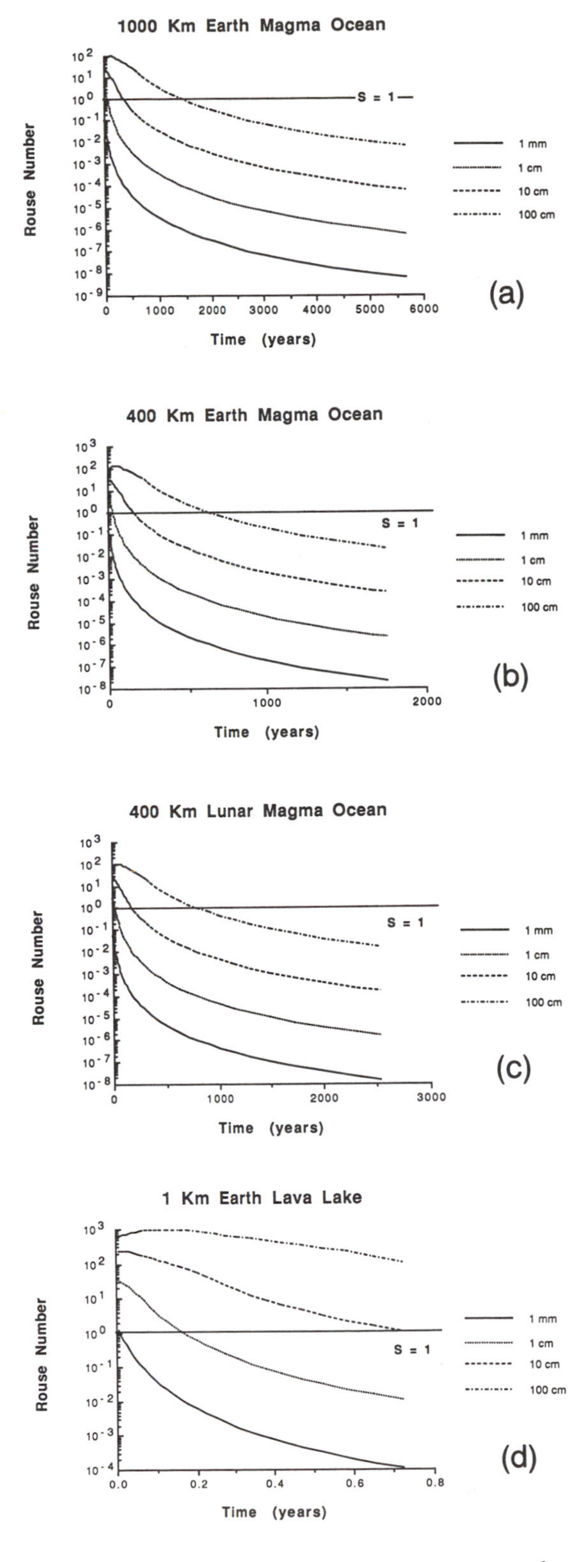

Fig. 6. *The Rouse number vs. time. As in Fig. 5, the 1000-km Earth and 400-km lunar magma ocean are similar in their ability to suspend crystals. The differences are due to the difference in time required to cool the magma oceans. The first several percent of crystallization occur very rapidly but the last 10% takes much longer to crystallize. The 1-km lava lake model shows that it is only capable of suspending crystals of the order of a millimeter before settling occurs. Perhaps more important is the disappearance of the inertial flow zone at only 32% crystallization.*

transfer from its surface. For the 1000-km terrestrial magma ocean, the boundary layer is of the order of a few millimeters thick. When the ocean is convecting as vigorously as predicted, it is unlikely that a chilled crust would remain coherent or become very thick because shear stresses generated by convection would easily tear sections apart. Convective velocities calculated for the 1000-km terrestrial magma ocean were 151.5 cm/sec when the ocean is completely liquid to 22.5 cm/sec at 50% crystallization. *Hofmeister* (1983) argued that the thickness of the chilled crust would reach a steady-state thickness of no more than a few meters.

The rather rapid cooling times computed in Fig. 6 are mainly a result of our assumption that the surface heat is radiated directly into space: The models correspond to magma oceans in which the hot magma has ready access to the surface, no permanent rigid crust forms to insulate the hot liquid below, and the atmosphere does not impede radiative loss. Although these assumptions may be good for the Moon, their validity is arguable for the Earth (however, for Earth the atmosphere may actually increase heat loss by convection). These rapid cooling times also imply that formation of a magma ocean is not easy—if melting does not take place on a timescale shorter than the cooling time, then no magma ocean can be expected. This indicates that the formation of a substantial magma ocean requires either a rapid assembly time, a large impact event, or very rapid core formation.

We also examined the longevity of the inertial flow zone. As noted above, each iteration in the calculation performed a test using equation (7) to see if the inertial flow zone still existed. The inertial flow zone disappeared at 46% crystallization in the 400-km lunar run, 47% in the 400-km Earth run, and 49% in the 1000-km Earth run. The uncertainty in the parameters is sufficiently large and the model is sufficiently crude that these small differences probably have no significance. Thus, there is no substantial difference in the longevity of the inertial flow zone between the Earth and Moon. The degree of crystallization at which the magma-crystal mush locks up is not precisely determined, but is around 48-52% (*Murase and McBirney,* 1973). Magma oceans deeper than a few hundred kilometers on both the Earth and Moon thus probably retained the inertial flow zone until the lockup point. The inertial flow zone disappeared at about 32% crystallization in the 1-km lava lake run. In this case crystals will likely remain suspended in the very early stages of a lava lakes's cooling history. At some point well before the lockup point, however, the inertial flow zone disappears, allowing crystals to separate from the magma as cooling continues with fractional crystallization occurring, as is commonly observed.

From these considerations, it appears that magma oceans on either the Earth or Moon may cool via equilibrium crystallization with no substantial difference between them. Although this result is new, and perhaps unexpected to the many geochemists who have speculated on the chemical differentiation of magma oceans, it is rather problematical in view of the accepted geochemical differences between the Earth and Moon. The simplifying approximations we have made have evidently gone too far, and some important feature has been left out of the analysis. In the next section we speculate that this feature is the effect of gravity on the liquidus and solidus temperature profiles, which in turn affect the mean crystal radius during the cooling process, and thus the ability of crystals to settle out of the convective flow.

GRAVITY, LIQUIDUS, AND SOLIDUS TEMPERATURE PROFILES AND THE SETTLING OF CRYSTALS

We have shown that magma oceans on either the Earth or the Moon can plausibly keep crystals in suspension. Additionally, we found that the effect of the acceleration of gravity on the combined process of convection and

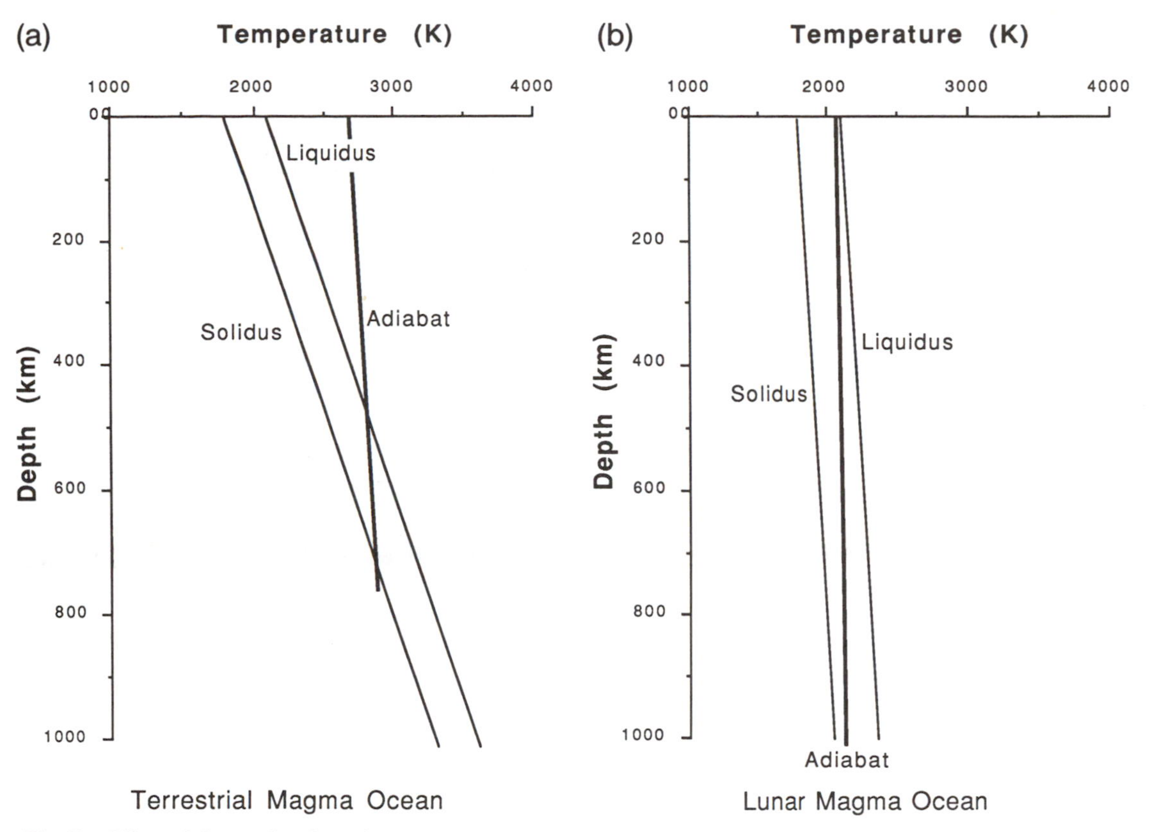

Fig. 7. *Effect of the acceleration of gravity on the slopes of the solidus, liquidus, and adiabatic temperature profiles.* **(a)** *The Earth profiles are plotted using 0.3 K/km for the adiabat and 1.5 K/km for the slope of the liquidus and solidus.* **(b)** *The Moon profiles are plotted using the same values divided by 6 and starting from the same surface temperature. The adiabats are placed to correspond to magma oceans of 600 km depth on both the Earth and Moon. The ratio of the slope of liquidus profile to the slope of adiabat is independent of gravity as required by the Clausius-Clapeyon equation and the equation for the adiabatic gradient (equation (2)). Note that the largest fraction of the Earth ocean is superliquidus while the entire lunar ocean is subliquidus.*

crystal settling is remarkably similar in the two bodies. We must then ask how these two bodies came to exhibit such divergent geochemical evolutionary paths, with the Earth showing few signs of differentiation while the Moon's geochemistry is dominated by early fractionation. Since the answer to this conundrum evidently does not lie in the mechanics of magma ocean convection or the physics of crystal settling, we must look to other, hitherto neglected, differences that depend upon gravity.

Besides the overall vigour of convection, gravity also affects the slopes of the solidus and liquidus temperature profiles, which in turn determine the way in which crystals are distributed in a magma ocean. Although we explicity neglected these slopes in the preceeding analysis, we now consider them in more detail. Figure 7 illustrates the basic effect. We plot the approximate adiabatic, liquidus, and solidus profiles for the Earth and Moon. Values of 0.3 K/km for the adiabat and 1.5 K/km for the slopes of the liquidus and solidus profiles are used for the Earth, typical of the upper few hundred kilometers. The difference in temperature between the liquidus and solidus is assumed constant at 300 K. The lower acceleration of gravity of the Moon causes the slopes dT/dz of the liquidus, solidus and adiabat to be lower than those on Earth by a factor of 1/6, although the *ratio* of the slope of the liquidus and solidus to the slope of adiabat remains constant between the two planets, because both are linear in gravity. The adiabats are placed on the diagram so that the two oceans have equal depths of 600 km, with the bottom of the ocean defined as the depth of the 50% crystallization surface. Note that the adiabat of the lunar magma ocean lies entirely between the liquidus and solidus. Also note that most of the volume of the terrestrial magma ocean lies above the liquidus. This difference has important consequences for the mean crystal size in a cooling magma ocean. Figure 8 schematically illustrates the effect.

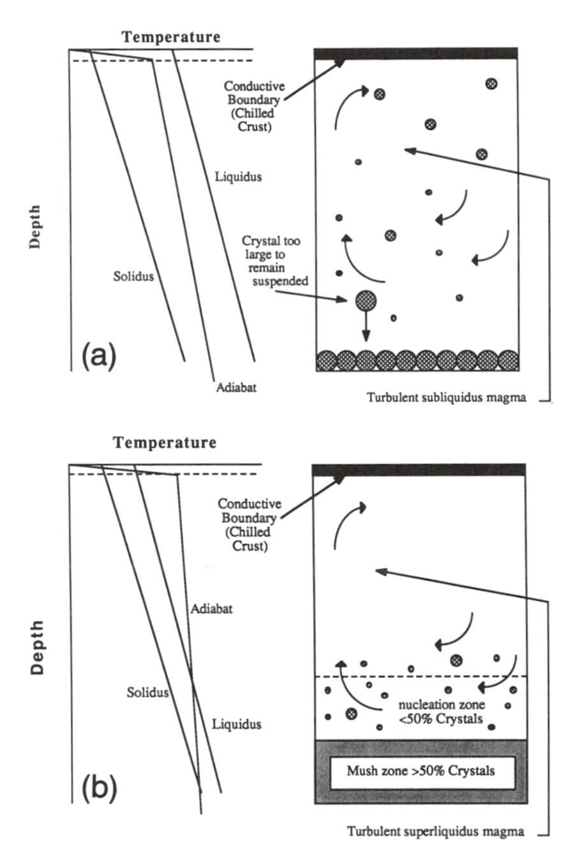

Fig. 8. *Schematic diagram of the liquidus and solidus temperature profiles for* **(a)** *the Earth and* **(b)** *the Moon, neglecting the phase changes on the Earth. Because an adiabat can lie entirely between the solidus and liquidus on the Moon, crystals continue to grow in size until they are large enough to settle out. However, on the Earth, crystals nucleate in a relatively narrow zone near the bottom of the ocean. Any crystals that are swept into the main body of the ocean remelt, thus keeping the mean crystal size small.*

Nucleating crystals that are suspended in a lunar magma ocean remain below the liquidus temperature. They continue to grow no matter where they are transported in the ocean. As time passes, the mean crystal size increases until the crystals are large enough to over come convective stirring and settle to the bottom (or float to the top). The thermal cycling of crystals as they circulate through the magma ocean may have enhanced the crystal size by the elimination of small crystals in favor of large ones, a phenomenon that is commonly exploited in laboratory crystallization experiments (M. J. Drake, personal communication, 1989). On the other hand, the largest fraction of a terrestrial magma ocean is superliquidus. Crystals forming in the narrow subliquidus region are swept by convection into the superliquidus region where they may melt completely. Because residence times in the subliquidus region are limited and growth times consequently short, the mean crystal size remains smaller than in the case of the Moon and probably remains under the threshold size for settling. Thus, a terrestrial magma ocean may have been able to keep the liquid and crystals in intimate contact and thus cooled via equilibrium crystallization, while a lunar magma ocean underwent fractionation.

The residence time of crystals in the subliquidus zones of the two magma oceans can be roughly estimated. Because the lunar magma ocean is nearly all subliquidus during the pertinent part of its cooling history, the residence time of crystals is the time required for the Moon to crystallize to the lockup point at about 50% crystallization. In our model, this time was on the order of 2500 years (Fig. 6c). If crystals grew for the entire cooling time at the linear crystal growth rate of 10^{-10} cm/sec cited by *Marsh* (1988), they would reach radii of about 8 cm. Still larger sizes would be attained if the surface temperature boundary condition discussed above was such as to greatly prolong the cooling of the magma ocean. The large crystals thus formed can segregate from the melt by density differences, resulting in the geochemical fractionations observed in the lunar samples. On the Earth, the thickness of nucleation zone is of the order of 100-km (the exact thickness depends on the depth of the magma ocean). The typical residence time of these crystals is on the order of the depth of the nucleation zone divided by the mean speed that crystals are swept along by the fluid. If crystals are swept at even one-half of the convective velocity, this residence time is about 36 hours at the beginning of crystallization and about 10 days at the end of crystallization, using the convective speeds noted above for the 1000-km magma ocean. At the crystal growth rate noted above, crystal radii would be only about 10^{-4} cm for the 10-day residence time, small enough to be easily suspended by convection. Crystals in a terrestrial magma ocean are thus unlikely to grow large enough to overcome convective stirring and fractionate from the magma.

These arguments are based on a simplification of the melting profiles. Figure 9 illustrates more realistic temperature profiles for the Earth and Moon. This figure shows the phase diagram of *Ito and Takahashi* (1986), with pressure being converted to depth on the Earth and Moon using the appropriate gra-

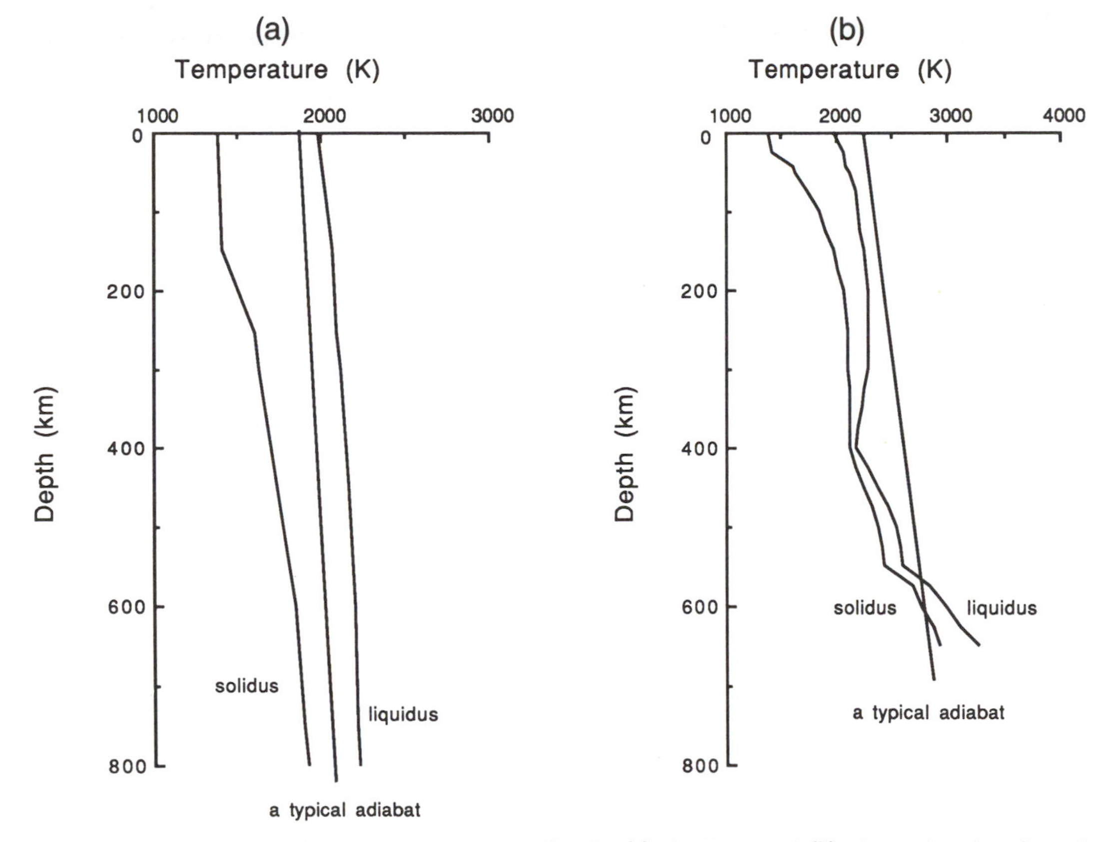

Fig. 9. *Realistic liquidus and solidus temperature profiles for* **(a)** *the Moon and* **(b)** *the Earth, taken from the Ito and Takahashi (1987) phase diagram with pressure coordinate converted to depth. Also plotted are typical adiabats in magma ocean of depth 600 km on the Earth and the Moon showing the approximate relationship to the solidus and liquidus. As the magma ocean cools the temperatures follow adiabats shifted to the left (lower temperature) of the ones plotted. At the times shown temperatures in the entire lunar magma ocean lie between the liquidus and solidus, but only a small fraction of the Earth's magma ocean is subliquidus.*

dients of 1 kbar/4 km depth on Earth and 1 kbar/20 km on the Moon. Also drawn are adiabats that would exist in a 600-km-deep magma ocean on Earth and the Moon.

On Earth, gravity causes the pressure to increase with depth relatively rapidly. At about 400 km there is a major shift in the slopes of the profiles due to perovskite, a high-pressure phase, becoming the liquidus phase. Likewise, the slopes take another large increase at a depth of about 550 km due to the mineral majorite garnet becoming the liquidus phase. Thus, for the Earth to have a magma ocean deeper than ~400 km, the major part of the ocean must be heated to above ~2200 K and is superliquidus as illustrated. On the Moon, the transition to perovskite does not happen until near its center, if it occurs at all. Thus, lunar magma oceans do not have to be as strongly heated to reach 50% crystallization.

The effect of gravity on the temperature profiles of the Earth and Moon noted above still exists even in the more realistic picture. The situation for the Moon is almost identical to that illustrated in Fig. 7 and 8. The phase changes below 400 km on the Earth cause the adiabat to cross the liquidus and solidus in a very narrow region. Cooling of a magma ocean less than about 550 km deep introduces additional complications. There are ocean depths that would have two subliquidus regions with different minerals being the liquidus phase. Even so, melting and recrystallization should occur that would also help keep the mean crystal size small. There is an obvious need for treating this problem in a more sophisticated manner than we have done here. Nevertheless, the concept presented above may hold the key to reconciling the evidence that the Earth must have been melted in its early history with the evidence that the Earth never fractionated.

CRYSTAL SETTLING VS. MAGMA PERCOLATION

If convection keeps the crystals and liquid in intimate contact on the bottom of a terrestrial magma ocean, the geochemical evolution of the system after the crystal fraction exceeds about 50% depends on leakage of the residual magma from the interstitial space between the crystals into the main body of the magma ocean. If this leakage is rapid compared to the advance of the 50% crystallization surface, the ocean will still fractionate and it will not matter that convection keeps crystals in suspension (see, e.g., *Langmuir,* 1989). However, if the upward advance of the 50% crystallization surface is faster than the percolation of residual magma, the crystals will remain in intimate contact with the residual liquid and the ocean will solidify via equilibrium crystallization. A crude estimate of these rise rates can be made as follows. In the model described above, the entire heat flux results from solidification and cooling of the magma. Using the values of $c_p = 1.09 \times 10^7$ ergs/g·K (0.26 cal/g), $L = 5 \times 10^8$ erg/g, and $\Delta T_{liquidus\text{-}solidus} = 400$ K in equation (21), it can be shown that the latent heat loss contributes about 10% to the total heat flux. This heat flux, $\dot{Q}''_s$, must be

$$\dot{Q}''_s = L\dot{m} \qquad (23)$$

where L is the latent heat and $\dot{m}$ is the mass converted from liquid to solid per unit time per unit area. This mass solidification rate can further be expressed as

$$\dot{m} = \rho\dot{z} \qquad (24)$$

where $\dot{z}$ is the upward velocity of the 50% crystallization surface. Using the values from the 1000-km Earth magma ocean model with a total heat flux of about 3.3×10^7 ergs/cm^2·sec and density of 3.3 g/cm^3, the velocity of the 50% crystallization line is about 0.002 cm/sec, or 0.69 km/year.

The rate at which the residual magma rises from of the interstitial space between crystal grains can be roughly estimated using Darcy's law

$$\vec{u} = -\sigma\vec{E} = -\frac{k}{\mu}\vec{E} \qquad (25)$$

where σ is the volume conductivity of the system, k is the permeability of the medium, μ is the fluid viscosity, $\vec{E}$ is the impelling force on the fluid (the buoyant force caused by the density difference between the residual magma and the crystals in this application), and $\vec{u}$ is the "Darcy velocity." The Darcy velocity is the volumetric flow rate per unit area, which equals the speed of fluid transfer through the porous medium per unit area (*Hubbert,* 1956; *Turcotte and Schubert,* 1982, equation 9-1, p. 383). Modeling the porosity of the mush as a cubic matrix surrounded by circular tubes that carry the fluid (*Turcotte and Schubert, 1982,* equation 9-207, p. 414) and using buoyancy as the driving force, the speed of the magma with respect to the porous mush (Δv) is

$$\Delta v = \frac{b^2 \phi \Delta \rho g}{24 \pi \mu} \qquad (26)$$

where b is the grain diameter, $\Delta\rho$ is the difference in density between the liquid and solid, ϕ is the melt fraction ($\sim$50% at the lock up point), g is the acceleration of gravity, and μ is the viscosity of the fluid. From Fig. 4, the viscosity of the magma at 50% crystallization is about 6000 p. The grain diameter of the crystals is difficult to predict. For the estimate, we assume it is as large as 1 mm to be conservative. We take $\Delta\rho$ to be 0.6 g/cm^3. A melt fraction of 0.5 corresponds to a permeability of 6.63×10^{-9} m^2 in equation (25), in the range of pervious sand (*Turcotte and Schubert,* 1982, p. 382, Table 9-1). Using these values, the rise speed of the residual magma is 6.63×10^{-6} cm/sec or 2.09×10^{-3} km/yr, much smaller than the advance of the 50% crystallization surface. By this argument, it appears that the residual magma cannot escape from the interstitial space. Thus, if the mush is formed by crystal suspension, the magma ocean should solidify in equilibrium. As the ocean solidifies, the Rayleigh number and rate of heat loss will decrease, also decreasing the rate of advance of the 50% crystallization surface. There is probably a crossover depth at which these two rates are comparable. If this depth is less than 100-200 km, the geochemical fractionation should not be seen on the Earth.

CONCLUSIONS

Clearly, the fractional crystallization of a magma ocean is a far more complex process than has previously been envisaged. The simple physical analysis presented in this paper indicates that it may be possible for a magma ocean to cool without fractional crystallization due to crystal suspension, contrary to the implicit assumption made by the majority of previous investigators. We have demonstrated several important principles that affect the ability of convecting systems to keep crystals in suspension.

Conclusions that can be drawn from this work are: (1) The ability of a magma body to suspend crystals, regardless of their size, probably depends critically on the existence of the inertial flow zone. Magma bodies must be of planetary scale for the inertial flow zone to exist through cooling to the rheological lockup point. Crystal settling in layered intrusions, magma chambers, and lava lakes is probably controlled by the theory of *Martin and Nokes* (1988), rather than by convective suspension. (2) Even though a higher acceleration of gravity increases vigor of convection, it aids crystal separation overall. (3) A magma ocean on either the Earth or Moon would have a difficult time fractionating unless the crystals become quite large, of the order of centimeters in radius. (4) As the ocean crystallizes, it is less likely to separate due to the increase in viscosity.

We showed, contrary to what might be naively expected, that the effect of the acceleration of gravity on convective vigor and crystal settling alone is insufficient to cause the geochemical disparity between the Earth and Moon. We then argued that magma oceans on the two bodies are likely to have cooled by different paths because of the effect of the planet's gravity on the liquidus and solidus temperature profiles.

The problem of convective stirring on a planetary scale is complex and needs more study. It may be the key to reconciling the evidence from the physics of planetary formation that predicts extensive early melting of the Earth and the geochemical evidence that indicates the Earth's upper mantle was never strongly fractionated. Examination of this problem from a physical point of view has turned up a number of surprising features and shown that little of our intuition gained from studies of subsolidus convection can be transferred to the problem of magma ocean convection. More study is needed on the physics of the upper conductive boundary layer and its control of the overall rate of magma ocean cooling, the effect of the complex interaction between liquidus, solidus and adiabatic profiles, and the phenomena controlling crystal size. Nevertheless, we believe that our study has revealed a number of important qualitative features of crystal fraction in planetary-scale magma oceans that will have to be reckoned with in any future studies of the problem.

Acknowledgments. We thank A. Ingersoll for steering us to Kraichnan's important paper on high Rayleigh number convection, M. Foley for introducing us to Vanoni's book on sedimentation, and M. Drake for many helpful suggestions as this work unfolded. Critical comments by anonymous referees aided in clarifying our presentation. This work was partially supported by NASA grant NAGW-428.

REFERENCES

Abe Y. and Matsui T. (1985) The formation of an impact generated H_2O atmosphere and its implications for the early thermal history of the Earth. *Proc. Lunar Planet. Sci. Conf. 16th,* in *J. Geophys. Res., 90,* C545-C559.

Agee C. B. (1988) Geochemical constraints on early Earth melting (abstract). In *Papers Presented to the Conference on Origin of the Earth,* p. 1. Lunar and Planetary Institute, Houston.

Agee C. B. and Walker D. (1988) Mass balance and phase density constraints on early differentiation of chondritic mantle. *Earth Planet. Sci. Lett., 90,* 144-156.

Bagnold R. A. (1965) An approach to the sediment transport problem from general physics. *U.S. Geol. Surv. Prof. Pap. 422 I.*

Benz W. and Cameron A. G. W. (1990) Terrestrial effects of the giant impact. In *Origin of the Earth,* this volume.

Binder A. B. (1986) The initial thermal state of the Moon. In *Origin of the Moon* (W. K. Hartmann, R. J. Phillips, and G. J. Taylor, eds.), pp. 425-434. Lunar and Planetary Institute, Houston.

Brandeis G. and Marsh B. D. (1989) The convective liquidus in a solidifying magma chamber: A fluid dynamic investigation. *Nature, 339,* 613-616.

Cameron A. G. W. and Benz W. (1988) Effects of the giant impact on the Earth (abstract). In *Papers Presented to the Conference on Origin of the Earth,* pp. 11-12. Lunar and Planetary Institute, Houston.

Chandrasekhar S. (1961) *Hydrodynamic and Hydromagnetic Stability.* Oxford Univ., Oxford. 652 pp.

Chu T. Y. and Goldstein R. J. (1973) Turbulent convection in a horizontal layer of water. *J. Fluid Mech., 60,* 141-159.

Drake M. J., Malvin D. J., and Capobianco C. J. (1988) Primordial differentiation of the Earth (abstract). In *Papers Presented to the Conference on Origin of the Earth,* p. 17. Lunar and Planetary Institute, Houston.

Elder J. W. (1976) *The Bowels of the Earth.* Oxford Univ., New York. 222 pp.

Garon A. M. and Goldstein R. J. (1973) Velocity and heat transfer measurements in thermal convection. *Phys. Fluids, 16,* 1818-1825.

Goldstein R. J. and Tokuda S. (1980) Heat transfer by thermal convection at high Rayleigh numbers. *Int. J. Heat Mass Transfer, 23,* 738-740.

Hartke G. J., Canuto V. M., and Dannevik W. P. (1988) A direct interaction approximation treatment of high Rayleigh number convective turbulence and comparison with experiment. *Phys. Fluids, 31,* 256-262.

Hofmeister A. M. (1983) Effect of a hadean terrestrial magma ocean on crust and mantle evolution. *J. Geophys. Res., 88,* 4963-4983.

Hubbert M. K. (1956) Darcy's law and the field equations of the flow of underground fluids. *Petroleum Trans. AIME, 207,* 222-239.

Ito E. and Takahashi E. (1987) Melting of peridotite at uppermost lower-mantle conditions. *Nature, 328,* 514-517.

Kato T., Ringwood A. E., and Irifune T. (1988) Experimental determination of element partitioning between silicate perovskites, garnets, and liquids: Constraints on the early differentiation of the mantle. *Earth Planet. Sci. Lett., 89,* 123-145.

Kaula W. M. (1979) Thermal history of Earth and Moon growing by planetesimals impacts. *J. Geophys. Res., 84,* 999-1008.

Kirk R. L. and Stevenson D. J. (1988) The role of differentiation in the stress histories of the terrestrial planets: Implications for the Moon and Mars (abstract). In *Lunar and Planetary Science XIX,* pp. 605-606. Lunar and Planetary Institute, Houston.

Kraichnan R. H. (1962) Turbulent thermal convection at arbitrary Prandtl number. *Phys. Fluids, 5,* 1374-1389.

Langmuir C. H. (1989) Geochemical consequences of in situ crystallization. *Nature, 340,* 199-205.

Langseth M. G., Keihm S. J., and Peters K. (1976) Revised lunar heat-low values. *Proc. Lunar Sci. Conf. 7th.,* pp. 3143-3171.

Laufer J. (1951) *Investigation of Turbulent Flow in a Two-Dimensional Channel.* National Advisory Committee for Aeronautics Report 1053.

Long R. R. (1975) Relation between Nusselt number and Rayleigh number in turbulent thermal convection. *J. Fluid Mech., 73,* 445-451.

Martin D. and Nokes R. (1988) Crystal settling in a vigorously convecting magma chamber. *Nature, 332,* 534-536.

Marsh B. D. (1988) Crystal capture, sorting, and retention in convecting magma. *Geol. Soc. Am. Bull., 100,* 1720-1737.

Matsui T. and Abe Y. (1986) Evolution of an impact induced atmosphere and magma ocean on the accreting Earth. *Nature, 319,* 303-305.

McBirney A. K. and Murase T. (1984) Rheological properties of magmas. *Annu. Rev. Earth Planet. Sci., 12,* 337-357.

McFarlane E. A. and Drake M. J. (1990) Element partioning and the early thermal history of the Earth. In *Origin of the Earth,* this volume.

Melosh H. J. (1990) Giant impacts and the thermal state of the early Earth. In *Origin of the Earth,* this volume.

Melosh H. J. and Kipp M. E. (1988) Giant impact theory of the moon's origin: Implications for the thermal state of the early Earth (abstract). In *Papers Presented to the Conference on Origin of the Earth,* pp. 57-59. Lunar and Planetary Institute, Houston.

Murase T. and McBirney A. R. (1973) Properties of some common igneous rocks and their melts at high temperature. *Geol. Soc. Am. Bull., 84,* 3563-3592.

Ohtani E. (1985) The primordial terrestrial magma ocean and its implication for stratification of the mantle. *Phys. Earth Planet. Inter., 38,* 70-80.

Palme H., Spettle B., Wänke H., Bischoff A., and Stöffler D. (1984) Early differentiation of the Moon: Evidence from trace elements in plagioclase. *Proc. Lunar Planet. Sci. Conf. 15th,* in *J. Geophys. Res., 89,* C3-C15.

Priestley C. B. H. (1954) Convection from a large horizontal surface. *Aust. J. Phys., 7,* 176-201.

Ringwood A. E. (1988) Early history of the Earth-Moon system (abstract). In *Papers Presented to the Conference on Origin of the Earth,* pp. 73-74. Lunar and Planetary Institute, Houston.

Rouse H. (1946) *Elementary Mechanics of Fluids.* Wiley, New York. 376 pp.

Safronov V. S. (1969) *Evolution of the Protoplanetary Cloud and Formation of the Earth and Planets.* Nauka, Moscow. Translated by the Israel Program for Scientific Translation (1972).

Safronov V. S. (1978) The heating of the Earth during its formation. *Icarus, 33,* 1-12.

Schubert G. (1979) Subsolidus convection in the mantles of terrestrial planets. *Annu. Rev. Earth Planet. Sci., 7,* 289-342.

Shaw H. R. (1972) Viscosities of magmatic silicate liquids: An empirical method of prediction. *Am. J. Sci., 272,* 870-893.

Solomon S. C. (1986) On the early thermal state of the Moon. In *Origin of the Moon* (W. K. Hartmann, R. J. Phillips, and G. J. Taylor, eds.), pp. 435-452. Lunar and Planetary Institute, Houston.

Stevenson D. J. (1981) Models of the earth's core. *Science, 214,* 611-619.

Sumer B. M. (1985) Recent developments on the mechanics of sediment suspension. In *Transport of Suspended Solids in Open Channels* (W. Bechteler, ed.), pp. 3-13. A. A. Balkema, Rotterdam.

Taylor S. R. (1986) Origin of the Moon: Geochemical considerations. In *Origin of the Moon* (W. K. Hartmann, R. J. Phillips, and G. J. Taylor, eds.), pp. 125-144. Lunar and Planetary Institute, Houston.

Taylor S. R. (1988) Planetary compositions. In *Meteorites and the Early Solar System* (J. F. Kerridge and M. S. Matthews, eds.), pp. 512-534. Univ. of Arizona, Tucson.

Threlfall D. C. (1975) Free convection in low-temperature gasoline helium. *J. Fluid Mech., 67,* 17.

Turcotte D. and Schubert G. (1982) *Geodynamics.* Wiley, New York. 450 pp.

Vanoni V. A., ed. (1975) *Sedimentation Engineering.* American Society of Civil Engineers, New York. 745 pp.

Verhoogen J. (1980) *Energetics of the Earth.* National Academy of Sciences, Washington, DC. 139 pp.

Wetherill G. W. (1976) The role of large bodies in the formation of the Earth and Moon. *Proc. Lunar Sci. Conf. 7th,* pp. 3245-3257.

Wetherill G. W. (1986) Accumulation of the terrestrial planets and implications concerning lunar origin. In *Origin of the Moon* (W. K. Hartmann, R. J. Phillips, and G. J. Taylor, eds.), pp. 519-550. Lunar and Planetary Institute, Houston.

Wood J. A. (1970) Petrology of the lunar soil and geophysical implications. *J. Geophys. Res., 32,* 6497-6513.

HEAT AND MASS TRANSPORT IN THE EARLY EARTH

G. F. Davies

Research School of Earth Sciences, Australian National University, GPO Box 4, Canberra, ACT 2601, Australia

Mass transport was responsible for establishing the main stratification of the Earth (core, mantle, and crust) and also was the dominant means of heat transport through most of the Earth's history. During early accretion, impacts supplied the heat input and also enhanced heat loss by stirring the near-surface. Initial core segregation probably began catastrophically when the Earth was about half of its present radius and homogenized the temperature near the melting temperature of the mantle. Thereafter, core segregation would have been rate-limited by accretion. Subsequent thermal evolution would have depended on whether the surface was insulated by a thick atmosphere or a buoyant crust, either of which might have permitted a magma ocean to accumulate. Magma extruding onto a cold surface is a potent heat removal mechanism: In this case accretion in much less than 1 m.y. would have been required for a magma ocean to accumulate, and a magma ocean generated by a late giant impact would cool to a mush in about 1000 years. Solid-state mantle convection would have begun soon after initial core segregation (or upon establishment of a cool surface). Mantle convection would have been enhanced by continuing core segregation and melt extraction of heat. At a mean temperature perhaps 100°C higher than at present, the mantle would have entered a regime in which a further increase by 100°C would have required increasing the heat input by one or two orders of magnitude because of heat extraction by melt. In the latter regime, intrasilicate density differences may have played an important role. A substantial magma ocean seems likely only in the cases of an insulated surface or a late giant impact. As the Earth cooled, there may have been one or more transitions in tectonic style, but how these might relate to the accumulation, chemistry, and structural style of the preserved continental crust is still not clear. The lack of clear evidence for any compositional stratification in the present mantle cautions against models involving substantial early stratification.

INTRODUCTION

Mass transport, or mass motion, in the Earth's interior was obviously involved with the establishment of the primary stratification of the Earth into crust, mantle, and core. Mass transport has also been the main agent of heat transport in the deep interior for most of the Earth's history. Here we will concentrate on mass transport as the agent of heat transport, since that has driven the tectonic evolution of the Earth, but some brief discussion of core-mantle separation and the differentiation of the continental crust will also be given.

It is now widely appreciated that mass transport in the mantle, in the form of mantle convection, is the key to understanding heat transport in the earth as it operates at present. There are two reasons why this is so. First, thermal conduction and radiation are very inefficient over length scales of thousands of kilometers (*Schatz and Simmons,* 1972; *Shankland et al.,* 1979), so the Earth would take many times its present age to cool significantly by conduction. Second, the advent of plate tectonics made it clear that the mantle is internally mobile, and the characteristic velocities of centimeters per year are capable of transporting heat much more efficiently over long distances. The second point was underscored by *Tozer's* (1965, 1972) argument that the temperature dependence of silicate rheology means that a large planet will almost inevitably become hot enough and soft enough to convect internally, and that the convection will then in fact strongly regulate the internal temperature.

In considering the early stages of the Earth's evolution, including its formation, it becomes clear that heat transport by mass motion, called advection, is likely to have been dominant then as well. However, three general types of advection can be identified, distinguished by the source of the mass motion: (1) impact of accreting material, (2) internal chemical buoyancy, and (3) internal thermal buoyancy. Each of these probably dominated at some stage in the early history of the Earth, and there may have been distinct substages in which the roles of different kinds of chemical buoyancy were important. It is also worth singling out melting as having a distinctive effect on heat transport, because although the sources of motion of the magma may be the same as for the solid phase, its much lower viscosity makes it potentially orders of magnitude more efficient in transporting heat.

A subject as remote from direct evidence as the formation of the Earth must be approached with some humility because, as will be plain from other papers in this volume, we are still very much at the stage of identifying processes that *may* have been dominant at various times. Nevertheless, a surprising amount can be said with reasonable confidence provided certain conditions applied. The challenge here is to find a way through the plethora of possible combinations of circumstances. I will use two devices for this purpose. First, I will try to concentrate on processes rather than events, and to be clear about the conditions under which they would be important. Second, I will outline initially what seems to be a plausible general sequence of events, while noting major and minor variations on this sequence. In this way, perhaps, some order will be brought to the presentation without unduly channeling readers' lateral thinking.

SOME ASSUMPTIONS, AND A REFERENCE SEQUENCE OF MAIN EVENTS

In the reference scenario it will be assumed that the terrestrial planets grow by collisional accretion in a cold, gas-free environment and that any atmosphere is thermally transparent so that the planetary surface is well below typical rock melting temperatures (cf. *Tajika and Matsui,* 1990). Approximately homogeneous accretion will be assumed (i.e., no major change in composition of the accreting material during the growth of the planet). Melts will be assumed to be less dense than their associated solids. Giant impacts, where the impactor is greater than a few percent of the mass of the growing planet, will not be considered in the first instance. However, a substantial fraction of unaccreted mass will be assumed to be in relatively large bodies.

With the above assumptions about the accretion environment, the Earth will accrete over a few tens of millions of years, in the initial stages as a cold mixture of rock and metal phases (*Safronov,* 1969). As the Earth grows, its increasing gravity will increase the impact velocity of accreting material, and the near-surface temperature will thereby be raised in proportion to the square of the radius of the Earth such that the temperature will be a few hundred degrees at a radius of 1000 km and melting temperatures will be reached at a radius of roughly 3000 km (50% uncertainty) (*Safronov,* 1978; *Kaula,* 1979; *Davies,* 1985). Once melting occurs in the outer part of the Earth, separation of metal and silicate will occur, most likely followed shortly by a catastrophic overturn in which the separated metal and the metal/silicate protocore exchange places (*Stevenson,* 1980; *Davies,* 1982). Segregation of metallic material into the core will then be rate-limited by accretion. Probably soon after initial core segregation the mantle will be hot enough to start convecting, and convection will then remain the dominant mechanism for heat removal from the deeper interior. Also, if large degrees of melting occur, melt migration will become the main heat removal mechanism in the outer few hundred kilometers of the Earth. Melting, initiated by impact heating, will be sustained thereafter by gravitational energy released internally by core segregation and by radioactivity. Melting continues to the present, but for much of the Earth's history its main cumulative effect has been to differentiate the continental crust rather than transport significant amounts of heat.

This scenario can be divided into four main phases. In the first, there is no internally driven mass motion and the interior is cold. The near-surface is stirred by impacts, so that impacts accomplish both heat deposition and some heat removal. In the second phase, internal motions are driven by the density difference between metal and silicate phases as the core segregates, at first catastrophically and then in a steady rain of newly accreted metal. In the third phase, internal motions are driven by thermal density differences. The third phase presumably takes over gradually from the second, and it may for some time continue to be affected by chemical or phase density differences within the silicates, as will be discussed later. The fourth phase does not separate clearly in time, but most of the extant continental crust dates from the last two-thirds of the Earth's history; this may be the most distinctive feature of this era, so it is distinguished on this basis. It is possible that phase four coincides more or less with the modern form of plate tectonics.

PHASE 1: IMPACT HEATING AND STIRRING

The theory of collisional accretion in a cold, gasless environment was developed by *Safronov* (1969) and *Wetherill* (1976), and it predicts a power-law distribution of accreting bodies, the largest of which could be between 0.1% and 10% of the mass of the growing Earth. *Safronov* (1978) pointed out that the effect of the large accreting bodies (tens of kilometers and greater) is to deposit heat at a significant depth below the surface from where it may not escape to the surface before being buried by further accretion. Thus a cold environment need not imply a cold accreted planet, as had been thought previously on the basis of a theory in which very small impactors were implicit (e.g., *Hanks and Anderson,* 1969).

Kaula (1979) presented elaborate calculations that confirmed this general conclusion. *Davies* (1985) incorporated constraints on the scaling of impact effects with impactor size due to work by *Holsapple and Schmidt* (1982), which Kaula's scalings had not satisfied, and also separately characterized the several important physical processes involved so that their relative importance could be more readily evaluated.

As well as depositing heat, an impact stirs the near-surface material. *Safronov* (1978) proposed treating the effect of successive impacts as a random vertical walk of the stirred material, which allows the calculation of an effective thermal diffusivity. *Davies* (1985) identified three important length scales in the impact heating/stirring problem: (1) the depth scale of heat deposition, (2) the depth scale of impact stirring, and (3) the depth scale of heat removal by diffusion. The latter, which we can call the cooling depth scale, is given by κ/v, where κ is the effective diffusivity and v is the upward velocity of the accreting surface, i.e., the rate of growth of the planetary radius. If the cooling depth scale is not significantly larger than the heat deposition depth scale, then a significant fraction of the deposited heat will be retained. It turns out that all three depth scales are independent of the planetary growth rate and proportional to the planet radius. It follows that in this regime the efficiency of heat retention is constant, with a value between about 40% and 70% (*Davies,* 1985).

The above discussion concerns the efficiency with which heat deposited by an impact is retained within the planet in the long term. In addition, we must consider the efficiency with which the kinetic energy of the impactor is converted into deposited heat during the course of the impact. It is not appropriate to delve into this complex topic here: It has been considered in some detail by *O'Keefe and Ahrens* (1977). Briefly, there is an initial conversion of kinetic energy into heat by the passage of a shock wave generated by the impact, and this may be more than 80% efficient. However, a lot of this heat may be contained in small fragments thrown out by the impact, and therefore may be lost by radiation before the fragments fall back. *Kaula* (1979) estimates a net efficiency of the total impact process of 10-20%.

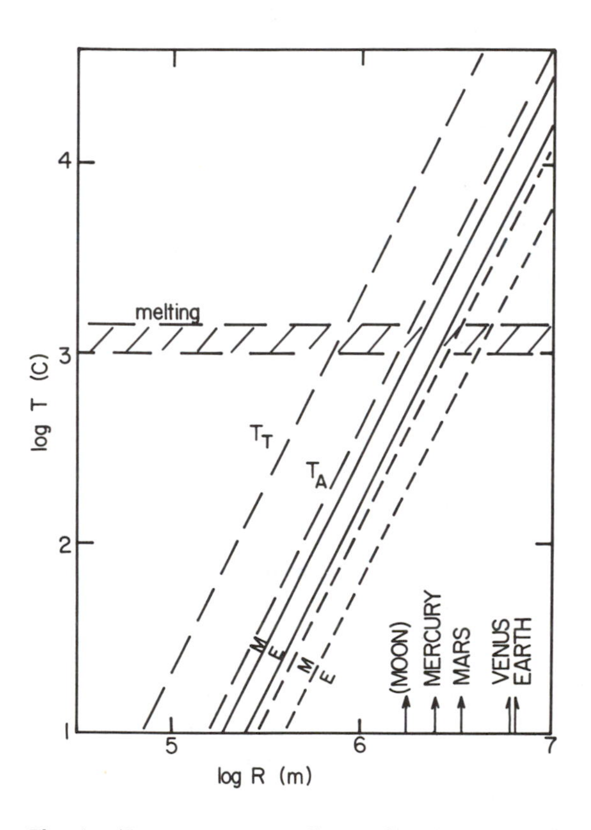

Fig. 1. *Temperature vs. radius profiles for terrestrial planets in Phase 1 of their growth (impact heating and stirring). The profiles do not apply above the shaded melting band. T_T is the temperature attained if all of the kinetic energy of accreting material is trapped in the planet, and T_A is the temperature attained if 20% of the kinetic energy is deposited and trapped. The solid lines bound the most likely case that 80% of kinetic energy is lost on impact and a further 30-60% is lost through impact stirring (for a final retention of 8-14%). The solid and short-dashed curves labeled "M" and "E" represent different assumptions about the scaling of energy deposition and impact stirring. From Davies (1985).*

Thus something like 5-15% of the kinetic energy of impactors may be deposited as heat and retained in the interior of the growing planet. Since there is enough gravitational energy initially available to heat the Earth by 40,000 K or more, this efficiency is quite enough to yield a hot Earth. A feature of *Safronov's* (1969) collisional accretional theory is that the kinetic energy of impactors grows as the square of the planet radius due to the stronger gravitational perturbations of the planetesimal swarm by the growing planet. The result is that temperature in the planet also increases as the square of the radius: After the impact stirring zone migrates out from a given radius, the temperature at that radius remains essentially constant (until Phase 1 of our scenario is completed). Quantitative results from *Davies* (1985) are shown in Fig. 1. Temperatures are low at the center of the planet and approach melting at a few thousand kilometers radius. Evidence for Mars being hot early in its history (*Chicarro et al.*, 1985; *Watters and Maxwell*, 1986; *Chen and Wasserburg*, 1986) is consistent with this result. The onset of melting would mark the end of Phase 1.

This analysis will not apply if a significant amount of interplanetary gas were present, since this transforms the accretionary dynamics (*Patterson and Spaute*, 1988). Nor would it apply if the surface of the Earth were kept very hot by a thick, opaque atmosphere (*Tajika and Matsui*, 1990), since this would prevent heat loss from the interior. This analysis would be inaccurate if the unaccreted material included bodies that were greater than a few percent of the Earth's (growing) mass, since the details of such individual impacts would come to dominate the thermal state in the later stages (*Melosh and Kipp,* 1988). If accreting bodies are all very small, less than about 1 km (e.g., *Patterson and Spaute,* 1988), the analysis applies but the results are very different: Temperatures are much lower.

PHASE 2: CORE SEGREGATION

Elsasser (1963) proposed that core segregation would begin when the metal phase in the outer parts of the Earth began to melt: He envisaged that the metal melt would collect into large blobs that would then sink through the metal/silicate interior. One potential difficulty with this proposal is that the cold interior might be strong enough to resist the intrusion of the metal blobs. *Stevenson* (1980) suggested that the metal might fracture the cold interior, while *Davies* (1982) noted that even the highest plausible material strengths could be overcome by sufficiently large metal blobs of the order of 100 km in radius. Either way it is plausible, on the basis of the above estimates for Phase 1, that core segregation would have begun when the Earth had reached about half of its final radius.

An initial core segregation of this type would have been a runaway process because a large amount of gravitational energy would have been converted to heat in the process, thus reducing the material strength that resisted further segregation. The total energy difference between a completely homogeneous Earth and the present Earth would be sufficient to heat the Earth by about 2300 K (*Flasar and Birch,* 1973). When the Earth had reached half its final radius, the available energy would have been equivalent to about 600 K. With the predicted parabolic temperature profile from Phase 1 and an outer temperature T_R of about 1700 K, the Earth would have had a mean temperature at the onset of core segregation of $0.6T_R$, or about 1000 K. Thus initial core segregation could have raised the mean temperature to about 1600 K, close to melting and high enough to reduce material strengths and viscosities. (Note that at this stage the central pressure would be only about 30 GPa and would not have greatly raised either melting temperatures or the adiabatic temperature.) The temperature is also likely to have been homogenized by the vigorous stirring accompanying core segregation.

This picture depends on the metal being able to segregate into large blobs. *Stevenson* (1990) has noted that surface tension effects of liquid iron in a solid silicate matrix may resist this, and suggests that other mechanisms may have been involved. He notes that metal droplets in a silicate liquid may be limited to about 1 cm in size, but nevertheless estimates that in an initially liquid, homogeneous Earth, the entire core-mantle separation might have taken only about 100 years.

In fact it is likely that initial melting of the metal phase would have been accompanied by substantial silicate melting: The melting temperatures are similar, and impact heating would have been quite heterogeneous. Thus it may be that the largest impacts of the period generated "lakes" of liquid metal/liquid silicate mixture from which substantial amounts of metal segregated before they cooled (on about the same timescale—see later discussion). Reheating by shock waves from later impacts may have allowed large bodies of metal to collect at the base of the outer hot zone.

The timescale of this initial core segregation is very difficult to estimate because it depends on the rheology of silicates at extreme stresses and moderate temperatures and pressures, which is unknown. If the main phase took a matter of days, then velocities of the order of 10 m/sec are implied; this seems plausible, but is only a guess. After this main phase the Earth would probably have been thermally homogenized because the sinking iron would have carried heat into the cool interior, as well as releasing gravitational energy. It is possible that the core material refroze during its descent or was subsequently refrozen by the increasing pressures at the center. The mantle seems likely to have been chemically homogenized by the vigorous stirring, because the metal-silicate density differences are much greater than the intrasilicate chemical and thermal density differences, and seems likely to be near its solidus.

After the initial catastrophic segregation there probably followed a period in which the near-surface was maintained between the solidus and liquidus by the competition between impacts and melt eruption (see later discussion) and the mantle was stirred and heated by the descent of newly accreted core material. The duration of this phase is difficult to estimate, but the combination of the buffering of the near-surface temperature, the deeper heating, and the gradual steepening with time of the solidus at depth, in response to the rising pressure, is likely to have made the temperature profile sufficiently unstable to have initiated convection in the mantle before the main phase of accretion was complete. This would mark the (gradual) transition from Phase 2 (core segregation) to Phase 3 (thermal convection in the mantle).

It is difficult to imagine that core segregation could have been delayed to near the end of accretion. If penetration of the metal/silicate protocore was more difficult than supposed above, then segregation might have been delayed, but melting near the surface would have produced a metal layer, resulting in a gravitationally highly unstable configuration of metal over metal + silicate (*Stevenson,* 1981). The protocore would have been displaced and presumably must then have deformed or fractured and been broken down. Heating from a late giant impact or heat retention by an opaque atmosphere would only have promoted segregation. It seems implausible that impact stirring of near-surface melt could have prevented metal separation from the melt.

Core segregation would be delayed if accretion were relatively cool. This would be possible if all accreting bodies were very small, as noted earlier. Core segregation would then depend on radioactive heating, and the timing would depend critically on the actual temperature reached during accretion. Billions of years would be required if the initial temperature were very low (*Davies,* 1980a), and this is probably precluded by lead isotopes, which require core formation to have occurred within about 100 m.y. of meteorite formation at 4.57 Ga (assuming that core segregation was the main event responsible for separating uranium from lead; *Oversby and Ringwood,* 1971; *Davies,* 1984).

It is likely, though not certain, that the metal and silicates chemically equilibrated near the Earth's surface in the above scenario and its main variations. However, it is plausible that they did not equilibrate under the different conditions prevailing in the deep interior of the nearly grown Earth, so that the mantle and core may well be out of chemical equilibrium, as has been noted by others (*Stevenson,* 1981; *Ahrens,* 1990).

THE ROLE OF MELT IN HEAT REMOVAL

The presence of melt is likely to greatly increase the efficiency of heat removal if the melt erupts onto a cold surface. This has a strong bearing on the presence or absence of a magma ocean, but seems to have been either controversial or often overlooked. It is therefore discussed here before we proceed to Phase 3.

A very crude approximation will illustrate the effect. Suppose melt is erupted onto a surface that is about 1000 K cooler than the interior, and then spreads and forms a chill crust. Heat will have to be conducted through the chill crust, so it will regulate the rate of heat loss. In a Hawaiian basaltic lava lake, the chill crust was maintained at only a few centimeters thick because it continuously foundered into the actively circulating magma (*Duffield,* 1972). Suppose that this is true in our primordial lava lake: The heat flux, q, out of the crust will then be

$$q = K\Delta T/d \qquad (1)$$

where K is the conductivity, ΔT is the temperature drop across the crust, and d is the thickness of the crust. With $K = 3$ W/mK, $\Delta T = 1000$ K, and $d = 30$ mm, we get $q = 10^5$ W/m^2.

Now supposing we have a global magma ocean, and let us ask at what rate this heat loss would freeze the magma ocean. If the latent heat of crystallization is of the order of $H = 500$ J/g (*Bottinga and Richet,* 1978) and the magma ocean has depth D and density ρ, then the latent heat of crystallization per unit area of the surface is

$$Q_H = H\rho D \qquad (2)$$

If the surface heat loss results in crystallization of the magma ocean, then we can equate the heat flux to the negative rate of change of Q_H

$$q = -Q_H/t = -H\rho D/t \qquad (3)$$

Taking $\rho = 3000$ kg/m^3, we can obtain a rate of decrease of D of 2 km/yr. Thus a magma ocean 100 km deep would freeze within about 50 years.

A slightly more sophisticated treatment of heat loss from a magma ocean, based on a parameterized convection formula, is given in Appendix A. There the heat flux is also estimated to be in the range 10^4-10^5 W/m^2. This implies a chill crust thickness and a freezing time of the order of 10 cm and 100 years, respectively. Given that ultramafic magmas have viscosities as low as 100 p (*Bottinga and Weill,* 1972), that the magma ocean would therefore be convecting very vigorously (*Tonks and Melosh,* 1990), and that the accretionary bombardment would be continuing, tending to keep the crust broken up, these results are not implausible.

The critical assumptions in the above treatment are that the surface is cold and the chill crust is denser than the melt so that it can founder. If there is an insulating thick atmosphere, then the first assumption would be invalid, and heat loss would be regulated by the atmosphere (*Tajika and Matsui,* 1990). If the crust is buoyant, then it would become much thicker and heat loss would be much lower. This seems to have been the situation on the Moon, where the buoyant anorthosite crust allowed a magma ocean to accumulate.

Now consider some implications of the rapid heat losses just estimated. First, unless heat can be supplied at a similar rate, a magma ocean will never accumulate. Possible heat sources are accretion and core segregation. Considering first the latter, the total energy released by core segregation would have been about 1.5×10^{31}J, but about one third of this might be absorbed in warming up the interior during and soon after the initial catastrophic overturn. Thus about 10^{31}J would have been available to cause melting, but its rate of release would have been limited by the rate of accretion. To match the lowest heat flux estimate of 3000 W/m^2, the main phase of accretion would have had to occur in less than 0.2 m.y. This is about 100 times the rate predicted by the Safronov theory.

If we consider accretion itself as the energy source, about 20 times as much energy is available, but the efficiency of its deposition and retention in the Earth might be only 5-15%, so we have about 1.5 to 5 times as much heat actually available to cause melting as for core segregation. Thus accretion over 1 m.y. (still a much shorter time than in the Safronov theory) might match the lowest heat loss estimate. Of course, if the heat loss were as high as 10^5 W/m^2, the accretion time would have to have been 30 k.y. or less. It is clear that in the standard Safronov theory, on which our reference scenario is based and in which the largest secondary bodies are less than 1% of the Earth mass, a global magma ocean would not accumulate. In fact, these results imply that on average only a few percent of the Earth's surface would need to be molten in order to remove all of the heat of accretion and core segregation.

There are several alternatives that would allow a magma ocean to accumulate. An insulating atmosphere has already been mentioned. A buoyant crust on the magma would have the same effect, though there is little evidence for one having existed on the Earth (*Drake et al.,* 1988), in contrast to the Moon (*Wood,* 1970). A variation on this is a "buried" magma ocean due to a reversal of liquid-solid relative densities at a depth of hundreds of kilometers in the mantle (*Agee,* 1988). Rapid accretion (in 1 m.y. or less) may be possible in the context of "runaway accretion," in which the primary bodies grow much faster and larger than secondary bodies (*Patterson and Spaute,* 1988; *Wetherill,* 1988), though in this case the theories are not as quantitatively developed as the Safronov type of theory. Finally, one or several late giant impacts could melt much of the Earth (*Melosh and Kipp,* 1988), yielding a transient global magma ocean that would cool to a crystal-liquid mush on a timescale of about 1000 years.

Aside from the question of a magma ocean, the above estimates imply that melting can extract heat very efficiently from a near-surface melt zone. This effect will strongly buffer temperatures throughout the Earth, except in extreme circumstances, such as a giant impact. It could also have a strong modulating effect on solid-state mantle convection, tending to make it more efficient, as we shall see.

PHASE 3: THERMAL MANTLE CONVECTION

The basic heat transport characteristics of thermal convection in the mantle have been established with the aid of "parameterized convection" theory, which is based on a simple relation between heat flux and temperature difference. Here, two points will be stressed: First, the rate of convection is a strong function of temperature, especially if melting is involved, and second, the likelihood that the mantle solidus is steeper than the adiabat makes it unlikely that all of the mantle would have been melted except in the immediate aftermath of a giant impact.

The emphasis here will be on convection in the mantle, based on the assumption that convection in the core will adjust its temperature quickly to any change in the mantle. In other words, the mantle will regulate the thermal evolution of the whole interior.

Pure Thermal Convection

Several authors more or less independently demonstrated that a simple relation between temperature difference and convective heat flux could be used to calculate the thermal evolution of a convecting mantle (*Sharpe and Peltier,* 1978, 1979; *Turcotte et al.,* 1979; *Davies,* 1980a; *Schubert et al.,* 1980; *Stacey,* 1980). The so-called parameterized convection relation is usually put in the form of a relation between Nusselt number, Nu, which measures heat flux, and Rayleigh number, Ra, which measures temperature difference

$$\mathrm{Nu} = a\,\mathrm{Ra}^p \quad (4)$$

where $\mathrm{Nu} = qD/K\Delta T$ and $\mathrm{Ra} = g\alpha D^3 \Delta T/\kappa\nu$ and a and p are constants. Here, q is the convected heat flux, D is the depth of fluid, K is conductivity, ΔT is the temperature difference across the fluid layer, α is the coefficient of thermal expansion, κ is the thermal diffusivity, and ν is the kinematic viscosity. This can be rewritten in a more directly relevant form as

$$q/q_0 = (T/T_0)^{1+p}(\nu/\nu_0)^{-p} \quad (5)$$

where the subscript 0 denotes a reference state. The form of equation (5) can be derived from a boundary layer theory first developed by *Turcotte and Oxburgh* (1967). This theory balances the driving thermal buoyancy forces against viscous resistance forces. It is found that with the Rayleigh number definition used above p = 1/3. A very simplified derivation of equation (5) is given in Appendix B, for comparison with the theory (below) including melt extraction of heat.

In the silicate mantle, the heat flux given by equation (5) is a strong function of temperature because the viscosity of silicates is a strong function of temperature: Viscosity decreases by nearly an order of magnitude for a 100 K increase in temperature. This is conveniently represented by an approximation that applies near mantle temperatures

$$\nu/\nu_0 = (T/T_0)^{-n} \quad (6)$$

where $n \approx 30$ at mantle temperatures (*Davies,* 1980a). It can be seen that most of the temperature dependence of q is due to the viscosity, and equations (5) and (6) can be combined into

$$q/q_0 = (T/T_0)^m \quad (7)$$

where $m = 1 + p(n+1) \approx 10$.

The implication of equation (7) is that a large change in heat flux can be accommodated by a small change in temperature. For example, a six-fold increase in heat input will accomplish only a 20% increase in temperature. This effect is illustrated in Fig. 2, which shows temperature histories calculated by *Davies* (1980b) assuming an initial temperature half of the present mean temperature and the release of core segregation heat over various periods ranging from 30 m.y. to 1 G.y. (the latter being unrealistic). The peak heat flux is about 60 times the present value, but the peak temperature is only about 50% greater than at present. Other effects will probably reduce the peak temperature further, as we will see. Figure 2 also shows that the thermal transient from core segregation is dissipated in about 0.5 G.y.

Christensen (1984) questioned this strong sensitivity to temperature on the basis of numerical models with a temperature-dependent viscosity. However, his models developed a static, cool upper boundary layer that is not a good analogue of the Earth's mobile lithosphere. *Gurnis* (1989) has shown explicitly that a mobile boundary layer (more like plates) recovers the strong temperature dependence of the heat flux.

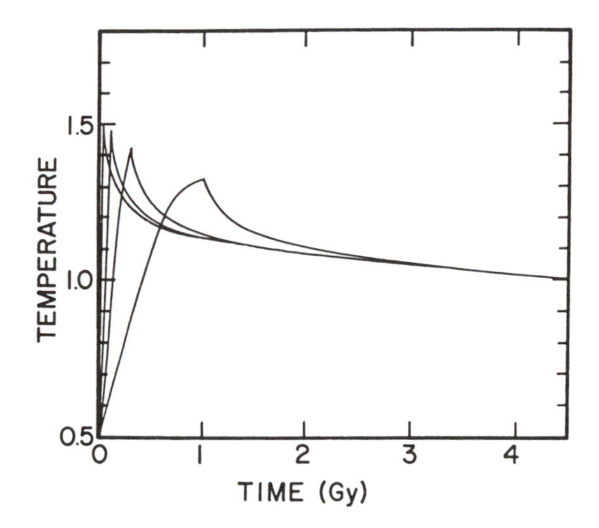

Fig. 2. *Temperature histories of the mantle undergoing pure thermal convection with heat input from core segregation and radioactivity. The different curves are for different durations of core segregation (the longer times being unrealistic): The end of core segregation is marked by the peak in each curve. Peak temperatures will be reduced by the stirring effect of sinking core material and by extraction of heat by melt (see text and Fig. 4). Temperatures are scaled to the present mantle temperature. From Davies (1980b).*

Thermal Plus Chemical Convection (Metal/Silicate Case)

Davies (1980b) noted that in the particular case of core segregation, chemical buoyancy forces would also be driving mantle flow due to the large density difference between metal and silicate. He supposed that an effective Rayleigh number could be defined in terms of the sum of the buoyancies and found that the heat flux would be enhanced by the factor r^p, where r is the ratio of total buoyancy to thermal buoyancy

$$r = 1 + d_c \Delta\rho / \rho d_t \alpha T \quad (8)$$

where d_c and d_t are the effective chemical and thermal boundary layer thicknesses and $\Delta\rho$ is the density difference between metal and silicate. Note that the buoyancy is the product of density difference and volume, and that for the chemical buoyancy the latter is expressed as an equivalent boundary layer thickness even though the geometry of the separated metal regions would probably not be simple boundary layers. With plausible ratios d_c/d_t between 1 and 10, values of r^p between 2.5 and 5 were found. This would reduce the required temperature by between 10% and 17%.

Actually, it is implicit in this formulation that the chemical and thermal buoyancies vary in proportion, but there is really no need for this assumption. In the pure thermal buoyancy case, as the convection goes faster the boundary layer thickness decreases because the fluid spends less time gaining or losing heat at the boundaries. This effect would not be present for chemical buoyancy, and so the effect of chemical buoyancy has probably been underestimated. A more careful analysis of the effect of metal/silicate buoyancy on mantle cooling is called for.

Thermal Convection with Heat Extraction by Melt

Figure 3 shows schematically the relationships between the solidus, liquidus, and adiabat in the mantle. In the absence of melting, the adiabatic gradient is only about 0.3 K/km, and the melting gradients are several times larger, on average. Because of the latent heat of melting, the adiabat is steeper in the melting region. With a latent heat of melting of 600 kJ/kg (*Bottinga and Richet,* 1978) and a specific heat of 700 J/kgC (*Stacey,* 1977), the latent heat is equivalent to a temperature change of about 900°C. The significance of this in relation to buoyancy forces can be seen from the fact that the average temperature deficit of oceanic lithosphere relative to the mantle (temperature about 1400°C) is about 700°C, and the negative buoyancy of this lithosphere

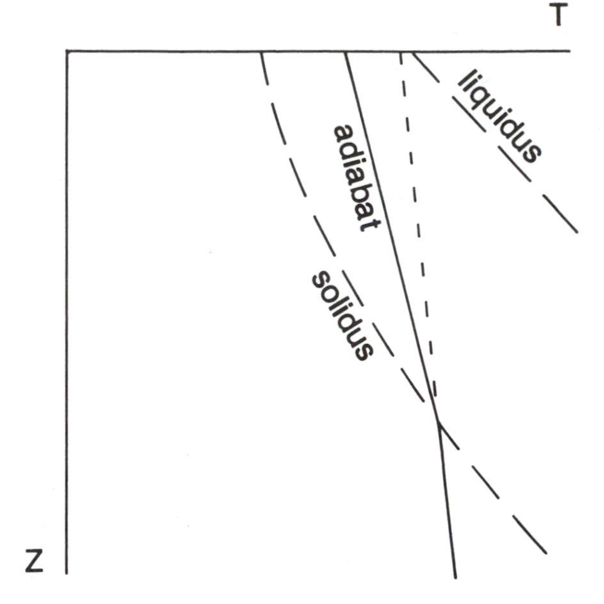

Fig. 3. *Schematic solidus, liquidus, and adiabat temperatures vs. depth, illustrating the steeper gradient of the adiabat in the partial melt zone due to the latent heat of fusion.*

is the main agent driving plate tectonics and mantle convection at present (*Davies,* 1988a,b).

Suppose that in a region of upwelling in the early Earth, material rising along an adiabat intersects the solidus at a depth of 300 km and the liquidus at the surface (the numbers here are only rough estimates to illustrate the phenomenon). Suppose that the liquid speads out on the surface and cools, losing both its latent heat and thermal heat. The net effect is that an amount of heat equivalent to an average temperature deficit of about 450°C (900°C at the surface, 0°C at the solidus at depth) has been removed from a thickness of 300 km of the mantle. With a thermal expansion coefficient of 3×10^{-5}, this implies an excess mass per unit area of the Earth's surface due to thermal contraction of about 1.3×10^{7} kg/m^2. The excess mass of old oceanic lithosphere, estimated from 3 km of subsidence relative to ocean rises, is about 7×10^{6} kg/m^2. Thus we see that extraction of heat by large degrees of melting can generate more negative buoyancy to drive mantle convection than occurs in the present oceanic lithosphere. Remember, however, that the mantle viscosity would also be much less viscous, perhaps by two or more orders of magnitude, in this early hot mantle: Solid state convection would go very much faster.

There is a crucial difference between the "melt" boundary layer just described and the usual thermal boundary layer: The thickness of the melt boundary layer is independent of the rate at which material flows through it, whereas the thickness of the thermal boundary layer gets smaller as the material velocity increases. The latter effect is because the depth to which cooling penetrates by conduction is proportional to the square-root of the time spent at the surface, which is less. If the heat input into a convecting layer with a normal thermal boundary layer is increased, the temperature rises, the density contrasts increase, the viscosity decreases, and the fluid flows faster, but the thermal boundary layer thickness decreases. The net effect is that the flow is faster, but not in full proportion to the other effects: This is why the exponent p in equations (4) and (5) is only 1/3. With a melt boundary layer, on the other hand, the thickness will actually increase because melting will begin at a greater depth, and the full effect of the viscosity decrease will apply: p will be greater than 1 and the exponent m in equation (7) will be greater than 30.

A simplified but complete derivation of these results is given in Appendix C. The resulting heat flux is

$$q = g\rho^2 C_p \alpha \beta^2 (\delta T)^4 / (8\eta) \qquad (9)$$

where $\delta T = T - T_m$, T_m is the base temperature of the adiabat for which melting first occurs, $\beta = d_m/\delta T$, and d_m is the depth at which melting first occurs.

Convection in this regime with a melt boundary layer will thus be extremely sensitive

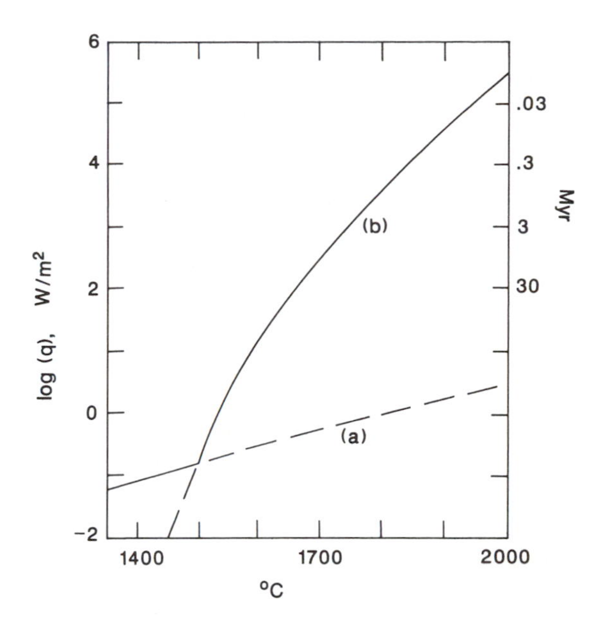

Fig. 4. *Heat flux-temperature relations for a convecting fluid with* **(a)** *a standard thermal diffusion boundary layer, and* **(b)** *a melt boundary layer. In the hotter regime of the melt boundary layer, both internal heat and latent heat of fusion are extracted by melt from a thick layer that is then negatively buoyant. The result, with temperature-dependent silicate rheology, is that heat transport is extremely sensitive to temperature. The scale on the right shows the accretion timescale that would yield the corresponding heat input rate on the left scale. Thus, for example, accretion within 30 m.y. would require a heat flux of 100 W/m^2 and the mantle would reach a temperature of about 1650° C.*

to temperature and, conversely, further temperature increases will be more strongly buffered. In Fig. 4, heat flux is plotted against temperature for both pure thermal convection, equation (7) with m = 10, and for convection with a melt boundary layer. The numbers used here are very approximate (Appendix C), but the character is clear. At a mean mantle temperature perhaps only 100°C above the present mantle temperature, the effect of the melt boundary layer becomes dominant, and the heat flux increases at a rate of more than an order of magnitude per 100°C at higher temperatures. Further implications of these results will be noted below.

Convection with Silicate Density Differences

Whenever melting occurs, there will inevitably be some local differentiation and stratification, and the resulting silicate layers will generally have densities different from the initial material. Indeed this is true at present, where the oceanic lithosphere is stratified into oceanic crust and depleted mantle. Depending on the particular composition, depth, and temperature, substantial density differences can occur (up to about 2%), although at present the net density anomaly of subducted lithosphere is less than 0.5% (*Ringwood and Irifune,* 1988).

Predicting density profiles will be left to those more expert than I. For the moment we may note that the densities will be complicated functions of the degree of melting, the depths of layers, and any crystal settling, among other things. For example, a basalt layer might be thick enough to convert to eclogite at its base, and thus there might be buoyant basalt overlying dense eclogite overlying slightly buoyant depleted mantle. The dynamic effects in turn are potentially complex, depending on densities, layer thicknesses, and available timescales. Thus if the net buoyancy of the pile were negative it would probably founder. On the other hand, the eclogite might drop out, leaving buoyant residues, or the foundering eclogite might pull the basalt after it, etc. As 100% melting is approached, the stratification may decrease except for the effects of crystal settling.

These possibilities will need careful evaluation, and we will do no more here than note the multiplicity of possibilities. However, before predictions of complicated mantle stratifications are undertaken, the lack of strong evidence for stratification of the present mantle should be kept in mind. This will be discussed in the section on present stratification of the Earth.

Some Implications

The combined effects of metal/silicate density differences and a melt boundary layer will be to strongly buffer temperatures on the high side, so that very high rates of energy input will be required to produce substantial melting in the mantle.

Peak temperatures reached during accretion (and core segregation) can be estimated from Fig. 4 as a function of the accretion timescale. Suppose a total energy of 5×10^{31}J is deposited in the Earth (about 15% efficiency; *Flasar and Birch,* 1973). With a surface area of the Earth of 5.1×10^{14}m^2 and an assumed timescale, the heat flux required to balance this input can be calculated. The timescales corresponding to various heat fluxes are shown on the vertical axis of Fig. 4, allowing the corresponding temperatures to be read off. In the standard Safronov theory, the main phase of accretion would occur within 10-30 m.y., the peak heat flux would be 100-300 W/m^2 and the peak temperature will be less than 1700°C. At this temperature, melting might begin at 200-300 km depth with about 50% melting at the surface, yielding over 100 km of komatiitic melt (e.g., *Wyllie,* 1971).

Note that this scenario does not imply a magma ocean. We have already seen that heat fluxes of this magnitude can be attained with only a few percent of the Earth's surface covered with melt. Melt will be produced in areas of mantle upwelling, but it will freeze before covering the whole surface. The accretion timescale would have to be about 0.3 m.y. or less before the surface were covered with melt. Even then, the depth of melt would be variable and might be quite thin over downwellings. A late giant impact could of course produce a real magma ocean.

A sketch of how the mantle might have looked in this regime is given in Fig. 5a, with a corresponding schematic geotherm in Fig. 5b. Because large volumes of magma extrude, the basalt/komatiite will be steadily depressed by later lava flows until it moves out of range of the spreading magma. Thus relatively cool crust will be carried down and the geotherm will be depressed in the manner shown in Fig. 5b. This form of geotherm was

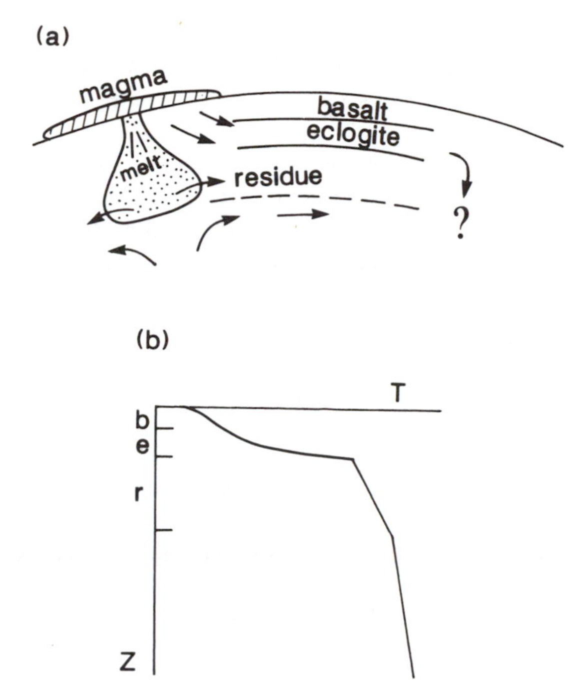

Fig. 5. **(a)** *Sketch of a possible tectonic regime with a thick melt boundary layer.* **(b)** *Geotherm through the center of* **(a)**. *Melting begins at great depth in upwelling regions, but if the melt extrudes onto a cold surface it freezes quickly without accumulating into a magma ocean. With large degrees of partial melting, great thicknesses of lava (notionally basalt, or "b") will accumulate, depressing the geotherm and possibly transforming to a higher-pressure assemblage like eclogite ("e"). A thick residual layer ("r") will underlie this. Depending on details of temperature and composition, some or all of the boundary layer will probably be negatively buoyant and founder, perhaps without traveling very far. It is also possible that some of the boundary layer might be positively buoyant and not founder with the rest, thus accumulating a buoyant layer.*

deduced by *O'Reilly and Davies* (1981) in the context of the resurfacing of Io. Below the crust the geotherm is lowered by the effects of latent heat extraction by melting.

As noted earlier, some or all of the cooler crust/residue zone is likely to be negatively buoyant and thus will founder. The stage at which this will occur depends on details of composition, density, the geotherm, and rheologies, and therefore is very difficult to predict. It seems likely, however, that the assemblage would not move very far laterally before foundering occurred, so the possibility exists that the basic tectonic scale, defined by the distance between upwelling and downwelling, would have been much less than in the present Earth. (It is the great strength of the present oceanic lithosphere that allows plates to persist at the surface despite being gravitationally unstable after only about 10 m.y.)

If the arguments developed here are basically correct, then something similar to this regime must have existed at some time early in Earth's history if the Earth was ever substantially hotter than at present. Trying to identify particular regimes with times in the Earth's history must await further quantification. The great efficiency of heat removal with a melt boundary layer may mean that this regime did not last more than 100 m.y. or so, but this assumption is conjectural at this point.

We have already looked at the possible effects of having compositional or latent heat effects that make the top of the mantle more negatively buoyant. The next step would be to look at mantle convection with a positively buoyant layer, such as a thick basaltic crust, on top. This might be relevant to the melt boundary layer regime just described and is also possibly relevant in the later thermal boundary layer regime (i.e., the present plate regime). It is notable that if the present plate regime is run backward, then there may have been a time, with a hotter mantle and faster plates, when the combination of thicker oceanic crust and shorter cooling time at the surface meant that the oceanic lithosphere was buoyant and would not subduct. This possibility is a contradiction, and means that it would not have been possible to form oceanic lithosphere in the modern sense, and some other tectonic regime would have prevailed. This other regime may have intervened between the putative melt boundary layer regime and the modern plate regime. Figure 6a shows a sketch of how this regime may have looked, and Fig. 6b is a sketch of the modern plate regime for comparison.

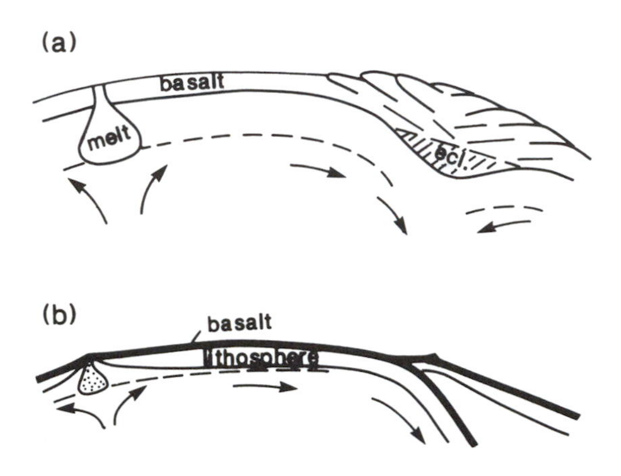

Fig. 6. *Sketch of a possible tectonic regime between the hotter melt boundary layer regime of Fig. 5 and the modern plate regime, sketched for comparison in* **(b)**. *Because of still large degrees of partial melt, the resulting thicker basaltic crust, and faster turnover, the "lithosphere" may have been too buoyant to subduct directly. Instead, the basalt may have accumulated in thick piles while the underlying material foundered. Probably a thick basaltic root would convert to denser eclogite, possibly foundering episodically as a result.*

PHASE 4: ACCUMULATION OF THE CONTINENTAL CRUST

Most of the existing continental crust is younger than 3 Ga (*DePaolo,* 1983), and the outstanding question is whether this means that little continental crust (broadly defined) existed prior to 3 Ga, or whether it existed but has been recycled into the mantle. Although the majority view seems to be that subduction zones play a crucial role in generating continental crust (*Taylor,* 1977), the mechanism of crustal differentiation is still quite obscure (e.g., *Hofmann,* 1988). This means that most of what can be said about the relationship between crustal accumulation and the physical processes of mass and heat transport that are the subject of this paper is quite conjectural.

A little light is shed by the calculations of *Gurnis and Davies* (1985, 1986), who pursued the consequences of assuming that the crustal accumulation rate is related in simple ways to the mantle convection rate and hence to the heat budget. In the simplest case with no recycling, most of the action would have occurred early, and most of the crust would be older than 3 Ga, which is clearly not the case. The present age distribution of the crust can be broadly reproduced if recycling is included, but substantial recycling rates persisting to the present are required (of order 1 km^3/yr): It is not clear whether such recycling rates are realistic. *Gurnis and Davies* (1986) showed that the final age distribution could be substantially changed just by assuming that younger crust is several times more vulnerable to erosion and recycling than older crust: It is even possible to produce an age peak near 2.5 Ga even though the crustal *production* rate is smoothly declining. The distinction between the crustal accumulation or growth curve and the present age distribution of the preserved crust is crucial to this discussion.

With this background, one can speculate a little. It is possible that continental crust generation is linked with plate tectonics in particular, and that the dearth of older crust reflects the absence of plate tectonics. Thus one might suppose that the great volume of crust dating from about 2.7 Ga reflects the main onset of plate tectonics. Plate tectonics is reasonably reliably traced back to about 2.2 Ga (*Hoffman,* 1989) and it is not implausible that it was operating at 2.7 Ga. Two variations on this are possible. One is that plate tectonics extend well back into the Archean (with the possible buoyancy of the oceanic lithosphere a potential problem, as noted above) and that the Archean-Proterozoic "boundary" results only from the fact that in the Archean the continental crust was smaller in volume and less geochemically mature. Another variation is that crustal production is possible in other tectonic regimes as well (e.g., Fig. 6a), and the Archean-Proterozoic transition reflects a number of changes, of which a tectonic transition is only one. Obviously other conjectures are possible.

With regard to the earliest Earth, then, we are left mostly with questions. Was there a layer we would identify as a crust? Was there material we would classify as continental crust? Was there a lot or a little? This aspect of mass transport in the early Earth will be fertile ground for future work.

PRESENT STRATIFICATION OF THE EARTH

Proposed models of the early Earth must obviously meet the constraint that they should produce something consistent with the present Earth. The main trouble, of course, is that our understanding of how the early Earth relates to the present Earth is also very deficient. Another problem is that our knowledge of the present Earth is also limited.

The main stratification into crust, mantle, outer core, and inner core is well established. Within the mantle, a mineralogical stratification is also well established, the seismological transition zone being explained as primarily due to transformations of upper mantle minerals to denser lower mantle phases. A rheological stratification is also plausible, though not well defined. The main argument currently is about internal compositional stratification of the mantle, and in particular whether there is an interface at the 670 km discontinuity. Those who would predict stratification of the early mantle should be cautioned by the fact that present mantle stratification is, at best, not clearly resolved.

One impetus for compositional layering comes at present from experiments on the equation of state of lower mantle material (*Knittle et al.,* 1986; *Knittle and Jeanloz,* 1987). Though impressive experimental achievements, these results are still pioneering and unconfirmed and the data scant. Other geochemical and isotopic evidence for mantle layering, though widely cited, is indirect and not compelling (*Davies,* 1984). On the other side, the positive geoid anomaly over subduction zones requires that subducted lithosphere is not compensated by depressions either at the Earth's surface or at the putative 670 km interface, or by buoyancy from any other source near 670 km, but can be quantitatively explained with a moderate viscosity increase at 670 km (*Richards and Hager,* 1984, 1988). Further, the fact that ocean rise profiles are well explained by the surficial cooling of the oceanic lithosphere argues against actively buoyant mantle rising under them, and this precludes a strong thermal boundary layer at 670 km, as would be required if this were a barrier to flow (*Davies,* 1988b, 1989). At present, the model in which 670 km is a viscosity interface but not a compositional interface has survived more empirical tests than the model in which it is a compositional interface.

CONCLUDING REMARKS

The main conclusions have been summarised in the abstract. Here I will merely observe that despite its remoteness, there really are things that can be sensibly said about the behavior of the earliest Earth. Although still sketchy and subject to revision, present ideas are much more detailed and realistic than those of one or two decades ago. Nevertheless, this volume gives a strong warning not to become too attached to particular ideas while observational constraints remain so meagre or indirect. The subject seems to be at an exciting stage: Major alternatives have been sketched out, and there is great scope for creative quantification. Such quantification must not be too elaborate at this stage, however.

APPENDIX A: COOLING RATE OF A MAGMA OCEAN

A relation of the form of equation (4) between Nusselt number and Rayleigh number seems to apply very generally to convecting fluids, including low-viscosity fluids in which the Reynolds number (measuring inertial effects) is not very small (e.g., *Tonks and Melosh,* 1990). The constants typically have values like $a = 0.1$ and $p = 1/3$, and the conclusions here are not sensitive to these values. With typical material properties, equation (1) yields a heat flux, q, of

$$q = 2(\Delta T)^{4/3} \nu^{-1/3} \qquad \text{(A1)}$$

where the constant is in S.I. units, $\Delta T = 1400°C$ is the temperature difference between the magma and the surface, and ν is the kinematic viscosity (viscosity/density) of the magma. Ultramafic and mafic magmas have relatively low viscosities in the range 10-10^5 Pa sec (*Bottinga and Weill,* 1972), so with a density of 3000 kg/m^3 we get a heat flux in the range 10^4 - 2×10^5 W/m^2, the higher value applying to the lower viscosity ultramafic magmas.

APPENDIX B: CONVECTIVE HEAT FLUX WITH A DIFFUSIVE THERMAL BOUNDARY LAYER

Consider the fluid layer sketched in Fig. B1: The layer has depth D and internal temperature T, and the top surface is kept at $T = 0°C$. A cold boundary layer will form next to the top surface whose thickness, d, is governed by the thermal diffusivity ($\kappa = K/\rho C_p$; see main text for definitions) and is given by

$$d = (D\kappa/v)^{1/2} \qquad \text{(B1)}$$

where v is the fluid velocity along the top surface and we are assuming the fluid stays at the surface for a horizontal distance D. This boundary layer turns downward and forms a column whose negative buoyancy force (density anomaly times volume) is

$$F = gDd\rho\alpha T/2 \qquad \text{(B2)}$$

T/2 being the average temperature deficit of the boundary layer. This driving force is balanced by viscous resistance. The viscous stress is equal to viscosity (η) times velocity gradient (2v/D), so the total viscous force, R, is

$$R = \eta(2v/D)D = 2\eta v \qquad \text{(B3)}$$

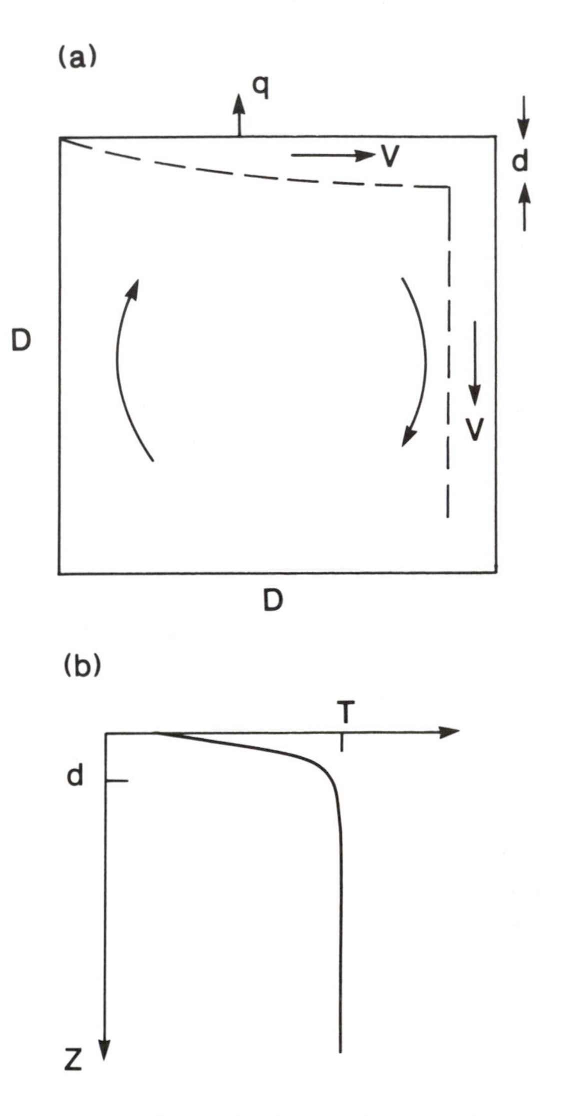

Fig. B1. (a) *Sketch of a viscous fluid layer undergoing conventional convection. A heat flux, q, conducts out the top, yielding a thermal boundary layer of thickness d. This moves to the right and descends with velocity v. The cell is assumed to be square for convenience of size D.* **(b)** *Temperature-depth profile through the top thermal boundary layer of* **(a)**.

Heat escapes out the top surface by conduction from the boundary layer, so the average heat flux is

$$q = KT/d \qquad (B4)$$

If we now equate F and R and use the resulting equation and equation (B1) to eliminate v and d, the result is

$$q^3 = gD\rho\alpha K^3T^4/(4\eta\kappa) \qquad (B5)$$

so that

$$q \propto T^{4/3}\eta^{-1/3} \qquad (B6)$$

which is equivalent to equation (5).

APPENDIX C: CONVECTIVE HEAT FLUX WITH A MELT BOUNDARY LAYER

Now consider the fluid layer sketched in Fig. C1a in which the fluid down to depth d is partially molten and it is assumed that the melt escapes to the surface and cools efficiently. A schematic geotherm, ignoring the adiabatic gradient, is shown in Fig. C1b: Above the depth d, the geotherm follows the solidus curve, which has a temperature of T_m at the surface. Suppose, therefore, that the depth d is given by

$$d = \beta(T - T_m) = \beta\delta T \qquad (C1)$$

where T is the internal fluid temperature and β is a constant. Now because of the latent heat of fusion, this "melt boundary layer" has an average temperature deficit of $(T - T_m)/2$. As in Appendix B, when this descends vertically it will exert a driving (negative) buoyancy force F

$$F = gDd\rho\alpha\delta T/2 \qquad (C2)$$

The viscous resistance force balancing this is given by equation (B3), as before. The rate of heat removal, Q, from the melt boundary layer

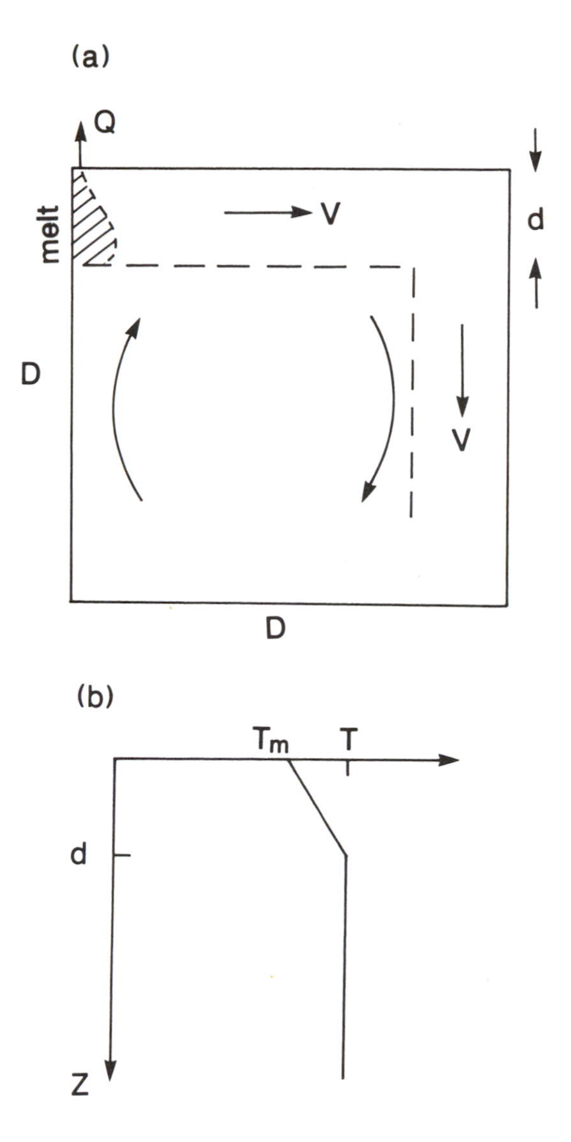

Fig. C1. **(a)** *Sketch of convection with a "melt boundary layer." Melting is assumed to begin at depth d in the upwelling region, and the melt extrudes and cools, removing internal and latent heat at a rate Q. The resulting cooler, denser (by assumption) residual moves to the right and descends with velocity v.* **(b)** *Temperature-depth profile through the upper boundary layer of* **(a)**. *Compare with Fig. B1.*

is given by the volume flux through the melt zone times the heat deficit per unit volume

$$Q = dv\rho C_P\delta T/2 \qquad (C3)$$

Averaging this over the top surface gives the average heat flux, q

$$q = dv\rho C_P \delta T/2D \qquad (C4)$$

As before, we can equate F and R and eliminate d and v to get

$$q = g\rho^2 C_P \alpha \beta^2 (\delta T)^4/(8\eta) \qquad (C5)$$

which is the same as equation (9).

REFERENCES

Agee C. B. (1988) Geochemical constraints on early earth melting (abstract). In *Papers Presented to the Conference on Origin of the Earth,* p. 1. Lunar and Planetary Institute, Houston.

Ahrens T. J. (1990) Earth accretion. In *Origin of the Earth,* this volume.

Bottinga Y. and Richet P. (1978) Thermodynamics of liquid silicates, a preliminary report. *Earth Planet. Sci. Lett., 40,* 382-400.

Bottinga Y. and Weill D. F. (1972) The viscosity of magmatic silicate liquids: a model for calculation. *Am. J. Sci., 272,* 438-475.

Chen J. H. and Wasserburg G. J. (1986) Formation ages and evolution of Shergotty and its parental planet from U-Th-Pb systematics. *Geochim. Cosmochim. Acta, 50,* 955-968.

Chicarro A. F., Schultz P. H., and Masson P. (1985) Global and regional ridge patterns on Mars. *Icarus, 63,* 153-174.

Christensen U. R. (1984) Heat transport by variable viscosity convection and implications for the earth's thermal evolution. *Phys. Earth Planet. Inter., 35,* 264-282.

Davies G. F. (1980a) Thermal histories of convective earth models and constraints on radiogenic heat production in the earth. *J. Geophys. Res., 85,* 2517-2530.

Davies G. F. (1980b) Exploratory models of the earth's thermal regime during core segregation. *J. Geophys. Res., 85,* 7108-7114.

Davies G. F. (1982) Ultimate strength of solids and formation of planetary cores. *Geophys. Res. Lett., 9,* 1267-1270.

Davies G. F. (1984) Geophysical and isotopic constraints on mantle convection: an interim synthesis. *J. Geophys. Res., 89,* 6017-6040.

Davies G. F. (1985) Heat deposition and retention in a solid planet growing by impacts. *Icarus, 63,* 45-68.

Davies G. F. (1988a) Role of the lithosphere in mantle convection. *J. Geophys. Res., 93,* 10451-10466.

Davies G. F. (1988b) Ocean bathymetry and mantle convection 1. Large-scale flow and hotspots. *J. Geophys. Res., 93,* 10467-10480.

Davies G. F. (1989) Effect of a low viscosity layer on long-wavelength topography, upper mantle case. *Geophys. Res. Lett., 16,* 625-628.

DePaolo D. P. (1983) The mean life of continents: estimates of continental recycling rates from Nd and Hf isotopic data and implications for mantle structure. *Geophys. Res. Lett., 10,* 705-708.

Drake M. J., Malvin D. J., and Capobianco C. J. (1988) Primordial differentiation of the earth (abstract). In *Papers Presented to the Conference on Origin of the Earth,* p. 17. Lunar and Planetary Institute, Houston.

Duffield W. A. (1972) A naturally occurring model of global plate tectonics. *J. Geophys. Res., 77,* 2543-2555.

Elsasser W. M. (1963) Early history of the earth. In *Earth Science and Meteoritics* (J. Geiss and E. D. Goldberg, eds.), pp. 1-30. North-Holland, Amsterdam.

Flasar F. M. and Birch F. (1973) Energetics of core formation: A correction. *J. Geophys. Res., 78,* 6101-6103.

Gurnis M. (1989) A reassessment of the heat transport by variable viscosity convection with plates and lids. *Geophys. Res. Lett., 16,* 179-182.

Gurnis M. and Davies G. F. (1985) Simple parametric models of crustal growth. *J. Geodynamics, 3,* 105-135.

Gurnis M. and Davies G. F. (1986) Apparent episodic crustal growth arising from a smoothly evolving mantle. *Geology, 14,* 396-399.

Hanks T. C. and Anderson D. L. (1969) The early thermal history of the earth. *Phys. Earth Planet. Inter., 2,* 19-29.

Hoffman P. F. (1989) Speculations on Laurentia's first gigayear (2.0 to 1.0 Ga). *Geology, 17,* 135-138.

Hofmann A. W. (1988) Chemical differentiation of the earth: the relationship between mantle, continental crust and oceanic crust. *Earth Planet. Sci. Lett., 90,* 297-314.

Holsapple K. A. and Schmidt R. M. (1982) On the scaling of crater dimensions. 2. Impact processes. *J. Geophys. Res., 87,* 1849-1870.

Kaula W. M. (1979) Thermal evolution of Earth and Moon growing by planetesimal impacts. *J. Geophys. Res., 84,* 999-1008.

Knittle E. and Jeanloz R. (1987) Synthesis and equation of state of (Mg,Fe)SiO_3 perovskite to over 100 gigapascals. *Science, 235,* 668-670.

Knittle E., Jeanloz R., and Smith G. L. (1986) Thermal expansion of silicate perovskite and the stratification of the earth's mantle. *Nature, 319,* 214-216.

Melosh H. J. and Kipp M. E. (1988) Giant impact theory of the moon's origin: Implications for the thermal state of the early earth (abstract). In *Papers Presented to the Conference on Origin of the Earth,* pp. 57-58. Lunar and Planetary Institute, Houston.

O'Keefe J. D. and Ahrens T. J. (1977) Impact-induced energy partitioning, melting and vaporization on terrestrial planets. *Proc. Lunar Sci. Conf. 8th,* pp. 3357-3374.

O'Reilly T. C. and Davies G. F. (1981) Magma transport of heat on Io. *Geophys. Res. Lett., 8,* 313-316.

Oversby V. M. and Ringwood A. E. (1971) Time of formation of the earth's core. *Nature, 234,* 463-465.

Patterson C. and Spaute D. (1988) Planetary accretion by runaway growth: Formation of the earth (abstract). In *Papers Presented to the Conference on Origin of the Earth,* p. 67. Lunar and Planetary Institute, Houston.

Richards M. A. and Hager B. H. (1984) Geoid anomalies in a dynamic earth. *J. Geophys. Res., 89,* 5987-6002.

Richards M. A. and Hager B. H. (1988) The earth's geoid and the large-scale structure of mantle convection. In *The Physics of the Planets* (S. K. Runcorn, ed.), pp. 247-272. Wiley, New York.

Ringwood A. E. and Irifune T. (1988) Nature of the 650-km seismic discontinuity: implications for mantle dynamics and differentiation. *Nature, 331,* 131-136.

Safronov V. S. (1969) *Evolution of the Protoplanetary Cloud and Formation of the Earth and Planets.* Nauka, Moscow. Translated by the Israel Program for Scientific Translation (1972).

Safronov V. S. (1972) *Evolution of the Protoplanetary Cloud and Formation of the Earth and Planets.* NASA Technical Translation TTF-667.

Safronov V. S. (1978) The heating of the earth during its formation. *Icarus, 33,* 8-12.

Schatz J. F. and Simmons G. (1972) Thermal conductivity of earth materials at high temperature. *J. Geophys. Res., 77,* 6966-6983.

Schubert G., Stevenson D., and Cassen P. (1980) Whole planet cooling and the radiogenic heat contents of the earth and moon. *J. Geophys. Res., 85,* 2531-2538.

Shankland T. J., Nitsan U., and Duba A. G. (1979) Optical absorption and radiative heat transport in olivine at high temperature. *J. Geophys. Res., 84,* 1603-1610.

Sharpe H. N. and Peltier W. R. (1978) Parameterized convection and the earth's thermal history. *Geophys. Res. Lett., 5,* 737-740.

Sharpe H. N. and Peltier W. R. (1979) A thermal history for the earth with parameterized convection. *Geophys. J. R. Astron. Soc., 59,* 171-203.

Stacey F. D. (1977) *Physics of the Earth,* 2nd ed. Wiley, New York.

Stacey F. D. (1980) The cooling earth: A reappraisal. *Phys. Earth Planet. Inter., 22,* 89-96.

Stevenson D. J. (1980) Lunar asymmetry and paleomagnetism. *Nature, 287,* 520-521.

Stevenson D. J. (1981) Models of the earth's core. *Science, 214,* 611-619.

Tajika E. and Matsui T. (1990) The evolution of the terrestrial environment. In *Origin of the Earth,* this volume.

Taylor S. R. (1977) Island arc models and the composition of the continental crust. In *Island Arcs, Deep Sea Trenches, and Back-Arc Basins,* pp. 325-335. Maurice Ewing Series 1, American Geophysical Union, Washington, DC.

Tonks W. B. and Melosh H. J. (1990) The physics of crystal settling and suspension in a turbulent magma ocean. In *Origin of the Earth,* this volume.

Tozer D. C. (1965) Thermal history of the earth. *Geophys. J. R. Astron. Soc., 9,* 95-112.

Tozer D. C. (1972) The present thermal state of the terrestrial planets. *Phys. Earth Planet. Inter., 6,* 182-197.

Turcotte D. L. and Oxburgh E. R. (1967) Finite amplitude convection cells and continental drift. *J. Fluid Mech., 28,* 29-42.

Turcotte D. L., Cooke F. A, and Willemann R. J. (1979) Parameterized convection within the Moon and terrestrial planets. *Proc. Lunar Planet. Sci. Conf. 10th,* pp. 2375-2392.

Watters T. R. and Maxwell T. A. (1986) Orientation, relative age and extent of the Tharsis Plateau ridge system. *J. Geophys. Res., 91,* 8113-8125.

Wetherill G. W. (1976) The role of large bodies in the formation of the moon. *Proc. Lunar Sci. Conf. 7th,* pp. 3245-3257.

Wetherill G. W. (1988) Accumulation of the Earth from runaway embryos (abstract). In *Papers Presented to the Conference on Origin of the Earth,* pp. 104-105. Lunar and Planetary Institute, Houston.

Wood J. A. (1970) Petrology of the lunar soil and geophysical implications. *J. Geophys. Res., 75,* 6497-6513.

Wyllie P. J. (1971) *The Dynamic Earth.* Wiley, New York.

THE PRIMARY SOLAR-TYPE ATMOSPHERE SURROUNDING THE ACCRETING EARTH: H_2O-INDUCED HIGH SURFACE TEMPERATURE

S. Sasaki[1]

Lunar and Planetary Laboratory, University of Arizona, Tucson, AZ 85721

In the solar nebula, a growing planet attracts ambient gas to form a solar-type atmosphere. The structure of this H_2-He atmosphere is calculated assuming the Earth was formed in the nebula. The blanketing effect of the atmosphere renders the planetary surface molten when the planetary mass exceeds 0.2 M_E (M_E being the present Earth's mass). Reduction of the surface melt by atmospheric H_2 should add a large amount of H_2O to the atmosphere: Under the quartz-iron-fayalite oxygen buffer, partial pressure ratio P_{H_2O}/P_{H_2} becomes higher than 0.1. Enhancing opacity and gas mean molecular weight, the excess H_2O raises the temperature and renders the atmosphere in convective equilibrium, while the dissociation of H_2 suppresses the adiabatic temperature gradient. The surface temperature of the proto-Earth can be as high as 4700 K when its mass is 1 M_E. Such a high temperature may accelerate the evaporation of surface materials. A deep totally-molten magma ocean should exist in the accreting Earth.

INTRODUCTION

Investigation of the Earth's initial thermal state is very important when reconstructing its history. If the primordial Earth was partially or totally molten, metal could have easily separated from homogeneously-accreted materials to give birth to the core; in addition, chemical fractionation in the cooling stage may have formed the present layered structure of the mantle. Very early core formation is postulated from lead isotope data (suggesting that core was formed within 2×10^8 yr after planetary accretion; *Vollmer,* 1977; *Davies,* 1984) and palaeomagnetic studies (within 1×10^9 yr; *McElhinny and Senanayake,* 1980). Noble-gas isotope data suggest the early Earth was active enough to form the atmosphere and ocean by degassing from the interior (*Hamano and Ozima,* 1978; *Staudacher and Allègre,* 1982). Recent studies of planetary formation suggest that the Earth was formed within 10^7 (in the solar nebula)-10^8 (in gas-free state) yr (*Hayashi et al.,* 1985; *Safronov,* 1969) and the growing Earth could be heated by capturing a large amount of gravitational energy that was released through planetesimal accretion.

The accretional energy, if trapped completely, could heat the Earth up to 4×10^4K on average. Most of this energy would radiate away from the planet unless the accretion is within the order of 1×10^5 yr (*Hanks and Anderson,* 1969). Some fraction of it should be buried beneath the surface at collision and may heat the planetary interior. While this "self-blanketing effect" could melt and differentiate the planetary interior (*Safronov,* 1978; *Kaula,* 1979; *Coradini et al.,* 1983; *Davies,*

[1]Now at Faculty of Science, Hiroshima University, Hiroshima, 730, Japan

1985), the surface temperature, which is determined by a balance of the accretional energy release and black-body radiation, should remain low ($\lesssim 400$ K) during the accretion.

An optically-thick primordial atmosphere around a protoplanet can enhance the surface temperature by its blanketing effect. Two types of protoatmospheres have been proposed: One is an impact-induced H_2O-CO_2 atmosphere (*Lange and Ahrens,* 1982) and the other is a solar-type H_2-He atmosphere of gravitationally-attracted solar nebula gas (*Hayashi et al.,* 1979). *Ringwood* (1970, 1975) earlier described the evolution of the accreting Earth if it were surrounded by a primordial atmosphere: Degassed H_2O and CH_4 would react to form a massive reducing atmosphere of H_2 and CO, which should be thinned by mixing with surrounding nebular gas. Earth-forming materials would be reduced during this process. Though Ringwood basically considered that rapid planetary accretion ($<10^5$ yr) should result in high surface temperature ($T > 2000$ K), he pointed out that evaporated materials might prevent free escape of the accretional energy.

A qualitative estimate of the blanketing effect by the impact-induced atmosphere was started by *Abe and Matsui* (1985), who based their calculations on gas-free (in-vacuum) planetary accretion (*Safronov,* 1969). The high opacity of H_2O molecules should delay the escape of accretional energy as well as insolation and may elevate the surface temperature. A feedback mechanism due to H_2O dissolution into the molten planetary surface may keep the surface temperature between silicate solidus and liquidus (*Abe and Matsui,* 1986; *Zahnle et al.,* 1988).

A protoplanet growing in the solar nebula attracts ambient nebular gas to form a gravitationally-bound solar-composition atmosphere, which can also enhance the temperature of the planetary surface via a blanketing effect. *Hayashi et al.* (1979) first calculated the structure of the solar-type atmosphere and showed that the surface temperature of the proto-Earth could be enhanced to as high as 4000 K. However, considering the decrease in temperature gradient due to opacity decrease by dust evaporation at high altitude (2-3 times planetary radius), *Mizuno et al.* (1982) showed that the surface temperature may be about 2300 K when the planetary mass is 1 M_E.

In the present study, we develop a numerical program of chemical equilibrium where change of mean molecular weight is fully taken into account. A large amount of H_2O is formed through reaction between the atmospheric H_2 and surface melt. The additional H_2O in the H_2-He atmosphere can oppose the effect of opacity drop by dust evaporation and the surface temperature could be as high as 4700 K. Such a high temperature has not been inferred except on the basis of the giant impact scenario (*Benz et al.,* 1987; *Melosh and Kipp,* 1988). Though a planet formed by giant impact should cool down quite rapidly, the atmosphere-induced high temperature would continue until escape of the atmosphere (possibly a few 10^6 yr). A schematic picture of the primary solar-type atmosphere that includes processes we will discuss below is shown in Fig. 1.

ASSUMPTIONS AND BASIC EQUATIONS

In calculating the structure of the primary atmosphere we have made the following assumptions: (1) The atmospheric structure is spherically symmetric and is in hydrostatic equilibrium. (2) The atmosphere is an ideal gas composed of H_2 and He. We also consider minor species such as H_2O and CO. The dissociation and change of chemical composition of gas species are also included. (3) A protoplanet is composed of silicate and metallic materials and its mean density $\bar{\rho}$ is 5000 kg/m^3. (4) The inner and outer boundaries of the atmosphere are given by the planetary surface $r_s = (3M_s/4\pi\bar{\rho})^{1/3}$ and the Hill sphere $r_h = (M_h/3M_\odot)^{1/3}a$, respectively.

Here r is the distance from the planetary center, M is mass, and a is the heliocentric distance, and subscripts s and h denote the planetary surface and the Hill sphere, respectively. Thus M_s is mass of a protoplanet. The outer boundary conditions are given by a model of the solar nebula: For the proto-Earth, we use the values $\rho_h = 1.34 \times 10^{-6}$kg/m^3 and $T_h = 280$ K, which are estimated based on *Hayashi et al.* (1985). (5) Luminosity (outward energy flux) L is given by released accretional energy. The primary atmosphere is thermally in steady state and the insolation and the gravitational energy released by gas accretion are neglected; the luminosity is constant throughout the atmosphere and we have $L = GM_s\dot{M}_s/r_s$ where $\dot{M}_s = 1\ M_E/\tau_{accr}$ is the mass accretion rate (τ_{accr} being typical accretion time).

The structure of a spherical atmosphere can be expressed by the equation of hydrostatic equilibrium

$$\frac{dP}{dr} = -\frac{GM}{r^2}\rho \qquad (1)$$

the equation of mass conservation

$$\frac{dM}{dr} = 4\pi r^2 \rho \qquad (2)$$

and the equation of state for an ideal gas

$$P = \rho \frac{k_B}{\mu m_H} T \qquad (3)$$

In the above equations, G is the gravitational constant, P is gas pressure, ρ is gas density, M is the mass contained in a sphere with radius r (planetary mass and gas mass), T is gas temperature, μ is mean molecular weight of gas, k_B is Boltzmann constant, m_H is mass of a hydrogen atom. The solar abundance of H_2 and He (*Anders and Ebihara,* 1982) gives $\mu = 2.28$. Change of chemical composition (e.g., increase in H_2O) or dissociation of gas species (e.g., dissociation of H_2) may change the mean molecular weight μ.

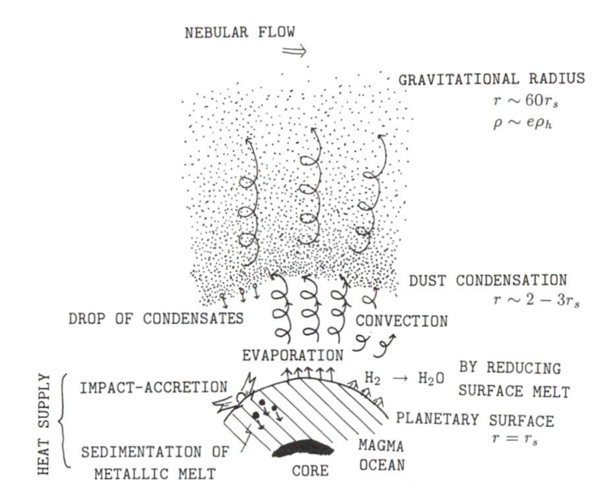

Fig. 1. *Schematic picture of the primary atmosphere when $M_s = 1\ M_E$.*

So long as the atmosphere is optically thick, temperature distribution is given by the equation of radiative equilibrium

$$\frac{16\sigma T^3}{3\kappa\rho}\frac{dT}{dr} = -\frac{L}{4\pi r^2} \qquad (4)$$

where σ is the Stefan-Boltzmann constant and κ(P, T) is opacity, which also depends on gas components. We adopt the gas opacity of *Mizuno* (1980) and the dust opacity of *Pollack et al.* (1985). The dust opacity is proportional to the dust abundance f (normalized to the interstellar value 3×10^{-3}kg/kg) and its value is about 0.1 m^2/kg when f = 1 and mean dust size $d = 1 \times 10^{-5}$m. While the temperature dependence of dust opacity is very weak except at evaporation and its pressure dependence is limited to change of evaporation temperature, the gas opacity (due mainly to pressure-induced absorption by H_2 and line absorption by H_2O) has a large pressure and temperature dependence.

When gas opacity is very dominant, the positive temperature dependence of gas opacity renders the entropy gradient negative. Moreover, when dust abundance is extremely high or accretion is very rapid, the atmosphere becomes convectively unstable (*Sasaki and Nakazawa,* 1990). In such cases, the above radiative temperature gradient is replaced by an adiabatic temperature gradient. Taking into account change in gas molecular weight, we can express the adiabatic gradient as

$$\frac{dT}{dr} = -(1 - \frac{1}{\gamma} + \delta)\frac{GM\mu m_H}{k_B r^2} \quad (5)$$

where $\gamma \equiv (d\ln P/d\ln\rho)_s$ is the adiabatic exponent and $\delta \equiv (d\ln\mu/d\ln P)_s$ is a parameter expressing mean molecular weight change along the adiabat. While the solar H_2-He mixture gives $\gamma = 1.42$ and $\mu = 2.28$, the actual γ and μ are given by a function of temperature, pressure, and chemical composition. Additional atmospheric constituents such as H_2O may increase gas molecular weight μ to steepen the above adiabatic gradient. On the other hand, when hydrogen molecules are dissociating, not only μ but also γ is reduced since density should vary more greatly than pressure through a phase change. Equation (5) shows that the decrease in γ, like μ, should decline the temperature gradient. In other words, release of dissociation energy, increasing apparent specific heat, should suppress the rise of gas temperature. We calculated γ according to *Hayashi and Nakano* (1963).

Integrating the above equations downward from the Hill sphere numerically by the Runge-Kutta method, we obtain the pressure, density, and temperature distributions of the primary atmosphere. We find that increases in P, ρ, and T are very small outside the gravitational radius $r_q = GM_s\mu m_H/k_BT$ where gravitational potential and thermal energy are balanced: A short analytical estimate from equations (1) and (3) shows that the pressure and density increase outside r_q by only a factor of $e \approx 2.7$. The gas outside r_q, which should suffer from flow in the nebula (*Takeda et al.,* 1985), cannot be said to belong to the planet. Hereafter we take the gravitational radius as the outer boundary of the atmosphere, especially in calculating the atmospheric mass. The size of the atmosphere r_q, which is 4×10^8m when $M_s = 1\ M_E$ and $T_h = 280$ K, is smaller than the Hill radius $r_h = 1.5 \times 10^9$m, but it is much larger (~60 r_s) than that of impact-generated atmosphere (whose thickness is about a few 10^5m).

ANALYTIC ESTIMATE OF TEMPERATURE

The study of planetary evolution requires an understanding of the surface temperature, which constrains the interior structure of the accreting planet. In the primary atmosphere, the temperature at the planetary surface can be obtained as the bottom temperature by integrating downward from the Hill radius. In order to compare with numerical results, we show here analytic estimates of atmospheric temperature.

Eliminating ρ and r from equations (1) and (4), we obtain the differential equation

$$\frac{dP}{dT} = \frac{64\pi\sigma GM}{3\kappa L}T^3 \quad (6)$$

Assuming $M = M_s$ (thus neglecting the atmospheric mass) and constant κ, integration of the above yields

$$P = \frac{16\pi\sigma GM_s}{3\kappa L}T^4 \quad (7)$$

where $P \gg P_q$ and $T \gg T_q$ are assumed (subscript q denoting a value at the gravitational radius). Using equation (3), substitution of equation (7) into equation (4) yields the temperature distribution

$$\frac{dT}{dr} = -\frac{GM_s\mu m_H}{4k_B}\frac{1}{r^2} \quad (8)$$

where both κ and L have been eliminated. Assuming that the mean molecular weight is constant, when $T \gg T_q$ and $r \ll r_q$, we have

$$T = \frac{GM_s\mu m_H}{4k_B}\frac{1}{r} \quad (9)$$

The bottom temperature of the atmosphere, i.e., the surface temperature of the protoplanet is expressed by

$$T_{r=r_s} = \frac{GM_s\mu m_H}{4k_B}\left(\frac{4\pi\bar{\rho}}{3M_s}\right)^{1/3} = 4.1 \times 10^3$$

$$\left(\frac{\mu}{2.28}\right)\left(\frac{\bar{\rho}}{5000\ \mathrm{kg/m^3}}\right)^{1/3}\left(\frac{M_s}{1M_E}\right)^{2/3} \mathrm{K} \tag{10}$$

When the whole atmosphere is in convective equilibrium, from equation (5) we may replace 1/4 in equations (9) and (10) by $(1 - 1/\gamma + \delta)$. This value is 0.3 if $\gamma = 1.42$ (solar H_2-He mixture) and μ is constant ($\delta = 0$). Then the analytic value of T_s is 20% higher than that in equation (10) (4.9×10^3K when parenthetical values are unities).

Further analytic estimates including opacity change are developed in *Sasaki and Nakazawa* (1990). Though the temperature equations (9) and (10) are independent of the absolute value of the opacity, spatial change of opacity may modify the temperature distribution: Downward κ increase (decrease) would enhance (lower) temperature.

EFFECT OF EXCESS H_2O

In previous works on the primary solar-type atmosphere, the atmospheric composition is assumed to be constant and determined by the solar abundance (*Hayashi et al.,* 1979; *Mizuno et al.,* 1982; *Mizuno and Wetherill,* 1984; *Nakazawa et al.,* 1985). During the accretion of protoplanets, however, some possible mechanisms supply excess water and other volatile species to the primary atmosphere. For example, impact degassing from accreting materials (*Lange and Ahrens,* 1982), infalling of icy materials, degassing from molten surface, and reduction of silicate or iron oxide may enhance the H_2O content of the primary atmosphere; changes in the H_2O content strongly affect the opacity and mean molecular weight of atmosphere.

In the early stage of planetary accretion when the surface temperature is lower than its melting temperature, the main source of additional H_2O is impact degassing from accreting planetesimals (*Lange and Ahrens,* 1982), even when the accretion proceeds in the solar nebula. After the surface temperature surpasses the decomposition temperature of hydrous minerals (~1000 K), H_2O may be outgassed rather completely (*Abe and Matsui,* 1986). The exact quantity of degassed H_2O depends on the H_2O abundance of accreting materials, which is rather uncertain. In a later stage of accretion, terrestrial planets might collect icy planetesimals such as comets that were formed in a distant region from the sun and transported inward due to radial diffusion (*Nakagawa et al.,* 1983); accretion of icy materials such as comets may also be a source of H_2O.

However, once the surface temperature reaches the melting temperature of silicate or metal (~1500 K) in the primary H_2-He atmosphere, surface materials should undergo reduction by H_2 and supply additional H_2O to the atmosphere. Dust suspended in the atmosphere can be also reduced to supply H_2O. From the viewpoint of the atmosphere, the atmospheric gas is *oxidized* by the molten planetary surface. As a result, irrespective of other mechanisms of water supply, H_2O abundance in the bottom atmosphere can be determined by chemical equilibrium at the planetary surface.

In dissociation equilibrium of H_2O, the partial pressure ratio P_{H_2O}/P_{H_2} ($= n_{H_2O}/n_{H_2}$) can be expressed by the oxygen partial pressure P_{O_2} and temperature determining the equilibrium constant. Moreover, P_{O_2} is also expressed as a function of temperature when a possible reduction reaction is postulated (such a reaction is called an "oxygen buffer"). Assuming that metallic iron exists at the planetary surface as raw materials for the planetary core, probable reactions of the oxygen buffer in the present case are as follows (*Nordstrom and Munoz,* 1985):

(SPF; silicon-periclase-forsterite buffer)

$$2MgO + Si + O_2 \longleftrightarrow Mg_2SiO_4$$

(QIF; quartz-iron-fayalite buffer)

$$2Fe + SiO_2 + O_2 \longleftrightarrow Fe_2SiO_4$$

(IM; iron-magnetite buffer)

$$3/2Fe + O_2 \longleftrightarrow 1/2Fe_3O_4$$

(IW; iron-wüstite buffer)

$$1.894Fe + O_2 \longleftrightarrow 2Fe_{0.947}O$$

Let us assume that one of the above reactions controls oxygen fugacity. We can estimate P_{O_2} directly from the differences of Gibbs' free energy of the products and reactants in the assigned buffer equation, assuming that the species (except O_2) in the equation are not in a gas phase and neglecting nonideality. Successively, we get P_{H_2O}/P_{H_2}. Calculated P_{H_2O}/P_{H_2} for each oxygen buffer is shown in Fig. 2 as a function of temperature.

Among the four oxygen buffers above, the SPF buffer, where iron compounds are reduced completely, is rather unrealistic because the present Earth interior is not so highly reduced and because the atmospheric H_2 is not abundant enough to reduce all of the iron compounds as discussed later. Moreover, considering the existence of silicate melt, we prefer the QIF buffer to the IM-IW buffer, but this choice is not very important since both buffers give similar P_{H_2O}/P_{H_2} (see Fig. 2). The addition of Mg in silicate should make only a small modification (more reducing) on the QIF buffer curve (*Holland,* 1984). Figure 2 shows that the QIF buffer gives much higher P_{H_2O}/P_{H_2} ($0.01 \sim 0.3$) than that of solar abundance gas (6×10^{-4} at 2000 K). The reaction rate is too slow to change P_{H_2O}/P_{H_2} significantly at low temperature ($T \ll T_m$). However, when the ambient temperature is higher than 1500 K, H_2O concentration is much enhanced, i.e., we have $P_{H_2O}/P_{H_2} = 0.1$ to 0.3 if the QIF oxygen buffer is achieved completely. As a result, the oxidization of the atmosphere (or reduction of the planetary surface) may enhance the H_2O abundance in the atmosphere more than 10^2 times relative to solar.

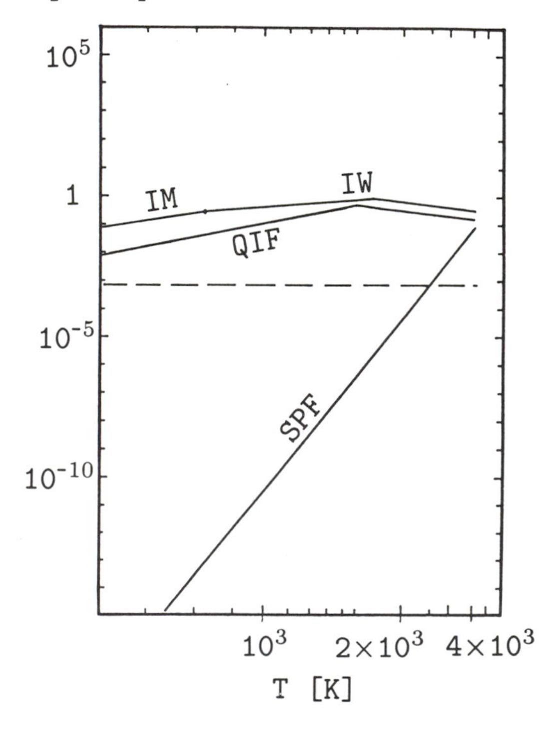

Fig. 2. *Partial pressure ratio P_{H_2O}/P_{H_2} determined by oxygen buffer systems: SPF (Si-MgO-Mg_2SiO_4), QIF (SiO_2-Fe- Fe_2SiO_4), IM (Fe-Fe_3O_4), IW (Fe-$Fe_{0.947}O$). The oxygen fugacity at each buffer system is estimated from thermodynamical data in Robie et al. (1979). Pressure, which is assumed to be 1×10^5Pa, does not largely affect results. The dashed horizontal line expresses the ratio determined by solar abundance (Anders and Ebihara, 1982).*

ATMOSPHERIC STRUCTURE WITH EXCESS H_2O

Let us now consider the atmospheric structure when additional water exists. For a parameter expressing H_2O abundance, we use the relative abundance of oxygen atoms to hydrogen atoms normalized to the solar abundance ratio, $R(O) = [N(O)/N(H)]/[N(O)/N(H)]_{solar}$, which is normalized to the solar abundance ratio: $[N(O)/N(H)]_{solar} = 7.4 \times 10^{-4}$. For a given R(O), T, and P, we determine the equilibrium abundances of H_2, H_2O, OH, H, O, CH_4, and CO and we then obtain the mean molecular weight and specific heat of the atmospheric gas. Mean molecular weight change due to species other than H_2 and H was neglected in the previous studies.

Since in the condition of enhanced H_2O the atmosphere becomes convective and stirred, as noted later, we assume that R(O) is constant in the lower region of the atmosphere. We assume that excess water exists in the region where $T > 1000$ K. Note that the excess H_2O may be supplied initially from dehydration of the planetary surface and 1000 K is just above decomposition temperature of hydrous minerals.

The relations between R(O) and relative partial pressures P_{H_2O}/P_{H_2}, P_{OH}/P_{H_2}, P_H/P_{H_2}, and P_O/P_{H_2} are displayed in Fig. 3 for gas density

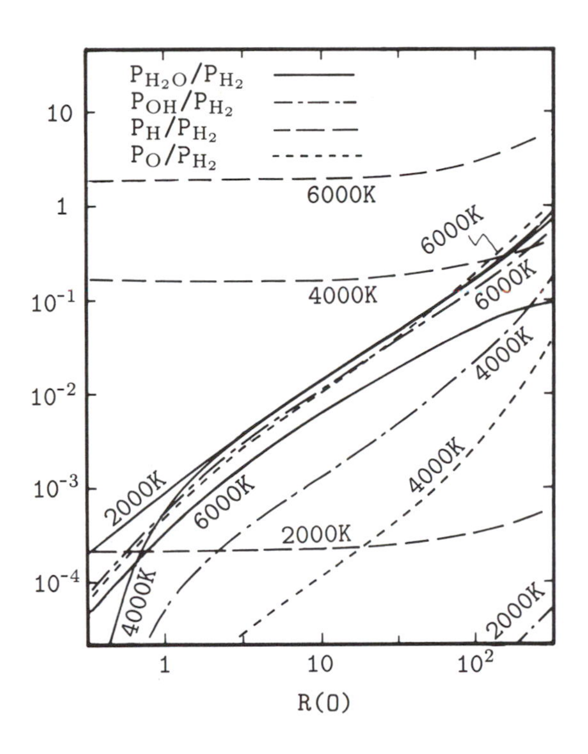

Fig. 3. *The relation between oxygen enhancement factor R(O) and partial pressure ratios P_{H_2O}/P_{H_2} (solid curves), P_{OH}/P_{H_2} (dotted-and-dashed curves), P_H/P_{H_2} (dashed curves), and P_O/P_{H_2} (dotted curves) for three different temperatures: 2000, 4000, and 6000 K. We assume $\rho = 1.0$ kg/m^3, which approaches the gas density at the planetary surface for the standard model in Table 1.*

1.0 kg/m^3 and various temperatures. The relative abundance of H_2O increases with R(O). When $R(O) \sim 10^2$, P_{H_2O}/P_{H_2} becomes slightly higher than 0.1, which corresponds to the ratio from the QIF oxygen buffer. This is valid in a wide range of temperature ($T < 5000$ K); hereafter we take 10^2 as a representative R(O), which describes the chemical condition at the molten planetary surface.

We calculated the atmospheric structure for various values of R(O). For other parameters we used values summarized in Table 1: planetary mass $M_s = 1\ M_E$, relative dust abundance $f = 1$, mean dust size $d = 1 \times 10^{-5}$m, and typical accretion time $\tau_{accr} = 1 \times 10^7$yr. Figure 4 illustrates calculated temperature distributions of the atmosphere for various values of R(O) as functions of the radius, and Fig. 5 shows changes in gas molecular weight, temperature, and pressure at the planetary surface against R(O).

When the atmosphere has the solar composition and the additional water is absent ($R(O) = 1$), as shown in Fig. 4, temperature gradient is declined in the dust-evaporation region where $r_v \sim (1$ to $1.5) \times 10^7$m owing to rapid opacity drop. The bottom temperature of the atmosphere, T_s, is depressed to be 2.3×10^3K, which is lower than the analytically-estimated value 4.1×10^3K (*Mizuno et al.*, 1982; *Nakazawa et al.*, 1985).

The additional H_2O enhances the bottom temperature of the atmosphere. When R(O) is small, mean molecular weight is changed very little and temperature increase is due mainly to increase in H_2O opacity. The large $d\kappa/dT$ of opacity due to H_2O molecules

TABLE 1. *Parameters used in the standard model.*

Planetary mass	M_s	$1\ M_E = 5.97 \times 10^{24}$ kg
Typical accretion time	τ_{accr}	1.0×10^7 yr
Planetary mean density	$\bar{\rho}$	5000 kg/m^3
Relative dust abundance	f	1
Mean dust size	d	1×10^{-5} m

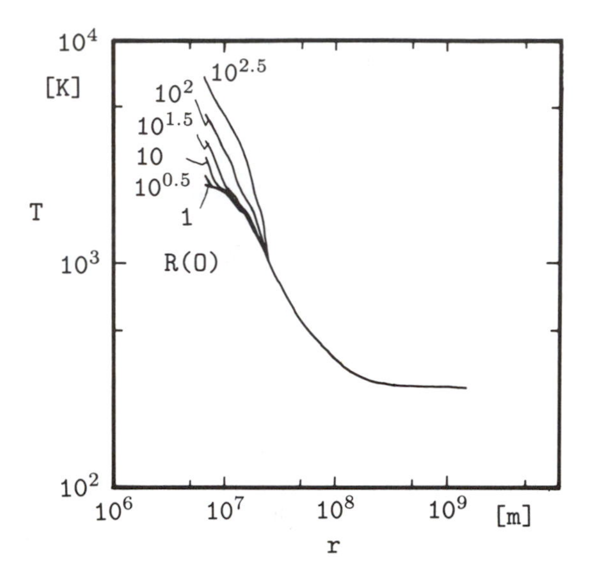

Fig. 4. *Temperature distribution of the primary atmosphere for various oxygen atom abundance relative to hydrogen atom: R(O) = 1, $10^{0.5}$, 10, $10^{1.5}$, 10^2, and $10^{2.5}$ Other parameters used are based on Table 1.*

renders the lowest region convective and enhances the bottom temperature. The excited convective motion may stir the atmosphere and transport H_2O as well as evaporated materials. When $R(O) > 10^{1.5}$, gas mean molecular weight starts increasing promptly, and when $R(O) = 10^2$ (corresponding to the QIF buffer), we get $\mu = 3.9$. The convective temperature gradient as well as radiative gradient is proportional to mean molecular weight; atmospheric temperature should be raised by mean molecular weight increase. When $R(O) = 10^2$, we get $T_s = 4.7 \times 10^3$K. But this is not higher than the analytically obtained value, 4.9×10^3K, for the adiabatic atmosphere at $\mu = 2.28$. Latent heat exchange at H_2 dissociation (decreasing γ) together with negative pressure dependence of mean molecular weight (decreasing δ) suppresses the adiabatic temperature gradient. When $R(O) = 10^2$, we have $\gamma = 1.2$ and $\delta = -0.036$; estimated $(\mathrm{dln}\ T/\mathrm{dln}\ P)_s = (1 - 1/\gamma + \delta)$ is 0.14, which is twice as small as the adiabat of a solar H_2-He gas, 0.30, and even smaller than the radiative gradient, 0.25. Thus the effect of enhanced molecular weight on the adiabatic temperature gradient is compensated by that of H_2 dissociation. In any event, however, the temperature depression due to dust disappearance in the lower region is denied by the excess H_2O; the surface temperature is greatly enhanced (by 2400°) compared with the case $R(O) = 1$. Under a more oxidized condition ($R(O) > 10^2$), the effect of increase in mean molecular weight by additional H_2O surpasses that of H_2 dissociation and gas temperature is raised greatly; we have $\mu = 7.5$ and $T_s = 6.9 \times 10^3$K when $R(O) = 10^{2.5}$.

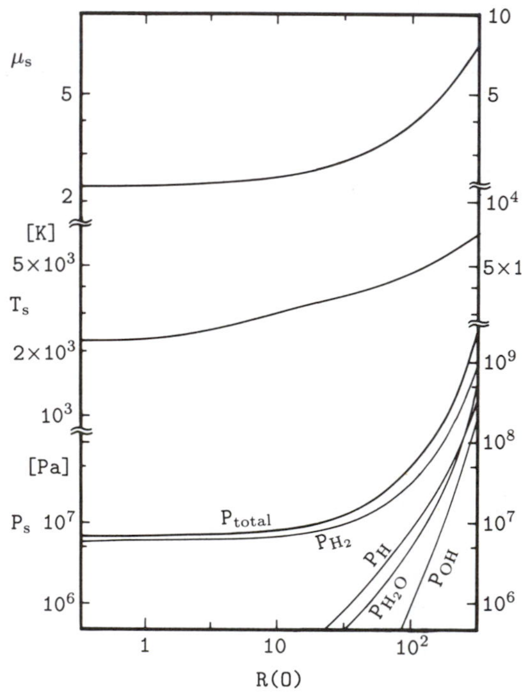

Fig. 5. *Gas molecular weight μ, temperature T_s, and pressure P_s at the planetary surface for various values of oxygen enhancement factor R(O). As for the pressure, not only the total value but also partial pressures (P_{H_2}, P_{H_2O}, P_{OH}, P_H) are shown for comparison. Other parameters used appear in Table 1. When $R(O) > 10^{1.5}$, abundance of H_2O molecules becomes large enough compared with H_2 to increase gas molecular weight appreciably.*

Pressure as well as density is suppressed by opacity increase but enhanced by mean molecular weight increase. Then the balance of both keeps pressure roughly the same at low R(O). Since dependence of pressure or density on gas molecular weight is large (P, $\rho \propto \mu^4$ based on the analytic estimate by *Sasaki and Nakazawa*, 1990), they increase greatly when R(O) > 30 as seen in Fig. 5. Figure 3 shows that when $\rho = 1\,kg/m^3$ ($P \sim 1 \times 10^7 Pa$), P_H is higher than P_{H_2} and both P_{OH} and P_O are enhanced instead of P_{H_2O} at very high temperature (≥ 6000 K). In the primary atmosphere, however, higher pressure accompanying high R(O) may prevent H_2 and H_2O from dissociation despite enhanced temperature as shown in Fig. 5; the surface value of dissociation degree does not change greatly ($P_H/P_{H_2} \sim 0.3$) in the range R(O) ≥ 10^2.

The mass of the attracted atmosphere is calculated from the density distribution; the atmospheric mass within the gravitational radius is $3.7 \times 10^{-4} M_E$ (R(O) = 1), $4.1 \times 10^{-4} M_E$ (R(O) = 10), $1.5 \times 10^{-3} M_E$ (R(O) = 10^2), and $4.5 \times 10^{-2} M_E$ (R(O) = $10^{2.5}$) when f = 1 and $M_s = 1\,M_E$. When R(O) ≥ 10^2, the addition of H_2O enhances the atmospheric mass as well as the pressure. When R(O) = 10^2, total mass of oxygen atom (including a form of H_2O or OH) in the atmosphere is $7 \times 10^{-4} M_E$, which can oxidize only 0.7% of core-forming metal. Thus the atmospheric H_2O mass is much smaller than metal or silicate determining oxygen state. Even at the extremely oxidized condition R(O) ≥ $10^{2.5}$, the atmospheric oxygen atom can be supplied if 30% of the core-forming metal was oxide before accretion onto the Earth.

SURFACE TEMPERATURE CHANGE DURING ACCRETION

Figure 6 illustrates the change of the surface temperature T_s during planetary accretion for various R(O). At first, when $M_s \geq 0.05\,M_E$, a radiative zone starts developing and T_s becomes approximately proportional to $M_s^{2/3}$, as suggested by equation (10). We have

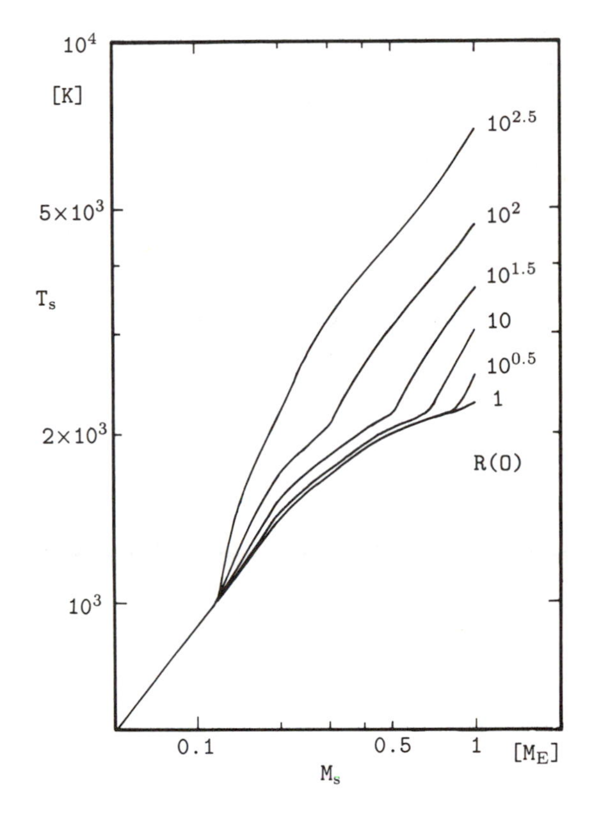

Fig. 6. *Changes of surface temperature during planetary accretion for various oxygen enhancement factor: R(O) = 1, $10^{0.5}$, 10, $10^{1.5}$, 10^2, and $10^{2.5}$. The excess oxygen is taken into account, when the temperature is higher than 1000 K. Mass accretion rate determining the outgoing energy flux is given by an equation $\dot{M}_s/M_E = \tau_0^{-1}(M_s/M_E)^{2/3}(q' - M_s/M_E)/(q' - 1)$ where $\tau_0 = 1 \times 10^7$yr and $q' = 1.2$ (Sasaki, 1987). The change in $\dot{M}$ is a factor of two.*

assumed that excess H_2O should appear when $T_s > 1000$ K, just above the decomposition temperature of hydrous minerals. The surface temperature reaches the melting temperature of silicates (~1500 K) when $M_s = 0.16$-$0.2\,M_E$. In the case R(O) = 1, the opacity drop due to dust evaporation in the lower atmosphere suppresses the surface temperature to as low as 2300 K (*Mizuno et al.,* 1982). When R(O) < 1, though not expressed in Fig. 6, an evolutionary curve of temperature is nearly the same as that of the case R(O) = 1. As seen in the previous section, the excess H_2O raises the surface temperature. So long as R(O) ≤ 10, the surface temperature increase is not appreciable until the last stage of accretion ($M_s \leq 0.8\ M_E$), because the depth of the non-dust region is too small for the temperature to increase appreciably. When $R(O) = 10^2$ under the QIF oxygen buffer, T_s becomes as high as 4000 K even at $M_s = 0.8\ M_E$ owing to an increase in mean molecular weight. And finally, when $M_s = 1\ M_E$, we have $T_s = 4700$ K. Under a more oxidized condition where $R(O) = 10^{2.5}$, we get $T_s = 5900$ K at $M_s = 0.8\ M_E$ and 6900 K at $M_s = 1\ M_E$.

As seen in Fig. 6, Venus is massive ($0.8\ M_E$) enough to have a high surface temperature due to the blanketing effect of the primary atmosphere. Formation theories of planets predict that Venus should have grown more rapidly than the Earth (*Safronov,* 1969; *Hayashi et al.,* 1985); therefore Venus is more likely to have developed a solar-type atmosphere. Figure 6 also shows that the surface temperature is below the melting temperature of silicate (1500 K) when a protoplanet is Mercury-size ($0.05\ M_E$) or Mars-size ($0.1\ M_E$). This result is not altered by changing the mass accretion rate, since the outgoing energy flux L should hardly affect the temperature distribution of the atmosphere (see equation (9)). Though a high surface temperature might cause the evaporation of silicate and explain the low silicate/metal ratio of Mercury (*Cameron,* 1985), the mass of Mercury is too small to produce the blanketing effect with a solar-type atmosphere. As for Mars, the blanketing effect produced by impact-generated H_2O would raise the surface temperature to the melting point if the accretion of Mars was rather rapid ($<1 \times 10^7$ yr) (*Abe,* 1986). Notwithstanding Abe's assumption of a gas-free accretion, the temperature rise caused by the impact-degassed H_2O could have occurred also in the presence of nebular gas. The blanketing effect by the impact-induced gas should be more likely on Mars since the H_2O abundance in Mars-forming planetesimals is expected to have been higher than that in Earth-forming ones.

DISCUSSIONS AND IMPLICATIONS

Effect of Evaporated Materials on Temperature

When a primitive atmosphere contains excess H_2O, the positive temperature dependence of H_2O opacity tends to keep the bottom region of the atmosphere in convective equilibrium. The evaporated materials can be transported upward by vigorous convection. Since the surface temperature is higher than the evaporation temperature of surface materials, the evaporation may enhance the abundance of heavy molecules in the atmosphere. Silicate or metallic vapor like H_2O vapor enhances temperature by increasing opacity and mean molecular weight, although these effects are neglected in our calculation. [In an analogy of formation of gas planets (*Stevenson,* 1984), a gas with very high mean molecular weight ($\mu > 10$) should collapse inward owing to its self-gravity, which would prevent such a heavy vapor from becoming the main constituent of the lower atmosphere.]

The transported materials in vapor phase recondense and increase dust abundance in the region where the gas temperature is low (<2000 K) (see Fig. 1). Fine dust particles ($\leq 10^{-6}$m), which can remain suspended within the gas, can be lifted by convection and fill the atmosphere; the primary atmosphere should become very opaque as a whole by the recondensed dust, even though the initial dust abundance was small. The typical transport time is $r_q^2/K \sim r_q^2/\ v_c\ell \sim 10^2\text{-}10^3$ yr, where K is eddy diffusion coefficient, v_c is velocity of convective motion, and ℓ is mixing length (*Hunten,* 1975); the dust transport in the atmosphere is very rapid compared with the planetary accretion.

Figure 7 shows the surface temperature for various values of dust abundance parameter f. Remember that if dust abundance is as high as the interstellar value ($f \sim 1$), the temperature gradient is suppressed in the dust evaporating (or condensing) region and the surface temperature is lowered when

R(O) = 1. Even when R(O) = 10, this is still effective so long as f ≥ 10. But under a higher R(O), relative temperature suppression by enhanced dust abundance would be smaller; the high temperature at the planetary surface should be kept irrespective of dust abundance.

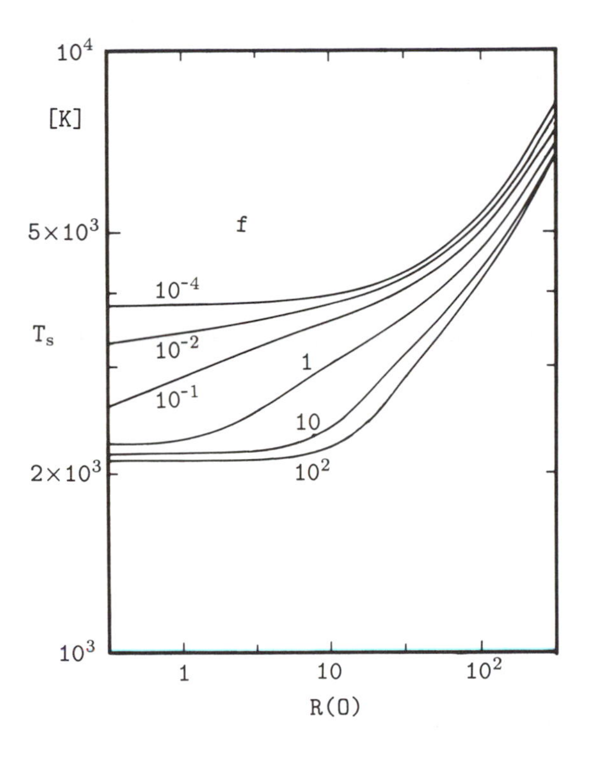

Fig. 7. *The surface temperature for various values of relative oxygen abundance R(O) and relative dust abundance f. Other parameters are based on Table 1.*

Protoplanetary Dust Ball

Even if the initial dust abundance is small, dust can be supplied by evaporated material from the hot planetary surface and fine dust can be transported outward in a relatively short timescale. Since the radius of the thick region should be the gravitational radius $r_q = \mu m_H GM/k_B T$, we may have the radius of the "dust ball" $r_{DB} \sim r_q \sim 4 \times 10^8 m \sim 60 r_s \sim$ (distance between the Earth and the Moon) ~ (half of the solar radius), which would be comparable to or larger than the radius of so-called brown dwarfs. This "dust ball" (a planet wearing an optically-thick extended atmosphere) may be observed in a orbit around some early stars, if a circumstellar nebula is optically thin or if the nebula as well as the primary atmosphere is in the dissipation stage. The lifetime of the "dust ball" should be determined by the accretion timescale of planets or escape timescale of the primary atmosphere: We have 10^6-10^7yr in the solar system. Astrometric observations would discriminate "dust balls" from brown dwarfs, since mass of "dust balls" (1-10 M_E) should be too small to cause gravitating effects on the main star.

Possibility of Satellite Formation from Evaporated Materials

In the H_2-He atmosphere, the outward transport of the evaporated materials is restricted by deficiency of gravitational energy and/or angular momentum. First, outward mass motion requires a work large enough to compensate the difference of gravitational potential. The kinetic energy of incoming planetesimals releases a large amount of heat, some of which would be used to push up the evaporated materials. Let us introduce energy conversion efficiency h, which is the heat fraction used for the upward mass transport. When accreted mass is Δm, the generated heat can heave mass $h\Delta m$ to an appreciable height ($r \gg r_s$). One may assume that the lunar mass ($\sim 0.01\ M_E$) would be supplied by evaporation if $\Delta m \sim 0.1\ M_E$ and $h \sim 0.1$.

However, once condensed dust particles become large enough to be free from gas suspension, they must sediment downward to evaporate again. In the H_2-He atmosphere, gravity is balanced with pressure gradient (see equation (1)) and the gas is not rotating globally, as opposed to a spin-out gas disk where gravity should act to balance the centrifugal force of the gas Keplerian motion. [In the model of a spin-out disk around a Jovian planet (*Pollack et al.,* 1990), satellite-forming materials, which were transported outward, should have Keplerian velocity.]

Convective velocity of the atmosphere should be negligibly small ($v_c \sim 10$ m/sec at the dust-condensing region from an order-of-magnitude estimate by mixing length theory) compared with the Keplerian velocity ($\sqrt{GM_s/r_v} \sim 6$ km/sec). Then, condensed particles do not have Keplerian velocity. They may grow by collision induced by turbulent motion and they should finally drop to evaporate again when their sedimentation velocity exceeds gas turbulent velocity [where the particle size ~ centimeters (*Mizuno and Wetherill,* 1984)]. Evaporated materials may enhance the dust abundance but deficiency of angular momentum hinders dust from accumulating to form a large body.

The formation of the Moon from the evaporated Earth materials is therefore doubtful in the primary solar-type atmosphere because of the lack of angular momentum in condensed materials. The Moon should probably have formed by another process: Giant impact and capture (*Wood,* 1986; *Stevenson,* 1987) are two such possibilities.

Magma Ocean

The high surface temperature produces a deep totally-molten magma ocean. Separation of metal from silicate is very rapid in the totally-molten zone and core formation should proceed simultaneously with accretion; downward motion of heavier metal should release a large amount of gravitational energy that can keep the interior molten and vigorously convective (*Sasaki and Nakazawa,* 1986).

As the surface temperature decreases owing to escape of the atmosphere, the magma ocean cools. So long as the gradient of melting temperature curve is steeper than that of the silicate adiabat (*Ohtani,* 1985) and convective heat transport in the magma ocean is fast enough to catch up with the surface temperature decrease, the totally-molten magma ocean should solidify from the bottom upward. The chemical fractionation, if any, should start from the high-pressure region (~100 GPa) where experimental data on the partitioning of trace elements are absent. If elemental fractionation in the lower-mantle pressure range is quite different from that of *Kato et al.* (1988), and admitting later mixing within the upper mantle region that possibly suffered from elemental fractionation depicted by *Kato et al.* (1988), the presence of the deep totally-molten magma ocean is not impossible; otherwise the cooling of the totally-molten ocean would be too rapid to cause any fractionation or chemical layering.

Atmospheric Escape and a Degassed Atmosphere

Since the noble gas abundances in the present terrestrial atmosphere are much lower than that of the solar-type gas, the primary H_2-He atmosphere probably escaped from the Earth. At the T-Tauri stage of the protosun, the strong solar EUV-FUV could blow off the primary atmosphere as well as the solar nebula (*Sekiya et al.,* 1980, 1981).

Let us consider the case that atmospheric escape started during accretion and the primary solar-type atmosphere escaped nearly completely. After that, degassing from impact materials may provide volatile species mainly of H_2O and CO_2. Contrary to the case of H_2-He atmosphere, the molten surface of the protoplanet *reduces* the degassed H_2O atmosphere to form H_2 (*Abe and Matsui,* 1986). The partial pressure ratio P_{H_2O}/P_{H_2} should be the same as that of the solar-type atmosphere if the same oxygen buffer is applicable to both cases. Since a large amount of H_2 is produced by reduction of H_2O, the degassed atmosphere should also suffer from intense UV irradiated from the T-Tauri sun (*Zahnle et al.,* 1988), but its hydrodynamic escape rate may be smaller than that of the H_2-He atmosphere because of abundant heavier CO, CO_2, and N_2. The degassed atmosphere should rather oxidize the surface materials through conversion of H_2O to H_2; iron oxide is formed from iron. Therefore the escape of the solar-type atmosphere, if it occurred during accretion, might change the effect of the atmosphere on planetary materials from reducing to oxidizing. But since bulk mass of the Earth materials is much larger than the atmospheric mass (as seen in the section on atmospheric structure), this would not upset the oxidation state of all metal or silicate and the change would be confined near the planetary surface.

Evaporation—Fractionation by Volatility

The depletion of volatile elements such as Na, K, and Rb in the present Earth and Moon (if formed from the Earth; *Wood,* 1986) may be caused by evaporation within the primary atmosphere. In the course of the escape and cooling of the atmosphere, less volatile or heavier species recondense and fall back to the Earth. The more volatile or lighter species should be dragged out of the Earth together with the atmosphere (*Sasaki and Nakazawa,* 1988). However, Fig. 6 shows that the the eucrite parent body ($\leq 10^3$km), which is considered to be also depleted in volatile elements, is too small to have a hot primary atmosphere.

Acknowledgments. I would like to thank K. Nakazawa and Y. Abe for valuable discussions; D. M. Hunten for discussions and checking a manuscript; H. Mizuno for sharing his opacity data; and J. W. Larimer and R. Boldi for their thoughtful and constructive reviews which suggested the importance of hydrogen dissociation. Numerical calculations were performed by S-810 and M-680 at the University of Tokyo, RVAX at the University of Arizona (with support by Grant No. NSF AST-85-14520 to D. M. Hunten), and Cyber 205 at the John von Neumann Center in Princeton.

REFERENCES

Abe Y. (1986) Early evolution of the terrestrial planets: accretion, atmosphere formation, and thermal history. D.Sc. dissertation, Univ. of Tokyo, Tokyo.

Abe Y. and Matsui T. (1985) The formation of an impact-generated H_2O atmosphere and its implications for the early thermal history of the Earth. *Proc. Lunar Planet. Sci. Conf. 15th,* in *J. Geophys. Res., 90,* C545-C559.

Abe Y. and Matsui T. (1986) Early evolution of the earth: Accretion, atmosphere formation, and thermal history. *Proc. Lunar Planet. Sci. Conf. 17th,* in *J. Geophys. Res., 91,* E291-E302.

Anders E. and Ebihara M. (1982) Solar-system abundances of the elements. *Geochim. Cosmochim. Acta, 46,* 2363-2380.

Benz W., Slattery W. L., and Cameron A. G. W. (1987) The origin of the moon and the single impact hypothesis II. *Icarus, 71,* 30-45.

Cameron A. G. W. (1985) The partial volatilization of Mercury. *Icarus, 64,* 285-294.

Coradini A., Federico C., and Lanciano P. (1983) Earth and Mars: early thermal profiles. *Phys. Earth Planet. Inter., 31,* 145-160.

Davies G. F. (1984) Geophysical and isotopic constraints on mantle convection: An interim synthesis. *J. Geophys. Res., 89,* 6017-6040.

Davies G. F. (1985) Heat deposition and retention in a solid planet growing by impacts. *Icarus, 63,* 45-68.

Hamano Y. and Ozima M. (1978) Earth-atmosphere evolution model based on Ar isotopic data. In *Advance in Earth and Planetary Science 3* (E. C. Alexander and M. Ozima, eds.), pp. 155-171. Center for Academic Publ., Tokyo.

Hanks T. C. and Anderson D. L. (1969) The early thermal history of the Earth. *Phys. Earth Planet. Inter., 2,* 19-29.

Hayashi C. and Nakano T. (1963) Evolution of stars of small masses in the pre-main sequence stages. *Prog. Theor. Phys., 30,* 460-474.

Hayashi C., Nakazawa K., and Mizuno H. (1979) Earth's melting due to the blanketing effect of the primordial dense atmosphere. *Earth Planet. Sci. Lett., 43,* 22-28.

Hayashi C., Nakazawa K., and Nakagawa Y. (1985) Formation of the solar system. In *Protostar and Planets II* (D. C. Black and M. S. Matthews, eds.), pp. 1100-1153. Univ. of Arizona, Tucson.

Holland H. D. (1984) *The Chemical Evolution of the Atmosphere and Oceans.* Princeton Univ., Princeton, New Jersey. 582 pp.

Hunten D. M. (1975) Vertical transport in atmospheres. In *Atmospheres of Earth and the Planets* (B. M. McCormac, ed.), pp. 59-72. Reidel, Dordrecht.

Kato T., Ringwood A. E., and Irifune T. (1988) Experimental determination of element partitioning between silicate perovskites, garnets and liquids: constraints on early differentiation of the mantle. *Earth Planet. Sci. Lett., 89,* 123-145.

Kaula W. M. (1979) Thermal evolution of Earth and Moon growing by planetesimal impacts. *J. Geophys. Res., 84,* 999-1008.

Lange M. A. and Ahrens T. J. (1982) The evolution of an impact-generated atmosphere. *Icarus, 51,* 96-120.

McElhinny M. W. and Senanayake W. E. (1980) Paleomagnetic evidence for the existence of the geomagnetic field 3.5 Ga ago. *J. Geophys. Res., 85,* 3523-3528.

Melosh H. J. and Kipp M. E. (1988) Giant impact theory of the moon's origin: Implications for the thermal state of the early earth (abstract). In *Papers Presented to the Conference on Origin of the Earth,* pp. 57-58. Lunar and Planetary Institute, Houston.

Mizuno H. (1980) Formation of the giant planets. *Prog. Theor. Phys., 64,* 544-557.

Mizuno H. and Wetherill G. W. (1984) Grain abundance in the primordial atmosphere of the Earth. *Icarus, 59,* 74-86.

Mizuno H., Nakazawa K., and Hayashi C. (1982) Gas capture and rare gas retention by accreting planets in the solar nebula. *Planet. Space Sci., 30,* 765-771.

Nakagawa Y., Hayashi C., and Nakazawa K. (1983) Accumulation of planetesimals in the solar nebula. *Icarus, 54,* 361-376.

Nakazawa K., Mizuno H., Sekiya M., and Hayashi C. (1985) Structure of the primordial atmosphere surrounding the early-Earth. *J. Geomag. Geoelectr., 37,* 781-799.

Nordstrom D. K. and Munoz J. L. (1985) *Geochemical Thermodynamics.* Benjamin Cummings Publ., Menlo Park, California. 477 pp.

Ohtani E. (1985) The primordial terrestrial magma ocean and its implication for stratification of the mantle. *Phys. Earth Planet. Inter., 38,* 70-80.

Pollack J. B., McKay C. P., and Christofferson B. M. (1985) A calculation of the Rosseland mean opacity of dust grains in primordial solar system nebulae. *Icarus, 64,* 471-492.

Pollack J. B., Lunine J. I., and Tittemore W. C. (1990) Origin of the Uranian satellites. In *Uranus.* Univ. of Arizona, Tucson, in press.

Ringwood A. E. (1970) Origin of the moon: the precipitation hypothesis. *Earth Planet. Sci. Lett., 8,* 131-140.

Ringwood A. E. (1975) *Composition and Petrology of the Earth's Mantle.* McGraw-Hill, New York. 618 pp.

Robie R. A., Hemingway B. S., and Fisher J. R. (1978) Thermodynamic properties of minerals and related substances at 298.15K and 1 bar (10^5 pascals) pressure and at higher temperatures. *U. S. Geol. Surv. Bull. 1452.*

Safronov V. S. (1969) *Evolution of the Protoplanetary Cloud and Formation of the Earth and Planets.* Nauka, Moscow. Translated by the Israel Program for Scientific Translation (1972).

Safronov V. S. (1978) The heating of the Earth during its formation. *Icarus, 33,* 1-12.

Sasaki S. (1987) Structure and evolution of the primordial H_2-He atmosphere surrounding the accreting Earth and the origin of the terrestrial noble gas. D.Sc. dissertation, Univ. of Tokyo, Tokyo.

Sasaki S. and Nakazawa K. (1986) Metal-silicate fractionation in the growing Earth: Energy source for the terresrial magma ocean. *J. Geophys. Res., 91,* 9231-9238.

Sasaki S. and Nakazawa K. (1988) Origin of isotopic fractionation of terrestrial Xe: hydrodynamic fractionation during escape of the primordial H_2-He atmosphere. *Earth Planet. Sci. Lett., 89,* 323-334.

Sasaki S. and Nakazawa K. (1990) Did a primary solar-type atmosphere exist around the proto-earth? *Icarus, 84,* in press.

Sekiya M., Nakazawa K., and Hayashi C. (1980) Dissipation of the primordial terrestrial atmosphere due to irradiation of the solar EUV. *Prog. Theor. Phys., 64,* 1968-1985.

Sekiya M., Hayashi C., and Nakazawa K. (1981) Dissipation of the primordial terrestrial atmosphere due to irradiation of the solar far-UV during T Tauri stage. *Prog. Theor. Phys., 66,* 1301-1316.

Staudacher T. and Allègre C. J. (1982) Terrestrial xenology. *Earth Planet. Sci. Lett., 60,* 389-406.

Stevenson D. J. (1984) On forming the giant planets quickly (superganymedean puff balls) (abstract). In *Lunar and Planetary Science XV,* pp. 822-823. Lunar and Planetary Institute, Houston.

Stevenson D. J. (1987) Origin of the moon—the collision hypothesis. *Annu. Rev. Earth Planet. Sci., 15,* 271-315.

Takeda H., Matsuda T., Sawada K., and Hayashi C. (1985) Drag on a gravitating sphere moving through a gas. *Prog. Theor. Phys., 74,* 272-287.

Vollmer R. (1977) Terrestrial lead isotopic evolution and formation time of the Earth's core. *Nature, 270,* 144-147.

Wood J. A. (1986) Moon over Mauna Loa: A review of hypotheses of formation of Earth's moon. In *Origin of the Moon* (W. K. Hartmann, R. J. Phillips, and G. J. Taylor, eds.), pp. 17-55. Lunar and Planetary Institute, Houston.
Zahnle K. J., Kasting J. F., and Pollack J. B. (1988) Evolution of a steam atmosphere during Earth's accretion. *Icarus, 74,* 62-97.

EARTH ACCRETION

T. J. Ahrens

Lindhurst Laboratory of Experimental Geophysics, Seismological Laboratory 252-21, California Institute of Technology, Pasadena, CA 91125

Accretion of the Earth is described in terms of the infall of volatile-, silicate-, sulfide-, and iron-bearing planetesimals. The shock pressures induced in the constituents of the planetesimals upon impact with the Earth determine their fate and the oxidation state associated with reaction of the coaccreting protoatmosphere. For planetesimal infall velocities of <0.9 km/sec, the accreting Earth has a radius of $\lesssim$1400 km, and the impacting planetesimals retain their full complement of volatiles, e.g., H_2O, CO_2, SO_2, NH_3, CH_4, and noble gases. This portion of the initially accreted Earth is the present source of the mantle-derived ^3He. The present rate of ^3He flux from the Earth's mantle is fully consistent with the existence of a planetesimal-derived primitive undifferentiated reservoir of an equivalent mass as a sphere of 1400 to 2000 km radius. Upon the onset of impact vaporization, at shock pressures of ~19 GPa, the water driven from the planetesimals by impact is able to react with metallic iron. This oxidation results in an increase in the iron silicate budget of the Earth. The composition of the mantle indicates that a maximum of 0.12 to 0.22 of the Earth's mantle accreted under oxidizing conditions. As more water is released by impact, the protoatmosphere inhibited infrared radiative cooling of the surface. As a consequence, impact energy is trapped under the protoatmosphere and the surface temperature rises to the melting point of wet silicates (~1400 K); surface temperature then becomes buffered by the melting reaction. Composition of the protoatmosphere is controlled by temperature and partial solubility in molten silicate. Iron silicates are reduced to metallic iron. Iron, both thermally and shock heated, and iron sulfide form the Earth's core. As accretion continues, the largest planetesimals impacting the Earth increase in diameter from several kilometers to several hundred kilometers and possibly to thousands of kilometers. We examine the constraints on impact-induced loss of the entire protoatmosphere as a result of giant impact of large planetesimals. A ~2000-km-diameter, 10^{37}-erg impactor is found to be sufficient to completely eject into space the protoatmosphere of the Earth. Impact of a larger 3000-km-diameter object is calculated to expel the protoatmosphere to speeds of >15 km/sec.

INTRODUCTION

Recent exploration of the solar system using spacecraft and Earth-based observations, observations of nearby stars, and the study of the Earth as a planet, combined with laboratory experimentation and theoretical modeling, have provided an increasingly detailed picture of the processes that form the Earth and the terrestrial planets. In this paper we employ recent laboratory data specifying the shock pressures required to devolatilize model minerals of the presumed Earth-forming planetesimals and complementary thermodynamic data to outline the sequence of conditions on the Earth's surface as it grew from infalling planetesimals. We assume the model of planetary formation developed by *Safronov* (1972), *Levin* (1972), *Wetherill* (1985), and *Greenberg et al.* (1978), in which gases of solar-like composition condensed to form dust, and then the dust (associated with the remaining gas) accreted to form larger objects—the so-called planetesimals. These in turn accrete into planets through their mutual interactions (Fig. 1). As the planets grew via

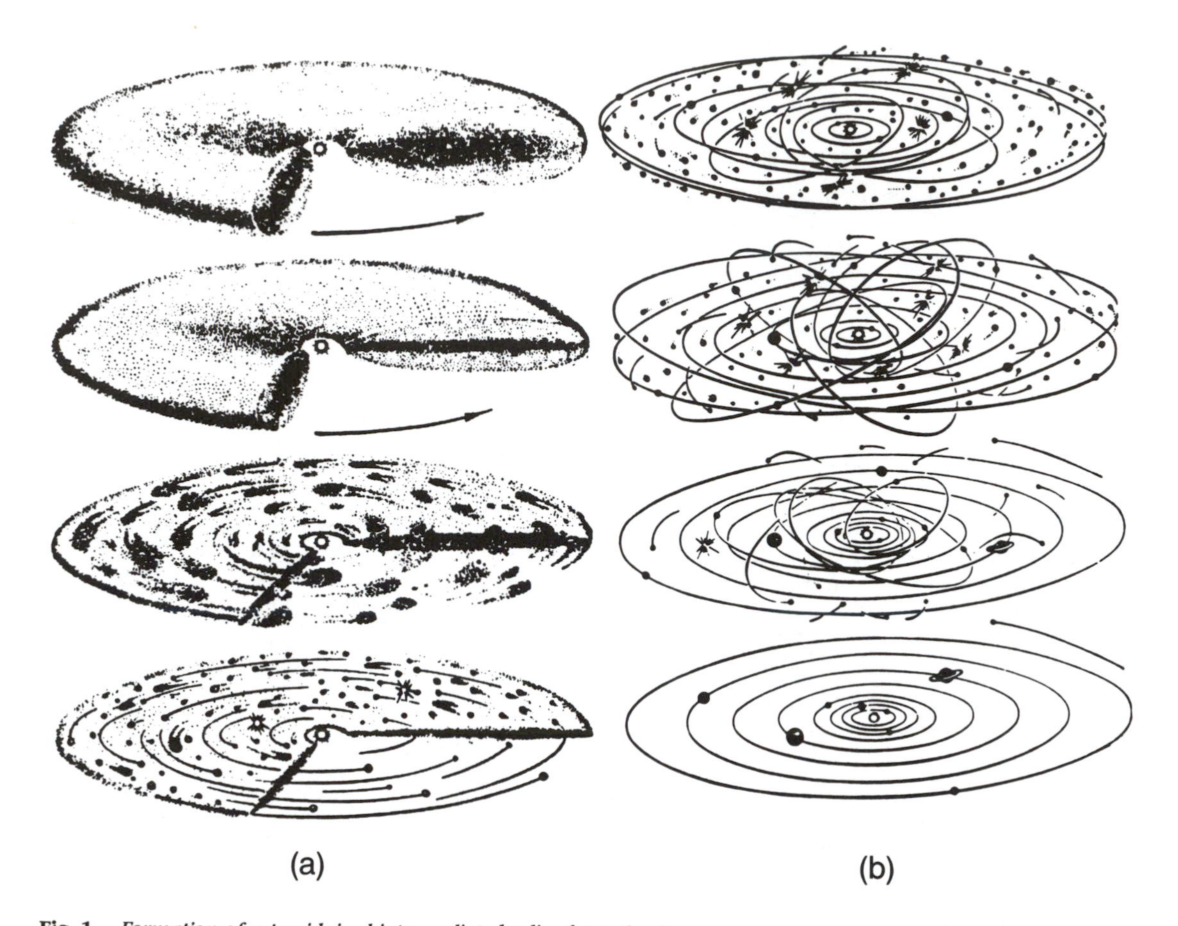

Fig. 1. *Formation of asteroid-sized intermediate bodies form the dust component of the solar nebula—a conditional variant assuming* **(a)** *instantaneous formation of the solar nebula and* **(b)** *a breakup of the dust disc caused by the onset of gravitational instability. There is gradual accretion of intermediate bodies into planets. The accretion of gas by giant planets is not shown. The initially flat system of intermediate bodies thickens due to their mutual gravitational perturbations. After Levin (1972).*

accretion, so did the planetesimals, so as the planets approached their final size they were impacted by larger and larger planetesimals. The last few impacts on the Earth and other planets were therefore major events.

Depending on the equation of state of the silicate planet and its escape velocity, the impact of a given composition projectile (planetesimal) at a certain velocity can give rise to net planetary accretion or erosion (by cratering) (*O'Keefe and Ahrens,* 1977b), as well as induce melting and vaporization (*O'Keefe and Ahrens,* 1977a). Recent numerical studies of large-body impact (*Benz et al.,* 1986, 1987, 1988) demonstrate how a few large planetesimals colliding with a terrestrial planet near the close of the accretional epoch can give rise to whole-planet partial devolatilization. Moreover, Benz et al. show that loss of sufficient ejecta will occur such that an orbiting disc of material may have temporarily been formed about the Earth, possibly accounting for the formation of the Moon. Also, major impacts have long been associated with the inclinations of the spin axes of the planets with respect to their orbital plane.

These are believed to be solely the result of major collisions suffered by the planets during the terminal phases of their accretion.

Strong supporting evidence for a Safronov-type scenario has come from the surprising similarity in the composition of the sun and the primitive undifferentiated meteorites in both major and minor nongaseous elements (Fig. 2). Moreover, the isotopic composition of the sun, meteorites, and planets are surprisingly uniform and demonstrate only minor planet-to-planet variations.

Studies of gravitational perturbations arising from commensurabilities within the zone of the main belt asteroids suggest that this region is the major source of the meteorites. Moreover, the similarity of noble gas abundance patterns in meteorites and terrestrial planetary atmospheres (Fig. 3) suggest that meteorites (and hence asteroids) and planets

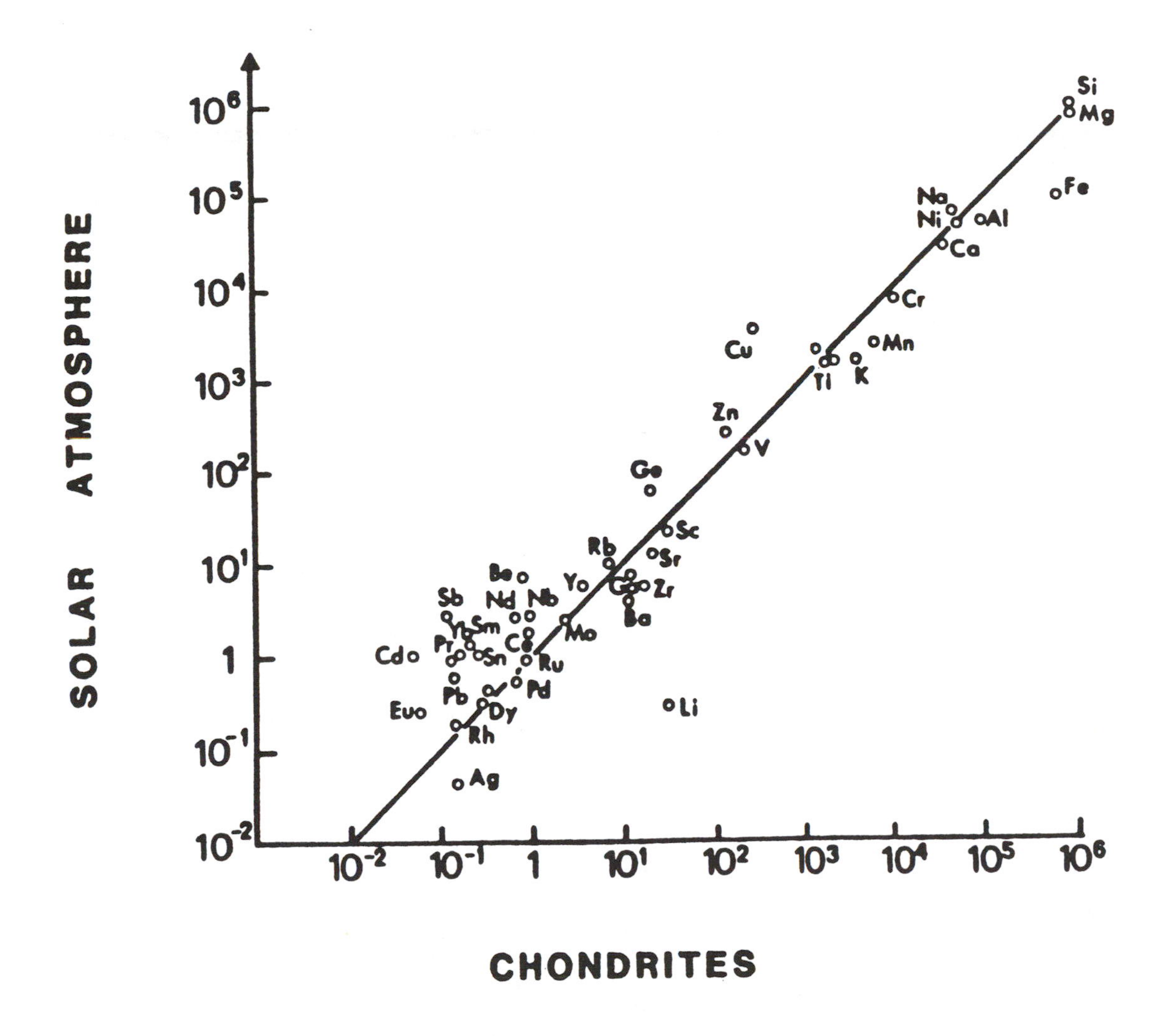

Fig. 2. *Atomic abundance of the elements in the solar photosphere vs. the abundance in chondritic meteorites. Plot is normalized with respect to 10^6 atoms of Si (after Allègre, 1982).*

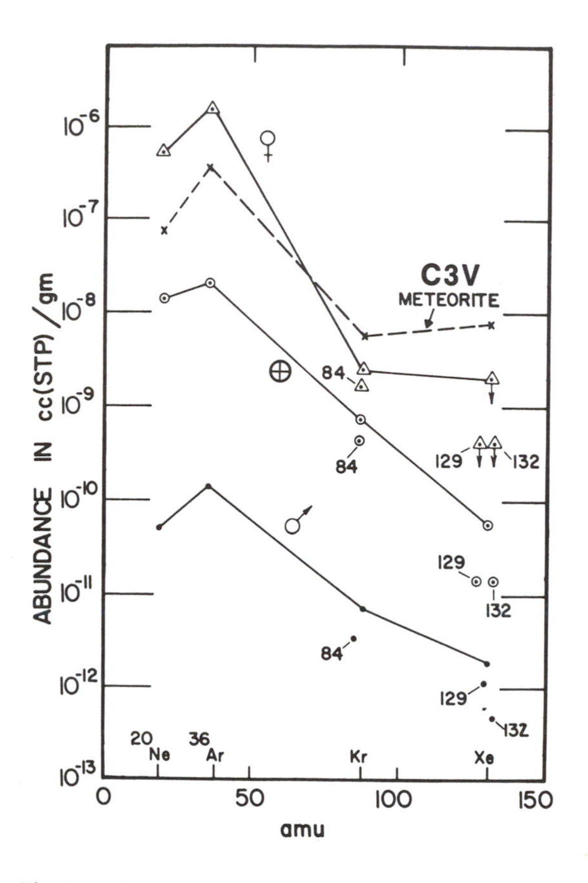

Fig. 3. *Abundances of noble gases on Venus, Earth, and Mars (indicated by symbols) and in class C3V carbonaceous chondrites (after Donahue and Pollack, 1983).*

originated from the same accumulation of matter orbiting the sun as that suggested by the Safronov model (Fig. 1).

Mankind's just-completed reconnaissance and, in some cases, more detailed exploration of all the planets and satellites of the solar system (with exception of the Pluto-Charon system) demonstrate that all the ancient (solid) terranes of the planets and their satellites provide physical evidence of planetesimal accretion—namely, extensive ancient cratered terranes.

Finally, the most convincing evidence that the Earth and the planets accreted via the planetesimal impact process described in the Safronov scenario has come from the optical and microwave images of several regions immediately surrounding nearby stars (e.g., Beta Pictoris, HL Tauri, and R Monocerotis), which demonstrate that gas and presumed dust-laden accretion discs are presently forming around these objects, and these structures presumably represent an early stage in the formation of planetary systems (*Beckwith et al.,* 1986; *Sargent and Beckwith,* 1987; *Smith and Terrile,* 1984).

In this paper, we assume that planetesimals similar to the primitive meteorites accreted to form the Earth. We apply recent laboratory data for the minimum shock pressure required to induce impact devolatilization of meteorites to infer the radius of the "pristine sphere" of undifferentiated material that comprised the "core" of the undifferentiated cold Earth. We then examine the consequences of further impact devolatilization on the production of oxidized iron now residing in the Earth's mantle (and possibly also in the core). We then consider the effect on the accreting Earth's mantle composition of a massive water-bearing protoatmosphere overlaying a magma ocean. With increasing time, the number of planetesimals impacting the Earth decreased, but their largest mass increased, possibly to approximately that of the Moon or Mars. Finally, we examine constraints on the size or energy of such impacts required to entirely eject from both the Earth and the present atmosphere, as well as the putative massive protoatmosphere.

THE ACCRETION PROCESS—APPLICATION OF LABORATORY IMPACT AND EQUATION-OF-STATE EXPERIMENTS

We develop a sequence that approximately describes the point in the accretion sequence of the Earth at which the impact heating of planetesimals gives rise to melting of various components within the Earth. Based on our present knowledge of the Earth's structure, the composition of meteorites, and the availability of shock wave data for terrestrial materials,

TABLE 1. *Simplified Earth constituent.*

	Percent	Mineral		Percent	Mineral*
Mantle + Crust	8	Anorthite	Mantle + Crust	23	Serpentine‡
	23	Enstatite†		23	Forsterite
	35	Forsterite†		15	Enstatite
				8	Anorthite
Core	34	Iron-Nickel	Core	34	Iron-Nickel

* After *Lange and Ahrens* (1984).
† Assumed 3% by weight, H_2O.
‡ Water content of 3% given explicitly via equivalent serpentine content.

Lange and Ahrens (1984) formulated a simple model of planetesimal composition for the objects that form the Earth (Table 1). In the present study, because of the paucity of thermodynamic data, we simplify this approach even more by employing data for a pure iron fraction to represent the bulk of the material in the Earth's core. In addition, we assume a gabbro component to simulate the effect of mantle and crustal silicates, and a serpentine component to approximate the water- and other highly volatile-bearing phases that were probably present in the Earth-forming planetesimals. We also assume that the iron core of the Earth first formed after quantities of iron and iron sulfide (FeS), both of which are abundant in planetesimals melted together upon impact on the growing Earth's surface. As indicated in Table 2, the shock pressures required to induce melting (incipient melting, IM) and complete melting (CM) of the iron and gabbro components can be inferred from shock wave equation-of-state data. For purposes of simplification, and again because of the lack of definitive thermodynamic data, we do not take into account the thermal or shock-induced incongruent vaporization of these media.

The shock pressures required to induce partial and complete melting of FeS upon release to ambient pressure were estimated using the entropy production criterion (*Ahrens and O'Keefe,* 1972) and the shock wave equation of state of *Brown et al.* (1984) for $Fe_{0.9}S$. We assume that the onset of melting occurs when an entropy gain of 1.22 J/g/K occurs, as shown in Table 3 and Fig. 4, corresponding to shock pressures of approximately 100 GPa. Complete shock melting is inferred to result from shock compression to 136 GPa.

TABLE 2. *Shock states and Earth radius for melting and vaporization.*

	Incipient Melting			Complete Melting		
	GPa	km/sec	km	GPa	km/sec	km*
Iron†	220	6.50†	3704	260	7.30	4160
Troilite	100	3.66	2085	136	4.62	2632
Gabbro†	43	1.86	1060	52	2.18	1242
Serpentine‡	19	0.91	519	60	2.45	1396

* If $V_{esc}/2 = u$ is assumed, then radius is multiplied by a factor of 2 for uniform Earth.
† *Ahrens and O'Keefe* (1977).
‡ Values given are for incipient and complete impact devolatilization (after *Tyburczy et al.,* 1986).

TABLE 3. *Shock entropy production in FeS.*

Density (g/cm^3)	Shock Pressure (GPa)	Hugoniot Temperature (K)	Entropy Gain (J/g/K)
5.5	10.4	400	0.167
6.5	40.8	900	0.627
7.0	64.8	1400	0.878
7.5	102	2650	1.240
8.0	158	5090	1.611

Probably the best constrained of any of the values given in Table 2 are those for the shock pressures required for incipient and complete vaporization of serpentine. The values are based on the direct experiments of *Tyburczy et al.* (1986).

The question of whether shock melting or vaporization of various components of planetesimals occur upon impact on an accreting planetary surface is not straightforward and is dependent on the assumed equation of state of planetesimals and the impacted Earth. An actual accreting surface on a planet is likely to have a porous regolith-like surface early in its history before shock devolatilization of volatile-bearing materials occurs. Later in the accretion process, the planet is likely to have a thick envelope of protoatmospheric gases, typically H_2 and H_20, overlying a magma ocean. Moreover, the minimum infall velocity of planetesimals onto a planet will just equal the planetary escape velocity. For a constant density object, the planetary escape velocity, V_{esc}, is proportional to planetary radius, r, thus

$$V_{esc} = (8 \pi G \rho/3)^{1/2} r \qquad (1)$$

Here G is the gravitational constant and ρ is the mean planetary density.

We assume, for simplicity, that the infall velocity always equals the escape velocity, V_{esc}, and the shock pressure induced upon impact, p, is related to impact velocity by

$$p = \rho_o u (C_o + Su) \qquad (2)$$

Here ρ_o is initial density, C_o is zero pressure bulk sound velocity, and S is the shock wave equation-of-state parameter. If the planetary surfaces are solid and identical in composition to the planetesimals (that is, having the same equation of state as the impacting planetesimals), we should assume $u = V_{esc}/2$. However, if the planetary surface is highly differentiated, and/or very porous (e.g., porosity of 30% to 50%), $u \cong V_{esc}$ is a better approximation. We employ this latter assumption in Fig. 5 to give the approximate proto-Earth radius for incipient melting (IM) and complete melting (CM) of each component. Values of r for both assumptions are given in Table 2. The sequence of Table 2 is qualitatively similar to that given, for example, by *Ringwood* (1979, Fig. 7.1, p. 111).

Recent experiments and theoretical work have better quantified the physics that take place during the latter stages of accretion. The shock wave equation-of-state parameters that are used to construct composite planetesimal properties are given in Table 4. These planetesimals are assumed to have been homogeneous in composition with time. We have also attempted to construct a whole-Earth average equation of state. If we were to assume that some degree of inhomogeneous accretion occurred, we would obtain some variation of

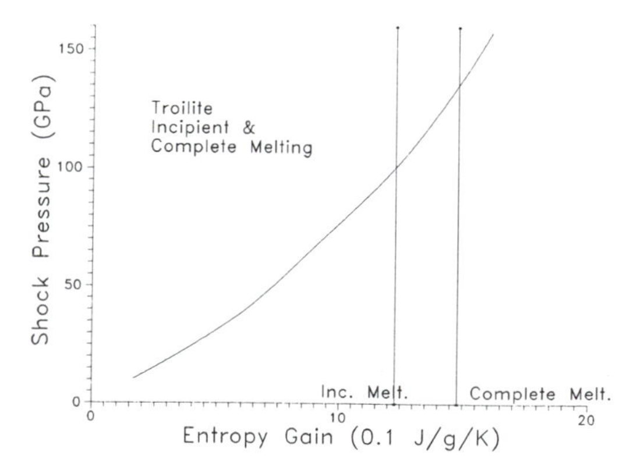

Fig. 4. *Entropy gain vs. shock pressure upon impact of FeS. Shock pressures required to induce incipient and complete melting are indicated.*

TABLE 4. *Average Earth equation of state.*

	Upper Mantle and Crust	Lower Mantle	Outer Core	Inner Core
Uncompresed density (g/cm^3)	3.31	3.99	6.73	7.60
Mass fraction of Earth	0.16	0.50	0.31	0.016
Zero presure bulk modulus (GPa)	131	195	127	398
Radial travel time at bulk sound velocity (sec)	111	313	523	168

the equation-of-state parameters with planetary size. Instead, for simplicity, we employ a value averaged according to the percent of the bulk sound speed travel time through the different regions of the Earth (upper and lower mantle and inner and outer core) (Table 4). The equation-of-state parameters obtained in this way are $\rho_0 = 5.75\ g/cm^3$, $C_o = 6.55$ km/sec, and $S = 1.6$. This density is slightly greater than the average Earth (compressed) density of $5.5\ g/cm^3$. The alternate approach, for example, would be to mass average the specific volume, which would yield an average uncompressed density, ρ_0, of $4.51\ g/cm^3$.

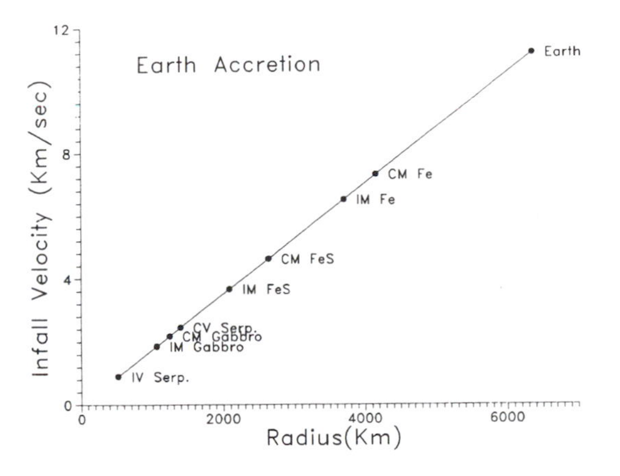

Fig. 5. *Radius (r) of the accreting Earth vs. assumed infall (escape) velocity (equation (1)). Minimum radii are plotted for incipient vaporization of water (and other gases) from serpentine (IV, Serp), and complete gas loss from serpentine (CV, Serp). Incipient and complete impact-induced melting of silicate is represented by the incipient and complete melting of gabbro (IM gabbro and CM gabbro). Similarly, the points where FeS and Fe undergo impact-induced melting are indicated.*

THE PRIMITIVE ACCRETION CORE OF THE EARTH

Recent impact devolatilization experiments on both serpentine and C3 carbonaceous chondrites (*Tyburczy et al.,* 1986) demonstrate that the onset of detectable volatilization of H_2O and other volatiles [CO_2, SO_2, NH_3, CH_4, and noble gases (?)] occurs at approximately 19 GPa and complete devolatilization occurs upon unloading from shock pressures greater than 60 GPa.

Thus we define the "primitive accretion core" as the radius, R_{pac}, of the Earth accreted before total volatilization occurs. According to the mean equation of state of the Earth chosen here, this core has a radius of ~1400 km, slightly smaller or comparable to the size of the Moon, which is ~1/100th the mass of the Earth. If, instead, various assumptions are made about the effective equation of state of serpentine, either crystalline or porous, estimates of R_{pac} range from 0.2 to 0.5 R_e, where R_e is the final Earth radius. Is this range of R_{pac} reasonable? One approach to answering this question is to try to constrain the mass of a hypothetical "primitive" reservoir of large ion lithophile, light rare elements, and noble gases, evidence for which is observed in molten rocks that appear to have deep mantle sources. For simplicity, we examine the ^3He flux because this isotope can rapidly diffuse from the ocean to the atmosphere. Then it is lost to space via Jeans' escape. Also, it is too chemically inert to be easily stored in the crust. Thus it probably is not efficiently subducted (or recycled) from crustal rocks. Moreover, ^3He is not produced by radioactive decay. The production of ^3He from cosmic-ray sources does not seriously contaminate estimates of the flux from the mantle. We only examine an order of magnitude mass of a reservoir since we observe the total flux of ^3He and we need to integrate this flux over geological time. Using the mantle flux of ^3He determined by *Craig et al.* (1975) of 4 ± 1 atom/cm^2/sec, we infer a value of 4.9×10^{12} mol of ^3He produced, temporarily stored in the ocean and atmosphere, and ultimately lost to space (uniformly) over geologic time (4.6 aeons). Since we may anticipate that the ^3He loss model will merely provide a lower bound to the initial ^3He inventory. This flux came from portions of the mantle that presumably have not experienced total gas loss since the material was accreted by the Earth. We assume that it comprises the primitive accretion core (pac). If this pac were a sphere, what is its characteristic size and how does this compare to R_{pac} inferred from the shock devolatilization experiments conducted on model planetesimals (i.e., the carbonaceous chondrites)?

One approach we may use to answer the above question is to examine the ^3He abundance within carbonaceous meteorites. *Ozima and Podesek* (1983) give the ^4He content of various gas-rich primitive meteorites as 10^{-4} to 10^{-3} cm^3 STP/g. When these values are taken with the $^3He/^4He$ solar system ratio of 1.43×10^{-4}, a primitive meteoritic abundance of 5.8×10^{-13} to 5.8×10^{-12} mol ^3He/cm^3 is obtained. Taking 0.2 and 0.5 R_e as R_{pac} with the above range of ^3He concentration yields values of 4.2×10^{12} to 6.6×10^{13} mol ^3He as the Earth's primordial budget. The lower bound of 4.2×10^{12} mol of ^3He compares favorably with $5 \pm 1 \times 10^{12}$ mol ^3He inferred above from the *Craig et al.* (1975) total present Earth flux.

THE IRON-WATER REACTION

In the early stages of Earth accretion, when $r = 0.2$ to $0.5\ R_e$, once complete release of water and other volatiles from volatile-bearing materials takes place upon shocking to above 60 GPa, chemical reactions between the gases in the atmosphere to the surface are expected to occur. Early consideration of the consequences of an impact-induced atmosphere on the ultimate composition of the Earth are summarized in *Ringwood* (1979). He pointed out that the reaction of metallic iron with water at low temperatures would give rise to an oxidation reaction

$$MgSiO_3 + Fe + H_2O \rightarrow 1/2\ Fe_2SiO_4 + 1/2\ Mg_2SiO_4 + H_2 \quad (3)$$

Presumably, during Earth accretion, all the H_2 produced via reaction (3) would be lost by Jean's escape. Later, *Lange and Ahrens* (1984) considered this reaction and pointed out that the total oxidized iron of the Earth's mantle, and possibly the core, limited the amount of the Earth that could have been homogeneously accreted. Lange and Ahrens show that the Earth cannot have accreted homogeneously with only reaction (3) taking place. Taking the 3% total water content of planetesimals they assume, it follows that the H_2O would be used up oxidizing Fe to $FeSiO_3$ or Fe_2SiO_4, and the Earth would be too rich in oxidized iron and would not retain the oceanic water budget. Also, *Jones* (1987) argued that a reaction such as that shown in equation (3) removed H_2O from the accreting mantle. *Lange and Ahrens* (1984) concluded (as did *Ringwood*, 1979; *Wänke et al.*, 1984; and *Drake*, 1987) that some inhomogeneous accretion must have taken place. Although Lange and Ahrens infer that the 6% of the Earth that accreted inhomogeneously occurred such that the metallic iron core accreted first, the recent thermodynamic calculations of *Holloway* (1988) (Fig. 6) point to initial oxidation of the surface materials when the surface temperatures are in the 400° to 600°C range. Holloway considered a silicate-free assemblage of C, H_2O, FeS, FeO, and Fe in the mole ratio 1:0.5:6:11:16 at various temperatures. According to his calculations, at low temperatures (<800°C) the solids on the surface of Earth become oxidized, and the coexisting fluid becomes reduced.

In terms of carbon- and silicate-bearing species, equation (3) can be written as

$$(1/2)C + MgSiO_3 + Fe + H_2O \rightarrow (1/2)Fe_2SiO_4 + (1/2)Mg_2SiO_4 + (1/2)CH_4 \tag{4}$$

We denote accretion in which iron becomes oxidized via reactions of the type indicated by equation (4) as Mode A, or oxidizing. *Ringwood* (1979) has used the same terminology. [Unfortunately, *Wänke et al.* (1984) denoted this type of accretion as Mode B and this has led to confusion.]

Thus, if the planetesimals contained no iron silicate, and depending on whether the mantle has a composition of $(Mg_{0.88}Fe_{0.12})_2SiO_4$ to $Mg_{0.8}Fe_{0.2}SiO_3$, a maximum of some 12 to 22 wt.% of the mass of the mantle could have been accreted via Mode A reactions. If the accreting planetesimals contained constant ratios of core to mantle elements these percentages will also correspond to percentages of the whole Earth. The 12% to 22% is an upper bound of how much of the Earth could have accreted at lower temperatures with an oxidizing atmosphere. Clearly, to the extent to which the Earth accreted some of its iron budget from iron-bearing silicates, the upper bound on the fraction of the Earth accreted via Mode A would be reduced.

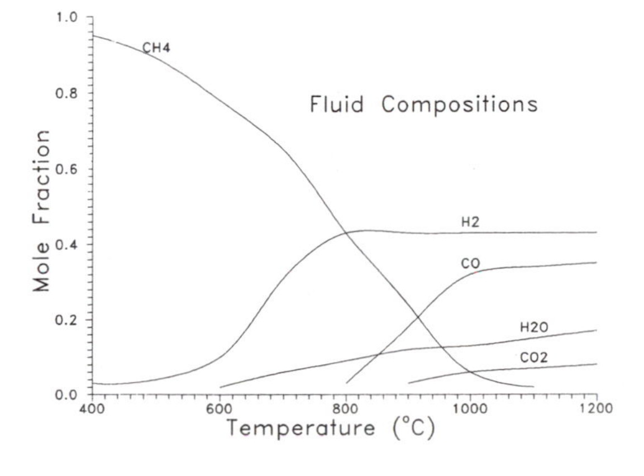

Fig. 6. *Mole fraction of gas (CH_4, H_2, H_2O, CO, and CO_2) in equilibrium with silicate-free hypothetical planetesimal mineral assemblage with a molar ratio C:1, H_2O:0.5; FeS:6.0; FeO:11.0; and Fe:16.0 (after Holloway, 1988).*

MODE B (REDUCING) ACCRETION THROUGH THE MAGMA OCEAN

The question of gas interaction with the solid Earth and the issue of homogeneous vs. heterogeneous accretion must be viewed in a different light since the physical models of Abe and Matsui (*Abe,* 1986, 1988a; *Abe and Matsui,* 1985, 1986, 1988; *Matsui and Abe,* 1986a) have been developed. They examined the earlier suggestion of a massive water protoatmosphere (*Lange and Ahrens,* 1982a) and the consequences of the initial experimental data (*Lange and Ahrens,* 1982b) regarding water devolatilization. They developed a quantitative model of the consequences of a massive water atmosphere blanketing the Earth and inducing a magma ocean during a large fraction of the accretion epoch. The Abe and Matsui model assumes that the impact rate of planetesimals required to maintain a magma ocean underlying a 10^2 to 10^3 bar protoatmosphere of water is ~300 W/m^2. This impact rate, which implies a total Earth accretion time of ~10^7 years, gives rise to a surface temperature buffered by the melting of silicates in equilibrium with the water of the atmosphere (1500 K). The 300 W/m^2 impact rate gives rise to a high surface temperature and causes melting of rocks as a consequence of the opacity of a H_2O- and CO_2-rich atmosphere to infrared radiation. This opacity traps most of the energy of planetesimal impact on the surface.

In the Abe and Matsui model, the amount of protoatmosphere is controlled by the solubility of H_2O in molten silicate on the surface of the magma ocean. The impact rate of 300 W/m^2 would correspond to something like a 10-km-diameter object, such as the Cretaceous-Tertiary extinction bolide, impacting the present Earth every 4.5 days. The preliminary thermodynamic calculations of Holloway (Fig. 6) for a simplified silicate-absent system imply that at the high surface temperatures required for magma ocean formation, the coexisting atmospheric gases will be H_2, CO, H_2O, and CO_2. Under these conditions, reduction of iron silicates would occur via reactions such as

$$CO + MgFeSiO_4 \rightarrow Fe + CO_2 + MgSiO_3 \quad (5)$$

Reaction (5) produces a metallic core and magnesium-rich mantle from an iron-silicate-rich, carbon-bearing planetesimal. Although the iron budget of the mantle is probably not solely determined by low-temperature oxidizing reactions such as those in equations (3) and (4), nor high-temperature reducing reactions such as equation (5), the latter reaction does limit how much of the Earth could have formed under Mode B (reducing) conditions. Thus, from the constraints on mantle composition given above (after equation (4)), we can infer that some 78% to 88% of the Earth can be accreted in Mode B. Core formation would begin according to the scheme of Fig. 5 once a magma ocean began to form in the radius range of 0.2 to 0.5 R_e. The temperature of the magma ocean would have been above the 1261 K eutectic in Fe-S sulfur system. Upon thermal or shock-induced melting of the FeS component of the planetesimals and melting of Fe, when a radius of ~4000 km is achieved, development of a core is expected in virtually all scenarios for Earth accretion. In addition, S and C may be among the light elements in the liquid core. The present outer core of the Earth has a density that is some 10% less than pure iron. We presume that the inner, solid core of the Earth, which appears to be nearly pure iron, formed later in time, as latent heat of crystallization is released at the inner core/outer core boundary.

LOSS OF THE PROTOATMOSPHERE

Both *Wetherill* (1985) and *Greenberg et al.* (1978) have demonstrated that, as the planets grew, so did the largest planetesimals that impacted them. Models indicate that the larger planetesimals at the start of accretion of the terrestrial planets had diameters of kilometers, whereas at the close of accretion, the largest planetesimals had accreted to the size of the smaller planets (e.g., Mercury and Mars) with diameters ~4000 to 6000 km.

Recently, *Benz et al.* (1986, 1987) and *Kipp and Melosh* (1986) constructed numerical models of the effects upon the Earth of large Moon-and Mars-sized body impacts. Their work was largely directed toward the question of whether sufficient ejecta from such an interaction could end up in orbit around the Earth such that the Earth's Moon could result from large impacts. Recently, partial loss or erosion of the Earth's present or protoatmosphere was considered in a series of papers by *Walker* (1986), *Zahnle et al.* (1988), *Ahrens and O'Keefe* (1987), *Ahrens et al.* (1989), and *Melosh and Vickery* (1989). The latter three papers concluded that for a sufficiently large impactor, all the atmosphere above the plane tangent to the Earth at the point of impact is ejected from the Earth. For the present atmosphere, 5.5×10^{-4} of the atmosphere can be lost by single impact of a minimum of 10^{30} ergs (e.g., Table IX, *Ahrens et al.,* 1989). For larger impacts, the blast wave theory of *Ahrens et al.* (1989) or the impact vaporization theory of *Melosh and Vickery* (1989) is not directly applicable.

To estimate the energy, and hence approximate planetesimal size, such that upon impact the entire planetary atmosphere is blown off, we employ a different approach than previous efforts and consider the shock wave that is entirely propagated within the Earth as sketched in Fig. 7.

The key calculation we wish to conduct is to relate the particle velocity of the Earth-atmosphere interface, u_{fs}, antipodal to a major impact to the atmospheric free-surface velocity, V_{esc}, which will result from being shocked first by the moving solid planet and then isentropically being released into space at greater than escape velocity. This is a conservative calculation as we use the density and pressure of the atmosphere at its base to conduct this calculation. Moreover, atmosphere covering the planet closer to the impact than the antipode is expected to achieve high velocity because it is shocked to higher pressures by the solid planet. We note that, in general, as a shock wave is propagated upward in an exponential atmosphere, because the density encountered by the traveling shock is decreasing, the shock velocity and particle velocity increases with altitude as discussed by *Zel'dovich and Raizer* (1967, Vol. 2, Chap. XII, section 5.25). Thus, we can safely neglect shock attenuation in the atmosphere, and our tactic is to assume the particle velocity at the Earth-atmosphere interface (the independent variable) and calculate the shock pressure induced in the gas by the outward surface of the Earth. The solid Earth therefore acts like a piston with velocity, u_{fs}, pushing on the atmosphere. The present calculational estimates can only be considered approximate because the effect of the Earth's core (which may defocus the shock) is neglected. Moreover, we also neglect the expected antipodal shock, focusing in the mantle, which is a consequence of the spherical shape. We employ equation (1.82) of *Zel'dovich and Raizer* (1967) in a strong shock approximation to calculate the shock pressure, p_1, at the base of the atmosphere from

$$u_{fs} = \{(\gamma + 1)^{1/2} + [(\gamma - 1)^2/(\gamma + 1)]^{1/2}\} (p_1 V_0/2)^{1/2} \quad (6)$$

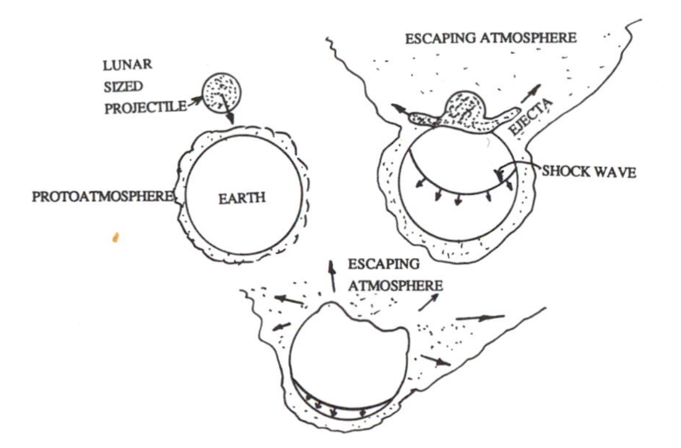

Fig. 7. *Sketch of lunar-sized planetesimal impacting the Earth. The protoatmosphere is blown away by the shock-wave-induced motion of the solid or molten planet.*

Here V_0 is the initial specific volume of the atmosphere (at its base), p_1 is the initially (unknown) shock pressure, and u_{fs} is the independent variable. For an air (1 bar) atmosphere, we assume a specific volume of 10^3 cm^3/g and a polytropic exponent, γ, of 1.4 ± 0.1.

For a water protoatmosphere, because of the limitations of available low-pressure and high-temperature thermodynamic data, we assume a 100-bar water protoatmosphere with a surface temperature of 1100 K and a specific volume of 49 cm^3/g. Because of the expected molecular disassociation upon shock compression, we again use the value of γ of 1.4 ± 0.1 for a nominal value. It turns out that only the piston velocity and the polytropic constant determine the gas expansion velocity. The total free expansion velocity V_{esc} is then given by

$$V_{esc} = u_{fs} + u_r \quad (7)$$

where u_r is given by the Riemann integral relation

$$u_r = \int_0^{p_1} \left(-\frac{\partial V}{\partial p} \right)_s^{1/2} dp \qquad (8)$$

For simplicity, we assume upon isentropic release from the shock state (p_1, V_1) that the shocked atmosphere behaves as a polytropic gas, with an equation of state of the form

$$V = (c/p)^{1/\gamma} \qquad (9)$$

Here c is a constant determined from p_1 (calculated from equation (6)) and for the strong shock approximation (equation (1.80), *Zel'dovich and Raizer,* 1967)

$$V_1 = V_o (\gamma - 1)/(\gamma + 1) \qquad (10)$$

Upon substituting V into equation (8) and integrating, we obtain

$$u_r = (c/\gamma)^{1/2} c^{(1-\gamma)/2\gamma} (p^{s+1})/(s+1) \qquad (11)$$

where

$$s = (\gamma - 1)/(2\gamma - 1) \qquad (12)$$

Calculated values for u_{fs} vs. V_{esc} from equation (7) are given in Table 5 and plotted in Fig. 8.

What then is the outward rock velocity for antipodal impact of a given energy on the Earth? The actual problem has not yet been worked out in detail and involves a formidable calculation that should take into account the internal structure of the Earth. We instead employ a scaled calculation of the antipodal response of the Moon to a large impact conducted by *Hughes et al.* (1977). Their calculation showed that for a 10^{32}-erg (46 km diameter, 11 km/sec, silicate impactor) impact on the Moon, the peak antipodal free-surface velocity is 11-75 m/sec. Scaling the energy to obtain a planet free-surface velocity of 3 km/sec to in turn obtain an atmospheric free-surface velocity of 11.2 km/sec, the Earth's escape velocity, requires an impactor of $\sim 10^{38}$ ergs for the Earth. By comparison, using the empirical formula of *Melosh* (1984) for the decay of a strong shock wave in rock, the pressure upon spherical spreading and irreversible work carried out on the silicates encompassed by a shock from a point explosion in a while space is given by

$$p = c_p \rho_p U/2 (a/r)^{1.7} \qquad (13)$$

where c_p, ρ_p, U, and a are the longitudinal elastic velocity, density, impact velocity, a, and radius of the projectile and r is radial distance from the source. The exponent in equation (13), 1.7, is very close to the value independently inferred by *Ahrens and O'Keefe* (1987) to be appropriate for stress wave decay from an 11-km/sec impact onto silicate. For a 10^{32}-erg impact, the antipodal point on the Moon is predicted to have a free-surface velocity of 2 m/sec, which is a factor of 6 to 35 less than the *Hughes et al.* (1977) calculation. Thus, Melosh's formula (which is for a whole space and not a sphere) appears to slightly understate the required free-surface velocity by a factor of 6 to 35. For a Moon-sized object impacting the Earth at 11 km/sec, with an energy of 6×10^{37} ergs, the antipodal shock pressure is 8 GPa (from equation (12)). This results in a free-surface velocity of 1.1 km/sec. If this velocity is increased by the factor of 6 correction, it can be seen from Fig. 8 that a

TABLE 5. *Relation of Earth free-surface velocity (u_{fs}) and particle velocity of overlying atmospheric gas upon being shocked and released to zero pressure (V_{exp}) for various values of polytropic exponent (γ).*

	Rock Free-Surface Velocity (km/sec) u_{fs} Protowater or Air Atmosphere	Gas Velocity (km/sec) V_{exp}
$\gamma = 1.3$	1	3.94
	2	7.89
	5	19.7
	10	39.4
$\gamma = 1.5$	1	3.45
	2	6.9
	5	17.2
	10	34.5

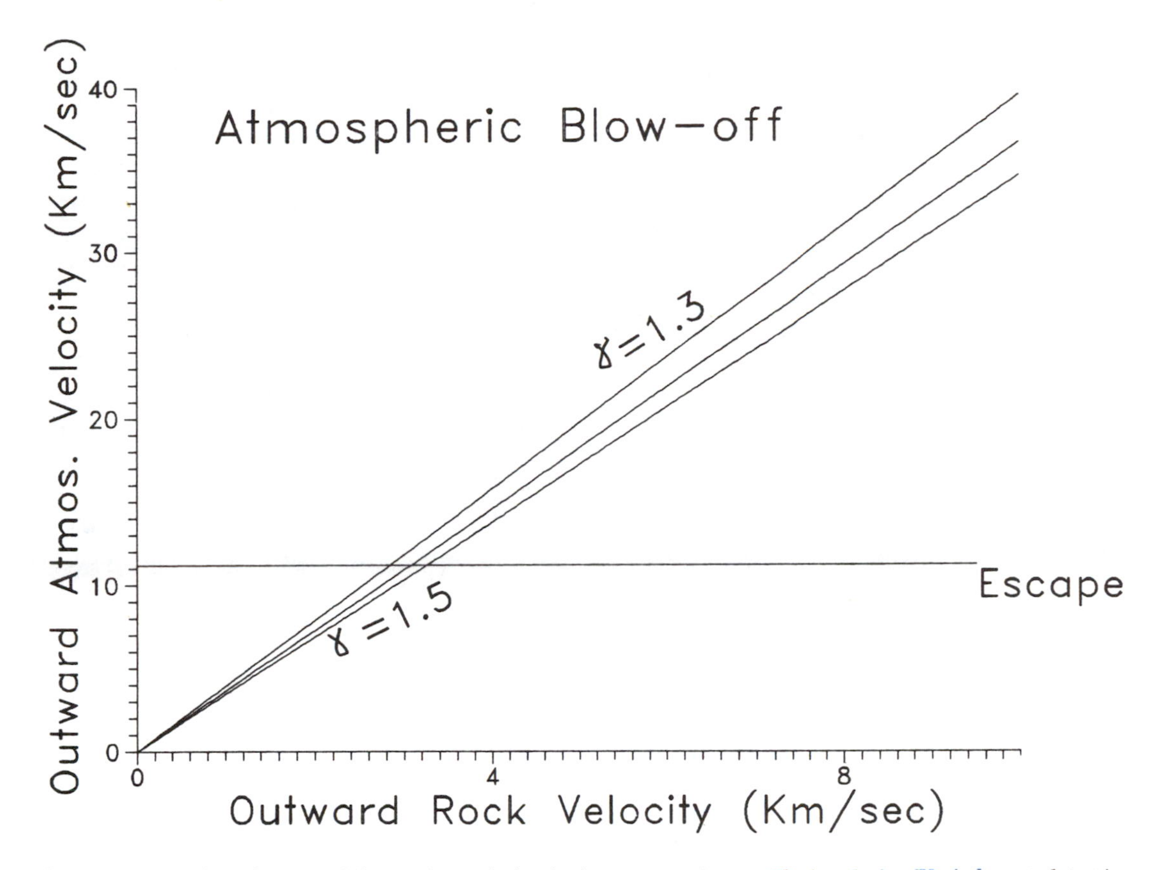

Fig. 8. *Relationship of outward free-surface velocity (u_{fs}) to outward atmospheric velocity (V_{esc}) for a polytropic atmosphere with various values of γ.*

collisional event of this magnitude will result in air or water atmosphere achieving velocities of ~25 km/sec, respectively, and thus be completely blown off. The minimum energy impact expected to just induce complete atmospheric escape is therefore found to be $\sim 10^{37}$ ergs.

CONCLUSIONS

It is likely that the Earth accreted via the infall of planetesimals of approximately the same overall composition as the nongaseous elements that are now inferred for its composition. Initially the planetesimals are thought to have been kilometers in diameter and the diameter of the largest planetesimal grew with time as the Earth accreted. Numerical models of the accretion process suggest that toward the end of the accretion epoch, several large planetesimals with diameters in the range ~4000 to 6000 km impacted the terrestrial planets.

We assume that planetesimals contained similar abundances of readily releasable volatiles such as H_2O, NH_3, CH_4, SO_2, CO_2, and noble gases in approximately the same proportion as observed in primitive chondrites. Experiments have determined that shock pressures of ~19 GPa are required to induce the onset of devolatilization of primitive carbonaceous meteorites. For a mean Earth equation of state, a core of material that has not undergone shock devolatilization is hypothesized to form an undifferentiated primordial reservoir containing approximately the initial complement of volatiles present in the planetesimals. The minimum radius of such an undifferentiated core would be ~1400 km if it contained the meteoritic abundance of ^3He of 6×10^{-13} to 6×10^{-12} mol/cm^3. This would yield a total Earth abundance of 4×10^{12} to 7×10^{13} mol of ^3He. This range of abundance is compatible with the present observed degassing rate of ^3He from the Earth of 4 ± 1 atoms/m^2/sec occurring over geologic time (4.6 aeons). As water is released from impacting planetesimals, the oxidation of metallic iron to $FeSiO_3$ and Fe_2SiO_4 adds iron silicate to the budget of the Earth. At this time the Earth's atmosphere is inferred to be highly reducing, containing mostly CH_4. Depending on the assumed mantle composition, 0.12 to 0.22 is the maximum fraction of the mantle that could have accreted in this mode. Following the notation of Ringwood, this is denoted as Mode A (accretion) (Fig. 9). Once several atmospheres of water are released and the infall of impacting planetesimals delivers >300 W/m^2 onto the Earth's surface, thermal blanketing occurs. The infrared-opaque atmosphere of H_2O, CO, and CO_2 prevents the shock-heated surface rocks from losing their energy via radiation to space. Extensive melting of wet silicates occurs on the surface. The surface temperature is bufferd by the melting point of wet silicates (~1500 K) and the gas content of the atmosphere is controlled by solubility in molten silicate. Most of the Earth is accreted in this so-called B mode. Iron silicates in the planetesimals are reduced to metallic iron and this iron and the iron and iron sulfide in the planetesimals goes into the Earth's core.

As indicated in Fig. 9, large impacts on the Earth occasionally ejected the accumulated primordial atmosphere temporarily, and for short periods of time thereafter the Earth accretes according to the oxidized A mode. Impacts of ~1000-km-diameter, ~10^{36}-erg planetesimals at the final Earth size are calculated to approximately cause the antipodal solid or molten planet free-surface to achieve velocities of 3 km/sec. This in turn is sufficient to drive a shock into the overlying 100-bar, H_2O-rich atmosphere such that it releases to a final velocity $\gtrsim 11$ km/sec and escapes the Earth. The impact of larger ~4000-km-diameter impactors is expected to drive off an air or water atmosphere at velocities of $\gtrsim 25$ km/sec. After a sufficiently large impact, if the accretion rate declined so that the impact-produced power input at the surface fell below ~300 W/m^2 (the power required to sustain a protoatmosphere), the protoatmosphere will not reform. Cooling of the magma ocean and the formation of a liquid-water ocean as described by *Matsui and Abe* (1986a) will occur. Accretion in the A mode continues and is still taking place on the

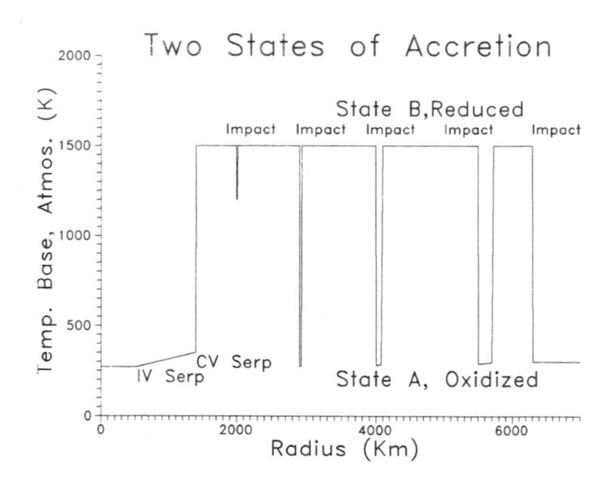

Fig. 9. *Diagrammatic sketch of the temperature state at the surface of the accreting Earth. A primitive accreted core containing most of the initial complement of gases within planetesimals accretes until complete vaporization of these gases occur upon impact. During the interval between initial and complete vaporization of serpentine, Mode A (oxidizing) accretion occurs. The loss of water and other infrared absorbing species to the atmosphere causes thermal blanketing and Mode B (reducing) accretion. Large impacts produce temporary loss of the protoatmosphere and temporary periods of Mode A accretion. Finally, the rate of impact of planetesimals declines to the point that generation of protoatmosphere and thermal blanketing does not occur and the magma ocean does not form again. The last vestiges of accretion are oxidizing (Mode A).*

Earth. Evidence of a small veneer of oxidized accreting material (accreted in Mode A) is recognized in the high sideerophile minor element abundances of the present upper mantle (e.g., *Brett,* 1984; *Drake,* 1987; *Ringwood,* 1978; *Schmitt et al.*, 1989; *Wänke et al.,* 1984).

Thus, we infer that as the Earth accreted, the surface condition went from oxidizing to reducing and again to oxidizing. Initially, the conditions on the Earth were such that surface materials were initially oxidized. Once impact heating produced a protoatmosphere, when the Earth was 0.2 to 0.5 of its present radius, the Earth accreted most of its mass (78% to 88%) under surface-reducing conditions, e.g., the iron in iron-bearing silicates were reduced to the metal and core formation readily occurred beneath a molten magma ocean. Toward the end of accretion, when the energy delivered to the Earth's surface by planetesimal impact decreased to less than $\sim$300 W/m^2, the protoatmosphere and magma ocean could not be sustained. Upper mantle, minor, and trace siderophile element abundances suggest that in this final oxidizing interval, the Earth accreted some 0.7% of its mass (*Schmitt et al.,* 1989). The Earth is very slowly accreting meteoritic and cometary material under oxidizing conditions at the present time.

Acknowledgments. The author appreciates the encouragement of the conference organizers, J. Jones and H. Newsom, to present this paper in written form. The paper has benefited from helpful comments proffered by J. Jones and two anonymous reviewers. This research was supported by NSF and NASA grants. This paper is Caltech Division of Geological and Planetary Sciences contribution no. 4800.

REFERENCES

Abe Y. (1986) Early evolution of the terrestrial planets: Accretion, atmosphere formation and thermal history. Ph.D. thesis, Univ. of Tokyo, Tokyo, Japan. 293 pp.

Abe Y. (1988a) Conditions required for formation of water ocean on an Earth-sized planet (abstract). In *Lunar and Planetary Science XIX,* pp. 1-2. Lunar and Planetary Institute, Houston.

Abe Y. (1988b) Partitioning of carbon between the proto-atmosphere and proto-mantle during accretion. In *Proceedings of the Erice Conference on Interaction of Solid Earth and Atmosphere.* Italian Phys. Soc., Rome, in press.

Abe Y. and Matsui T. (1985) The formation of an impact-generated H_2O atmosphere and its implications for the early thermal history of the Earth. *Proc Lunar Planet. Sci. Conf. 15th,* in *J. Geophys. Res., 90,* C545-C559.

Abe Y. and Matsui T. (1986) Early evolution of the Earth: Accretion, atmosphere formation and thermal history. *Proc. Lunar Planet. Sci. Conf. 17th,* in *J. Geophys. Res., 91,* E291-E302.

Abe Y. and Matsui T. (1988) Evolution of an impact-generated H_2O-CO_2 atmosphere and formation of a hot proto-ocean on Earth. *J. Atmos. Sci., 45,* 3081-3101.

Ahrens T. J. and O'Keefe J. D. (1972) Shock melting and vaporization of lunar rocks and minerals. *The Moon, 4,* 214-249.

Ahrens T. J. and O'Keefe J. D. (1977) Equations of state and impact-induced shock-wave attenuation on the moon. In *Impact and Explosion Cratering* (D. J. Roddy, R. O. Pepin, and R. B. Merrill, eds.), pp. 639-656. Pergamon, New York.

Ahrens T. J. and O'Keefe J. D. (1987) Impact on the Earth, ocean, and atmosphere. *Intl. J. Impact Eng., 5,* 13-32.

Ahrens T. J., O'Keefe J. D., and Lange M. A. (1989) Formation of atmospheres during accretion of the terrestrial planets. In *Origin and Evolution of Planetary and Satellite Atmospheres* (S. K. Atreya, J. B. Pollack, and M. S. Matthews, eds.), pp. 328-385. Univ. of Arizona, Tucson.

Allègre C. J. (1982) Cosmochemistry and primitive evolution of planets. In *Formation of Planetary Systems* (A. Brahic, ed.), pp. 283-364. Centre National d'Etudes Spatiales, Paris.

Beckwith S., Sargent A. I., Scoville N. Z., Masson C. R., Zuckerman B., and Phillip A. T. G. (1986) Small-scale structure of the circumstellar gas of H. L. Tauri and R. Monocerotis. *Astrophys. J., 309,* 755-761.

Benz W., Slattery W. L., and Cameron A. G. W. (1986) The origin of the Moon and the single impact hypothesis, I. *Icarus, 66,* 515-535.

Benz W., Slattery W. L., and Cameron A. G. W. (1987) Origin of the Moon and the single impact hypothesis, II. *Icarus, 71,* 30-45.

Benz W., Slattery W. L., and Cameron A. G. W. (1988) Collisional stripping of Mercury mantle. *Icarus, 74,* 516-528.

Brett R. (1984) Chemical equilibration of the Earth's core and upper mantle. *Geochim. Cosmochim. Acta, 48,* 1183-1188.

Brown J. M., Ahrens T. J., and Shampine D. L. (1984) Hugoniot data for pyrrohite and the Earth's core. *J. Geophys. Res., 89,* 6041-6048.

Craig H., Clarke W. B., and Beg M. A. (1975) Excess ^{3}He in deep water on the East Pacific Rise. *Earth Planet. Sci. Lett., 26,* 125-132.

Donahue T. M. and Pollack J. B. (1983) Origin and evolution of the atmosphere of Venus. In *Venus* (D. M. Hunter, L. Colin, T. M. Donahue, and V. I. Moroz, eds.), pp. 1003-1036. Univ. of Arizona, Tucson.

Drake M. J. (1987) Siderophile elements in planetary mantles and the origin of the Moon. *Proc. Lunar Planet. Sci. Conf. 17th,* in *J. Geophys. Res., 92,* E377-E386.

Greenberg R., Wacker J. F., Hartmann W. K., and Chapman C. R. (1978) Planetesimals to planets: Numerical simulation of collisional evolution. *Icarus, 35,* 1-26.

Holloway J. R. (1988) Planetary atmospheres during accretion: The effect of C-O-H-S equilibria (abstract). In *Lunar and Planetary Science XIX,* pp. 499-500. Lunar and Planetary Institute, Houston.

Hughes H. G., App F. N., and McGetchin T. R. (1977) Global effects of basin-forming impacts. *Phys. Earth Planet. Inter., 15,* 251-263.

Jones J. H. (1987) Core formation and accretion of the Earth volatiles (abstract). *Eos Trans. AGU, 68,* 1337-1338.

Kipp M. E. and Melosh H. J. (1986) Short note: A preliminary study of colliding planets. In *Origin of the Moon* (W. K. Hartmann, R. J. Phillips, and G. J. Taylor, eds.), pp. 643-648. Lunar and Planetary Institute, Houston.

Lange M. A. and Ahrens T. J. (1982a) The evolution of an impact-generated atmosphere. *Icarus, 51,* 96-120.

Lange M. A. and Ahrens T. J. (1982b) Impact-induced dehydration of serpentine and the evolution of planetary atmospheres. *Proc. Lunar Planet. Sci. Conf. 13th,* in *J. Geophys. Res., 87,* A451-A456.

Lange M. A. and Ahrens T. J. (1984) FeO and H_2O and the homogeneous accretion of the Earth. *Earth Planet. Sci. Lett., 71,* 111-119.

Levin B. J. (1972) Origin of the Earth. In *The Upper Mantle* (A. R. Ritsema, ed.), pp. 7-30. Elsevier, Amsterdam.

Matsui T. and Abe Y. (1986a) Evolution of an impact-induced atmosphere and magma ocean on the accreting Earth. *Nature, 319,* 303-305.

Matsui T. and Abe Y. (1986b) Impact induced atmospheres and oceans on Earth and Venus. *Nature, 322,* 526-528.

Melosh H. J. (1984) Impact ejection, spallation, and the origin of meteorites. *Icarus, 59,* 234-260.

Melosh H. J. and Vickery A. M. (1989) Impact erosion of the primordiaL martian atmosphere. *Nature, 338,* 487-489.

O'Keefe J. D. and Ahrens T. J. (1977a) Impact induced energy partitioning and melting, and vaporization on terrestrial planets. *Proc. Lunar Sci. Conf. 8th,* pp. 3357-3375.

O'Keefe J. D. and Ahrens T. J. (1977b) Meteorite impact ejecta: Dependence on mass and energy lost on planetary escape velocity. *Science, 198,* 1249-1251.

Ozima M. and Podosek F. A., eds. (1983) *Noble Gas Geochemistry.* Cambridge Univ., New York. 367 pp.

Ringwood A. E. (1978) Composition of the core and implications for the origin of the Earth. *Geochim J., 11,* 111-136.

Ringwood A. E., ed. (1979) *Origin of the Earth and Moon.* Springer-Verlag, Berlin. 295 pp.

Safronov V. S. (1969) *Formation of the Protoplanetary Cloud and Formation of the Earth and Planets.* Nauka, Moscow. Translated by the Israel Program for Scientific Translation (1972).

Sargent A. I. and Beckwith S. (1987) Kinematics of the circumstellar gas of HL Tauri and R Monocerotis. *Astrophys. J., 323,* 294-305.

Schmitt W., Palme H., and Wänke H. (1989) Experimental determination of metal/silicate partition coefficients for P, Co, Ni, Cu, Ga, Ge, Mo, and W, and some implications for the early evolution of the Earth. *Geochim. Cosmochim. Acta, 53,* 173-185.

Smith B. A. and Terrile R. J. (1984) A circumstellar disk around β-Pictoris. *Science, 226,* 1421-1424.

Tyburczy J. A., Frisch B., and Ahrens T. J. (1986) Shock-induced volatile loss from a carbonaceous chondrite: Implications for planetary accretion. *Earth Planet. Sci. Lett., 80,* 201-207.

Walker C. G. (1986) Impact erosion of planetary atmospheres. *Icarus, 68,* 87-98.

Wänke H., Dreibus G., and Jagoutz E. (1984) Mantle chemistry and accretion history of the Earth. In *Archean Geochemistry* (A. Kröner, G. Hanson, and A. Goodwin, eds.), pp. 1-24. Springer-Verlag, Berlin.

Wetherill G. W. (1985) Occurrence of giant impacts during the growth of the terrestrial planets. *Science, 228,* 877-879.

Zahnle K. J., Kasting J. F., and Pollack J. B. (1988) Evolution of a steam atmosphere during Earth's accretion. *Icarus, 74,* 62-97.

Zel'dovich Y. B. and Raizer Y. P., eds. (1967) *Physics of Shock Waves and High-Temperature Hydrodynamic Phenomena.* Academic, New York. 916 pp.

CORE FORMATION

This section contains three papers specifically discussing core formation in the Earth. Several other papers that are closely concerned with core formation have been placed in other sections. Taken together, these papers represent much of the current thinking about the formation of the Earth's core. The papers in this section are:

D. J. Stevenson: *Fluid Dynamics of Core Formation*
R. J. Arculus, R. D. Holmes, R. Powell, and K. Righter: *Metal-Silicate Equilibria and Core Formation*
H. E. Newsom: *Accretion and Core Formation in the Earth: Evidence from Siderophile Elements*

These papers are concerned with both the physics and chemistry of core formation, and the influence of the Giant Impact is evident. Early models of Elasser (1963, in *Earth Science and Meteoritics,* J. Geiss and E. D. Goldberg, eds., pp. 1-30, North-Holland) and Stevenson (1981, *Science, vol. 214,* 611-619) supported the idea that metal would accumulate and sink as large diapirs during the later stages of accretion. This model is rejected by Stevenson on the grounds of theoretical and experimental evidence against percolation of metallic Fe-rich liquid through a mostly solid matrix, due to the high surface tension of the metallic liquid. Stevenson discusses new models involving the possibility of core formation during a magma ocean epoch. He also raises the possibility that percolation of Fe-rich liquid could occur under higher pressure regimes in the Earth, leading to other possible models.

Arculus and coauthors discuss some of the chemical implications of the redox state of the Earth during core formation. They conclude that the chondritic Co/Ni ratio observed in the Earth's upper mantle is difficult to explain by metal-silicate equilibrium, because of the strong dependence on oxygen fugacity of the partitioning behavior of Fe, Co, and Ni, which results in dramatic changes in Ni/Co ratios over small ranges in oxygen fugacity.

The abundance of siderophile elements (elements with an affinity for Fe-metal) are depleted in the Earth's mantle relative to assumed initial chondritic abundances, and therefore provide clues to the process of core formation in the Earth. A general assumption is that the siderophile element data obtained for the upper mantle apply to the whole mantle. Some workers, however, suggest that the lower mantle may have a different composition from the upper mantle (e.g., Anderson, 1989, *Science, vol. 243,* 367-370). The paper by Newsom discusses the observed depletions of siderophile elements in the Earth's mantle in terms of quantitative theories of core formation. Newsom's paper identifies critical elements that do not fit different models for accretion and core formation in the Earth. Future studies of the abundances and partitioning behavior of these elements may provide important new constraints on the formation of the Earth. While no theory is completely successful, the heterogeneous accretion theory explains such difficult problems as the chondritic Ni/Co ratio of the mantle. This theory is the least constrained by pressure effects or the extent of melting of the mantle during core formation because it depends on successive accretion of veneers on a siderophile-depleted mantle.

Papers elsewhere in the book also address problems of core formation. Taylor and Norman (section 1) suggest that the siderophile trace element content of the mantle and core could in part reflect metal-silicate equilibrium at low pressure in small planetesimals before accretion.

In dramatic contrast, Ringwood (section 3) emphasizes the possibility that the siderophile element abundances in the Earth's mantle are due to partitioning at very high pressures into an oxygen-rich core; this abundance pattern is uniquely terrestrial. Ringwood also discusses high-pressure experimental data that suggest that the depletion of V, Cr, and possibly Mn in the Earth's mantle can be explained by equilibrium at high pressures. Davies (section 3) suggests that perhaps early melting processes could allow metal to accumulate into sufficiently large bodies at shallow depths, thus permitting the metal to sink toward the core. Ahrens (section 3) discusses the possibility that the oxidation state of the mantle and the fate of accreting metal is controlled by the process of accretion. Sims and coworkers (section 5) discuss how the abundances of siderophile elements in the Earth's mantle are obtained and new data for As, Sb, Mo, and W are presented. Their revised depletion of W relative to chondritic initial abundances is consistent with the depletion of Co and Ni as predicted by the heterogeneous accretion theory.

What is not discussed at length in this book is the current thinking about the present state of the Earth's core and the core mantle boundary. For a review of this topic, two articles by R. Jeanloz may be useful: Jeanloz (1989) In *Mantle Convection* (W. R. Peltier, ed.), Gordon and Breach; and Jeanloz (1990) *Annual Reviews of Earth and Planetary Science.*

The formation of the Earth's core remains an enigma, further complicated by the possibility of a giant impact. The heterogeneous accretion theory is the least sensitive to quantitative data for partition coefficients and is quite successful. However, additional partitioning data at both high and low pressures are critically needed to test some of the other core formation models, especially for the elements that are not consistent with current models. Improvements are also needed in our understanding of the abundances of siderophile elements in the Earth's mantle.

FLUID DYNAMICS OF CORE FORMATION

D. J. Stevenson

Division of Geological and Planetary Sciences, California Institute of Technology, Pasadena, CA 91125

Past discussions of core formation are incorrect or incomplete because they assume that metallic iron-rich liquid is able to migrate through a mostly solid silicate matrix by percolation, prior to macrosegregation and diapiric descent. Experimental and theoretical considerations suggest that percolation is largely prevented because of the high surface tension of iron. Two alternative views of core formation are offered. One assumes that percolation is possible in the deep mantle (where perovskite is the major phase). Iron is then supplied to the deep mantle by Rayleigh-Taylor instabilities of a silicate-iron suspension in the shallow mantle, and drains efficiently from the deepest mantle into the core by Darcy flow. The other model assumes complete or nearly complete melting of all or part of the mantle. Despite vigorous convection, iron droplets approximately one centimeter in radius are predicted and settle rapidly by Stokes flow, either to the core or into a layer or ponds that provide iron-rich diapirs that can descend to the core. These stories generally suggest very efficient core formation in the sense that the typical residence time of metallic iron in the mantle is orders of magnitude shorter than the formation time of Earth ($\sim 10^8$ years). Good chemical equilibrium between mantle and core phases is also predicted in many cases. Geochemical constraints and implications relevant to these scenarios are discussed but are largely inconclusive. The tentative inference of rapid core formation on Mars suggests a magma ocean and iron rainout.

INTRODUCTION

The Earth's core is mostly iron, mostly liquid, and mostly insoluble in the overlying mantle. Its existence could be due to either a "bottom-up" formation (accumulation of the core as an iron planet, followed by the addition of a silicate shell) or a "top-down" formation (accumulation of a roughly homogeneous planet, followed by separation and downward migration of an iron-rich liquid). The extremely heterogeneous or bottom-up formation cannot be disproved on purely observational grounds but is improbable for several reasons. First, the difference in condensation temperatures of iron and major silicate phases in the solar nebula is smaller than the *variations* in temperature (vertically or radially over a fraction of an astronomical unit, e.g., *Boss et al.,* 1989). It is therefore unreasonable to build an iron body by selective condensation. Second, the mechanical sorting of strong materials (e.g., iron) from weak materials during planetesimal collisions (cf. *Wasson,* 1988) can only work on a small scale, less than tens to hundreds of kilometers, if it works at all. Strength differences are irrelevant at the scale of the Earth. The vigor of the planetary accumulation process can be expected to intermingle bodies of varying iron content, largely rehomogenizing earlier heterogeneities. Third, the outer, liquid iron core of the Earth requires an alloying constituent that acts as an antifreeze, since all reasonable geotherms reach the core-mantle boundary at a temperature less than the freezing point of *pure* iron. Plausible antifreeze ingredients (oxygen, sulfur, hydrogen, carbon) would not be available in abundance if the core formed from the bottom up because these ingredients are not a sufficiently large part of the early, refractory iron-rich condensate of the solar nebula. These arguments are discussed further elsewhere (*Ringwood,* 1979, 1984; *Jacobs,* 1987, pp. 81-126) and reflect the general acceptance of the top-down core formation process. Accordingly, the core formation event is fluid dynamical, and hence the title of this contribution.

The fundamental question can be stated thus: How was the separation of iron from silicates accomplished? The answer to this question has important implications for the nature of the accretion process, the thermal state of the early Earth, the compositions of core and mantle, and the origin of geomagnetism. As explained below, there is reason to believe that past answers to the question have been incorrect or inadequate. The most commonly cited story of core formation is some embellishment of an idea by *Elsasser* (1963), who proposed that large blobs of iron (perhaps even hundreds of kilometers in size) would form in the upper mantle and migrate downward because of their high density and the deformable (viscous) nature of the mantle. This diapiric mode of descent is a Rayleigh-Taylor instability, and depends on the ability of iron to macrosegregate into a global layer or a large "magma chamber." In 1981, I proposed an alternative involving a fundamental symmetry breaking ($\ell = 1$ mode in spherical harmonics), whereby a cold primordial silicate-rich core moved sideways through an overlying layer of liquid iron (*Stevenson,* 1981). This very fast instability then leads to fracturing of the primordial core and rapid, downward migration of iron, perhaps along fractures or perhaps by the buoyant rise of core fragments replaced by iron fluid. However, this model, like the Elsasser model, assumes macrosegregation of iron at some early stage. It becomes indistinguishable from the Elsasser picture at later stages of accretion, when the interior is necessarily hot by the release of core formation energy alone. A third mode of descent, called the sinking layer model (*Vityazev and Mayeva,* 1976) involves the erosion of a colder, silicate-rich core by overlying liquid iron at a rate determined by downward heat diffusion. It is slower than other possibilities (*Stevenson,* 1981) and does not merit attention because the fastest process wins.

More recent work has addressed quantitative aspects of the Elsasser model (*Turcotte and Emerman,* 1983) and Stevenson model (*Davies,* 1982; *Ida et al.,* 1987).

The main problem with the "giant blob" mode of core formation in a mostly solid Earth lies in the microscale issue of whether liquid iron can percolate through a solid silicate matrix. If porous flow is inhibited then the macrosegregation of iron will be prevented or incomplete. This is a *surface tension* problem and is discussed in some detail in the next section. The importance of this issue has been independently recognized by *Urakawa et al.* (1987) for Earth, and by *Taylor* (1989) for the different but equally relevant context of meteorite parent bodies. However, the quantitative aspects of the problem have not been addressed before. There is another equally important reason to address anew the core formation dynamics: Current views of Earth accretion involve giant impacts and suggest that very high temperatures are reached (*Benz and Cameron,* 1990), sufficient to ensure complete melting and partial vaporization of the entire Earth. Although these models are motivated by the problem of lunar formation rather than by constraints imposed by known properties of Earth, the requirement of complete melting is difficult to avoid for any of the current, quantified versions of Earth accretion (e.g., *Wetherill,* 1985; *Stevenson,* 1987). Complete melting (or at least a magma ocean a few hundred kilometers deep) may also arise from the accumulation effect of small, impacting bodies, provided there is a dense steam atmosphere to act as a greenhouse (*Abe and Matsui,* 1985). This depends on assumptions concerning the spectrum of impactor masses (*Stevenson,* 1988; *Rintoul and Stevenson,* 1988; *Abe,* 1988). In any event, we must confront the possibility that a completely molten state is an appropriate environment in which to consider a large part of the core formation process. This is again an environment in which the microphysics of surface tension plays a role: The formation of large blobs may be prevented, for much the same reason that the hydrodynamics of raindrops in the Earth's atmosphere limits droplet growth.

It might be supposed that geochemical constraints establish the nature of core formation—indeed they might, if we understood them well enough. Strontium and lead isotopic constraints (*Oversby and Ringwood,* 1971; *Vollmer,* 1977; *Vidal and Dosso,* 1978; *Allègre et al.,* 1982) generally support rapid core formation, though they allow and perhaps even suggest an extended period during which the final "dregs" of the core are accumulated. For example, Allègre and coworkers favor 85‰ of the core formation within 50-200 Ma and the remaining 15‰ throughout subsequent evolution (perhaps including the present time). It is worth emphasizing (since the geochemists often fail to) that these conclusions are model-dependent. A different set of constraints arises from a consideration of trace elements. The mantle abundances of strongly siderophile elements, though low, are higher than expected for a silicate-metal equilibrium process, suggesting either a component (~1% of the mass of the mantle) of late accreting material that was never in equilibrium with metallic iron (*Chou,* 1978; *Jagoutz et al.,* 1979; *Morgan et al.,* 1981; *Schmitt et al.,* 1989) or the stranding in the mantle of a small amount of core-forming fluid (*Jones and Drake,* 1986) that was presumably later oxidized. It would seem likely that both are relevant, and core formation scenarios must be compatible with some combination of these factors. Other trace elements (e.g., Sc, Zr, Hf, rare earths) are fractionated during freezing of perovskite from a mantle-wide magma ocean (*Kato et al.,* 1988), and the suggested consequences of this for the oldest terrestrial crust have not been detected (*Ringwood,* 1988), leading him to conclude that the mantle was never totally molten. This inference cannot be accepted until confirmed by a detailed assessment of the dynamics of a partially solidified mantle, an exercise that is considerably more difficult than most appreciate and has not yet been seriously attempted. [*Tonks and Melosh* (1988) deal with only one aspect of this problem concerning the early phase of crystals suspended in a convecting melt.] Major element chemistry could also offer some clues concerning core formation. For example, the experimental indications of iron-silicate interactions under core-mantle boundary conditions (*Knittle and Jeanloz,* 1989) together with mantle compositional evidence might be used to establish how much of the mantle has equilibrated with the core at high pressure. However, the combined inferences of geochemistry are more promising than highly constraining at the present.

Geophysical inferences derived from data (as distinct from theoretical speculation) are even weaker. Evidence of a geomagnetic field exists in some of the oldest rocks (*McElhinny and Senanayake,* 1980), suggesting the existence of a substantial core prior to 3.5 Ga. However, our knowledge of core dynamics is not sufficiently quantitative that we can invert paleointensity data into useful assertions about core size or the existence of an inner core (see discussion in *Stevenson et al.,* 1983). Measured consequences of tidal evolution of the Earth-Moon system place a strict upper bound on the time-dependence of the Earth's moment of inertia (e.g., *Lambeck,* 1980, Chapter 10) and hence the present-day rate of core growth, but this bound is too large to be useful. Of course, there is much geophysical data on the nature of the *present* core (*Jacobs,* 1987, pp. 81-126).

Another set of potentially important constraints arise from consideration of Mars. The inferred "late veneer" is smaller than on Earth

(*Treiman et al.,* 1987) and lead isotope data indicate that the core formed early (*Chen and Wasserburg,* 1986), assuming that the SNC meteorites are a good indicator of martian crust. It is particularly relevant to our later discussion that Mars is a small planet with no high-density silicate perovskite.

Here, we proceed with the theoretical exercise of understanding core formation, largely unencumbered by tight observational constraints but mindful of the need to be compatible with basic physics and the less basic (but often strong) implications of accretion dynamics. The next section categorizes and describes the properties of core formation possibilities, with an emphasis on those that have received little attention previously. Later in this paper we deal with specific models of the percolative mode of core formation and with specific models of the rainstorm mode. We then conclude with an overview of these results and their geophysical and geochemical implications.

MODES OF DIFFERENTIATION

It is natural to think of heavier material separating from lighter material, but counterexamples abound: the Earth's troposphere, much of the interiors of giant planets and stars, and the Earth's mantle where grains of varying composition and density commingle. The difficulties of differentiation deserve emphasis because they are often insufficiently appreciated.

There are five broadly defined modes of possible differentiation:

1. Solids settling through solids. Here, the solid is assumed capable of creep and the problem is not the difficulty of settling, but the time it takes to create an embryo that can settle.

2. Liquid core fluid settling through a solid matrix. Here, the main issues are permeability (the ability of the melt to percolate) and macrosegregation. Subsequent migration by cracks or diapirs also are relevant here.

3. Solids settling up through the core fluid. This could occur at a sufficiently large volume fraction of metallic melt ($\gtrsim 30‰$) but pertains to only a small fraction of the time or volume. Here, it will be treated as a special case of (2) above. It may be more relevant to later evolution of the outer core.

4. Solid, metallic particles settling through silicate melt. This is implausible, since it requires a reversal of melting points relative to the current situation at the core-mantle boundary. It is discussed no further here.

5. Settling of liquid, metallic droplets from a completely liquid emulsion. Here, the problem is settling in the presence of turbulence and fluid deformation.

Settling of Solids from Solids

For simplicity, imagine a single solid embryo of the dense, iron-rich phase that grows by diffusion from the surrounding, initially iron-rich solid medium. The simplification of a single embryo avoids the complications of competition among embryos, an essential part of understanding grain growth in a polydisperse granular medium, a process known as Ostwald ripening (e.g., *Smith,* 1972). The following calculation is therefore excessively optimistic. If the medium is at rest, then the embryo radius can be as large as $R \sim (Dt)^{1/2}$ after an elapsed time t, where D is the diffusivity. However, settling occurs at a velocity given by Stokes formula

$$v_s \simeq \frac{2}{9} g \Delta\rho R^2/\eta \quad (1)$$

where g is the gravitational acceleration and $\Delta\rho$ is the density contrast between the embryo and the mean surroundings that have a viscosity η. Provided $v_s \lesssim D/R$, the motion has little effect on the diffusion. Once $v_s \gtrsim D/R$, a diffusive boundary layer of thickness $\delta \sim (DR/v_s)^{1/2}$ is present and the embryo grows according to

$$\frac{dR}{dt} \simeq \frac{D}{\delta} \quad (2)$$

Combining these equations, we conclude that

$$R \sim (Dt)^{1/2} \quad , \quad R < R_o$$
$$R \sim R_o(Dt/R_o^2)^2 \quad , \quad R > R_o \qquad (3)$$
$$R_o \equiv (9D\eta/2g\Delta\rho)^{1/3}$$

where R_o is the critical embryo size at which $v_s \sim D/R$. Once a time $\sim R_o^2/D$ has elapsed, the embryo has this critical size and the subsequent growth is very rapid (quadratic in time) because of the formation of a progressively thinner boundary layer. Substituting $g = 10^3$ cm/sec, $\Delta\rho \sim 4.5$ g/cm^3, we find

$$R_o \simeq 10^2 \text{ cm} \left(\frac{\eta}{10^{21}\text{ P}}\right)^{1/3} \left(\frac{D}{10^{-12}\text{ cm}^2/\text{sec}}\right)^{1/3} \qquad (4)$$

The scaling choices for η and D are nominal; actual values can deviate greatly. We can now calculate the distance H that a particle settles in time t

$$H \simeq R_o(t/\tau)^2 \qquad , \; t \lesssim \tau$$
$$H \simeq R_o\left[\frac{4}{5} + \frac{1}{5}(t/\tau)^5\right] \; , \; t \lesssim \tau \qquad (5)$$
$$\tau \equiv R_o^2/D$$

The interesting case is $t \gtrsim \tau$, for which the result may be written

$$H \sim (10 \text{ cm})\,(t/\tau)^5$$
$$\tau \sim (3 \times 10^8 \text{yr}) \left(\frac{\eta}{10^{21}\text{P}}\right)^{2/3} \left(\frac{10^{-12}\text{cm}^2/\text{sec}}{D}\right)^{1/3} \qquad (6)$$

The striking feature of this result is the extreme sensitivity to parameter choices. For $D \sim 10^{-12}$ cm^2/sec and $\eta \sim 10^{21}$ P, not too different from the present Earth (though the value for D may be optimistic), the settling is negligible. Reduction of the viscosity to 10^{18} P yields $\tau \sim 3$ Ma and core formation in ~ 100 Ma! However, this conclusion is misleading because such a medium will convect and the resulting strain field inhibits the growth of embryos by stretching them out (*McKenzie,* 1979; *Olson et al.,* 1984; *Gurnis,* 1986). This same process prevents unlimited growth of crystals in the present mantle. As already noted, this crude analysis also neglects the competition among embryos, which greatly reduces the growth because the chemical potential gradient driving the diffusion is determined by small surface tension effects rather than the large effect implicit in the naive formula $R \sim (Dt)^{1/2}$. Moreover, the substitution of the low viscosities in equation (6) is incompatible with the assumption of a completely solid system. In summary, the settling of solids from solids is not a significant process unless there is a fast nondiffusive process of supplying embryos. Such a process surely requires melt migration and therefore falls in the same category as the next differentiation mode that we discuss.

Settling of Liquid from Solid

Consider a partially molten medium in which the solid component is mantle material and the melt is rich in metallic iron. This requires a temperature between that of a low melting point iron alloy (such as the Fe-FeS eutectic system at low pressure) and the solidus of the mantle. However, much of the discussion may still apply if the mantle component is mildly supersolidus (i.e., the metallic melt is still the dominant fluid phase), with caveats as explained below. In a partial melt, there is always a small but finite solubility of the solid phase in the melt, and a textural evolution will take place whereby the solid grain-liquid interfaces evolve in shape and the melt locally redistributes. This is governed by minimization of surface energy. Strictly speaking, the true minimum of energy corresponds to arbitrarily large grains, but this takes infinite time and is generally prevented by nonthermodynamic considerations (cf. the discussion above concerning Ostwald ripening and straining by convection). But on a geologically short timescale, a "local" energy minimum is accomplished whose properties depend solely on the surface tensions of the participating phases and the volume fraction of melt.

The simplest consequence of this rapid process is a "textural equilibrium," in which two grains in contact with melt meet at an angle θ_D called the dihedral angle (Fig. 1). In the limit of small melt fraction, the triangular cross-section of a melt pocket or tubule will accordingly have positive or negative curvature if the dihedral angle is less than or greater than 60°, respectively. (Here, "positive" means that the center of curvature is inside the grain.) It is not hard to see that if $\theta_D < 60°$, then the curvature is also positive along the direction perpendicular to the plane of the triangular cross-section, allowing interconnection among the melt pockets and the formation of a spongelike topology. Conversely, $\theta_D > 60°$ implies the accumulation of melt into disconnected pockets, at least at low melt fraction.

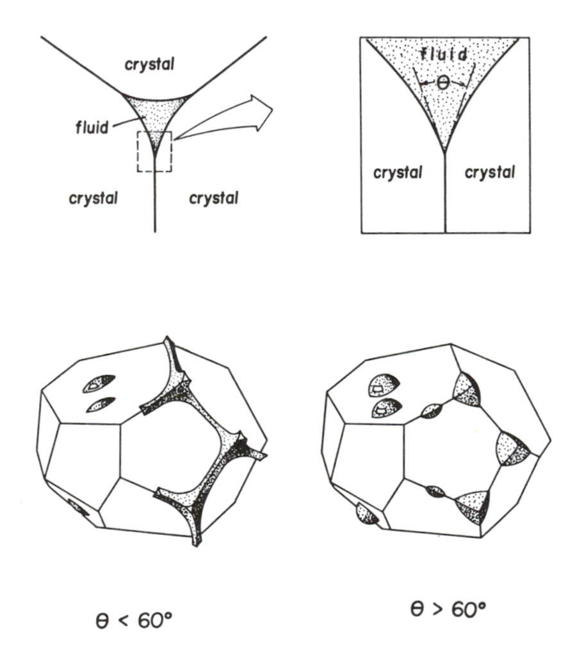

Fig. 1. *Upper diagrams illustrate the definition of the dihedral angle θ for surface equilibrium between solid grains and melt. Lower diagrams show a cutaway view of the resulting distribution of melt on a grain surface for θ below and above the critical value of 60°. Percolation at low melt fraction is only possible if $\theta <$ 60°. The isolated droplets depicted at $\theta < 60°$ should eventually disappear.*

There are extensive discussions of this phenomenon, in connection with metallurgy and partial melting phenomena in the present mantle (e.g., *Bulau et al.,* 1979; *McKenzie,* 1984; *von Bargen and Waff,* 1986). Although the basic concepts are well understood, the three-dimensional aspect together with real world complications (impurities, anisotropic surface tension, incomplete textural equilibrium, deviatoric stress, etc.) render predictions suspect. The presence of more than one solid phase also changes melt topology (*Toramaru and Fujii,* 1986). Figure 2 is nevertheless an attempt to delineate different topologies of two-phase media. The interconnected *sponge* topology occupies $\theta_D < 60°$ at small melt fraction f and larger θ_D at progressively larger melt fraction because of the merging of melt pockets, as indicated by the models of *von Bargen and Waff* (1986). At large θ_D and low f, we have the *Swiss cheese* topology (recall that this kind of cheese is a poor sponge: The holes, made by expanding gas, do not interconnect except when they occupy a large volume fraction of the cheese). At sufficiently large f, established to be ~20-30%, a *meatball* topology may develop in which grains or aggregates of grains could be completely surrounded by melt. This critical melt fraction has a poorly determined dependence on θ_D and delineates the matrix regime (a "solid" in the usual sense) from a dense suspension. The distinction is actually somewhat imprecise, since even a dense suspension has a yield stress (*Ryerson et al.,* 1988) and may support some kinds of shear waves (*Stevenson,* 1989a).

What is θ_D for the metallic melt in contact with major mantle phases (e.g., olivine, perovskite)? There do not appear to be direct measurements, but there are a number of indirect indications. *Walker and Agee* (1988) report experiments on the partial melting of a sample of the Allende meteorite in which the iron-rich sulfide melt exhibits immobility in the olivine cumulate. *Taylor* (1989) also describes experiments where an Fe-rich melt

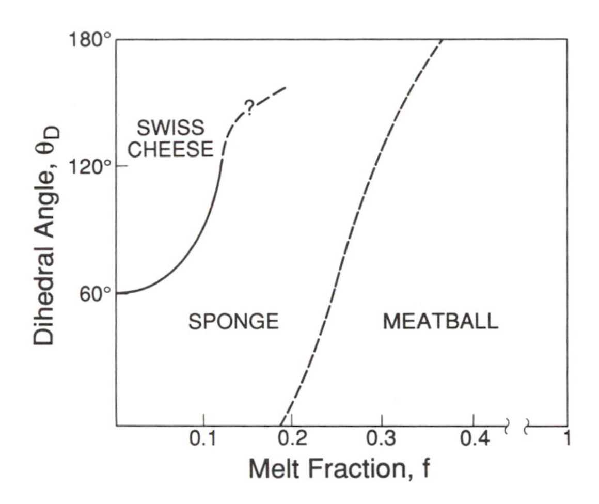

Fig. 2. *Topological phase diagram for all dihedral angles and melt fraction. In the sponge domain, the melt and solid are both fully connected. In the Swiss cheese domain, the melt pockets are largely disconnected. In the meatball domain, the grains are coated in melt. Dashed lines indicate uncertain boundaries.*

is unable to migrate except at very high melt fraction, and discusses the implications for meteorite parent bodies. *Fukai and Suzuki* (1986) show a photomicrograph of a recovered specimen of high-pressure experiments on the Fe-hydrous silicate system, showing large "balls of iron" surrounded by olivine, pyroxene, and garnet. All these observations are consistent with $\theta_D > 60°$, possibly $\theta_D \gg 60°$.

We can understand this at the level of fundamental physics. Force equilibrium at a grain-grain vertex requires

$$\theta_D = 2\cos^{-1}(\gamma_{ss}/2\gamma_{sl}) \qquad (7)$$

where γ_{ss} is the interfacial surface energy at grain-grain contact and γ_{sl} is the grain-metallic liquid interfacial energy. In the olivine-basalt system, $\gamma_{sl} \simeq 500$ erg/cm^2, $\gamma_{ss} \simeq 900$ erg/cm^2, and $\theta_D \approx 30°$-$45°$ (*Cooper and Kohlstedt,* 1982). The absolute surface tension of liquid silicates is around 300 erg/cm^2 (*Walker and Mullins,* 1981), less than but not enormously different from γ_{sl}. The absolute surface tension of olivine crystals is likely to be similar to γ_{ss} (see *Jurewicz and Watson,* 1984, for a similar argument concerning quartz). These estimates have a considerable uncertainty, typically ±30%, so we can only establish a rough idea of what to expect. The absolute surface tension of ultrapure liquid iron is about 1800 erg/cm^2 (*Dyson,* 1963); the high value is typical of a dense metal and due to the surface energy of the itinerant electrons. However, iron is unusually susceptible to surface activity. For example, less than 1% by mass of sulfur can drop the surface tension to ~1000 erg/cm^2 (*Dyson,* 1963) and oxygen can be almost as important (*Popel' et al.,* 1975). At higher mixture levels, the surface tension reaches a plateau value of around 800 erg/cm^2. Based in part on the aforementioned systematics, it is plausible that γ_{sl} for olivine-iron is not too different than γ for the iron alloy relative to vacuum, because of the enormous electronic difference between these two media and the high surface tension of metals. If we choose $\gamma_{ss} = 900$ erg/cm^2 and $\gamma_{sl} = 1000$ erg/cm^2, we find $\theta_D \simeq 125°$, compatible with the cited laboratory observations. The behavior of θ_D with γ_{sl} is shown in Fig. 3. If $\gamma_{sl} = 2000$ erg/cm^2 then θ_D rises to 154°, and γ_{sl} must drop to ~600 erg/cm^2 in order that θ_D drop to the critical value of 60°. Despite the optimistic comments of *Urakawa et al.* (1987), interconnection of melt pockets does not seem likely at low melt fraction and pressure.

Of course, some interconnection occurs at high enough melt fraction. Based on the numerical calculations of *von Bargen and Waff* (1986), the critical melt fraction f_c below which pinch-off takes place is roughly

$$f_c \sim 0.009(\theta_D - 60)^{1/2}, 60 < \theta_D \lesssim 100° \qquad (8)$$

At $f > f_c$, percolation may be possible while at $f < f_c$, the pockets of iron become disconnected. The pinch-off melt fraction is also shown in Fig. 3. However, there is another issue at $f > f_c$, $\theta_D > 60°$, which was analyzed by *Stevenson* (1986). In these circumstances, the melt fraction undergoes negative diffusion. In

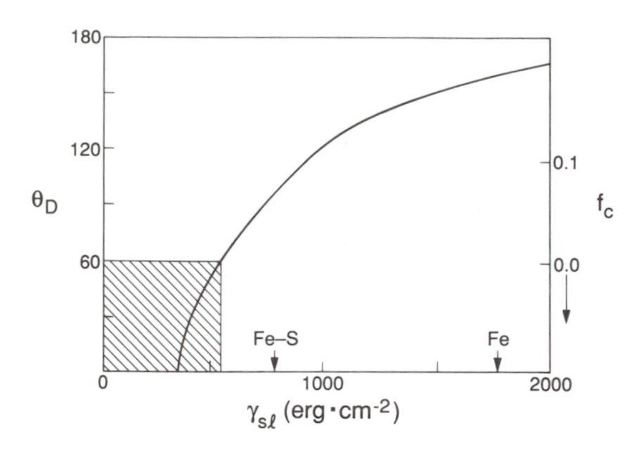

Fig. 3. *Behavior of dihedral angle (left axis) and critical melt fraction for percolation (right axis) as a function of the interfacial energy between the iron-rich melt and olivine grains. The hatched region ($\theta_D < 60°$) corresponds to percolation at low melt fraction. The surface tension values of Fe-S melt and pure Fe melt are shown for comparison. This diagram illustrates the unlikelihood that the metallic liquid will percolate in the Earth's upper mantle, at least at low melt fraction. In the text, it is argued that large-scale percolation is unlikely even at quite high melt fraction, because of negative diffusion of the melt fraction.*

other words, the melt prefers to aggregate into spherical pools surrounded by regions of melt-free solid. This happens because of the positive curvature of the surface free energy vs. melt fraction relationship (and is directly analogous to the thermodynamics of immiscibility). The process is rapid for low viscosity iron: The formation of centimeter-sized pools should occur in approximately minutes for a medium where $f \sim 0.1$, $\theta_D \sim 100°$, for example. This further exacerbates the problem of interconnection.

It is also possible that pressure plays a substantial role because of a possibly large pressure-dependence of θ_D, as has been suggested for the mantle (*von Bargen and Waff,* 1988), and it is very likely that the textural relationship between iron and perovskite differs substantially from that between iron and olivine. Experiments by F. Guyot (personal communication, 1988) suggest penetration of iron melt into solid perovskite, although it is unlikely that this is an equilibrium (i.e., low stress) experiment. The experiments by *Knittle and Jeanloz* (1989) might indirectly suggest penetration of metallic iron into perovskite since they imply less incompatibility of the two phases than is typical at low pressures, but this is only a speculation.

One possible deficiency of this analysis is that it always assumes textural equilibrium. In reality, this takes a finite time to achieve and is continually perturbed by the convective straining of the matrix. If we define a textural equilibrium time τ_{te} and a characteristic convective straining time $\tau_c(\sim \dot{e}^{-1}$ where $\dot{e}$ is the convective strain field) then the surface tension equilibrium result applies provided $\tau_{te} \gg \tau_c$. This probably applies in the present mantle (*McKenzie,* 1984) where $\tau_{te} \sim 10^3$-10^4 years and $\tau_c \sim 10^7$-10^8 years but the inequality may not be as well satisfied for early Earth where $\tau_c \sim 10^5$ years (*Stevenson,* 1989c). The problem of melt distribution in a rapidly strained solid has received considerable attention by this author but without quantitative resolution.

From this discussion, the following conclusions are indicated:

1. Dispersed metallic melt in the upper mantle is unable to percolate large distances because $\theta_D > 60°$. Even at high melt fraction, it will tend to aggregate into small droplets by negative "diffusion." These droplets can grow but only by the slow *solid state* diffusive process (cf. section on settling of solids from solids). It is unlikely that the melt fraction can drop below a quite large value (i.e., $f \sim 0.1$-0.2).

2. The behavior of θ_D in the deep mantle cannot be predicted with any confidence, but there are plausibility arguments in favor of a smaller value, allowing percolation. (This would be relevant to Earth but not Mars, where pressures are far lower.)

3. If $\theta_D > 60°$ everywhere, then efficient core formation cannot occur by the migration of metallic melt through the solid mantle matrix.

4. If $\theta_D < 60°$ in the deep mantle, then metallic depletion can occur from that region. This will create a top-heavy system because of the retention of iron in the mantle above. A diapiric instability ensues, delivering iron to regions where it can percolate. This can lead to efficient core formation, as illustrated schematically in Fig. 4.

One other factor merits mention. If there are two immiscible fluid phases (liquid metal and liquid silicate) then the behavior can be very complicated. This is a situation encountered in the oil industry, where oil is driven out of porous rock by a fluid (steam, water). In analogy with oil, the high surface tension metallic fluid can form "ganglions," relatively immobile, nonwetting domains that are surrounded by the mobile, wetting phase (e.g., *Adler and Brenner,* 1988). Depending on the relative amounts of metal and silicate melt, you can have the metal immobilizing the silicate or even forcing buoyant silicate downward, or you can have silicate melt advecting small amounts of iron upward. This is not well understood.

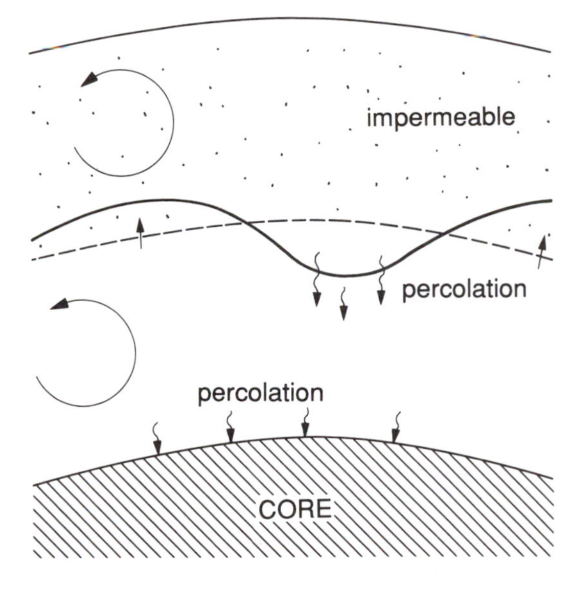

Fig. 4. *Cartoon illustrating percolative core formation from the lower mantle, fed by diapiric instabilities of impermeable, iron-bearing upper mantle.*

Even if the melt percolates, it may not macrosegregate. One way to accumulate melt is by having the melt pond at an impermeable barrier, e.g., subsolidus material. This is implicitly the idea in the scenarios advocated by *Elsasser* (1963) and *Stevenson* (1981). There may also be some form of *spontaneous* macrosegregation into veins, dikes, or magma chambers. There is reason to believe that this process occurs beneath midocean ridges (*Nicolas,* 1986; *Sleep,* 1988; *Scott and Stevenson,* 1989; *Stevenson,* 1989b). To the extent that we have any understanding of this process, there is no reason to believe that it could not also work for downward migrating metallic melt. It can arise either through magma fracturing or through a spontaneous fluid dynamical instability. In any event, this is likely to enhance rather than reduce the rate of core accumulation. Later in this paper I will present a specific model of percolative core growth.

Fluid Migrating through Fluid

Following a giant impact, we can expect the formation of an *emulsion*: a mixture of liquid silicate and liquid iron, in which the iron droplets could have a wide range of sizes. Despite extensive analysis of impact physics (e.g., *Melosh,* 1989), the expected spectrum of droplet sizes is hard to predict since it depends on poorly understood small-scale ("turbulent") shear instabilities driven by the postimpact hydrodynamic flow. To get a rough idea, suppose that some fraction of the impact energy is channeled into turbulence that adopts a Kolmogorov cascade of motions (cf. *Tennekes and Lumley,* 1972) with a dissipation timescale of order L/v_o ($L \sim$ excavation flow field dimensions $\sim 10^3$ km; $v_o \sim$ velocity of largest motions $\sim$ few km/sec). The local strain field is accordingly dominated by the smallest scale turbulent motions $v \sim v_o(\ell/L)^{1/3}$ for which $v\ell/\upsilon \sim 10^3$ and $\upsilon \sim 1$ cm^2/sec is the kinematic viscosity of the (ultrabasic) liquid silicate (*Kushiro,* 1980; *Bottinga and Weill,* 1972). We conclude that $\ell \sim 1$ cm, $v \sim 10^3$ cm/sec for this example. The strain rate is accordingly $10^3\,sec^{-1}$ and the associated deviatoric stress is $\sim 10^3$ dynes/cm^2. An iron blob of radius R cannot retain its integrity unless γ/R is greater than this stress, where $\gamma \sim 10^3$ dynes/cm is the interfacial surface tension between liquid iron and liquid silicate. Accordingly, the largest iron droplets are ~1 cm following the rapid (~hours) emulsification event of a large impact. The correctness of this estimate is not crucial for the following discussion, but instructive nonetheless.

Droplets subsequently evolve through collision, disruption, and settling. Diffusive growth (i.e., Ostwald ripening) is likely to be relatively unimportant because of the ease with which droplets can encounter each other. (Contrast this with the solid mixture problem discussed previously.) Consider, for example, a droplet of radius R falling through a medium containing droplets substantially smaller than itself. The Stokes velocity is given by equation (1) with $\eta \sim 1\text{-}10$ P. Accordingly, the droplet could grow at a rate

$$\frac{dR}{dt} \sim 10^2 f R^2$$

where f is the iron fraction in smaller droplets. This is spectacularly fast, if no other process intervenes. For example, at $f \sim 10^{-2}$ an initially 1 cm droplet could grow to a meter in size after an elapsed time of less than minutes! However, this is incorrect since large droplets self-destruct: If the deviatoric stress ($\sim \eta v/R$) exceeds the surface tension restoring pressure ($\sim \gamma/R$) then the droplet will fission. The predicted maximum droplet size is ~1 cm, somewhat comparable to our postimpact estimate. In the language of fluid dynamicists, this corresponds to a state of near unity Bond number. The corresponding Stokes velocity is of the order of a few tens of centimeters per second. Clearly, smaller droplets are readily swept up and the droplet distribution must be strongly peaked near the maximum possible size.

We must also check whether convective shear might limit the size even more. According to simple mixing length theory, the convective velocity v_c is given by

$$v_c \sim 0.1(F/\rho)^{1/3} \qquad (10)$$

where F is the heat flow, ρ is the fluid density, and the largest eddy scales are assumed to approach the pressure scaleheight (*Clayton,* 1968, p. 256; *Stevenson,* 1979). This assumes turbulent flow. The heat flow is sensitive to surface boundary conditions (dense atmosphere or naked magma ocean or chill crust or . . . ?). A value in excess of 10^5 erg/cm^2 sec is unlikely for an extended period because it is sufficient to eliminate the entire buried energy of Earth formation in 3×10^7 years (*Stevenson,* 1989c). The corresponding convective velocity of ~20 cm/sec suggests a Kolmogorov cascade extending down to eddies $\sim 6 \times 10^4$ cm with characteristic velocities ~2 cm/sec. The associated strain rate $\sim 3 \times 10^{-5}\,sec^{-1}$, too small to affect centimeter-sized droplets. This is hardly surprising since the thermal convection is a much more tranquil environment than the postimpact excavation flow. The centimeter-sized droplets settle at approximately a few tens of centimeters per second, and the largest scale convective eddies have comparable characteristic velocities. Can droplets accumulate into a core under these circumstances? Consider, first, the problem of an emulsion that extends all the way to the planet center. (This may never have happened, but we should consider it.) At first sight, it would seem that settling cannot take place, since the Stokes velocity at constant droplet size goes to zero linearly in r, the distance from the center, while the convective velocity is either finite or (in local mixing length theory) $\approx F^{1/3} \approx r^{1/3}$, as $r \to 0$. However, the tendency toward some settling creates a stable density gradient, which will inhibit and eventually stop the convection. To see this, let $4\pi r^2 N(r,t)dr$ be the number of droplets between r, r + dr at time t. Then

$$\frac{\partial N}{\partial t} = \frac{1}{r^2}\frac{\partial}{\partial r}(r^2 V_s N) + \frac{\kappa}{r}\frac{\partial^2(rN)}{\partial r^2} \quad (11)$$

where $V_s \propto r$ is the Stokes velocity of settling and κ is the eddy diffusivity. For κ, we use the product of convective velocity and largest eddy size ($\sim 10^8$ cm), and assume it is constant. The steady-state solution is

$$N(r) = N_o \exp(-V_o r^2/2\kappa R_p) \quad (12)$$

where $V_s \equiv V_o r/R_p$ is assumed and R_p is the radius of the planet. The constant N_o depends on the total amount of iron present. In order that convection persist, the thermal buoyancy must exceed the stabilizing influence of the iron droplet concentration gradient. If ΔT is the superadiabatic excess, between planet center and surface, then the approximate criterion is

$$\alpha \Delta T \gtrsim \bar{f}\,\frac{\Delta\rho}{\bar{\rho}} \cdot \left(\frac{V_o R_p}{\kappa}\right), \quad \kappa \gtrsim V_o R_p$$

$$\gtrsim \bar{f}\,\frac{\Delta\rho}{\bar{\rho}} \cdot \left(\frac{V_o R_p}{\kappa}\right)^{3/2}, \quad \kappa \lesssim V_o R_p \quad (13)$$

where α is the coefficient of thermal expansion, $\bar{f}$ is the mean iron volume fraction, $\Delta\rho$ is the density difference between iron and silicate, and $\bar{\rho}$ is the mean density of the planet. We also have the requirement $F \approx \bar{\rho} C_p \kappa \Delta T/R_p$ where F is the heatflow and C_p is the specific heat. These equations together imply an upper bound to $\bar{f} \sim 10^{-7}$ for $F \sim 10^5$ erg/cm^2 sec and $V_o \sim 3$ cm/sec (a conservative choice; a larger value is possible). Clearly, even a rather small amount of iron can stifle convection and cause core formation to proceed. With $\kappa = 0$, equation (11) predicts a formation timescale of only ~ 10-10^2 years despite the possible smallness of the Stokes velocity near $r = 0$. (Actually, droplets might grow larger near $r = 0$; the equilibrium droplet size behaves like r^{-1}!)

We turn now to the more relevant, later case of a well-defined core with an overlying magma ocean. Although convective velocities may exceed Stokes velocities, there is still a downward drift of iron droplets. In a nice set of experiments on crystal settling, *Martin and Nokes* (1988) show that sedimentation can take place despite vigorous convection, essentially because the convective velocities must go to zero in the bottom boundary layer, while the Stokes velocity persists. This observation should be even more relevant for the settling of droplets than the settling of crystals since there is less likelihood of entrainment back into the flow. New droplets can be created at the core-mantle boundary by wave breaking ("surf"), driven by convective disturbances. The appropriate measure of core-mantle boundary stability in this turbulent regime is obtained by comparing the Kolmogorov cascade of convective velocities $v_c(\ell) \sim 20(\ell/R_p)^{1/3}$ cm/sec with gravity wave velocities at the interface assuming wavelengths comparable to a convective eddy of size ℓ, i.e., $v_w \sim (g\ell\Delta\rho/\bar{\rho})^{1/2}$. According to *Linden* (1973), the entrainment velocity (volume of new droplets created per unit interface area per unit time) is of order $v_c(v_c/v_w)^n$ with $n = 3$. This is much smaller than the Stokes velocity and hence negligible even at the smallest eddies ($\sim 10^4$-10^5 cm). According to *Long* (1975), $n = 2$, but this also yields a negligible effect.

Application of *Martin and Nokes* (1988) is then straightforward and predicts core growth at a rate

$$\frac{dM_{core}}{dt} = 4\pi R_c^2 \rho_{Fe} f v_s \quad (14)$$

where R_C = core radius and v_s = Stokes velocity of droplets. Although the volume fraction of iron can be a function of height, it cannot be much *smaller* near the core-mantle boundary than further up, since this would be highly unstable (cf. *Trubitsyn and Kharybin,* 1987). It follows that a characteristic residence time of a droplet in the mantle is of order $R_c/v_s \sim 10$-10^2 years. In the section on core formation by rainfall, in particular, quantitative models are presented to illustrate this. A cartoon of this scenario is shown in Fig. 5.

PERCOLATIVE CORE FORMATION

In this section the previous generic discussion is converted into specific models and possible predictions. These are intended to be merely illustrative; the parameters are too poorly known to do otherwise. We envisage an Earth with a mostly solid mantle and with an instantaneous core radius $R_c(t)$ and surface radius $R_E(t)$. The mantle is split into two regions $R_c < r < R_t$ (the percolative zone) and $R_t < r < R_E$ (the nonpercolative zone). We shall call these the "lower" and "upper" mantle, respectively, but they need not correspond to the current definitions of these regions. We anticipate that core formation is efficient by seeking a steady-state model with the following features (see Fig. 4): (1) iron-bearing material is delivered to the nonpercolative zone by impact; (2) Rayleigh-Taylor (diapiric) instabilities deliver material from the iron-rich upper mantle to the percolative lower mantle; and (3) iron percolates steadily from this lower mantle to the core.

Let $f_u(t)$ and $f_l(t)$ be the upper and lower mantle mean iron volume fractions, respectively, and let dM/dt be the mean rate of accretion of which a fraction ϕ is metallic iron. Then we have

$$\frac{4}{3}\pi\rho_{Fe}\frac{d}{dt}[(R_E^3 - R_t^3)f_u] = \phi\frac{dM}{dt} - 4\pi R_t^2 f_u \rho_{Fe} V_{RT} \tag{15}$$

$$\frac{4}{3}\pi\rho_{Fe}\frac{d}{dt}[(R_t^3 - R_c^3)f_l] = 4\pi R_t^2 f_u \rho_{Fe} V_{RT} - 4\pi R_c^2 \rho_{Fe} V_D \tag{16}$$

$$V_{RT} \simeq \frac{2}{9} g\,(f_u - f_l)\,\Delta\rho\,(R_E - R_t)^2/\eta_m \tag{17}$$

$$V_D = k(f_l) g \Delta\rho/\eta_{Fe} \tag{18}$$

where V_{RT} is the Rayleigh-Taylor velocity, V_D is the Darcy law percolative flux (volume of iron per unit area per unit time), k is the permeability, and η_m, η_{Fe} are the viscosities of mantle (solid) and liquid iron, respectively. Of course, $V_{RT} \neq 0$ only if $f_u > f_l$. We shall assume a powerlaw $k(f) = k_o f^n$ and two possible choices of n. *McKenzie* (1984) favors n = 3, while theoretical simulations suggest n = 2, at least at low f (*von Bargen and Waff,* 1986). The value of k_o depends quadratically on grain size. We adopt a conservative grain size of 0.1 cm (tending to lower core formation efficiency) and use $k_o = 10^{-5}$ cm^2. If we write $dM/dt = M/\tau_a$, where τ_a is an "instantaneous accretion time," then the steady-state solutions to the above equations when the Earth is approaching its final mass give

$$f_u\,(f_u - f_l) \simeq 2 \times 10^{-6}\left(\frac{\eta_m}{10^{20}P}\right)\left(\frac{10^7 yr}{\tau_a}\right) \tag{19}$$

$$f_l^n \simeq 1 \times 10^{-7}\left(\frac{10^7 yr}{\tau_a}\right) \tag{20}$$

For any reasonable parameter choices, both f_u and $f_l \ll 1$, which is consistent with the idea of a quasisteady state of efficient core formation. For example, if $\eta_m = 10^{20}$ P, $\tau_a = 3 \times 10^7$ yr, and n = 2 (the most reasonable choice at low f), then $f_u \approx 9 \times 10^{-4}$, $f_l \approx 1.8 \times 10^{-4}$, and a given element of accreted metallic iron takes $\sim 10^4$ yr to reach the core. (This "residence time" is of order $\bar{f}\tau$.) If n = 3, the most unfavorable case imaginable, then $f_u = 3.4 \times 10^{-3}$, $f_l = 3.2 \times 10^{-3}$, and the residence time is a few tens of thousands of years.

It is possible that there is a pinch-off (zero permeability) at some small finite $f_l = f_c$ in the lower mantle, if $\theta_D > 60°$ even in that region (cf. equation (8)). Unless f_l is very small, the substitution of $k = k_o(f_l^n - f_c^n)$ leads to the prediction that both that both f_l and f_u are only slightly in excess of f_c. It is possible that this

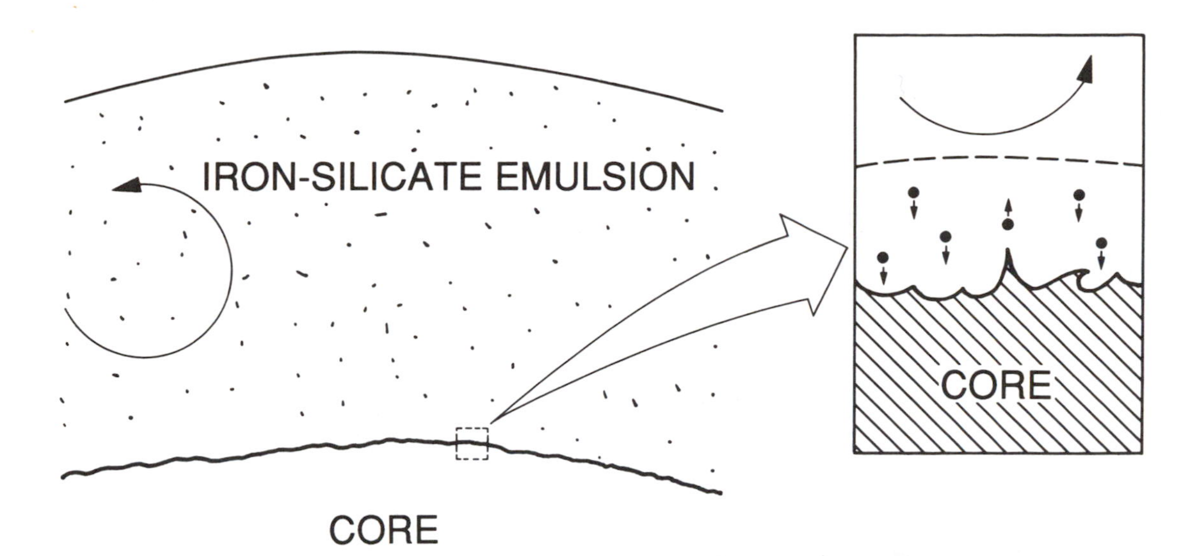

Fig. 5. *Cartoon of core formation by rainout. Insert shows a small portion of the core-mantle boundary. Below the dashed line, which delineates a viscous boundary layer, most iron droplets settle into the core. New upwardly mobile iron droplets are infrequently made by wavebreaking events ("surf").*

trapped iron is relevant to the problem of mantle siderophiles (*Jones and Drake,* 1986) although that appears to require special pleading. For example, $5 \times 10^{-3} < f_c \lesssim 1.5 \times 10^{-2}$ requires $60.5° \lesssim \theta_D \lesssim 63°$, an implausibly narrow range.

We can also use equations (15) through (18) to assess the decline of f_u, f_l as the accretion decays. Eventually, steady-state must fail. Suppose that we assume a half-life for the accretion flux of 3×10^7 years. We then obtain

$$\frac{df_u}{dt_7} \simeq 0.1 \exp(-t_7/3) - 5 \times 10^4 f_u (f_u - f_l) \quad (21)$$

$$\frac{df_l}{dt_7} \simeq 5 \times 10^4 f_u (f_u - f_l) - 10^6 f_l^n \quad (22)$$

where t_7 is the elapsed time in units of 10^7 years. At sufficiently long times, the asymptotic behavior of these equations for $n = 2$ is given by $f_u \sim 2.6 \times 10^{-5}/t_7$ and $f_l \sim 5.6 \times 10^{-6}/t_7$. In this model, accretion is negligible for $t_7 \gtrsim 40$ (i.e., more recent than ~4.1 Ga). These asymptotically small values of f_u and f_l are uninteresting in the sense that they have no important geochemical or geophysical consequence. Clearly, percolative core formation can be very efficient, even when only part of the mantle participates.

One other interesting complication deserves mention. Suppose that the lower mantle is intrinsically more dense that the upper mantle under conditions where neither region has metallic iron. It follows that the above solutions apply except that the negative buoyancy in V_{RT} must be increased by exactly the amount needed to offset this density stabilization. In equations (19), (21), and (22), this means that f_u - f_l is replaced by f_u - f_l - Δf, where $\rho_{Fe}\Delta f$ is the intrinsic density difference between upper and lower mantles. If we suppose that Δf is large (e.g., $\sim 10^{-2}$) then the previous steady-state example ($\eta_m = 10^{20}$ P, $\tau_a = 3 \times 10^7$ yr, $n = 2$) now yields $f_u \approx \Delta f$, $f_l \approx 1.8 \times 10^{-4}$. The revised solution to the asymptotic behavior (modified form of equations (21) and (22)) becomes $f_u \approx \Delta f$, $f_l \approx 2 \times 10^{-6}/t_7$. This is another way of "stranding" iron in the mantle that would seem very plausible, except that it is incompatible with some suggestions that the current lower mantle is ***enriched*** in iron (as magnesiowüstite).

CORE FORMATION BY RAINFALL

We turn now to specific models of the migration of liquid droplets through liquid. For this, we imagine a fully molten mantle that may be vigorously convecting and overlies a well-established liquid core. Figure 5 illustrates the scenario. The governing equation for the droplet number density N is

$$\frac{\partial N(z,t)}{\partial t} = \frac{1}{r^2}\frac{\partial}{\partial r}(r^2 V_s N) + \frac{\kappa}{r}\frac{\partial^2}{\partial r^2}(rN) + F(r,t) \qquad (23)$$

This equation is the same as equation (11) except for the addition of a source term. The boundary conditions are no net flux through the planet surface by Stokes flow and eddy diffusion

$$\left[V_s N + \kappa \frac{\partial N}{\partial r}\right]_{r=R_E} = 0 \qquad (24)$$

and no diffusive flux through the core mantle boundary

$$\left[\kappa \frac{\partial N}{\partial r}\right]_{r=R_c} = 0 \qquad (25)$$

We anticipate that core formation is easy and seek a quasisteady state in which the source term, assumed spatially uniform, balances the rainout. The appropriate steady-state solution for the total iron content is complicated in general, but simple for the limiting cases $\kappa \ll V_s R$ and $\kappa \gg V_s R$. (We assume $V_s =$ constant because gravity does not vary much in the mantle.) In the high eddy diffusion limit, the iron melt fraction f(r) is given by

$$f(r) \simeq \frac{0.1}{V_s \tau_a}\left(R_E + \frac{R_c^3}{R_E^2}\right) - \frac{0.1}{\kappa \tau_a}\left(\frac{r^2}{2} - \frac{R_c^3}{r} + \frac{R_c^2}{2}\right) \qquad (26)$$

where τ_a is the accretion time. For $v_s \sim 3$ cm/sec (a conservatively low value) and $\tau_a \sim 3 \times 10^7$ years, we find $f \sim 10^{-7}$. In the low eddy diffusion limit (which may correspond to no convection, since a large stabilizing density gradient develops), we find

$$f(r) \simeq \frac{0.3}{V_s \tau_a}\left(\frac{R_E^3}{r^2} - r\right) \qquad (27)$$

with necessarily similar mean values of f(r), but a much stronger gradient. We can find the minimum heat flow necessary to create a high diffusion state (equation (26)) by noting that we must have $\alpha \Delta T \gtrsim 0.1(R_E^2/\kappa\tau_a)\Delta\rho/\bar{\rho}$, where ΔT is the superadiabatic excess. Since $V_o R \sim 10^9$ cm²/sec, the smallest acceptable κ for this convective case is of that order, implying $\Delta T \gtrsim 10^{-3}$ K. The heat flow must thus be at least $\bar{\rho} C_p \kappa \Delta T / R_E \sim 10^5$ erg/cm² sec. This is only just possible on average, as discussed earlier. We conclude that core formation by rainfall might produce stable stratification and choke thermal convection. Of course, newly accreted material tends to be assimilated high in the mantle and may preserve some compositionally driven "convection" (strictly just Rayleigh-Taylor instabilities).

One more feature of rainfall dynamics is the extent to which droplets chemically equilibrate with the mantle through which they migrate. Each droplet can "process" a silicate volume of order $2\pi R\delta H$ where R = droplet radius, $\delta \sim (RD/V_s)^{1/2}$ is the diffusive boundary layer around a droplet, and H is the distance fallen. The ratio of processed volume to droplet volume is thus $\sim (D/RV_s)^{1/2}(H/R) \sim 10(D/10^{-7}\ \text{cm}^2/\text{sec})^{1/2}$ for $H \sim 10^8$ cm, $V_s \sim$ few cm/sec. For reasonable diffusivities in liquid silicates (*Hofmann,* 1980), this is greater than unity, implying that the droplets, though instantaneously a small volume fraction, are able to chemically "process" the entire mantle over the period of core formation.

IMPLICATIONS

I have sought in this paper to identify and quantify processes rather than develop an all-encompassing story of core formation. This would seem to be a wise strategy in view of the fact that the processes advocated here differ in major and qualitatively important ways from most previous scenarios. In one respect, this is not such an important development since the ideas advocated here lead to rapid or efficient core formation, a feature shared by previous explanations. However, the chemical implications of the various mechanisms can be markedly different. In this respect, the most striking aspect of the models presented here is chemical equilibration. In the diapiric mode of core formation, it has been argued (e.g., *Stevenson,* 1981) that the core-forming fluid equilibrates only with the uppermost mantle because it descends through the lower mantle in large, fast-moving blobs. In the percolative and rainfall models, descent may be fast but no element of core-forming fluid is ever more than ~1 cm away from mantle material, allowing good equilibration to occur. Another feature of the models presented here is excellent drainage: The "equilibrium" volume fraction of metallic iron in the mantle is always small (usually much less than 0.1%), and drainage to even smaller values is possible once accretion ceases. Of course, there may be ways of stranding larger amounts of iron in the mantle, by oxidizing it or by having $\theta_D > 60°$ everywhere, but these have an ad hoc character to them and often require restrictive assumptions. For example, the stranding of 1% iron by pinching off the percolation channels requires a value of θ_D that is fortuitously close to (yet above) 60°. However, it is always possible that the models described here are unduly simplistic, just as simple models for the drainage of basaltic melt from peridotite may overlook the existence of a "percolation threshold" at a few percent partial melting (*Toramaru and Fujii,* 1986) because of the effects of multiple phases.

Although we have chosen not to develop a complete chronological story of core formation, it seems worthwhile to provide a sketchy outline of such a story, if only to highlight the issues and concepts developed here. This story is cast in the context of current ideas of Earth accumulation including giant impacts (e.g., *Davies,* 1990; *Benz and Cameron,* 1990; *Stevenson,* 1989c). We imagine an embryo Earth that grows from planetesimals that are already often differentiated into core and mantle through some severe early heating event (presumably ^{26}Al). Despite poor percolation at low pressure, a core can begin to grow in this proto-Earth once the mass exceeds about the mass of Mars, since the accretional heating is then sufficient to produce melting. This earliest phase of core formation may not differ too much from that discussed in *Stevenson* (1981) but represents only a minor fraction of the total core formation process. As accretion continues, the mantle retains a large amount (~10%) of metallic iron as disconnected droplets until one of two things happen: Complete melting by a giant impact or sufficient pressure to create percolation in the lowermost mantle occur; plausibly both occur at various times. Following a partial or complete melting event, rainout of iron can be fast, even faster than freezing by convective cooling. The iron settles directly into the core, or accumulates at a rheological or freezing horizon in the lower mantle, where it undergoes further, diapiric instabilities. A combination of processes may operate; they are all fast. This combination or compromise picture is shown in Fig. 6. In Mars, where complete melting of the mantle may not occur and perovskite does not exist, the "ponding" of iron and subsequent diapiric instabilities are most likely.

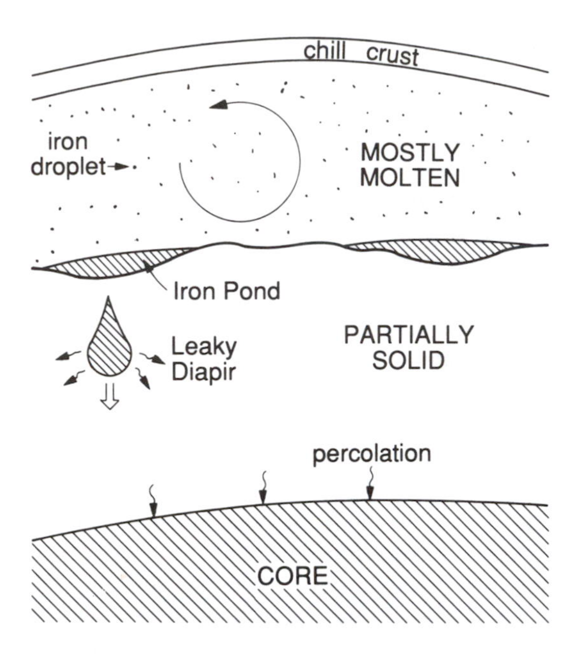

Fig. 6. *Something for everyone: A collage of core-formation mechanisms that may not all happen at the same time but that could each play a role.*

As accretion diminishes, the Earth cools and only the percolative mode can remain, fed by diapiric instabilities of the more iron-rich "upper" mantle (cf. Fig. 4). If percolation cannot occur then a substantial amount of iron alloy should be stranded, probably greater than what is compatible with mantle siderophiles, so it seems likely that percolation was, in fact, possible. All but an extremely small amount of core alloy can find its way to the core on a geologically short timescale. There does not appear to be a physically plausible way of delaying a substantial fraction of core formation to a much later time, such as a few hundred million years after accretion.

We have already briefly discussed Mars, a planet where the core appears to have formed quickly (*Chen and Wasserburg,* 1986) but where the mantle may be iron-rich (*Goettel,* 1981). In the spirit of comparative planetology, we end with a brief comment about the remaining terrestrial planets. Mercury evidently has a large iron core and may still have some form of magnetic field generation by convection in a fluid shell (*Schubert et al.,* 1988). The most plausible scenarios for the iron-rich character of Mercury involve severe heating, perhaps by a giant impact (*Cameron et al.,* 1988). This would cause complete melting and allow core formation by rainout. Venus presumably underwent a similar evolution to Earth, although it could be speculated that the largest giant impact on Venus was less traumatic than for Earth, conceivably reducing the role of rainout. The Moon has a core of uncertain size (*Hood,* 1986); most probably formed during a period of high degree of melting. It has sometimes been argued that the deep interior of the Moon began cold (e.g., *Solomon,* 1986) but these arguments are suspect (*Kirk and Stevenson,* 1989).

Acknowledgments. I thank the reviewers, especially N. Sleep, for useful comments. Support from NSF grant EAR 8816268 is gratefully acknowledged. This paper is contribution no. 4789 from the Division of Geological and Planetary Sciences, California Institute of Technology.

REFERENCES

Abe Y. (1988) Surface of a terrestrial planet growing by planetesimal impacts. *Proc. 21st ISAS Lunar Planet. Symp.*, pp. 225-231.

Abe Y. and Matsui T. (1985) Evolution of an impact-induced atmosphere and magma ocean on the accreting Earth. *Nature, 319,* 303-305.

Adler P. M. and Brenner H. (1988) Multiphase flow in porous media. *Annu. Rev. Fluid Mech., 20,* 35-59.

Allegre C., Dupré B., and Brévort O. (1982) Chemical aspects of the formation of the core. *Philos. Trans. R. Soc. London, 306* 49-59.

Benz W. and Cameron A. G. W. (1990) Terrestrial effects of the giant impact. In *Origin of the Earth,* this volume.

Boss A. P., Morfill G. E., and Tscharnuter W. M. (1989) Models of the formation and evolution of the solar nebula. In *Origin and Evolution of Planetary and Satellite Atmospheres* (S. K. Atreya, J. B. Pollack, and M. S. Matthews, eds.), pp. 35-77. Univ. of Arizona, Tucson.

Bottinga Y. and Weill D. F. (1972) The viscosity of magmatic silicate liquids: A model for calculation. *Am. J. Sci., 272,* 438-475.

Bulau J. R., Waff H. S., and Tyburczy J. A. (1979) Mechanical and thermodynamical constraints on fluid distribution in partial melts. *J. Geophys. Res., 84,* 6102-6108.

Cameron A. G. W., Fegley B. Jr., Benz W., and Slattery W. L. (1988) The strange density of Mercury: Theoretical considerations. In *Mercury* (F. Vilas, C. R. Chapman, and M. S. Matthews, eds.), pp. 692-708. Univ. of Arizona, Tucson.

Chen J. H. and Wasserburg G. J. (1986) Formation ages and evolution of Shergotty and its parent planet from U-Th-Pb systematics. *Geochim. Cosmochim. Acta, 50,* 955-968.

Chou C.-L. (1978) Fractionation of siderophile element ratios in the Earth's upper mantle. *Proc. Lunar Planet. Sci. Conf. 9th,* pp. 219-230.

Clayton D. D. (1968) *Principles of Stellar Evolution and Nucleosynthesis.* McGraw-Hill, New York. 256 pp.

Cooper R. F. and Kohlstedt D. L. (1982) Interfacial energies in the olivine-basalt system. *Adv. Earth Planet. Sci., 12,* 217-228.

Davies G. F. (1982) Ultimate strength of solids and the formation of planetary cores. *Geophys. Res. Lett., 9,* 1267-1269.

Davies G. F. (1990) Heat and mass transport in the early Earth. In *Origin of the Earth,* this volume.

Dyson B. F. (1963) The surface tension of iron and some iron alloys. *Trans. Metall. Soc. AIME, 227,* 1098-1102.

Elsasser W. M. (1963) Early history of the Earth. In *Earth Science and Meteorites* (J. Geiss and E. Goldberg, eds.), pp. 1-30. North-Holland, Amsterdam.

Fukai Y. and Suzuki T. (1986) Iron-water reaction under high pressure and its implication in the evolution of the Earth. *J. Geophys. Res., 91,* 9222-9230.

Goettel K. A. (1981) Density of the mantle of Mars. *Geophys. Res. Lett., 8,* 497-500.

Gurnis M. (1986) The effects of chemical density differences on convective mixing in the Earth's mantle. *J. Geophys. Res., 91,* 11407-11419.

Hofmann A. W. (1980) Diffusion in natural silicate melts: A critical review. In *Physics of Magmatic Processes* (R. B. Hargraves, ed.), pp. 385-418. Princeton Univ., Princeton, New Jersey.

Hood L. L. (1986) Geophysical constraints on the lunar interior. In *Origin of the Moon* (W. K. Hartmann, R. J. Phillips, and G. J. Taylor, eds.), pp. 361-410. Lunar and Planetary Institute, Houston.

Ida S., Nakagawa Y., and Nakazawa K. (1987) The Earth's core formation due to Rayleigh-Taylor instability. *Icarus, 69,* 239-248.

Jacobs J. A. (1987) *The Earth's Core,* 2nd ed. Academic, New York. 416 pp.

Jagoutz E., Palme H., Baddenhausen H., Blum K., Cendales M., Dreibus G., Spettel B., Lorenz V., and Wänke H. (1979) The abundances of major, minor and trace elements in the Earth's mantle as derived from ultramafic nodules. *Proc. Lunar Planet. Sci. Conf. 10th,* pp. 2031-2050.

Jones J. H. and Drake M. J. (1986) Geochemical constraints on core formation in the Earth. *Nature, 322,* 211-228.

Jurewicz S. R. and Watson E. B. (1984) Distribution of partial melt in a felsic system: The importance of surface energy. *Contrib. Mineral. Petrol., 85,* 25-29.

Kato T., Ringwood A. E., and Irifune T. (1988) Experimental determination of element partitioning between silicate perovskites, garnets and liquids: Constraints on early differentiation of the mantle. *Earth Planet. Sci. Lett., 89,* 123-145.

Kirk R. L. and Stevenson D. J. (1989) The competition between thermal contraction and differentiation in the stress history of the Moon. *J. Geophys. Res., 94,* 12133-12144 .

Knittle E. and Jeanloz R. (1989) Simulating the core-mantle boundary: An experimental study of high-pressure reactions between silicates and liquid iron. *Geophys. Res. Lett., 16,* 609-612.

Kushiro I. (1980) Viscosity, density, and structure of silicate melts at high pressures, and their petrological applications. In *Physics of Magmatic Processes* (R. B. Hargraves, ed.), pp. 93-120. Princeton Univ., Princeton, New Jersey.

Lambeck K. (1980) *The Earth's Variable Rotation.* Cambridge Univ., Cambridge. 449 pp.

Linden P. F. (1973) The interaction of a vortex ring with a sharp density interface: A model for turbulent entrainment. *J. Fluid. Mech., 60,* 467-480.

Long R. R. (1975) The influence of shear on mixing across density interfaces. *J. Fluid. Mech., 70,* 305-320.

Martin D. and Nokes R. (1988) Crystal settling in a vigorously convecting magma chamber. *Nature, 332,* 534-536.

McElhinny M. W. and Senanayake W. E. (1980) Paleomagnetic evidence for the existence of the geomagnetic field 3.5 Ga ago. *J. Geophys. Res., 85,* 3523-3528.

McKenzie D. P. (1979) Finite deformation during fluid flow. *Geophys. J. R. Astron. Soc., 58,* 689-715.

McKenzie D. (1984) The generation and compaction of partially molten rock. *J. Petrol., 25,* 713-765.

Melosh H. J. (1989) *Impacting Cratering: A Geologic Process.* Oxford Univ., New York. 245 pp.

Morgan J. W., Wandless G. A., Petrie R. K., and Irving A. J. (1981) Composition of the Earth's upper mantle—I. Siderophile trace elements in ultramafic nodules. *Tectonophysics, 75,* 47-67.

Nicolas A. (1986) A melt extraction model based on structural studies in mantle peridotites. *J. Petrol., 27,* 999-1022.

Olson P., Yuen D. A., and Balsiger D. (1984) Mixing of passive heterogeneities by mantle convection. *J. Geophys. Res., 89,* 425-436.

Oversby V. M. and Ringwood A. E. (1971) Time of formation of the Earth's core. *Nature, 234,* 463-464.

Popel' S. I., Tsarevskii B. V., Pavlov V. V., and Funsan Ye. L. (1975) Joint effects of oxygen and sulfur on the surface tension of iron. *Izv. Akad. Nauk SSR Met., 4,* 54-58.

Ringwood A. E. (1979) *Origin of the Earth and Moon.* Springer-Verlag, New York. 294 pp.

Ringwood A. E. (1984) The Earth's core: Its composition, formation and bearing upon the origin of the Earth. *Proc. R. Soc. London, A395,* 1-46.

Ringwood A. E. (1988) Early history of the Earth-Moon system (abstract). In *Papers Presented to the Conference on Origin of the Earth,* pp. 73-74. Lunar and Planetary Institute, Houston.

Rintoul D. and Stevenson D. (1988) The role of large infrequent impacts in the thermal state of the primordial Earth (abstract). In *Papers Presented to the Conference on the Origin of the Earth,* pp. 75-76. Lunar and Planetary Institute, Houston.

Ryerson F. J., Weed H. C., and Piwinskii A. J. (1988) Rheology of subliquidus magmas. I, Picritic compositions. *J. Geophys. Res., 93,* 3421-3436.

Schmitt W., Palme H., and Wänke H. (1989) Experimental determination of metal/silicate partition coefficients for P, Co, Ni, Cu, Ga, Ge, Mo, and W and some implications for the early evolution of the Earth. *Geochim. Cosmochim. Acta, 53,* 173-185.

Schubert G., Ross M. N., Stevenson D. J., and Spohn T. (1988) Mercury's thermal history and the generation of its magnetic field. In *Mercury* (F. Vilas, C. R. Chapman, and M. S. Matthews, eds.), pp. 429-460. Univ. of Arizona, Tucson.

Scott D. R. and Stevenson D. J. (1989) A self-consistent model of melting, magma migration and buoyancy-driven circulation beneath mid-ocean ridges. *J. Geophys. Res., 94,* 2973-2988.

Sleep N. H. (1988) Tapping of melt by veins and dikes. *J. Geophys. Res., 93,* 10255-10272.

Smith A. L., ed. (1972) *Particle Growth in Suspensions.* Academic, New York. 306 pp.

Solomon S. C. (1986) On the early thermal state of the Moon. In *Origin of the Moon* (W. K. Hartmann, R. J. Phillips, and G. J. Taylor, eds.), pp. 435-452. Lunar and Planetary Institute, Houston.

Stevenson D. J. (1979) Turbulent thermal convection in the presence of rotation and a magnetic field: A heuristic theory. *Geophys. Astrophys. Fluid Dyn., 12,* 139-169.

Stevenson D. J. (1981) Models of the Earth's core. *Science, 214,* 611-619.

Stevenson D. J. (1986) On the role of surface tension in the migration of melts and fluids. *Geophys. Res. Lett., 13,* 1149-1152.

Stevenson D. J. (1987) Origin of the Moon—the collision hypothesis. *Annu. Rev. Earth Planet. Sci., 15,* 271-315.

Stevenson D. J. (1988) Greenhouses and magma oceans. *Nature, 335,* 587-588.

Stevenson D. J. (1989a) Temperature gradient, crystallinity and rigidity of the outer core. *Eos Trans. AGU, 70,* 1212.

Stevenson D. J. (1989b) Spontaneous small-scale melt segregation in partial melts undergoing deformation. *Geophys. Res. Lett., 16,* 1067-1070.

Stevenson D. J. (1989c) Formation and early evolution of the Earth. In *Mantle Convection* (W. Peltier, ed.), pp. 817-873. Gordon and Breach, New York.

Stevenson D. J., Spohn T., and Schubert G. (1983) Magnetism and thermal evolution of the terrestrial planets. *Icarus, 54,* 466-489.

Taylor G. J. (1989) Metal segregation in asteroids (abstract). In *Lunar and Planetary Science XX,* pp. 1109-1110. Lunar and Planetary Institute, Houston.

Tennekes H. and Lumley J. L. (1972) *A First Course in Turbulence.* MIT, Cambridge. 300 pp.

Tonks W. B. and Melosh H. J. (1988) A well stirred magma ocean: Implications for crystal settling and chemical evolution (abstract). In *Papers Presented to the Conference on the Origin of the Earth,* pp. 93-94. Lunar and Planetary Institute, Houston.

Toramaru A. and Fujii N. (1986) Connectivity of melt phase in a partially molten peridotite. *J. Geophys. Res., 91,* 9239-9252.

Treiman A. H., Jones J. H., and Drake M. J. (1987) Core formation in the shergottite parent body and comparison with the Earth. *Proc. Lunar Planet. Sci. Conf. 17th,* in *J. Geophys. Res., 92,* E627-E643.

Trubitsyn V. P. and Kharybin V. (1987) The convective instability of a sedimentation regime in the mantle. *Izv. Earth Phys., 23,* 638-645.

Turcotte D. L. and Emerman S. H. (1983) Dissipative melting as a mechanism for core formation. *Proc. Lunar Planet. Sci. Conf. 14th,* in *J. Geophys. Res., 88,* B91-B96.

Urakawa S., Kato M., and Kumazawa M. (1987) Experimental study on the phase relations in the system Fe-Ni-O-S up to 15 GPa. In *High Pressure Research in Mineral Physics* (M. H. Manghani and Y. Syono, eds.), pp. 95-111. Terra Scientific, Tokyo.

Vidal P. and Dosso L. (1978) Core formation—continuous or catastrophic? Sr and Pb isotope geochemistry constraints. *Geophys. Res. Lett., 5,* 169-171.

Vityazev A. V. and Mayeva S. V. (1976) Model of the early evolution of the Earth. *Izv. Acad. Sci. USSR Phys. Solid Earth, 12,* 79-82.

Vollmer R. (1977) Terrestrial lead isotopic evolution and formation time of the Earth's core. *Nature, 270,* 144-146.

von Bargen N. and Waff H. S. (1986) Permeabilities, interfacial areas and curvatures of partially molten systems: Results of numerical computations of equilibrium microstructures. *J. Geophys. Res., 91,* 9261-9276.

von Bargen N. and Waff H. S. (1988) Wetting of enstatite by basaltic melt at 1350°C and 1.0 to 2.5-GPa pressure. *J. Geophys. Res., 93,* 1153-1158.

Walker D. and Agee C. B. (1988) Ureilite compaction. *Meteoritics, 23,* 81-91.

Walker D. and Mullins O. Jr. (1981) Surface tension of natural silicate melts from 1200°-1500° C and implications for melt structure. *Contrib. Mineral. Petrol., 76,* 455-462.

Wasson J. T. (1988) The building stones of the planets. In *Mercury* (F. Vilas, C. R. Chapman, and M. S. Matthews, eds.), pp. 622-650. Univ. of Arizona, Tucson.

Wetherill G. W. (1985) Occurrence of giant impacts during the growth of the terrestrial planets. *Science, 228,* 877-879.

METAL-SILICATE EQUILIBRIA AND CORE FORMATION

R. J. Arculus

Department of Geology and Geophysics, University of New England, Armidale, New South Wales 2351, Australia

R. D. Holmes

Department of Geological Sciences, University of Michigan, Ann Arbor, MI 48109-1063

R. Powell

Department of Geology, University of Melbourne, Parkville, Victoria 3052, Australia

K. Righter

Department of Geological Sciences, University of Michigan, Ann Arbor, MI 48109-1063

The segregation of metal from silicate was of major physical and chemical importance in the early development of the Earth, but many of the details of this process are obscured by subsequent events. Important clues can be gathered from the nature of metal-silicate equilibria preserved in the disrupted fragments of differentiated meteorites, but under the higher pressure conditions prevailing in the Earth, the operation of different metal partitioning and separation processes are manifested in a set of unique geochemical characteristics. With respect to differentiated meteorites, clear evidence from thermodynamic calculations and experiments shows that the compositions of coexisting metal and silicate in the various pallasite groups are relict from equilibrium siderophile element partitioning events. However, the inconsistency of MgO/(MgO + FeO) (Mg#, a function in part of the distribution of Fe between silicate and metal components) and siderophile element abundances in silicate components of the postulated Eucrite Parent Body (EPB) [eucrites, diogenites, main group pallasites (MGP), and Group IIIAB irons] precludes straightforward analysis of the geochemistry and physics of metal formation and core separation. Nevertheless, metal separation in the absence of a large gravitational field clearly occurred. The Mg# of the Earth's upper mantle is unlike any major meteorite group or the Mg# of material drawn from the same O isotope reservoir (Moon and enstatite chondrites), but shows some overlap with, for example, chondritic silicate inclusions in the IAB irons, the olivine of the MGP, and some other rare chondrites. It is not obvious, however, that the terrestrial Mg# necessarily reflects the redox equilibria prevailing in the solar accretion disk, given the evidence for chemical fractionation of the Earth's upper mantle with respect to chondrites of various lithophile elements (e.g., Mg/Si, Ca/Al, and Nb/U). Furthermore, the fO_2-dependence of Fe, Ni, and Co partitioning between coexisting silicate and metal results in dramatic variations in Ni/Co ratios over small ranges in fO_2. The preservation of chondritic Ni/Co ratios in the Earth's upper mantle is not readily explained by segregation of small amounts of highly siderophile elements during later accretionary stages of relatively oxidized materials. Separation of liquid metal from solid silicate at low pressures is unlikely due to the high surface tension of the metal. Increased solubility of O in liquid Fe at high pressures (>6 GPa) results in a reduction of surface tension by a factor of 5, sufficient to allow complete wetting and permeation of a solid silicate matrix. In the case of the low pressure (<<6 GPa) EPB and Moon, metal separation possibly required the transient involvement of an extensively molten silicate state. Metal production and destruction near the surface of the Earth may also have involved molten silicate, but additional separation of liquid metal from and equilibrium with solid silicate (as a porous medium) at high pressures in the mantle is probable.

INTRODUCTION

Two of the major aims of geochemists are the determination of the present distribution of elements in the Earth and an inversion of this distribution backwards in time to some original condition. On an active planet like the Earth, these are clearly complex tasks, especially given the limitations of the extent of direct sampling that we enjoy. Interpretation of the meteoritical and lunar records is particularly important in these regards, in guiding our understanding of the kinds of chemical and physical processes that occurred during planetesimal/protoplanetary origins and early evolution. Nevertheless, the overall disequilibrium that must prevail, for example, between the crust, mantle, and core in the present Earth, and the effects of very high pressures on silicate-metal equilibria (*Ringwood,* 1966; *Stevenson,* 1981) must be kept in mind when pursuing meteorite analogs.

The major distinctive chemical reservoirs of the Earth are the core, mantle, crust, ocean, and atmosphere. In terms of mass, the first two are predominant, although a number of elements are strongly concentrated in the near-surface reservoirs (e.g., *Taylor and McLennan,* 1985), and analysis of the trace element and isotopic systematics of these accessible regions is crucial to understanding terrestrial evolution as a whole. Segregation of metal and coalescence in the core of the planet can result in a major release of energy, and impart a distinct chemical stamp to protomantle-forming silicates and oxides, depending on the mechanisms involved (*Davies,* 1984; 1988; *Stevenson,* 1985).

The questions addressed in this paper are how did core formation proceed, and are there any interpretable geochemical fingerprints remaining from the process? A variety of physical models have been proposed, and further progress depends on the degree to which observed chemical characteristics can be reconciled with one or more of these models.

Some end-member possibilities exist that are not necessarily espoused in their pure form at this time, but have previously been postulated and serve to focus analysis and discussion. All relate directly to the origin of the Earth as follows:

1. All metal-silicate redox equilibria were established in the solar accretion disk; core formation in planets like the Earth resulted from melting of the disseminated metal phase and separation from silicate in accreted materials without further chemical reactions (e.g., *Taylor,* 1988a,b).

2. The Earth grew from oxidized (volatile-rich) CI-type chondritic materials; reduction of FeO proceeded after the planet reached sufficient size within a blanketing, hot atmosphere created by evaporation of accreting debris, that extensive redox reactions could occur (e.g., *Ringwood,* 1966). Metal segregation by core-bound diapirism (*Elsasser,* 1963) occurred in a possible disequilibrium manner.

3. The Earth accreted from a geochemically varied population of planetesimals. Core chemistry was strongly modified by high-pressure metal-oxide-silicate equilibria, and later accretion may have experienced proportionally less metal separation than earlier stages (e.g., *Ringwood,* 1979; *Wänke,* 1981).

The procedure followed here is to outline the meteoritical evidence for metal production and segregation, examine some accepted and controversial facets of the Earth's mantle geochemistry relating to core formation, and explore some of the physical aspects of the process. The goal of a satisfactory hypothesis for the original state and early evolution of the Earth is the aim of this book. It will become clear that we have some distance to travel before realizing this goal in terms of the core formation event.

THE METEORITE RECORD

Evidence for variation in redox conditions in materials forming the solar accretion disk has been reviewed by *Rubin et al.* (1988). A range of molar MgO/(MgO + FeO) (Mg#) exists in chondritic materials (Fig. 1) that can be attributed to equilibria between metal-silicate dust and nebular gas in the most simple terms by

$$\underset{\text{in olivine}}{Fe_2SiO_4} = \underset{\text{in pyroxene}}{FeSiO_3} + \underset{\text{in metal}}{Fe} + \underset{\text{in gas}}{1/2O_2} \quad (1)$$

or more realistically in terms of dominant gas species

$$Fe_2SiO_4 + H_2 = FeSiO_3 + Fe + H_2O \quad (2)$$

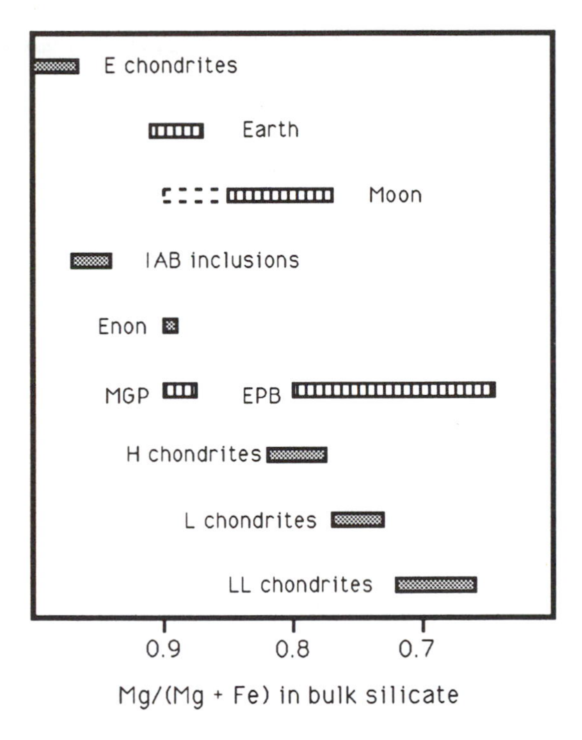

Fig. 1. *Estimated Mg/(Mg + Fe) (atomic) in the bulk silicate portion of the upper mantle of the Earth, the lunar mantle and various chondrite types. Sources of data are the Basaltic Volcanism Study Project (1981), Warren (1985b; dashed extension toward high Mg#s for the Moon), Kallemeyn and Wasson (1985) for the howardite-eucrite-diogenite (HED)-Main Group pallasite (MGP)-related reservoir Enon, Dreibus and Wänke (1980), Stolper (1977), and Jones (1984).*

The Mg# of silicates in chondritic materials is then a function of the local temperature, fH_2O/fH_2 (controlled in part by the proportion of solids:gas) and the degree to which diffusional equilibrium is maintained (*Nagahara,* 1986). Note that in a system with chondritic proportions of Mg:Fe:Si, a positive correlation exists between the modal abundance of metal and orthopyroxene.

The situation is more complex in that variations of bulk chondritic chemistry [e.g., Fe/Si (Fig. 2), Mg/Si and Al/Si (Fig. 3), and O isotopes (*Clayton et al.,* 1976)] point to the formation of different meteorite types in distinct chemical reservoirs.

The observed Mg#s for the Earth's upper mantle, lunar mantle, eucrite parent body (EPB), and main group pallasites (MGP) are also shown in Fig. 1. It can be seen that within the range of chondritic Mg#s, only the silicate inclusions trapped in IAB irons and some individual chondrites probably related to the IABs and EPB-MGP groups, respectively (on the basis of O isotope similarities), are close to the terrestrial upper mantle and the olivine of the MGP. In terms of simple redox character therefore, melting and separation of the metal phase of these IAB inclusions without further chemical equilibration with the silicate phases, might result in a silicate terrestrial mantle of appropriate Mg#. On the basis of O isotopes, only the enstatite chondrites are an appropriate building block for the Earth (e.g., *Smith,* 1982) but their (variably) highly reduced character (*Lusby et al.,* 1987) and strongly fractionated bulk chemistry (Fig. 3) require at least one other major silicate component for an appropriate match to be obtained with the chemistry of the Earth's upper mantle.

The primary conclusion reached by many is that a straightforward match of the chemistry of the Earth's mantle cannot be obtained within the chondritic spectrum, involving a core formation process through simple separation of metal from silicate (cf. model 1 above). However, the extensive variability in the chemistry of the solar accretion disk and

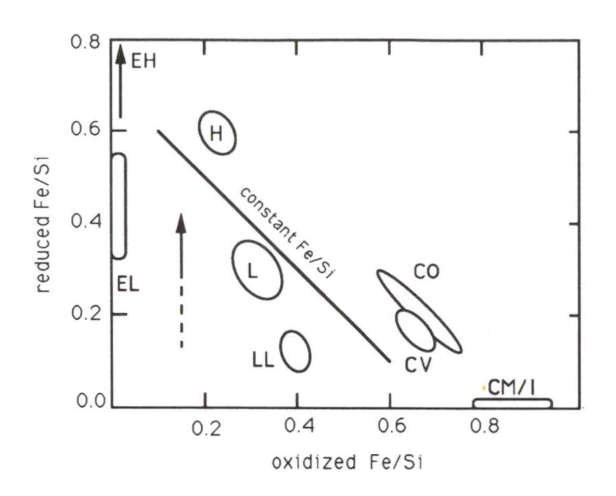

Fig. 2. *Variation of reduced Fe (FeS + Fe metal)/Si vs. oxidized Fe/Si (mole ratios) in chondrite groups after Larimer and Wasson (1988). The EH group plots at values of reduced Fe/Si > 0.94. The arrow with a dashed extension represents the value for oxidized Fe/Si of the Earth's upper mantle. The value for reduced Fe/Si for the Earth is uncertain.*

the development of a wide range of redox states can be interpreted to mean that every planetary body is the product of a unique blend of planetesimal input and development; limited radial mixing and homogenization is implied (*Taylor,* 1988a; *Weidenschilling,* 1988). Furthermore, the Mg# of the Earth's upper mantle may not be particularly diagnostic with respect to the character of the building blocks involved, nor be an unmodified average Mg# of these input components.

It is appropriate to turn to the differentiated meteorite population (achondrites and irons) to examine some of the possible chemical complications that arose during metal segregation. These materials are after all, our only *direct* samples of the processes of core formation, even if developed under low static pressures.

In this context, the IAB-IIICD and associated silicate inclusions (*Scott and Wasson,* 1975; *Bild,* 1977; *Kracher,* 1985) and the howardite-eucrite-diogenite (HED)-mesosiderite-MGP-IIIAB irons (*Wasson,* 1974; *Buseck,* 1977; *Scott,* 1977a) are the most important classes of material. It is possible that all of this latter group of differentiated meteorites are disrupted fragments of a single entity, the EPB. One undifferentiated chondrite (Enon, Fig. 1) was also derived from the HED-MGP-IIIAB reservoir, and may plausibly represent the kind of material from which the differentiated types developed (*Kallemeyn and Wasson,* 1985).

In terms of silicate-metal equilibria, the pallasites are particularly interesting. First of all, and perhaps coincidentally, the Mg# of the MGP olivine is close to that estimated for the Earth's upper mantle (Fig. 1). Second, a variety of thermodynamic calculations and electrochemical measurements show that the olivine present in pallasites was in chemical equilibrium with the metal phase at relatively high temperatures (*Olsen and Fredriksson,* 1966; *Holmes and Arculus,* 1983; *Brett and Sato,* 1984; *Righter et al.,* 1990). For example, thermodynamic data are available for calculation of the T-fO_2 relations of olivine-metal-phosphide-phosphate assemblages in the following equilibria

$$\underset{\text{in metal}}{2Fe + Si} + 2O_2 = \underset{\text{in olivine}}{Fe_2SiO_4} \qquad (3)$$

$$\underset{\text{in olivine}}{3Mg_2SiO_4} + \underset{\text{in schreibersite}}{4Fe_3P} + 8O_2 =$$

$$\underset{\text{in farringtonite}}{2Mg_3(PO_4)_2} + \underset{\text{in olivine}}{3Fe_2SiO_4} + \underset{\text{in metal}}{6Fe} \qquad (4)$$

The low concentration of Si in pallasite metal (<15 ppm) means that equation (3) is a lower limit for the MGP T-fO_2 relations (*Righter et al.,* 1990). However, a consistent variation of the distribution of Fe and Ni between olivine and metal as a function of intrinsic fO_2 for olivine separates has been demonstrated for the Eagle Station Trio, the anomalous pallasite Springwater, and the MGP (*Righter et al.,* 1990). Previous electrochemical measurements of fO_2 for MGP Brenham and Salta are also within the vicinity of the

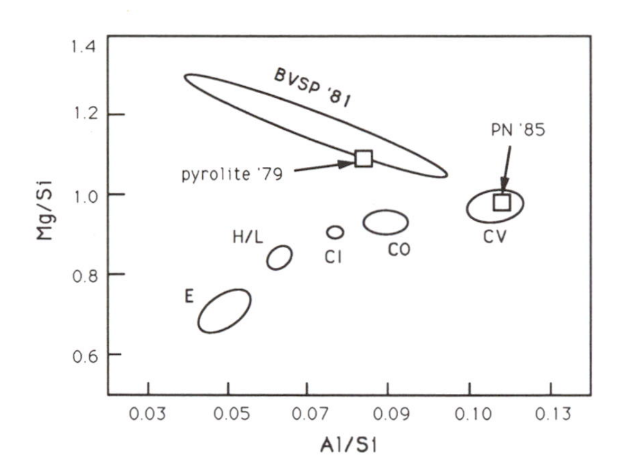

Fig. 3. *Variation of Mg/Si vs. Al/Si (weight ratios) in major chondrite groups and the Earth's upper mantle [analyses of spinel lherzolites from Basaltic Volcanism Study Project (1981)], with estimates of primitive upper mantle (boxes) indicated for Ringwood (1979) (=pyrolite'79), and Palme and Nickel (1985) (=PN'85). Figure modified after Larimer (1979) and Palme and Nickel (1985).*

iron-wüstite (IW) and quartz-iron-fayalite (QIF) buffers (*Holmes and Arculus,* 1983; *Brett and Sato,* 1984).

A first-order observation is that during segregation of core-forming metal in these small pallasite parent bodies, equilibrium with surrounding silicate was locally maintained. In addition, the coincident O isotope character of chromite from IIIAB irons and the MGP olivine (*Clayton et al.,* 1986) can be used to support the propositions that metal and silicate were derived from a *common* reservoir, and that pallasites do not represent randomly assembled components from dispersed regions of the solar accretionary disk.

However, the association of MGP and IIIAB irons also poses a number of physical and chemical problems with respect to understanding core formation processes and possible chemical fingerprints in mantle silicates: (1) the relative lack of separation of olivine from metal despite large density differences (*Wahl,* 1965) and the development of specific olivine-metal textures (*Buseck,* 1977; *Scott,* 1977b); (2) the rarity (and probably secondary nature) of orthopyroxene in pallasites is not expected for an equilibrium distribution of metal and silicate in a system with initial chondritic proportions of Mg-Fe-Si-O [see equilibrium (1)]; and (3) the difference in Mg# of the MGP olivine compared with the estimated Mg# of the source (i.e., the possible mantle of the EPB) of the eucrites must indicate that simple crust (HED)-mantle-(MGP)-core (IIIAB) relations for the EPB are inappropriate.

In fact as pointed out by *Buseck* (1977), pallasites are highly differentiated assemblages that may represent cumulate olivine in association with fractionated liquid metal-sulfide [by differentiation of the IIIAB solid core (*Scott and Wasson,* 1975)]. The olivine is deformed and was partially intruded by liquid metal (*Scott,* 1977b; *Ohtani,* 1983).

What was the character of the coexisting silicate melt if the olivine represents a cumulate (or restite) phase? Other experimental and analytical data are relevant to these aspects of possible melting and core formation processes in the EPB.

Stolper (1977) determined a point of liquidus multiple saturation at 10^5 Pa for noncumulate eucrites at ~1150°-1190°C and $-\log_{10} fO_2 \sim 13.3$. This is close to the calculated fO_2-T relations for the MGP and similar fO_2-T relations have been determined electrochemically for the diogenites and mesosiderites by *Hewins and Ulmer* (1984). The multiple saturation point represents a peritectic between melt and residual olivine-low Ca pyroxene-plagioclase-spinel-metal (i.e., a possible EPB mantle mineralogy). A reaction relationship of olivine and metal in the residue with coexisting melt was proposed by *Stolper* (1977), consistent with relations in the simple system Mg-Fe-Si-O (*Bowen and Schairer,* 1935; *Morse,* 1980).

On the other hand, the strong depletions but positive correlations of siderophilic W and Mo with respect to incompatible lithophiles in the eucrites is firm evidence for *prior* separation of core-forming metal from the eucrite source (*Palme and Rammensee,* 1981; *Newsom,* 1985). Despite the low fO_2 at time of melting, it would appear that metal was absent at the time of eucrite formation.

The peritectic melting relationships for the eucrites at 10^5 Pa have been confirmed (*Beckett and Stolper,* 1987; *Longhi and Pan,* 1988), but it appears that variable degrees of melting of at least two different sources are required to account for the range in eucrite chemistry (*Delaney,* 1987; *McSween,* 1989). The orthopyroxene-rich diogenites are plausibly cumulates from eucrite magmas (*Longhi and Pan,* 1988), and the howardites can be modeled as polymict breccias composed of both eucrite and diogenite clasts (*Dreibus and Wänke,* 1980; *Warren,* 1985a).

If some eucrites represent primary magmas, it is possible to infer the chemistry of the mantle from which they were derived (*Stolper,* 1977; *Consolmagno and Drake,* 1977; *Jones,* 1984), although this "mantle" is conspicuously absent from the spectrum of recovered meteorite types. The important features are (1) the relatively Fe-rich ($\sim Fo_{65}$) composition and (2) the chondritic lithophile abundances. Although the FeO-rich character of this source has been criticized on the basis of the calculated high FeO/MnO (*Dreibus and Wänke,* 1980), it is also important to realize that the composition of any core-forming metal previously extracted from such a source should bear a distinctive chemical stamp. Calculations are presented below for the chemical variation of olivine-metal assemblages during progressive reduction, but the important point here is that the metal in equilibrium with Fo_{65} olivine is relatively Ni-rich ($\sim Fe_{0.7}Ni_{0.3}$), and unlike that of the IIIAB-MGP metal. Conversely, the olivine (cumulate?) of the MGP is far more magnesian ($\sim Fo_{89}$) than the olivine in equilibrium with the parent eucrite magmas, although similarly forsteritic olivine is present in some howardites (e.g., Kapoeta; see *Delaney et al.,* 1980). In fact, *Kracher and Wasson* (1982) have pointed out that the calculated bulk composition of the IIIAB irons is closely similar in terms of Fe-Ni chemistry to the H chondrites.

Thus we are faced with a complex series of core-formation and melting events that do not appear to be simply linked. There is evidence that the EPB was formed ~4.56 Ga ago, and the eucrite formation event occurred at about 4.54 Ga (see *Tilton,* 1988, for review). Popular heat sources for the initiation of melting are currently believed to be the incorporation of ^{26}Al and/or electromagnetic induction (*Hewins and Newsom,* 1988). Evidence for early formation of planetesimal cores has also been inferred from the incorporation of "live" ^{107}Pd (*Kelly and Wasserburg,* 1978; *Kaiser and Wasserburg,* 1983).

It is possible that core-formation (IIIAB-MGP) occurred simultaneously with the eucrite-generation period, but that a radial variation in fO_2 and degree of metal extraction existed that precluded any equivalence between a eucrite-related and metal (core)-depleted residual mantle. It is also worth emphasizing that despite the similarity of O isotope characteristics, there is no assurance that the HED-IIIAB-MGP are all derived from a single body with a coherent geochemical evolution. Indeed, if the intact asteroid Vesta is the source of the HED (*Drake,* 1979) then the IIIAB-MGP must be derived from another fragmented parent body.

Turning to the disrupted products of other meteorite parent bodies, the process of core formation may have been preserved at an earlier stage of development in the IAB-IIICD iron groups than in the complex achondrite-pallasite-IIIAB assemblage. In these samples, a sulfur-rich core appears to have separated from a chondritic silicate + residual metal-bearing mantle, and partially back intruded the silicates to form brecciated assemblages (*Kracher,* 1985). Some minor degree of partial melting of the silicates was invoked by Kracher to account for the segregation of sulfur-rich metal (see below), but generally the chondritic proportions of siderophiles and lithophiles in the IAB inclusions means that extensive melting and differentiation did not occur (*Bild*, 1977).

The surprising feature of this parent body is the lack of extensive, highly siderophile element removal from the IAB inclusions, despite the high temperatures (~1200-1500 K) and presence of small amounts of melt inferred (*Bild*, 1977; *Kracher*, 1985).

Although the differentiated meteorites represent the end products of a complex variety of processes, and no simple picture of core segregation from silicate has emerged, it is important to bear in mind the following conclusions: (1) multiple small ($\lesssim$10-100-km radius) parent bodies did experience metal segregation and/or core formation with distinctive fractionated chemistries; (2) a large gravitational field was clearly not necessary for metal segregation; and (3) it is unlikely that large, persistent, blanketing coaccretionary atmospheres were present (see below).

METAL-SILICATE EQUILIBRIA

The importance for core formation studies of the quantification of Fe-Ni-Co and other siderophile partitioning between metal and silicate has prompted numerous studies (e.g., *Palme and Rammensee*, 1981; *Newsom and Drake*, 1982, 1983; *Jones and Drake*, 1983). In this section, an example is given of the compositional relations of coexisting metal and silicate during progressive reduction of an oxidized assemblage with CI proportions of siderophile elements. With these data, it is possible to estimate the composition of metal (proportions of Fe, Ni, and Co) that could have been in subsolidus equilibrium with a silicate assemblage of specific Mg#, and to explore the departure from metal-silicate equilibrium of individual siderophiles represented in the upper mantle of the Earth (see below). The calculations apply equally to the progressive oxidation of metal, if it is assumed that during the late growth of the Earth incoming metal was variably oxidized and retained in the mantle (*Morgan*, 1986; *Morgan et al.*, 1980, 1984; *Wänke*, 1981; *Wänke et al.*, 1984).

It should be recognized that these types of partitioning calculation represent only one facet of the problem. The extensive studies of H. Wänke, H. Palme, M. J. Drake, and colleagues of siderophile (and chalcophile) element partitioning between liquid and solid silicate with liquid and solid metal are vital for predictions of the chemical characteristics of planets undergoing melting concurrent with core separation (see *Jones and Drake*, 1986, and other papers in this volume for review).

At low pressures in the stability field of olivine, progressive reduction can be modeled by an iterative procedure employing thermodynamic relationships for olivine-metal equilibria and appropriate mass balance constraints. For example, the following equilibria can be written to describe the distribution of a given siderophile element between metal and silicate

$$2Fe + SiO_2 + O_2 = Fe_2SiO_4 \quad (5)$$

$$2Ni + SiO_2 + O_2 = Ni_2SiO_4 \quad (6)$$

$$2Co + SiO_2 + O_2 = Co_2SiO_4 \quad (7)$$

$$\Delta G^{o}_{(P,T)} = -RTlnK =$$
$$(a_{Fe_2SiO_4}) / [(a_{Fe})^2 (a_{SiO_2}) (fO_2)] \quad (8)$$

where a_i refers to the activity of component i in a silicate or metal phase, and similarly for equations (6) and (7).

In fact, in a CI (or peridotite) assemblage, a_{SiO_2} is buffered by coexisting olivine and orthopyroxene

$$Fe_2SiO_4 + SiO_2 = 2FeSiO_3 \quad (9)$$

At equilibrium in a metal-silicate system with buffered a_{SiO_2}, the fO_2 defined by equations (5)-(7) must be a constant. Adopting the standard free energy of formation data for Fe_2SiO_4, Ni_2SiO_4, and Co_2SiO_4 of *O'Neill* (1987a,b), a-X relations of *Seifert and O'Neill*

(1987) for olivine, and *Fraser and Rammensee* (1982) for metal and mass balance constraints of the form

$$N_{Fe}^{total} = N_{Fe}^{metal} + N_{Fe}^{olivine} \qquad (10)$$

$$X_{Fe}^{metal} + X_{Ni}^{metal} + X_{Co}^{metal} = 1 \qquad (11)$$

where N = number of moles and X is mole fraction, it is possible to determine the equilibrium distribution of the siderophile metals between silicate and metal as a function of fO_2. Examples of the results of these types of calculation for the silicate and metal phase at 1300 K are shown in Figs. 4 and 5. Ni-rich metal saturation occurs at about 2 log units more oxidized than IW (see also *O'Neill and Wall,* 1987). The Co content of the metal initially increases as reduction proceeds and then both Ni and Co become progressively diluted in the metal phase as the proportion of Fe increases. Reduction of 99% of available Fe occurs at some 3 log units below the IW buffer.

Some features of Figs. 4 and 5 are noteworthy. First, nonchondritic ratios of Ni/Co in metal and silicate are produced in the interval between complete oxidation and ~90% reduction. Incomplete oxidation of metal containing chondritic relative abundances of siderophile elements cannot be expected to give rise to chondritic relative abundances in the silicates and oxides generated during the oxidation event (cf. *O'Neill and Wall,* 1987; *Morgan,* 1986; *Wänke,* 1981). Second, the amount of Ni-rich metal (e.g., $Fe_{0.7}Ni_{0.3}$) that can be produced in equilibrium with iron-rich olivine (Fo_{65}) is small (<3%). If the EPB for example, had experienced core separation prior to the generation of the eucrites, overall extremely small quantities of Ni-rich metal (unlike that present in MGP and IIIAB irons) were removed.

Similar arguments have been advanced in analysis of the possible composition of the lunar core by *Newsom* (1989). In this case, a reduction in the Ni/Co ratio from about 20 (the chondritic ratio that is also preserved in the Earth's upper mantle) to about 3-5 in the source regions of mare basalts (*Delano,* 1986) requires very minor further reduction below the level of metal saturation (Fig. 4) under relatively low pressure conditions. Recall, however, that partitioning of siderophiles between silicate and metal is a function of temperature and the presence of silicate melt as well as liquid and/or solid metal (*Jones and Drake,* 1986).

Using the same general approach outlined in equations (5)-(11), *Holmes and Arculus* (1983) were able to show that under no low-pressure condition could the upper mantle of the Earth have equilibrated with an Fe-Ni-Co metal phase in a single-stage process.

In the following sections, the mechanism of metal formation and removal in a growing planet are further examined after some discussion of aspects of the chemical characteristics of the Earth's mantle, with particular reference to the evidence for melt-solid and metal-silicate fractionation.

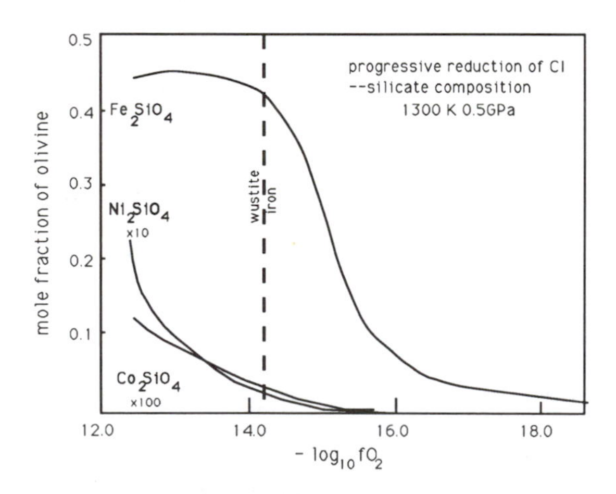

Fig. 4. *Variation of Fe_2SiO_4, Ni_2SiO_4 (×10), and Co_2SiO_4 (×100) contents of olivine in an initial CI bulk composition undergoing progressive reduction (or oxidation) as a function of fO_2. Calculations completed for 1300 K and 0.5 GPa. Mg_2SiO_4 forms the balance of the olivine composition.*

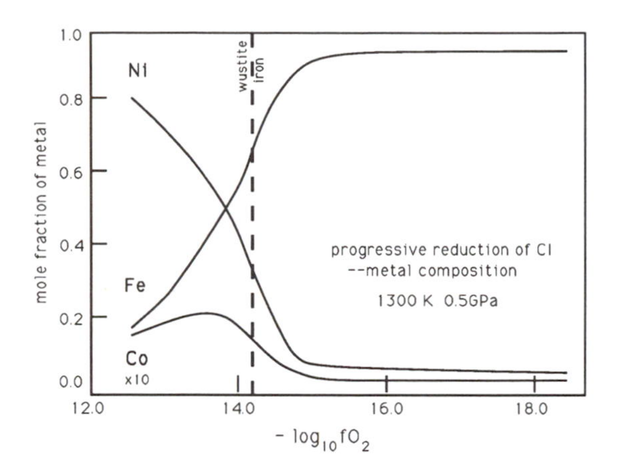

Fig. 5. *Variation of Fe, Ni, and Co (×10) content of metal in an initial CI bulk composition undergoing progressive reduction (or oxidation) as a function of fO_2. Calculations completed for 1300 K and 0.5 GPa.*

ASPECTS OF MANTLE CHEMISTRY

The chemical constitution of the upper mantle of the Earth and the persistence of this composition into the lower mantle continues to be actively debated. For example, the implications of a superchondritic Mg/Si ratio in the upper mantle (~1.0, e.g., *Palme and Nickel,* 1985) have been discussed at length (e.g., *Ringwood,* 1966, 1979, 1990; *Liu,* 1979, 1982a; *Agee and Walker,* 1988; *Herzberg and Feigenson,* 1988). Mechanisms such as preferential volatilization of Si during accretion, solution of Si in the core, preferential partitioning of Si in a perovskite-structured lower mantle phase during high-pressure solid-melt equilibria, selective olivine flotation in an internal magma ocean, and relative enrichment of Si in CI chondrites have all been proposed.

While a number of refractory lithophile elements appear to be present in chondritic relative proportions in the upper mantle, Ca/Al appears to be higher (~1.2) than chondritic (*Palme and Nickel,* 1985). Either addition of clinopyroxene or more probably removal of garnet on a global scale from the upper mantle has been proposed to account for this feature (*Palme and Nickel,* 1985). More recently, *Herzberg and Feigenson* (1988) have suggested near-solidus fractionation of majorite plus modified spinel is the critical process. However, it seems difficult to reconcile the absence of strong fractionations of Sc, Zr, and Hf vs. the rare earth elements with the former presence of an internal magma ocean that experienced high pressure phase (garnet + perovskite-structures) separation (*Ringwood,* 1990).

Ringwood (1979) has argued that the physical properties of the lower mantle do not require any change in chemical composition compared with the upper mantle. On the other hand, a number of workers have advocated a more SiO_2-rich (e.g., *Liu,* 1979, 1982b), FeO-poor and SiO_2-rich (*Watt and Ahrens,* 1982), and FeO-rich (*Anderson,* 1988) composition. Resolution of this question relates to the behavior of the transition zone as a somewhat leaky chemical boundary layer (*Silver et al.,* 1988), and, ultimately, determining the extent of core-mantle chemical exchange. The suggestion of *Anderson* (1988) that the Earth is characterized by the recently revised solar abundances of higher Fe, Ca, and Ti creates a major problem in accounting for differences in these respects between the Earth and the meteorite population.

Other evidence points to the existence of distinct, long-lived chemical reservoirs in the mantle that experienced early collective loss of continental crust-forming elements (*Hofmann,* 1988) but variable early degassing (*Staudacher and Allègre,* 1985).

The point of this discussion is that we are faced with an unknown degree of chemical variation through the mantle at present, and considerable evidence for a substantial amount of chemical differentiation through time. Nevertheless, it is still a worthwhile exercise to examine the distribution of siderophile elements in the accessible upper mantle for any indication of metal-silicate equilibration. It is possible, after all, that convective regimes in the mantle have not always been of the same form, and that whole-mantle convection was operative early in the Earth's history. Furthermore, if metal-silicate equilibration occurred predominantly at low pressures and stratification of the mantle was produced early, then the upper mantle may be expected to show signs of such processes.

The abundance patterns of siderophile elements in the upper mantle and the extent to which the relative fractionations reflect equilibrium with a metallic phase have been addressed in numerous studies. A recent assessment of these abundance is shown in Fig. 6. The critical observation of *Ringwood* (1966) that the siderophiles, Ni and Co, while depleted relative to chondritic abundances (~×0.2), are nevertheless overabundant compared with the type of metal-silicate partitioning described above, has prompted numerous subsequent attempts to find an acceptable mechanism for their partial depletion but chondritic relative abundance (Fig. 6). It has subsequently become apparent that the most siderophilic noble metals (Os, Ir, Ru, Rh, Pt, Pd, and Au) (PGEs) while 10× more depleted than Ni and Co, are also present in chondritic relative proportions (*Chou,* 1978; *Jagoutz et al.,* 1979; *Morgan et al.,* 1980, 1981). Explanations for these patterns include a change in distribution coefficient (silicate-metal) as a function of pressure (cf. *Brett,* 1971; *Ringwood,* 1971), a late but well mixed veneer of chondritic materials to account for the PGEs (*Chou,* 1978), partial retention of chalcophilic siderophiles including Ni, Co, W and Mo in the upper mantle during core separation (*Arculus and Delano,* 1981), selective oxidation of late accreting debris and removal of only the most noble metals (*Morgan et al.,* 1981; *Wänke,* 1981), equilibrium with an Fe-S-O metallic phase (*Brett,* 1984), and partial retention of metal in the upper mantle (*Jones and Drake,* 1986).

Wänke and Dreibus (1984) and *Ringwood* (1984) have also emphasized the stronger depletions of Cr, Mn, and V in the Earth's upper mantle (and the Moon) compared with

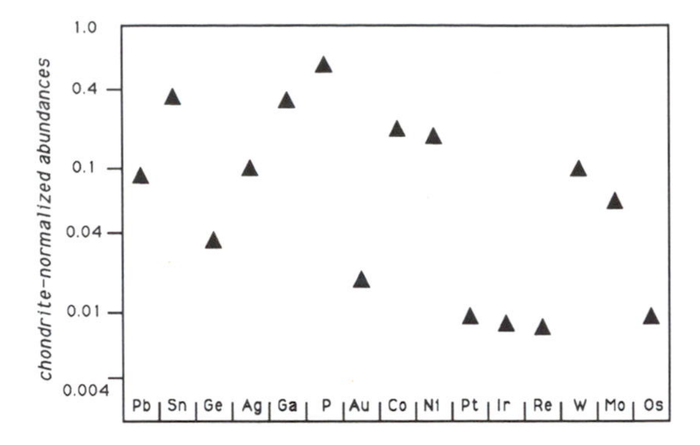

Fig. 6. *Abundances of representative siderophile elements in the Earth's upper mantle normalized to CI chondrites. Refractoriness generally increases from left to right. Data from Chou (1978), Jagoutz et al. (1979), Morgan et al. (1980), and Jones and Drake (1986).*

the siderophile-depleted EPB, and argue for an increased siderophile tendency at very high pressures for these elements. Whether it is possible to account for the lower Mg# of the lunar mantle compared with the Earth's upper mantle while simultaneously invoking a high-pressure (=Earth deep interior) metal-silicate partitioning event for the lunar mantle remains to be documented. In other words, the mechanism for reducing the Mn, Cr, and V is supposedly the same as that which caused FeO loss to the Earth's core, and yet the lunar mantle is richer in FeO than the Earth's. A valiant attempt by *Jones and Drake* (1986) to fit the abundances of siderophile elements of varying distribution coefficient with a model of metal separation concurrent with partial melting of the silicates, showed that "increasing the number of elements to be modeled has led to progressively poorer results," possibly implying that the present abundances are *not* controlled by *low-pressure,* liquid (metal and silicate) vs. solid (metal and silicate) partitioning.

The appropriate conclusions to be drawn from all this are that the present abundance pattern of siderophile elements in the upper mantle may well have resulted from several processes operating throughout the growth of the Earth, and it is time to examine more complex sequences of events. These need to be based on concurrent progress in physical models of the early Earth, and some aspects of current activity in this area are reviewed next.

METAL PRODUCTION AND DESTRUCTION

As outlined in the introduction, metal production was widespread in the solar accretionary disk, and some degree of metal-silicate fractionation is apparent between different chondrite groups on a small scale and the terrestrial planets on a larger scale (*Larimer and Wasson,* 1988). It can be assumed that the local H_2O/H_2 ratio was the most important variable with respect to metal production and stability.

There is also considerable evidence for reduction of FeO by C within chondritic and achondritic bodies (*Wasson et al.,* 1976; *Brett and Sato,* 1984; *Berkley et al.,* 1980; *Goodrich and Berkley,* 1986; *Rubin et al.,* 1988). Carbon is an effective reducing agent of Fe oxides at low pressures and high temperatures (Fig. 7) as everyday industrial experience can tell. In addition, the extreme sensitivity of the C-CO-CO_2-O_2 buffer to ambient pressure means that graphite can become an important reductant during magma ascent (*Sato,* 1978, 1979). A correlation between the redox state of a given silicate-metal assemblage and pressure as fixed by the graphite-CO-CO_2 surface is predicted (*Brett and Sato,* 1984).

Over a large depth range in the mantle however, ΔG_r^os for equilibria such as

$$2FeO + C = 2Fe + CO_2 \qquad (12)$$

are large and positive, indicating the stability of Fe^{2+}-bearing silicates in equilibrium with graphite or diamond, as commonly observed.

Extrapolation of thermodynamic data to very high pressures has indicated that a change from endothermic to exothermic conditions for equilibria such as (12) occur, at pressures above about 20 GPa (*Kuskov and Khitarov,* 1978; *Kuskov,* 1981; *Sato,* 1984). Thus, two zones of a large, growing terrestrial planet can be envisaged where production of Fe metal is possible.

Under subsolidus conditions at pressures >1.5 GPa, CO_2 reacts with silicates to form partially carbonated assemblages (*Wyllie,* 1978; *Eggler,* 1978)

$$Mg_2SiO_4 + CO_2 = MgCO_3 + MgSiO_3 \qquad (13)$$

so that the important equilibria at high pressures are of the general type

$$3Fe_2SiO_4 + C = 2Fe + FeCO_3 + 3FeSiO_3 \qquad (14)$$

Whether the Earth accreted sufficient C to render these types of metal production reactions volumetrically significant is unclear. However, if the assumption is made that the initial stages of accretion were the coolest

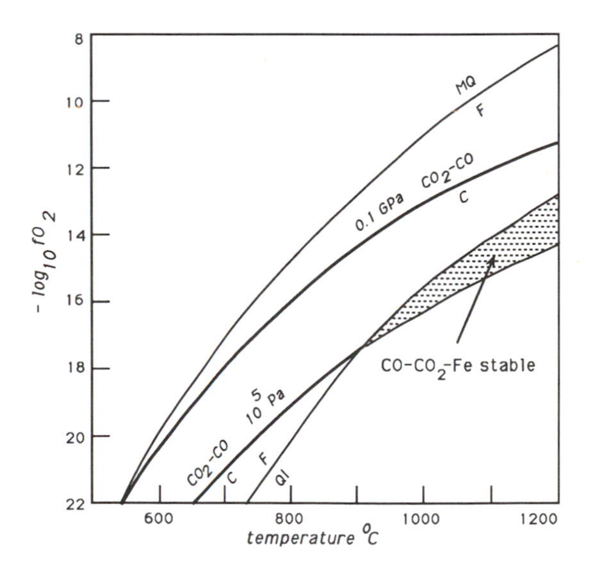

Fig. 7. *Oxygen fugacities (in bars) as functions of temperature for representative solid buffers (QIF is SiO_2 + 2Fe + O_2 = Fe_2SiO_4; FMQ is $3Fe_2SiO_4 + O_2 = 2Fe_3O_4 + 3SiO_2$) and the carbon-$CO_2$-CO surface at 10^5Pa (1 bar) and 0.1 GPa (1 kbar). Only at high temperatures and low pressures (dotted region) is the reduction of FeO (in silicates) by C possible.*

allowing potential volatiles (H_2O, S, C, ^3He) to be trapped, then subsequent growth and pressure increase could result in important reduction of silicates by C. The proportion of reduced Fe initially accreted to final complement in the core would then provide a measure of indigeneously accreted C, if this were the only reductant available. However, the metallization of FeO and solution in Fe at high pressures (*Knittle and Jeanloz,* 1986; *McCammon et al.,* 1983; *Kato and Ringwood,* 1989) is another possible mechanism for independent "metal" production.

Destruction of metal in an impact-induced, thermally blanketing atmosphere has also been suggested by *Abe and Matsui* (1985, 1986). The extent of oxidation of incoming Fe is a function of H_2O/H_2 and T [see reaction (2)], and the size range of incoming or impact-produced Fe particles. *Abe and Matsui* (1986) argue that with a H_2O content of incoming planetesimals >0.1 wt.%, and an accretion time of $\sim 5 \times 10^7$a, a steam atmosphere and an external magma ocean is produced in which both Fe and H_2O may react. Given a mass excess of Fe with respect to H_2O, the fO_2 of the magma ocean would be low and controlled by equilibria such as that represented by the synthetic, relatively low pressure (>0.3 GPa) ferrosilite-iron-fayalite buffer.

Ahrens (1988, 1990) has also suggested that the critical region of metal production/destruction is at the Earth's surface during accretionary growth, possibly with a secular change in redox conditions. The critical questions are (1) Was the accretion rate sufficient to form a steam atmosphere and consequently a surface magma ocean? (2) What were the proportions of C-H-O in the atmosphere-magma ocean system? and (3) Did the oxidation rate of incoming Fe by H_2O exceed the low-pressure reduction rate of FeO by C?

Answers to some of these questions may have to be partly sought in terms of atmospheric-early gas loss models (e.g., *Pepin,* 1988) and impact erosion of early atmospheres (*Walker,* 1986). However, the postulated magma ocean may be a critical feature for core development in another manner. The physical separation of core-forming metal from silicate has tacitly been assumed in the past to occur once a sufficient volume of (liquid) metal had accumulated (e.g., *Elsasser,* 1963; *Ringwood,* 1966; *Stevenson,* 1985). In fact, a variety of Stokes Law calculations have shown that metal agglomerations have to exceed >100 m in radius for realistically short descent times given the viscosity of mantle silicates (*Taylor,* 1989). Analytical interest in the separation of basaltic melts from peridotite residues through flow in porous media has recently been marked (*Waff and Bulau,* 1979; *McKenzie,* 1984; *Scott and Stevenson,* 1986). The properties of the matrix and the degree of surface wetting (Fig. 8) have been emphasized. It appears that the high surface tension (π) of pure liquid Fe (1.8 Nm^{-1}, *Iida and Guthrie,* 1988) compared with those of solid/liquid silicates ($\sim$0.3 Nm^{-1}) results in coalescence of liquid Fe droplets at multiple grain intersections, with dihedral angles (ϕ) >60° (Fig. 8). Experimental evidence for the distribution of FeS in meteoritic matrices has shown the type of predicted isolation and dissemination (*Walker and Agee,* 1988).

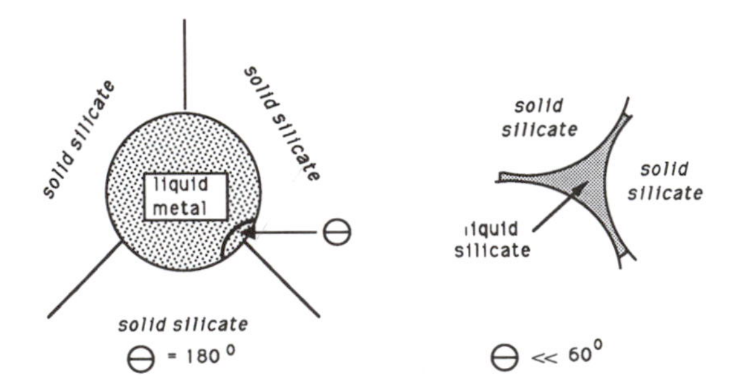

Fig. 8. *Schematic representation of the variation of the dihedral angle (ϕ) for liquids of high surface tension (e.g., metal at low pressure) and relatively low surface tension (e.g., silicate) in contact with solid silicates. Only for $\phi < 60°$ is melt connectivity established in a porous matrix.*

Industrial experience in iron and steelmaking shows that separation of liquid metals from silicates takes place in time periods of $<10^3$ sec over distances of ~ 10 m when the silicate is also molten. *Stevenson* (1988) and *Taylor* (1989) have accordingly argued that the critical stage in core separation is one in which sufficiently large agglomerations of liquid metal can accumulate to overcome the viscous resistance of solid silicates, and that these agglomerations can only realistically be produced in the presence of liquid silicate. Hence the possible significance of an early external magma ocean, or the role of a putative interior magma ocean in the Earth.

A difficulty arises when separations of comparatively small fractions of metal appear to be required, such as for the EPB or the Moon (*Newsom,* 1984; *Ringwood,* 1986; *Wänke and Dreibus,* 1986). Is it possible that core formation can only take place with the lubricating help of a liquid silicate? The presence of small amounts of silicate melt in the IAB parent body was invoked on these grounds by *Kracher* (1985), and the presence of a lunar magma ocean may also have been critical for metal separation.

The effect of other solutes on the surface tension of liquid Fe are shown in Fig. 9. In order of increasing effect, C-P-N-S-O can reduce the surface tension by a factor of 2 in the case of S at ~ 1 wt.% addition to ~ 0.6 Nm^{-1}. However, experimental evidence cited above suggests that this is still insufficient at low pressures to wet the surfaces of solid silicates. Further reduction of surface tension to the range 0.2-0.6 Nm^{-1} is required. For higher pressures, it can be shown from the differential of the Gibbs Free Energy for a phase

$$dG = VdP + SdT + \pi dA + \Sigma \mu_i dn_i \qquad (15)$$

where A is the surface area and other symbols have their usual meaning, that

$$(\partial \pi / \partial P)_{T,A,n_i} = (\partial V / \partial A)_{T,P,n_i} \qquad (16)$$

but this merely states the reduction in G is favored by coalescence of isolated droplets into larger ones, and a metastable dissemination is likely to persist in the absence of some form of silicate grain boundary migration.

It is clear however, that increased solubility of O in liquid Fe in the range 6-15 GPa results in a further significant lowering of surface tension of the metal phase, such that ϕ is reduced below 60°, and complete wetting of oxide and probably silicate grain boundaries occurs (*Urakawa et al.,* 1987). Thus, at sufficiently high pressures, the larger planets can experience liquid metal-solid silicate *equilibrium* and *separation.* Herein lies a possible resolution to the conflict between the need to invoke the agglomeration of large pods of liquid metal close to the growing Earth's surface, and disequilibrium descent to a core vs. the possible high-presssure equilibrium between liquid metal and solid silicate recognized in terms of the Cr, Mn, and V

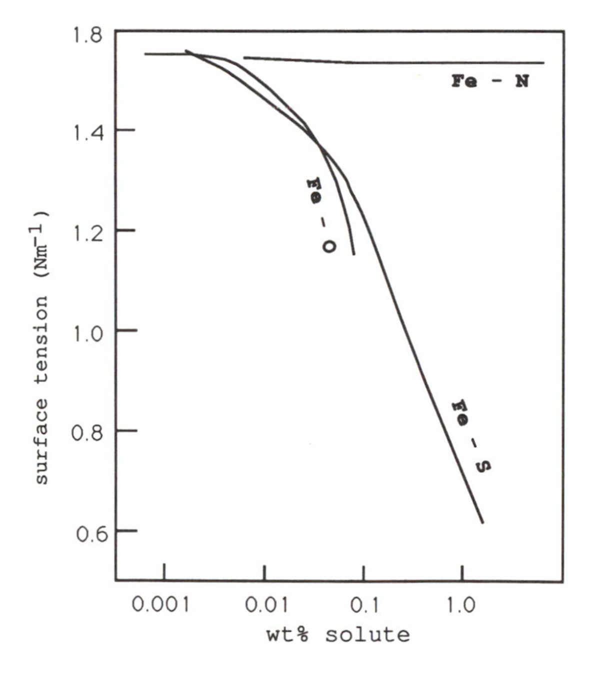

Fig. 9. *The effect on the surface tension of pure liquid Fe of additions of O, S, and N at liquidus temperatures (modified after Iida and Guthrie, 1988).*

depletion of the upper mantle, and the dissolution of O in the metal phase (see *Ringwood,* 1990).

In terms of other potential mechanisms of metal destruction, it is appropriate to recall the possible oxidation of metallic Fe in the upper mantle through reaction with CO_2 (or carbonate). Such a mechanism clearly depends on the rate of metal descent and the abundance of oxidized carbon. Selective oxidation would not be expected to result in chondritic relative abundances of Ni and Co (cf. *Morgan et al.,* 1981; *Wänke,* 1981) unless coincidentally the Ni/Co ratio of the upper mantle had been previously set superchondritic by olivine flotation. Subsequent selective oxidation of metal (with a greater proportion of Co than Ni being oxidized (see Figs. 4 and 5) might fortuitously reestablish a chondritic Ni/Co ratio (cf. *O'Neill and Wall,* 1987).

If the current predictions for the fate of a core of a major impacting body are correct, and a "rain" of metal droplets from this core were deposited on the surface of the Earth subsequent to the ejection of Moon-bound silicate (*Cameron and Benz,* 1988), selective oxidation of the metal may have been possible. But it is difficult at this point to quantify such a process.

In terms of current proximity to metal saturation of the upper mantle, there is some evidence that the sources of midocean ridge basalts are between the magnetite-wüstite and iron-wüstite buffers (*Christie et al.,* 1986; *Bryndzia et al.,* 1989). However, other regions of the upper mantle are clearly more oxidized (*Haggerty and Tompkins,* 1983; *Arculus,* 1985; *Ulmer et al.,* 1987) and in fact on a global scale, the fO_2 varies from near (Ni-rich) metal saturation to carbon-carbonate equilibria (*Eggler,* 1983; *Arculus and Delano,* 1987; *O'Neill and Wall,* 1987).

UPPER MANTLE Mg#—COINCIDENCE?

Earlier we addressed the possibility that the Mg# of the upper mantle might be a diagnostic parameter in terms of origin, by way of coincidence with a chondritic silicate-oxide input or reflecting a core-forming process analogous to a differentiated (achondritic)-metal assemblage (*Arculus and Powell,* 1988). However, it appears that despite the worldwide homogeneity of this value (*Palme and Nickel,* 1985), a number of perturbations to some initial more diagnostic condition are possible, and a unique inversion to an original state is not practical at this stage.

There are a few aspects of this ratio that are worth brief mention. The first is that Mg# = 0.9 corresponds to a maximum (negative) deviation of molar volume on the olivine join forsterite-fayalite (*Newton and Wood,* 1980), but any significant energy well deriving from this property would require concurrent reduction-mass transport through a large pressure range (>1 GPa). Second, Fo_{90} corresponds approximately to the composition in equilibrium over a considerable temperature range of the orthorhombic olivine structure with the β- and spinel-structures (e.g., *Bina and Wood,* 1987). Nevertheless, a plausible mechanism for solid state density stratification during the growth of the Earth that could effectively partition Mg into the shallower (lower density) uppermost mantle is lacking (cf. *Liu,* 1982). And finally, if *Anderson* (1988) is correct in proposing the lower mantle has a lower Mg#, then the more immediate task is to account for radial heterogeneity within the Earth rather than supposing the Mg# of the upper mantle necessarily records any information as to accretionary origins.

SUMMARY AND FUTURE DIRECTIONS

The processes of metal formation and segregation during the early growth of the Earth are by no means fully understood. There is ample meteoritical evidence for variations in redox conditions in the solar accretion disk, and in proportions of metal:silicate, widespread melting and coalescence of large volumes of metal in the probable absence of persistent, thermally blanketing atmospheres and large gravitational fields. In addition, it is clear that silicate-metal reactions in these systems reached heterogeneous equilibrium in terms of stable partitioning of various siderophile elements.

Apart from melting and coalescence of accreted metal, it is unclear what further redox equilibria were important for metal production or destruction in the Earth. At least two major depth ranges could have been important for further metal formation by C reduction (near surface and at $P > 20$ GPa), but this must have been dependent to some extent on competition with Fe + H_2O (and solution of H in Fe) reactions, particularly within any large, hot, early atmosphere (e.g., *Ringwood,* 1979; *Abe and Matsui,* 1985).

Metallization of FeO at very high pressures and solution of FeO in Fe has been experimentally demonstrated (*Knittle and Jeanloz,* 1986; *Ohtani and Ringwood,* 1984; *Kato and Ringwood,* 1989); the implications are that the signatures of high-pressure silicate-metal equilibria might be recognizable in the present upper mantle, that ready separation of liquid metal from solid silicate occurred, and that O is the major light element in the core (see *Ringwood,* 1979 for review).

A major question concerns the mechanics of core separation at low pressures (e.g., *Stevenson,* 1988; *Cooperman and Kaula,* 1985). It appears that the high interfacial energies of metal-silicate contacts prevents the gravitational separation of metal unless sufficiently large volumes are accumulated to overcome the viscous resistance of solid silicates and oxides. Accordingly, there is considerable interest in the possibility that initial accumulation of metal took place in the presence of silicate magma. The range of pressure over which reduction occurs in the surface tension of liquid metal concomittant with increased solution of light elements in the metal, sufficient to wet solid silicate interfaces, is not fully explored, but probably exceeds the central pressures attained in the Moon and EPB.

Evidence for widespread melting in the achondritic meteorite-iron populations means that efficient metal segregation under such conditions did occur (*Hewins and Newsom,* 1988). There is, however, considerable debate as to the existence and location of widespread melting in the early Earth. The lack of trace lithophile element fractionation in the present upper mantle has been argued by *Ringwood* (1988) to preclude extensive internal melting and high-pressure phase separation, despite the likelihood of massive ($T \gg$ solidus throughout much of the mantle) energy input during a giant Moon-forming impact (*Cameron and Benz,* 1988). Nevertheless, significant major lithophile element fractionation is preserved in the present upper mantle (high Mg/Si and Ca/Al) that may be accounted for by olivine-spinel structured silicate-majorite fractionation (*Agee and Walker,* 1988; *Herzberg and Feigenson,* 1988).

If a relatively shallow, albeit transient development of a magma ocean or "Great Lakes" occurred (cf. *Warren,* 1985b) on the growing Earth (*Abe and Matsui,* 1986), then separation and initial agglomeration of metal is probable. A complication is the variable oxidation of siderophiles by any CO_2 or carbonate at pressures greater than a few MPa. And there is clearly no direct evidence from the distribution of siderophiles in the present-day upper mantle for the operation of such processes.

In the event of downward motion of sufficiently large volumes of accumulated metal, an additional large energy release is inevitable (e.g., *Davies,* 1984), and whether the Earth escaped massive internal melting from the operation of this process alone is of long-standing interest. Even in the absence of silicate melting, it is probable that the deeper portions of the Earth's mantle acted as a porous medium for metal percolation, and extensive high-pressure equilibration between silicate and metal occurred.

It is clear that further progress in understanding these events will be aided immeasurably by high-pressure experimentation on melt-solid equilibria and trace-element partitioning, and the effects of these conditions on metal-silicate interfacial energies. Core formation is inextricably linked with other major fractionation events in the early Earth.

Acknowledgments. Research on metal-silicate equilibria has been supported at the University of Michigan by NSF grant EAR 8417131. J. Walker has been a constant source of stimulating discussion with respect to the conditions prevailing in the early Earth, and we owe a great debt to A. E. Ringwood and S. R. Taylor for the advice and intellectual challenges offered during sojourns in Canberra. Thanks are also extended to R. Brett and J. Delano for their constructive reviews, and to H. Newsom for his forbearance.

REFERENCES

Abe Y. and Matsui T. (1985) The formation of an impact-generated H_2O atmosphere and its implications for the early thermal history of the Earth. *Proc. Lunar Planet Sci. Conf 15th,* in *J. Geophys. Res., 90,* C545-C559.

Abe Y. and Matsui T. (1986) Early evolution of the Earth: Accretion, atmosphere formation, and thermal history. *Proc. Lunar Planet. Sci. Conf. 17th,* in *J. Geophys. Res., 91,* E291-E302.

Agee C. B. and Walker D. (1988) Static compression and olivine flotation in ultrabasic silicate liquid. *J. Geophys. Res., 93,* 3437-3449.

Ahrens T. J. (1988) Composition of the Earth's core reflects conditions at the surface during accretion (abstract). In *Papers Presented to the Conference on Origin of the Earth,* pp. 2-3. Lunar and Planetary Institute, Houston.

Ahrens T. J. (1990) Earth accretion. In *Origin of the Earth,* this volume.

Anderson D. L. (1988) Bulk chemistry and compositional stratification of the Earth (abstract). In *Papers Presented to the Conference on Origin of the Earth,* pp. 4-5. Lunar and Planetary Institute, Houston.

Arculus R. J. (1985) Oxidation status of the mantle: Past and present. *Annu. Rev. Earth Planet. Sci., 13,* 75-95.

Arculus R. J. and Delano J. W. (1981) Siderophile element abundances in the upper mantle: Evidence for a sulfide signature and equilibrium with the core. *Geochim. Cosmochim. Acta, 45,* 1331-1343.

Arculus R. J. and Delano J. W. (1987) Oxidation status of the upper mantle: Present conditions, evolution, and controls. In *Mantle Xenoliths* (P. H. Nixon, ed.), pp. 589-598. Wiley, Chichester.

Arculus R. J. and Powell R. (1988) Development of $Mg/(Mg+Fe)(Mg^*) = 0.9$ in the Earth's mantle (abstract). In *Papers Presented to the Conference on Origin of the Earth,* p. 6. Lunar and Planetary Institute, Houston.

Basaltic Volcanism Study Project (1981) *Basaltic Volcanism on the Terrestrial Planets.* Pergamon, New York. 1286 pp.

Beckett J. R. and Stolper E. (1987) Constraints on the origin of eucritic melts: An experimental study (abstract). In *Lunar and Planetary Science XVIII,* pp. 54-55. Lunar and Planetary Institute, Houston.

Berkley J. L., Taylor G. J., Keil K., Harlow G. E., and Prinz M. (1980). The nature and origin of ureilites. *Geochim. Cosmochim. Acta, 40,* 1429-1437.

Bild R. W. (1977) Silicate inclusions in group IAB irons and a relation to the anomalous stones Winona and Mt. Morris (Wis). *Geochim. Cosmochim. Acta, 41,* 1439-1456.

Bina C. R. and Wood B. J. (1987) Olivine-spinel transitions: Experimental and thermodynamic constraints and implications for the nature of the 400-km seismic discontinuity. *J. Geophys. Res., 92,* 4853-4866.

Bowen N. L. and Schairer J. F. (1935) The system $MgO\text{-}FeO\text{-}SiO_2$. *Am. J. Sci., 29,* 151-217.

Brett R. (1971) The Earth's core: Speculations on its chemical equilibrium with the mantle. *Geochim. Cosmochim. Acta, 35,* 203-221.

Brett R. (1984) Chemical equilibration of the Earth's core and upper mantle. *Geochim. Cosmochim. Acta, 48,* 1183-1188.

Brett R. and Sato M. (1984) Intrinsic oxygen fugacity measurements on seven chondrites, a pallasite, and a tektite and the redox state of meteorite parent bodies. *Geochim. Cosmochim. Acta, 48,* 111-120.

Bryndzia L. T., Wood B. J., and Dick H. J. B. (1989) The oxidation state of the Earth's sub-oceanic mantle from oxygen thermobarometry of abyssalt spinel peridotites. *Nature, 341,* 526-527.

Buseck P. R. (1977) Pallasite meteorites-mineralogy, petrology and geochemistry. *Geochim. Cosmochim. Acta, 41,* 711-740.

Cameron A. G. W. and Benz W. (1988) Effects of the giant impact on the Earth (abstract). In *Papers Presented to the Conference on Origin of the Earth,* pp. 11-12. Lunar and Planetary Institute, Houston.

Chou C.-L. (1978) Fractionation of siderophile elements in the earth's upper mantle. *Proc. Lunar Planet. Sci. Conf. 9th,* pp. 219-230.

Christie D. M., Carmichael I. S. E., and Langmuir C. H. (1986) Oxidation states of mid-ocean ridge basalt glass. *Earth Planet. Sci. Lett., 79,* 397-411.

Clayton R. N., Mayeda T. K., Prinz M., Nehru C. E., and Delaney J. S. (1986) Oxygen isotope confirmation of a genetic association between achondrites and IIIAB iron meteorites (abstract). In *Lunar and Planetary Science XVII,* p. 141. Lunar and Planetary Institute, Houston.

Clayton R. N., Onuma N., and Mayeda T. K. (1976) A classification of meteorites based on oxygen isotopes. *Earth Planet. Sci. Lett., 30,* 10-18.

Consolmagno G. J. and Drake M. J. (1977) Composition and evolution of the eucrite parent body: Evidence from rare earth elements. *Geochim. Cosmochim. Acta, 41,* 1271-1282.

Cooperman S. A. and Kaula W. M. (1985) Was core formation violent enough to homogenize the early mantle? (abstract). In *Workshop on the Early Earth: The Interval from Accretion to the Older Archean* (K. Burke and L. D. Ashwal, eds.), pp. 17-19. LPI Tech. Rpt. 85-01, Lunar and Planetary Institute, Houston.

Davies G. F. (1984) Geophysical and isotopic constraints on mantle convection: An interim synthesis. *J. Geophys. Res., 89,* 6017-6040.

Davies G. F. (1988) Thermal histories of convective Earth models and constraints on radiogenic heat production in the Earth. *J. Geophys. Res., 85,* 2517-2530.

Delaney J. S. (1987) The basaltic achondrite planetoid (abstract). In *Lunar and Planetary Science XVII,* pp. 166-167. Lunar and Planetary Institute, Houston.

Delaney J. S., Nehru C. E., and Prinz M. (1980) Olivine clasts from mesosiderites and howardites: Clues to the nature of achondritic parent bodies. *Proc. Lunar Planet. Sci. Conf. 11th,* pp. 1073-1087.

Delano J. W. (1986) Abundances of cobalt, nickel and volatiles in the silicate portion of the Moon. In *Origin of the Moon* (W. K. Hartmann, R. J. Phillips, and G. J. Taylor, eds.), pp. 231-247. Lunar and Planetary Institute, Houston.

Drake M. J. (1979) Geochemical evolution of the eucrite parent body: Possible nature and evolution of asteroid 4 Vesta? In *Asteroids* (T. Gehrels, ed.), pp. 765-782. Univ. of Arizona, Tucson.

Dreibus G. and Wänke H. (1980) The bulk composition of the eucrite parent asteroid and its bearing on planetary evolution. *Z. Naturforsch., 35a,* 204-216.

Eggler D. H. (1978) The effect of CO_2 upon partial melting of peridotite in the system Na_2O-CaO-Al_2O_3-MgO-SiO_2-CO_2 to 35 kb, with an analysis of melting in a peridotite-H_2O-CO_2 system. *Am. J. Sci., 278,* 305-343.

Eggler D. H. (1983) Upper mantle oxidation state: Evidence from olivine-orthopyroxene-ilmenite assemblages. *Geophys. Res. Lett., 10,* 365-368.

Elsasser W. M. (1963) Early history of the Earth. In *Earth Science and Meteoritics* (J. Geiss and E. Goldberg, eds.), pp. 1-30. North-Holland, Amsterdam.

Fraser D. G. and Rammensee W. (1982) Activity measurements by Knudsen cell mass spectrometry—the system Fe-Co-Ni and implications for condensation processes in the solar nebula. *Geochim. Cosmochim. Acta, 46,* 549-556.

Goodrich C. A. and Berkley J. L. (1986) Primary magmatic carbon in ureilites: Evidence from cohenite-bearing metallic spherules. *Geochim. Cosmochim. Acta, 50,* 681-691.

Haggerty S. E. and Tompkins L. A. (1983) Redox state of Earth's upper mantle from kimberlitic ilmenites. *Nature, 303,* 295-300.

Herzberg C. T. and Feigenson M. D. (1988) Experimental constraints on the formation of the Earth's upper and lower mantle (abstract). In *Papers Presented to the Conference on Origin of the Earth,* pp. 30-31. Lunar and Planetary Institute, Houston.

Hewins R. H. and Newsom H. E. (1988) Igneous activity in the early solar system. In *Meteorites and the Early Solar System* (J. F. Kerridge and M. S. Matthews, eds.), pp. 73-101. Univ. of Arizona, Tucson.

Hewins R. H. and Ulmer G. C. (1984) Intrinsic oxygen fugacities of diogenites and mesosiderite clasts. *Geochim. Cosmochim. Acta, 48,* 1555-1560.

Hofmann A. W. (1988) Chemical differentiation of the Earth: The relationship between mantle, continental crust, and oceanic crust. *Earth Planet. Sci. Lett., 40,* 297-314.

Holmes R. D. and Arculus R. J. (1983) Metal-silicate redox reactions: Implications for core-mantle equilibrium and the oxidation state of the upper mantle. In *Conference on Planetary Volatiles* (R. O. Pepin and R. O'Connell, eds.), pp. 77-80. LPI Tech. Rpt. 83-01. Lunar and Planetary Institute, Houston.

Iida T. and Guthrie R. L. (1988) *The Physical Properties of Liquid Metals.* Clarendon, Oxford. 288 pp.

Jagoutz E., Palme H., Baddenhausen H., Blum K., Cendales M., Dreibus G., Spetel B., Lorenz V., and Wänke H. (1979) The abundances of major, minor and trace elements in the Earth's mantle as derived from primitive ultramafic nodules. *Proc. Lunar Planet Sci. Conf. 10th,* pp. 2031-2050.

Jones J. H. (1984) The composition of the mantle of the eucrite parent body and the origin of eucrites. *Geochim. Cosmochim. Acta, 48,* 641-648.

Jones J. H. and Drake M. J. (1983) Experimental investigations of trace element fractionation in iron meteorites, II: The influence of sulfur. *Geochim. Cosmochim. Acta, 47,* 1199-1209.

Jones J. H. and M. J. (1986) Geochemical constraints on core formation in the Earth. *Nature, 322,* 211-228.

Kaiser T. and Wasserburg G. J. (1983) The isotopic composition and concentration of Ag in iron meteorites and the origin of exotic silver. *Geochim. Cosmochim. Acta, 47,* 43-58.

Kallemeyn G. W. and Wasson J. T. (1985) The compositional classification of chondrites: IV. Ungrouped chondritic meteorites and clasts. *Geochim. Cosmochim. Acta, 49,* 261-270.

Kato T. and Ringwood A. E. (1989) Melting relationships in the system Fe-FeO at high pressures: Implications for the composition and formation of the Earth's core. *Phys. Chem. Minerals, 16,* 524-538.

Kelly W. R. and Wasserburg G. J. (1978) Evidence for the existence of ^{107}Pd in the early solar system. *Geophys. Res. Lett., 5,* 1079-1082.

Knittle E. and Jeanloz R. (1986) High-pressure metallization of FeO and implications for the Earth's core. *Geophys. Res. Lett., 13,* 1541-1544.

Kracher A. (1985) The evolution of partially differentiated planetesimals: Evidence from iron meteorite groups IAB and IIICD. *Proc. Lunar Planet. Sci. Conf. 16th,* in *J. Geophys. Res., 90,* C689-C698.

Kracher A. and Wasson J. T. (1982) The role of S in the evolution of the parental cores of the iron meteorites. *Geochim. Cosmochim. Acta, 46,* 2419-2426.

Kuskov O. L. (1981) The role of oxidation-reduction reactions in the Earth's early history. In *Evolution of the Earth* (R. J. O'Connell and W. S. Fyfe, eds.), pp. 196-209. AGU Geodynamic Series 5, AGU, Washington, DC.

Kuskov O. L. and Khitarov N. I. (1978) Redox conditions and thermal effects of chemical reactions in the undifferentiated Earth. *Geokhimiya, 4,* 93-117.

Larimer J. W. (1979) The condensation and fractionation of refractory lithophile elements. *Icarus, 40,* 446-456.

Larimer J. W. and Wasson J. T. (1988) Siderophile element fractionation. In *Meteorites and the Early Solar System* (J. F. Kerridge and M. S. Matthews, eds.), pp. 416-435. Univ. of Arizona, Tucson.

Liu L. (1979) Calculations of high-pressure phase transitions in the system MgO-SiO_2 and implications for mantle discontinuities. *Phys. Earth Planet. Inter., 19,* 319-330.

Liu L. (1982a) Distribution of the chemical elements in the Earth with some implications. *Geochem. J., 16,* 179-198.

Liu L. (1982b) Speculations on the composition and origin of the Earth. *Geochem. J., 16,* 287-310.

Longhi J. and Pan V. (1988) Phase equilibrium constraints on the howardite-eucrite-diogenite association. *Proc. Lunar Planet. Sci. Conf. 18th,* pp. 459-470.

Lusby D., Scott E. R., and Keil K. (1987) Ubiquitous, high-FeO silicates in enstatite chondrites. *Proc. Lunar Planet Sci. Conf. 17th,* in *J. Geophys. Res., 92,* E679-E695.

McCammon C., Ringwood A. E., and Jackson I. (1983) A model for the formation of the Earth's core. *Proc. Lunar Planet. Sci. Conf. 18th,* in *J. Geophys. Res., 88,* A501-A506.

McKenzie D. (1984) The generation and compaction of partially molten rock. *J. Petrol. 25,* 713-765.

McSween H. Y. Jr. (1989) Achondrites and igneous processes on asteroids. *Annu. Rev. Earth Planet. Sci., 17,* 119-140.

Morgan J. W. (1986) Ultramafic xenoliths: Clues to Earth's late accretionary history. *J. Geophys. Res., 91,* 12375-12387.

Morgan J. W., Wandless G. A., Petrie R. K., and Irving A. J. (1980) Earth's upper mantle: Volatile element distribution and origin of siderophile element content. *Proc. Lunar Planet. Sci. Conf. 11th,* pp. 740-742.

Morgan J. W., Wandless G. A., Petrie R. K., and Irving A. J. (1981) Composition of the Earth's upper mantle-I. Siderophile trace element abundances in ultramafic nodules. *Tectonophysics, 75,* 47-67.

Morse S. A. (1980) *Basalts and Phase Diagrams.* Springer-Verlag, New York. 493 pp.

Nagahara H. (1986) Reduction kinetics of olivine and oxygen fugacity environment. (abstract). In *Lunar and Planetary Science XVII,* pp. 595-596. Lunar and Planetary Institute, Houston.

Newsom H. E. (1984). The lunar core and the origin of the Moon. *Eos Trans. AGU, 65,* 369-370.

Newsom H. E. (1985) Molybdenum in eucrites: Evidence for a metal core in the eucrite parent body. *Proc. Lunar Planet Sci. Conf. 15th,* in *J. Geophys. Res., 90,* C613-C617.

Newsom H. E. (1989) The nickel content of the lunar core (abstract). In *Lunar and Planetary Science XX,* pp. 784-785. Lunar and Planetary Institute, Houston.

Newsom H. E. and Drake M. J. (1982) Constraints on the Moon's origin from the partitioning behavior of tungsten. *Nature, 297,* 210-212.

Newsom H. E. and Drake M. J. (1983) Experimental investigation of the partitioning of phosphorus between metal and silicate phases: Implications for the Earth, Moon and eucrite parent body. *Geochim. Cosmochim. Acta, 47,* 93-100.

Newton R. C. and Wood B. J. (1980) Volume behavior of silicate solid solutions. *Am. Mineral., 65,* 733-745.

Ohtani E. (1983) Formation of olivine textures in pallasites and thermal history of pallasites in their parent body. *Phys. Earth Planet. Inter., 32,* 182-192.

Ohtani E. and Ringwood A. E. (1984) Composition of the core, I: Solubility of oxygen in molten iron at high temperatures. *Earth Planet. Sci. Lett., 71,* 85-93.

Olsen E. and Fredriksson K. (1966) Phosphates in iron and pallasite meteorites. *Geochim. Cosmochim. Acta, 30,* 459-470.

O'Neill H. St. C. (1987a) Quartz-fayalite-iron and quartz-fayalite magnetite equilibria and the free energy of formation of fayalite (Fe_2SiO_4) and magnetite (Fe_3O_4). *Am. Mineral., 72,* 67-75.

O'Neill H. St. C. (1987b) Free energies of formation of NiO, CoO, Ni_2SiO_4 and Co_2SiO_4. *Am. Mineral., 72,* 280-291.

O'Neill H. St. C. and Wall V. J. (1987) The olivine-orthopyroxene-spinel oxygen geobarometer, the nickel precipitation curve, and the oxygen fugactiy of the Earth's upper mantle. *J. Petrol., 28,* 1169-1191.

Palme H. and Nickel K. G. (1985) Ca/Al ratio and composition of the Earth's upper mantle. *Geochim. Cosmochim. Acta, 49,* 2123-2132.

Palme H. and Rammensee W. (1981) The significance of W in planetary differentiation processes: Evidence from new data on eucrites. *Proc. Lunar Planet. Sci. 12B,* pp. 949-964.

Pepin R. O. (1988) On the loss of Earth's primordial atmosphere (abstract). In *Papers Presented to the Conference on Origin of the Earth,* pp. 60-69. Lunar and Planetary Institute, Houston.

Righter K., Arculus R. J., Delano J. W., and Paslick C. (1990) Electrochemical measurements and redox equilibria in pallasite meteorites: Implications for the eucrite parent body. *Geochim. Cosmochim. Acta,* in press.

Ringwood A. E. (1966) Chemical evolution of the terrestrial planets. *Geochim. Cosmochim. Acta, 30,* 41-104.

Ringwood A. E. (1971) Core-mantle equilibrium: Comments on a paper by R. Brett. *Geochim. Cosmochim. Acta, 35,* 223-230.

Ringwood A. E. (1979) *Origin of the Earth and Moon.* Springer-Verlag, New York. 295 pp.

Ringwood A. E. (1984) Composition and origin of the Moon. In *Origin of the Moon* (W. K. Hartmann, R. J. Phillips, and G. J. Taylor, eds.), pp. 673-698. Lunar and Planetary Institute, Houston.

Ringwood A. E. (1988) Early history of the Earth-Moon system (abstract). In *Papers Presented to the Conference on Origin of the Earth,* pp. 73-74. Lunar and Planetary Institute, Houston.

Ringwood A. E. (1990) Earliest history of the Earth-Moon system. In *Origin of the Earth,* this volume.

Rubin A. E., Fegley B., and Brett R. (1988) Oxidation state in chondrites. In *Meteorites and the Early Solar System* (J. F. Kerridge and M. S. Matthews, eds.), pp. 488-511. Univ. of Arizona, Tucson.

Sato M. (1978) Oxygen fugacity of basaltic magmas and the role of gas-forming elements. *Geophys. Res. Lett., 5,* 447-449.

Sato M. (1979) The driving mechanism of lunar pyroclastic eruptions inferred from the oxygen fugacity behavior of Apollo 17 orange glass. *Proc. Lunar Sci. Conf. 10th,* pp. 311-325.

Sato M. (1984) The oxidation state of the upper mantle: Thermochemical modeling and experimental evidence. *Proc. 27th Intl. Geol. Congr., 11,* pp. 405-433.

Scott D. R. and Stevenson D. J. (1986) Magma ascent by porous flow. *J. Geophys. Res., 91,* 9283-9286.

Scott E. R. D. (1977a) Pallasites—metal compositions, classification and relationships with iron meteorites. *Geochim. Cosmochim. Acta, 41,* 349-360.

Scott E. R. D. (1977b) Formation of olivine-metal textures in pallasite meteorites. *Geochim. Cosmochim. Acta, 41,* 693-710.

Scott E. R. D. and Wasson J. T. (1975) Classification and properties of iron meteorites. *Rev. Geophys. Space Phys., 13,* 527-546.

Seifert S. and O'Neill H. St. C. (1987) Experimental determination of activity-composition relations in Ni_2SiO_4-Mg_2SiO_4 and Co_2SiO_4-Mg_2SiO_4 olivine solid solutions at 1200 K and 0.1 MPa and 1573 K and 0.5 GPa. *Geochim. Cosmochim. Acta, 51,* 97-104.

Silver P. G., Carlson R. W., and Olson P. (1988) Deep slabs, geochemical heterogeneity, and the large-scale structure of mantle convection: Investigation of an enduring paradox. *Annu. Rev. Earth Planet. Sci., 16,* 477-541.

Smith J. V. (1982) Heterogeneous growth of meteorites and planets, especially the Earth and Moon. *J. Geol., 90,* 1-125.

Staudacher T. and Allègre C. J. (1982) Terrestrial xenology. *Earth Planet. Sci. Lett., 60,* 389-406.

Stevenson D. J. (1981) Models of the Earth's core. *Science, 214,* 611-619.

Stevenson D. J. (1985) Thermal, dynamic and compositional aspects of the core-forming Earth. In *Workshop on the Early Earth: The Interval from Accretion to the Older Archean* (K. Burke and L. D. Ashwal, eds.), pp. 76-78. LPI Tech. Rpt. 85-01, Lunar and Planetary Institute, Houston.

Stevenson D. J. (1988) Fluid dynamics of core formation (abstract). In *Papers Presented to the Conference on Origin of the Earth,* pp. 87-88. Lunar and Planetary Institute, Houston.

Stolper E. M. (1977) Experimental petrology of the eucritic meteorites. *Geochim. Cosmochim. Acta, 41,* 587-611.

Taylor G. J. (1989) Metal segregation in asteroids (abstract). In *Lunar and Planetary Science XX,* pp. 1109-1110. Lunar and Planetary Institute, Houston.

Taylor S. R. (1988a) Planetary compositions. In *Meteorites and the Early Solar System* (J. F. Kerridge and M. S. Matthews, eds.), pp. 512-534. Univ. of Arizona, Tucson.

Taylor S. R. (1988b) Geochemical implications of planetesimal accretion (abstract). In *Papers Presented to the Conference on Origin of the Earth,* pp. 91-92. Lunar and Planetary Institute, Houston.

Taylor S. R. and McLennan S. M. (1985) *The Continental Crust: Its Composition and Evolution.* Blackwell, Oxford. 312 pp.

Ulmer G. C., Grandstaff D. E., Weiss D., Moats M. A., Buntin T. J., Gold D. P., Hatton C. J., Kadik A., Koseluk R. A., and Rosenhauer M. (1987) The mantle redox state: An unfinished story? *Geol. Soc. Am. Spec. Pap., 215,* 5-23.

Urakawa S., Kato M., and Kumazawa M. (1987) Experimental study on the phase relations in the system Fe-Ni-O-S up to 15 GPa. In *High Pressure Research in Mineral Physics* (M. Manghnani and Y. Syono, eds.), pp. 95-111. AGU Monograph 39, AGU, Washington, DC.

Tilton G. R. (1988) Age of the Solar System. In *Meteorites and the Early Solar System* (J. F. Kerridge and M. S. Matthews, eds.), pp. 259-275. Univ. of Arizona, Tucson.

Waff H. S. and Bulau J. R. (1979) Equilibrium fluid disribution in an ultramafic partial melt under hydrostatic stress conditions. *J. Geophys. Res., 84,* 6109-6114.

Wahl W. (1965) The pallasite problem. *Geochim. Cosmochim. Acta, 29,* 177-181.

Walker D. and Agee C. B. (1988) Ureilite compaction. *Meteoritics, 23,* 81-91.

Walker J. C. G. (1986) Impact erosion of planetary atmospheres. *Icarus, 60,* 87-98.

Wänke H. (1981) Constitution of terrestrial planets. *Philos. Trans. R. Soc. London, A303,* 287-302.

Wänke H. and Dreibus G. (1986) Geochemical evidence for the formation of the Moon by impact-induced fission of the proto-Earth. In *Origin of the Moon* (W. K. Hartmann, R. J. Phillips, and G. J. Taylor, eds.), pp. 649-672. Lunar and Planetary Institute, Houston.

Wänke H., Dreibus G., and Jagoutz E. (1984) Mantle chemistry and accretion history of the Earth. In *Archaean Geochemistry* (A. Kröner, G. N. Hanson, and A. M. Goodwin, eds.), pp. 1-24. Springer-Verlag, New York.

Warren P. H. (1985a) Origin of howardites, diagenites and eucrites: A mass balance constraint. *Geochim. Cosmochim. Acta, 49,* 577-586.

Warren P. H. (1985b) The magma ocean concept and lunar evolution. *Annu. Rev. Earth Planet. Sci., 13,* 201-240.

Wasson J. T. (1974) *Meteorites—Classification and Properties.* Springer-Verlag, New York. 316 pp.

Wasson J. T., Chou C.-L., Bild R. W., and Baedecker P. A. (1976) Classification of an elemental fractionation among ureilites. *Geochim. Cosmochim. Acta, 40,* 1449-1458.

Watt J. P. and Ahrens T. J. (1982) The role of iron partitioning in mantle composition, evolution and scale of convection. *J. Geophys. Res., 87,* 5631-5644.

Weidenschilling S. J. (1988) Formation processes and time scales for meteorite parent bodies. In *Meteorites and the Early Solar System* (J. F. Kerridge and M. S. Matthews, eds.), pp. 368-371. Univ. of Arizona, Tucson.

Wood J. A. and Morfill G. E. (1988) A review of solar nebula models. In *Meteorites and the Early Solar System* (J. F. Kerridge and M. S. Matthews, eds.), pp. 329-347. Univ. of Arizona, Tucson.

Wyllie P. J. (1978) Mantle fluid and compositions buffered in peridotite-CO_2-H_2O by carbonates, amphibole and phlogopite. *J. Geol., 86,* 687-713.

ACCRETION AND CORE FORMATION IN THE EARTH: EVIDENCE FROM SIDEROPHILE ELEMENTS

H. E. Newsom

Department of Geology and Institute of Meteoritics, University of New Mexico, Albuquerque, NM 87131

Several theories for accretion and core formation in the Earth are evaluated based on the depletions of siderophile elements. Because no theory successfully predicts the abundances of every element, the identification of key elements that do not fit a particular theory based on current assumptions is of great importance. These key elements require extra effort in the future to determine, for example, if their estimated mantle abundances are incorrect, or if their partition coefficients are especially sensitive to pressure. The model calculations include estimates of the uncertainties in the observed depletions as well as partition coefficients. The simplest model is the equilibrium S-rich core formation model. The calculated mantle depletions are least consistent with the observed depletions for V, W, Ni, Ge, and Au. Calculated abundances for the inefficient core formation theory best match the observed depletions with the assumption that approximately 2.5 wt.% S-rich metallic liquid and 0.04 wt.% solid metal are left behind in the mantle after core formation, and the metal and sulfide are subsequently oxidized and incorporated into mantle silicates. Key discrepancies remain for this theory, with inconsistencies between the calculated and observed abundances of V, Ge, and Au. The current calculations for the heterogeneous accretion model require that segregation of solid metal ceased after approximately 90% to 95% of the Earth had accreted. The calculations for W are now consistent with the new abundance data for the depletion of W. For the heterogeneous accretion theory, discrepancies still exist between the actual and calculated abundances of Ga and Ge. The calculated depletions of As and Sb are too great for S-rich core formation and heterogeneous accretion; however, their partition coefficients are not adequately constrained. A clear choice for the most successful theory is still not evident, although future investigations of the key elements identified in this study may establish a preference.

INTRODUCTION

Because the earliest history of the Earth is not preserved in known rock samples, evidence of the accretion and early evolution of the Earth must be sought in geochemical clues that persist in spite of the later evolution of the Earth. An important class of elements in this regard are the siderophile and chalcophile elements, which are strongly partitioned under reducing conditions into Fe-metal or Fe-sulfide, respectively. Major depletion of siderophile elements relative to lithophile elements (elements having an affinity for silicates), due to partitioning into Fe-metal, occurred during the accretion of the Earth (Fig. 1). The volatile siderophile elements have an additional depletion relative to CI chondrites due to volatility (Fig. 1). At some early time, the upper mantle became too oxidizing for Fe-metal to be stable. Later events under more oxidizing conditions resulted in partitioning of the siderophile elements only among silicate and oxide phases, such that their behavior is similar to other lithophile elements. In contrast, geochemical processes involving sulfides that fractionate chalcophile elements could have operated through geologic history.

An important assumption of this work is that we can estimate the abundances of the siderophile elements in the Earth's present mantle, and in the original primitive mantle (mantle + crust), from samples available at the Earth's surface. The available samples will probably allow reasonable estimates of the composition of the upper mantle, but the lower mantle remains a problem. The lower mantle may have the same trace element composition as the upper mantle if the whole mantle has been involved in convection and mixing (e.g., *Davies,* 1990). Some evidence exists, however, suggesting that the lower mantle has a different composition from the upper mantle (*Anderson,* 1989). Even if the lower mantle is different in major element composition from the upper mantle composition inferred from mantle nodules, the results discussed below would be valid if the trace element abundances are the same in both reservoirs, although this would be unlikely.

Several explanations have been proposed to account for the unexpectedly large siderophile element abundances in the Earth's primitive mantle. The following theories will be discussed quantitatively: (1) equilibrium S-rich core formation, where the mantle is assumed to be in equilibrium with a core consisting of S-rich metallic liquid, (2) inefficient core formation, in which metal and sulfide are left behind in the mantle, and (3) heterogeneous accretion, in which multiple stages of accretion and core formation are involved. Current data seem to strengthen the heterogeneous accretion model, but the issue is not conclusively resolved.

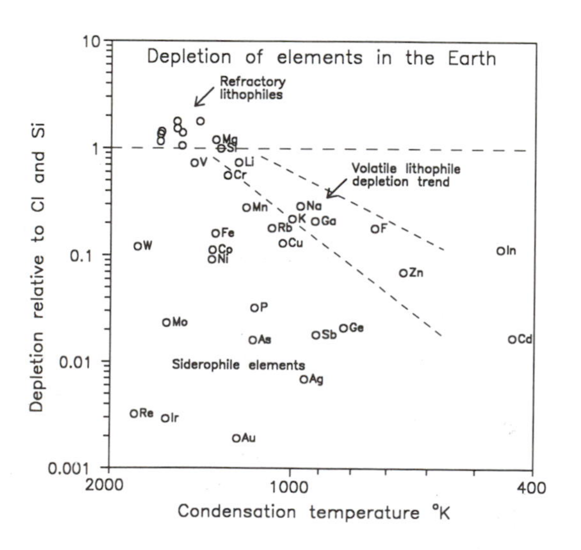

Fig. 1. *Lithophile and siderophile element depletions normalized to CI chondrites and Si (see text) in the Earth's mantle as a function of volatility. The depletions are primarily from Wänke et al. (1984) with additional data from Newsom et al. (1989) and Sims et al. (1990). The volatility of different elements is represented by their condensation temperature at a pressure of 10^{-4} atm (Wasson, 1985). The sloping dashed lines roughly outline the depletion trend for volatile lithophile elements in the Earth's mantle. Siderophile elements beneath this trend are depleted due to both their volatility and their siderophile or chalcophile nature.*

SIDEROPHILE ELEMENT DEPLETIONS IN THE EARTH'S MANTLE

We will begin by discussing the available data on the abundances and depletions of siderophile elements in the silicate portion of the Earth (also called the "primitive mantle"). An important part of this effort is the attempt to include estimates of the uncertainty in the abundances. Understanding the uncertainties is critical for comparison with model calculations for different accretion and core formation theories. Several different sources of data are available for determining primitive mantle abundances of siderophile elements. Lherzolite mantle nodules found in many basalts have proven to contain remarkably consistent abundances of compatible siderophile elements (elements retained in mantle silicates during partial melting), such as Co, Ni, and Ge (e.g., *Jagoutz et al.,* 1979). However, the abundances of incompatible elements such as W, P, Sb, and As are variable in nodules, and a different approach involving ratios of siderophile elements to incompatible lithophile elements in mantle-derived magmas has proven more satisfactory (e.g., *Rammensee and Wänke,* 1977; *Sims et al.,* 1990).

Normalizations and Volatile Corrections

The basic approach in discussing the abundances of siderophile elements in the primitive mantle is to compare the abundances with the presumed initial abundances in the material from which the solar system formed. The usual measuring stick is the composition of CI chondrites, the class of meteorites closest in composition to the sun (*Anders and Grevesse,* 1989). However, three different normalizations are commonly used in making comparisons with CI chondrites. The simplest approach is to consider the ratio of mantle abundances to CI abundances (e.g., *Jones and Drake,* 1986). However, the concentrations of refractory elements in the Earth's mantle are enriched relative to the abundances in the CI chondrites by a factor of about 2.7 (*Wänke et al.,* 1984). Therefore, with this normalization the initial abundances for refractory siderophile elements start out at a mantle/CI ratio of about 2.7.

A second normalization, to CI chondrites and Si, is also commonly used (e.g., *Wänke et al.,* 1984; Fig. 1), and is derived from an astronomical convention. This normalization is such that the abundance of Si in the mantle falls at 1.0. However, because Si is somewhat depleted relative to refractory elements in the Earth's mantle, due to volatility, this normalization results in refractory elements falling at an enrichment factor of about 1.3 (*Wänke et al.,* 1984).

The third normalization, to CI chondrites and refractory elements, is such that the initial abundances of refractory elements in the Earth fall at 1.0 (e.g., Figs. 2-4 and Fig. 6). The absolute depletions of refractory siderophile elements, presumably due to core formation, can therefore be obtained directly from the figure. An additional advantage of this normalization is that the abundances of many of the siderophile elements being considered are best known relative to the abundances of refractory elements such as Ce. Therefore, in Fig. 2, which is normalized to CI chondrites and refractory elements, the depletions of the incompatible refractory siderophile elements Mo and W are calculated by dividing the Mo/Ce ratio in the primitive mantle, for example, by the Mo/Ce ratio in the CI chondrites. For the compatible siderophile elements whose depletions are known from absolute abundances in mantle nodules (Mn, V, Cr, Co, Ni, Ge, Ag, and the highly siderophile elements; Fig. 2), the abundances are divided by the

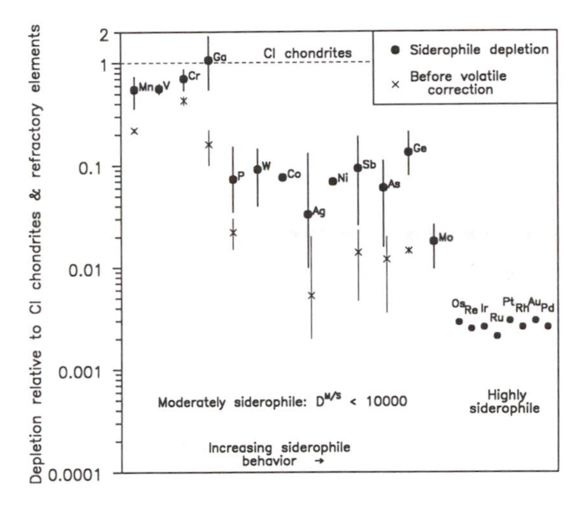

Fig. 2. *Depletion of siderophile elements in the Earth's primitive mantle (silicate portion of the Earth) normalized to mean CI chondrites (Anders and Grevesse, 1989) and refractory elements (see text). The moderately siderophile elements are arranged roughly in order of increasing siderophile behavior (with the exception of Sb and As, to better illustrate the Ge/Mo ratio), where the siderophile behavior is measured by the bulk metal/silicate partition coefficients during equilibration between olivine, silicate-melt, and metal at low degrees of partial melting. The depletions for the highly siderophile elements are from Chou et al. (1983); however, the uncertainties in the depletions are not shown (see Fig. 4 for the uncertainties on Au, Re, and Ir). For the volatile siderophile elements, Mn, Cr, Ga, P, Ag, Sb, As, and Ge, their actual mantle abundances and uncertainties are indicated by a cross. Because the depletion of these elements is due to both volatility and siderophility, their depletion has been corrected by removing the portion due to volatility (see text), with their corrected depletion indicated by a filled circle. The depletion of the refractory siderophile elements V, W, Co, Ni, and Mo, and essentially all of the highly siderophile elements, is due entirely to their siderophile nature and is also indicated by filled circles.*

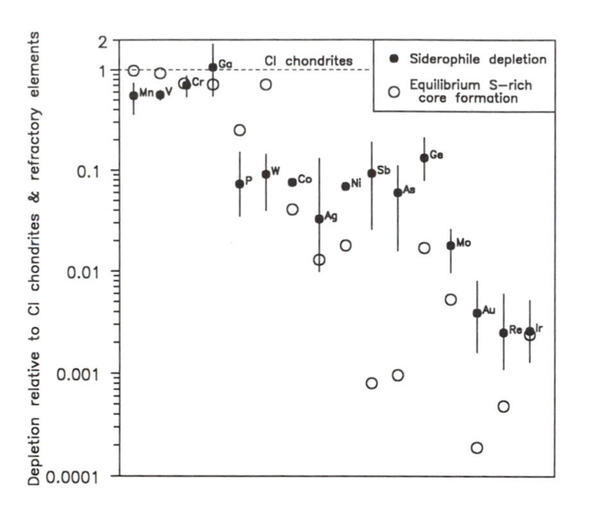

Fig. 3. *Calculated depletions are shown for the equilibrium S-rich core formation theory (Brett, 1984), compared to the observed volatile-corrected depletions (Table 1, Fig. 2). The calculated depletions (open circles) are the normalized concentrations in the mantle silicates for equilibrium between 31 wt.% S-rich metallic liquid (≈ 25 wt.% S) representing the present core material and a silicate mantle partially melted to 50%, using partition coefficients in Table 2 and assuming CI chondritic initial abundances (Anders and Grevesse, 1989).*

TABLE 1. *Depletions of siderophile elements in the Earth's primitive mantle.*

Element*		Median	High	Low	Refs.
Mn	Obs.	0.22	0.23	0.21	f
	Corr.	0.55	0.73	0.36	
V	Obs.	0.56	0.62	0.50	f
Cr	Obs.	0.43	0.47	0.43	f
	Corr.	0.70	0.86	0.54	
Ga	Obs.	0.16	0.22	0.10	b
	Corr.	1.1	1.8	0.55	
Ge	Obs.	0.015	—	—	b
	Corr.	0.13	0.21	0.080	
As	Obs.	0.012	0.020	0.0047	a
	Corr.	0.060	0.11	0.016	
Sb	Obs.	0.014	0.023	0.0047	a
	Corr.	0.093	0.19	0.026	
Ag	Obs.	0.0053	0.020	0.0020	b
	Corr.	0.033	0.13	0.010	
P	Obs.	0.022	0.030	0.015	c
	Corr.	0.073	0.15	0.035	
Co	Obs.	0.076	—	—	b
Ni	Obs.	0.069	—	—	b
W	Obs.	0.091	0.14	0.040	a
Mo	Obs.	0.018	0.026	0.0098	d
Au	Obs.	0.0030	0.0060	0.0013	e
	Corr.	0.0039	0.0080	0.0016	
Re	Obs.	0.0026	0.0060	0.0011	e
Ir	Obs.	0.0026	0.0052	0.0013	e

*Observed values (Obs.) are normalized to mean CI chondrites and refractory elements (*Anders and Grevesse,* 1989). The values listed as "Corr." are the volatile corrected depletions (see text).

References: (a) *Sims et al.* (1990); (b) *Wänke et al.* (1984); (c) *Weckwerth* (1983); (d) *Newsom and Palme* (1984); (e) *Chou et al.* (1983); (f) *Drake et al.* (1989).

concentrations in CI chondrites and by the enrichment of refractory elements in the mantle relative to CI chondrites. The estimates for the enrichment of refractory elements in the mantle range from a factor of 2.3 (for Ce, *Taylor and McLennan,* 1985) to a factor of 2.76 (for Al, *Newsom and Palme,* 1984; and for Ce, *Wänke et al.,* 1984), to a factor of 2.9 (for Al, *Palme and Nickel,* 1985). The value of 2.76 used by *Newsom and Palme* (1984) is adopted here, which is also consistent with the value for Ce from *Wänke et al.* (1984). The uncertainty in the enrichment factor based on the range of estimates adds an uncertainty of approximately 10-15% to the depletions of Co, Ni, Ge, and Ag, relative to the depletions of the other elements. This additional uncertainty is very small in terms of the absolute depletions, however.

The observed depletions of the volatile siderophile elements (Mn, Cr, Ga, P, Ag, Sb, As, and Ge) are plotted in Fig. 2. However, these elements are depleted due to both their siderophile behavior and to the general volatile element depletion in the Earth. Figure 1 shows the increasing depletion of volatile lithophile elements as a function of volatility, as measured by condensation temperatures. The observed trend allows a correction to be made to the volatile siderophile elements by subtracting the volatile depletion observed for lithophile volatile elements, which have similar condensation temperatures. The corrected siderophile depletions in Fig. 2 and Table 1 therefore represent the depletion due only to their siderophile behavior. Unfortunately, the lithophile volatile element trend is not very well defined, especially at lower condensation temperatures; this significantly increases the uncertainties of the corrected depletions. A very small volatile depletion for Au is listed in Table 1, but for clarity only the volatile corrected value is indicated in Figs. 2-4 and Fig. 6.

The reasons for large uncertainties in the depletions of a few elements (Fig. 2, Table 1) requires comment. The abundances of W, As, and Sb have been determined by *Sims et al.* (1990). These elements show good correlations with incompatible elements in midocean ridge basalts and ocean island rocks during normal igneous events, but their abundance in

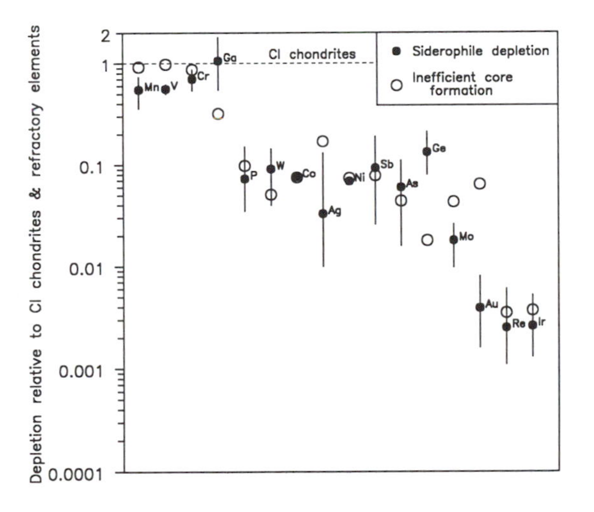

Fig. 4. *Calculated depletions for the inefficient core formation theory. The depletions for this theory (open circles) are calculated by first assuming simple equilibrium during core formation among 58.9 wt.% solid silicate, 6.6 wt.% liquid silicate, 19.2 wt.% solid metal, and 15.3 wt.% S-rich metallic liquid to establish the concentrations of siderophile elements in the different phases. The assumption is then made that 2.5 wt.% S-rich metallic liquid and 0.004 wt.% Fe-metal are left behind in the mantle to be subsequently destroyed by oxidation (Jones and Drake, 1986).*

the crust was enhanced, even relative to other incompatible elements, probably by hydrothermal processes (*Sims et al.,* 1990). A similar situation has been noted for Pb (*Newsom et al.,* 1986; *Hofmann et al.,* 1986; and *Hofmann,* 1988). The abundances for W, As, and Sb are therefore calculated using a mass balance between a crustal reservoir and a depleted mantle reservoir, introducing considerable uncertainties (*Sims et al.,* 1990). The large uncertainty for the abundance of Ag in the mantle is due to the great spread in Ag concentrations in mantle nodules (*Chou et al.,* 1983).

ACCRETION AND CORE FORMATION THEORIES

Several theories have been proposed to explain the abundances of siderophile elements in the Earth. Some of these include (1) partitioning into an FeO-rich iron metal phase at high pressures during core formation (*Ringwood,* 1979); (2) partitioning into a S-rich metal phase during core formation (*Brett,* 1984); (3) inefficient core formation, involving the retention of Fe-metal and sulfides during core formation (*Mitchell and Keays,* 1981; *Arculus and Delano,* 1981; *Jones and Drake,* 1986); and (4) heterogeneous accretion, involving multiple stages of accretion and core formation (*Chou,* 1978; *Wänke,* 1981; *Wänke et al.,* 1984). We will examine each of these in turn before discussing the comparative success of each model.

High-Pressure Partitioning of Siderophiles into Metal Alloys

The possibility that the Earth's siderophile element depletion pattern is simply due to partitioning into metal alloys (containing oxygen or silicon) at extremely high pressures (*Ringwood,* 1966a, 1979) cannot be directly tested due to the lack of metal/silicate partition coefficients at the required pressures, although a few preliminary studies are beginning to be published (*Seifert et al.,* 1988;

Ringwood, 1990). In addition, theoretical studies of core formation are divided as to whether low-pressure or high-pressure equilibration dominated. *Stevenson* (1981) and *Sasaki and Nakazawa* (1986) have suggested that core formation proceeds by segregation and accumulation of metal into relatively large blobs at shallow depths in the accreting Earth. By the time the metal blobs have sunk to high-pressure regions, their small ratio of surface area to mass and their rapid rate of sinking will prevent equilibration with surrounding silicates at high pressures. Recently, however, *Stevenson* (1990) has discussed the possibility that core formation occurs by segregation of metal at depth, from a metal-silicate emulsion during convection of a semiliquid mantle. The emulsion is the result of mixing due to large impacts during accretion.

Partitioning of Siderophiles into a Sulfur-rich Core

The possibility that partitioning into a Fe-S-O core was responsible for the siderophile element pattern in the Earth has been discussed by *Brett* (1984). The calculated depletion pattern for this theory is shown in Fig. 3, using the partition coefficients in Table 2. A best fit was found for equilibrium between the S-rich metallic liquid and a mantle 50% partially melted. In contrast to other theories, Ga is well modeled by this theory. The calculated abundances of the elements Mn, Cr, P, Co, Ag, Mo, and Re are within a factor of two of the observed abundances and can be considered successfully modeled.

Elements with well-known abundances and well-known partition coefficients that are not well modeled include V, W, Ni, Ge, and Au. The partition coefficients for Sb and As are not adequately constrained, which could be responsible for the poor fit for these elements. A late veneer could not only solve the problems with the low calculated abundances of Au and Re, but would also presumably provide chondritic relative abundances of the other highly siderophile elements at the required depletion level.

The assumption of equilibrium under more reducing conditions than used for the calculated results in Fig. 3 will provide a better match for Mn, V, P, and W, but the fits will worsen for the rest of the elements. Similarly, varying the degree of partial melting helps the fit for some elements, but worsens the fit for others.

A significant difficulty for this model is the large content of S required in the Earth's core (greater than or equal to 25 wt.%). Recent results of *Ahrens and Jeanloz* (1987) indicate that the physical parameters for the core are best matched with a S content of 11 ± 2 wt.%. As *Brett* (1984) mentions, however, reactions between the lower mantle and the core-forming metal could have modified the S content of the core.

Inefficient Core Formation

The inefficient core formation theory (*Mitchell and Keys,* 1981; *Arculus and Delano,* 1981; *Jones and Drake,* 1986) assumes that complete scavenging of metal and sulfide from the Earth's mantle during core formation would be extremely difficult to accomplish. Therefore, retention of small amounts of metal and sulfide can quantitatively explain the siderophile element abundances in the mantle. Several physical mechanisms that could lead to metal retention are discussed by *Stevenson* (1990). In a recent study of this idea, *Jones and Drake* (1986) investigated the resulting mantle abundances for different oxygen fugacity conditions of metal-silicate equilibrium, for different degrees of partial melting, and for different amounts of solid and metallic liquid left in the mantle. *Jones and Drake* (1986) found a best fit to the data as follows: They first assumed simple equilibrium among 58.9 wt.% solid silicate, 6.6 wt.% liquid silicate, 19.2 wt.% solid metal, and 15.3 wt.% S-rich metallic liquid. These amounts correspond to

TABLE 2. *Partition coefficients calculated for 1250°-1275°C and log fO_2 = -12.4.*

Element	$D^{LM/LS}$	$D^{SM/LS}$	$D^{SM/LM}$	$D^{SS/LS}$	Refs.
Mn	0.032	0.0078	0.06	1	k
V	0.13	0.0078	0.06	1	k
Cr	0.79	0.071	0.09	1.2	k
Ga	0.53	3.2	6.0	0.40	c, e
Ge	110	930	8.7	1.0	h, c
As	1,000	2,300	2.3	0.010	g, i
Sb	1,200	1,100	0.90	0.010	g, i
Ag	100	1.0	0.010	0.40	b
P	2.9	5.0	1.7	0.020	a
Co	89	210	2.3	3.0	b
Ni	1,700	2,200	1.3	10	h
W	0.40	8.0	20	0.010	j
Mo	180	450	2.5	0.010	h, b
Re	2,000	170,000	83	0.010	b, d
Ir	20,000	1,700,000	83	50	b
Au	10,000	13,000	1.3	1.0	c, f

*Partition coefficient abbreviations are LM = S-rich metallic liquid, LS = silicate liquid, SM = solid Fe-metal, and SS = solid (crystalline) silicate.

References: (a) *Newsom and Drake* (1983); (b) *Jones and Drake* (1986); (c) *Jones and Drake* (1983); (d) *Walker et al.* (1988); (e) *Drake et al.* (1984); (f) *Rammensee* (1978); (g) Sims et al., unpublished data; (h) *Schmitt et al.* (1989); (i) *Klöck and Palme* (1988); (j) *Newsom and Drake* (1982); (k) *Drake et al.* (1989).

10% partial melting of the mantle, a metal core (with 10 wt.% S) comprising 32 wt.% of the Earth (*Basaltic Volcanism Study Project,* 1981), and a very small amount of solid metal (0.04 wt.%) and S-rich metallic liquid (2.5 wt.%) to be left behind in the mantle. The temperature was fixed at about 1250°-1275°C, which is the temperature required for the presence of S-rich metallic liquid. The best fit was found for partition coefficients appropriate for the conditions at log fO_2 = -12.35. Thus the abundance of siderophile elements in the mantle is the equilibrium concentration in the silicate portion, with the addition of the siderophiles contained in the 2.5 wt.% S-rich metallic liquid (25 wt.% S), and 0.04 wt.% solid metal left behind and oxidized into the mantle. This calculation has been repeated, with the latest depletion data (Table 1) and partition coefficient data (Table 2), with the results illustrated in Fig. 4, confirming the results of *Jones and Drake* (1986).

The calculated depletions agree very well for many elements, with the exceptions of V, Ge, and Au. The calculated depletion of V is less than a factor of two from the observed abundance, but the partition coefficients and the depletion of V are well known (*Drake et al.,* 1989). The calculated depletion of Ge is a factor of 4 lower than the observed depletion of Ge (Fig. 4). The abundance of Ge is very constant in mantle nodules (*Jagoutz et al.,* 1979) and the relevant partition coefficients have been well determined (solid-metal/silicate-liquid, *Schmitt et al.,* 1989; solid-metal/liquid-metal, *Jones and Drake,* 1983; solid-silicate/silicate-liquid, *Malvin and Drake,* 1987). The calculated abundance of Au is probably at least a factor of 10 too high (Fig. 4), in spite of the significant scatter in the abundance of Au in mantle-derived magmas and mantle nodules (*Chou et al.,* 1983). The partition coefficients for Au, however, are not extremely well known (Table 2). Unlike Re and Ir, the partition coefficient between solid-metal and metallic-liquid for Au is near unity (*Jones and Drake,* 1983), such that the metallic-liquid contains a significant amount of the Au remaining after core formation.

The inefficient core formation theory also allows a constraint to be placed on the amount of S in the mantle, contributed by the S-rich metallic liquid. The calculation illustrated in Fig. 4 results in a mantle abundance of approximately 6000 ppm S (*Jones and Drake,* 1986), which is quite high, even though the actual S abundance in the mantle is highly uncertain (*Sun,* 1982). Estimates range from approximately 10 ppm S, based on mantle nodules (*Wänke et al.,* 1984), up to approximately 1000 ppm S, based on Archean komatiites (*Sun,* 1982).

Heterogeneous Accretion

The siderophile elements can be divided into two groups: the moderately siderophile elements with metal/silicate partition coefficients less than 10,000 and the highly siderophile elements with metal/silicate partition coefficients greater than 10,000 (*Newsom and Palme,* 1984). As can be seen from Fig. 2, the highly siderophile elements (which include Os, Re, Ir, Ru, Pt, Rh, Au, and Pd) are depleted in the mantle by factors of 300 to 400 relative to their assumed initial CI abundances, whereas most of the moderately siderophile elements are depleted by factors of 5-20 after corrections for volatile depletion. This pattern is in contrast to the depletion patterns for siderophile elements in the Moon (*Newsom and Taylor,* 1989) and the parent body of the eucrite meteorites (*Hewins and Newsom,* 1988), which show steadily increasing depletions as a function of increasing siderophile nature.

This bimodal pattern of siderophile element depletions has led to a theory for the heterogeneous accretion of the Earth (*Ringwood,* 1966b; *Turekian and Clark,* 1969; *Wänke,* 1981) involving multiple stages of accretion and core formation. This theory, represented schematically in Fig. 5, assumes that the Earth began accreting from a reduced component that contained iron metal and sulfides. As the proto-Earth grew, core formation depleted the siderophile elements in the mantle to levels significantly below the presently observed abundances. After much of the Earth had accreted, the accreting material became more oxidized, such that segregation of iron metal essentially ceased and the moderately siderophile elements were able to build up in the mantle to a similar level. During this second stage, however, a small amount of Fe-metal or S-rich metallic liquid continued to segregate, resulting in depletions of Mo and the highly siderophile elements below the Co and Ni level and below their present abundances. The final stage of accretion, sometimes called the late veneer, occurred after more than 99% of the Earth had accreted and all metal segregation had ceased. Most of the observed abundance of the highly siderophile elements contained in the mantle were therefore contributed by this veneer material (*Chou et al.,* 1978, 1983; *Sun,* 1982).

An example of a model calculation for this theory is presented in Fig. 6. Figure 6a illustrates the depletions expected for the first stage of accretion after 93% of the Earth had accreted. The partition coefficients used for this calculation (Table 3) are for conditions significantly more reducing (log $fO_2 = -14.3$) than for the conditions assumed for the S-rich core theory or the inefficient core formation theory. The reducing conditions were chosen to deplete Mn, V, and Cr by the required

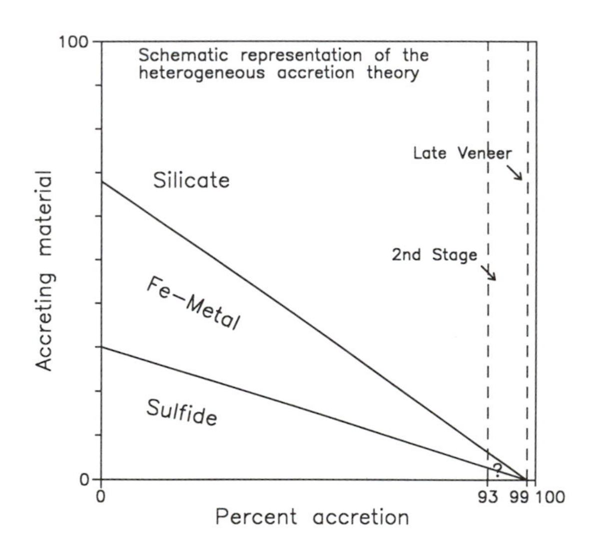

Fig. 5. *Schematic representation of the heterogeneous accretion theory. The composition of accreting material is represented as a function of the percent accretion of the Earth. For clarity, the width of the second stage and the late veneer, beginning at roughly 93% and 99% accretion respectively, are expanded. The fraction of core material accreting is shown as steadily decreasing, although an abrupt switch in metal content is not ruled out by the data. As discussed in the text, very small amounts of either metal or sulfide or both are assumed to continue to segregate to the core during the second stage of accretion. The final stage of accretion, the late veneer, is assumed to consist of entirely oxidized silicate material.*

amount, because the abundance of these elements is controlled by the first stage of accretion. Figure 6b illustrates the depletions after the addition of ~7% of chondritic material, which raises the abundances of most of the moderately siderophile elements to nearly their present abundances. Figure 6c indicates the results of segregation of 0.08 wt.% of Fe-metal, required to deplete the highly siderophile elements. This depletion must occur at a relatively low degree of partial melting in order to deplete Mo while not

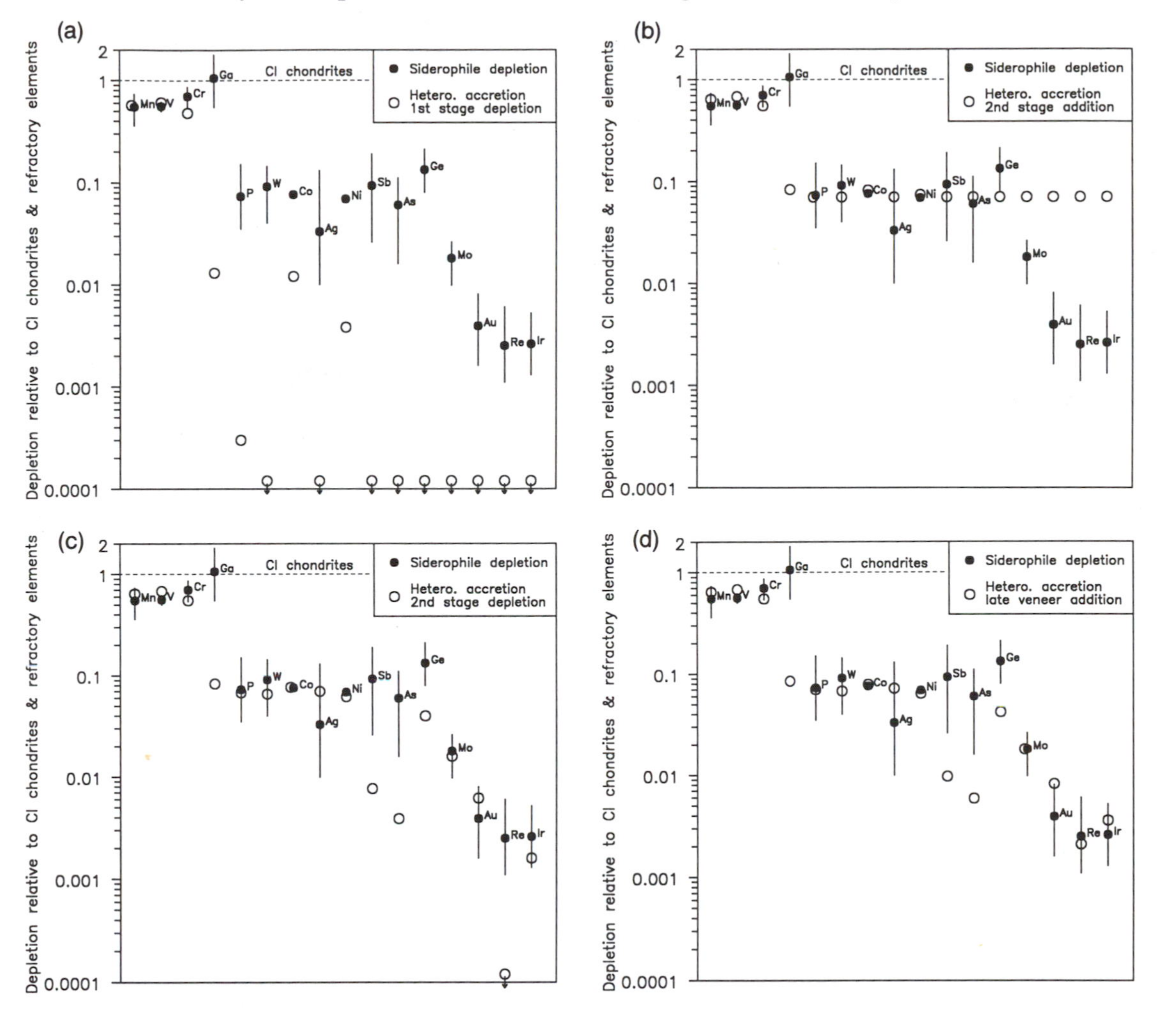

Fig. 6. *Results of a model calculation for the heterogeneous accretion theory.* **(a)** *Calculated depletions for the first stage of core formation until approximately 93% of the Earth had accreted. The depletions in the silicate mantle (open circles) were calculated by assuming equilibrium among 58.9 wt.% solid silicate, 6.6 wt.% liquid silicate, 19.2 wt.% solid metal, and 15.3 wt.% S-rich metallic liquid. Partition coefficients appropriate to an oxygen fugacity of log fO_2 = −14.3 (Table 3) were used in order to achieve the observed depletion of V.* **(b)** *Calculated abundances after the addition of the next 7% of material during the second-stage accretion of more oxidized material. The depletion of volatile siderophile elements in the material added is assumed to be the same as that observed for similarly volatile lithophile elements in the Earth's mantle (Fig. 1).* **(c)** *Calculated abundances after segregation of 0.08 wt.% Fe-metal during the second phase of accretion.* **(d)** *Calculated depletions for the final stage of the heterogeneous accretion model, where the addition of the late veneer (0.2 wt.%) brings up the abundances of highly siderophile elements to their observed level.*

significantly depleting Ge. Molybdenum and most of the highly siderophile elements can also be depleted by the segregation of approximately 0.1 wt.% of S-rich metallic liquid instead of Fe-metal; however, Ir is not sufficiently depleted by the segregation of S-rich liquid. Figure 6d represents the final abundances after addition of 0.2% of chondritic material (the late veneer), required to bring the highly siderophile elements to their present abundances.

The best constraints on this model are provided by the refractory siderophile elements such as V, Co, Ni, W, and Mo. The abundances of the compatible siderophile elements V, Co, Ni, and Ge are extremely well known from mantle nodules (Fig. 2). Previous work on the abundance of W by *Newsom and Palme* (1984) and *Newsom et al.* (1986), based on oceanic mantle-derived samples, suggested that W was more depleted than Co and Ni, which is inconsistent with this theory. However, new crustal data for W (*Sims et al.*, 1990) clearly indicates that the crust is enriched in W relative to the depleted mantle, such that the mass balance results in a revised depletion, indistinguishable from that of Co and Ni. The depletion of Mo has been confirmed by *Sims et al.* (1990) by additional work on Mo abundances in crustal reservoirs. The depletions of other moderately siderophile elements including As, Sb, Ag, and P are also consistent with the depletion of Co and Ni, but they have much larger errors due to uncertain mantle abundances and volatility corrections.

Several elements are not consistent with the calculated abundances for the heterogeneous accretion theory. The calculated abundances of Sb and As are lower than the measured abundances. The partition coefficients for these elements, however, are the least well known. The metal-silicate Ds are from an unpublished single reconnaissance measurement by K. W. Sims, R. H. Jones, P. D. Noll Jr., and H. E. Newsom. The solid-metal/liquid-metal partition coefficient is based on analysis of separate metal and sulfide grains in a terrestrial native iron deposit from Disko Island, Greenland (*Klöck and Palme,* 1988). This partition coefficient is of the greatest importance, because the large depletion of Sb and As relative to Co and Ni is due to the high liquid-metal/silicate-liquid partition coefficient (i.e., chalcophile behavior) calculated from the other two Ds.

The calculated Ge abundance is also below the measured mantle value, but the discrepancy is not as bad as for the inefficient core formation theory. The overabundance is hard to understand, however, since the partition coefficients for Ge are well known. A similar large discrepancy is observed for the calculated depletion of Ga.

A significant difficulty with the heterogeneous accretion theory is the problem of segregating very small amounts of metallic-liquid from the mantle during the later stages of accretion. Segregation of metal from an

TABLE 3. *Partition coefficients calculated for 1250°-1275°C and log fO_2 = -14.3.*

Element	$D^{LM/LS}$	$D^{SM/LS}$	Refs. from Table 2
Mn	0.32	0.032	k
V	2.5	0.15	k
Cr	5	0.45	k
Ga	17	100	e
Ge	5,700	50,000	h
P	470	800	h
Co	260	600	h
Ni	3,800	5,000	h
W	500	10,000	h
Mo	20,000	50,000	h

Partition coefficients between solid metal and S-rich metallic liquid and between solid silicate and silicate liquid are assumed to be identical to those reported in Table 2. Partition coefficients between Fe-metal and silicate liquid and between S-rich metallic liquid and silicate liquid for As, Sb, Ag, Re, Ir, and Au are poorly constrained, but are assumed great enough to cause depletion below 0.0001 times chondritic abundances (Fig. 6a).

impact-produced magma ocean is possible; however, segregation of small amounts of metal is even more difficult if low degrees of partial melting are required to explain the depletion of Mo relative to Ge as described above.

The S abundance for the heterogeneous accretion model is largely constrained by the amount of the oxidized component and the late veneer. The new estimate that the oxidized component need only represent the last 7% of accretion reduces the amount of S brought in during this stage to as little as 1400 ppm, depending on the volatile depletion of the accreting material. This is much closer to the estimates of <1000 ppm S in the mantle, but is still somewhat high.

The data presented in this paper, therefore, strengthen the evidence for a heterogeneous accretion event as the explanation for the abundance of the moderately siderophile elements in the mantle, although important questions remain. Some additional aspects and implications for these theories are discussed below.

DISCUSSION

The results of the calculations described above lead to several suggestions for future work. In order to distinguish between the equilibrium S-rich core formation theory, the inefficient core formation theory, and heterogeneous accretion models, several key siderophile elements must be intensively studied. The low-pressure partitioning behavior of As and Sb are still essentially unknown, and experimental studies are already underway. The behavior of V, Ga, W, Ni, Ge, and Au are especially important. In the case of the S-rich core formation theory the calculated depletions of V, W, Ni, Ge, and Au do not match observed abundances, for the inefficient core formation theory the calculated depletions of V, Ge and Au do not match the observed abundances, while in the case of the heterogeneous accretion theory the calculated abundances of Ga and Ge do not match the observed abundances. The high-pressure partitioning behavior of V, Ga, W, Ni, Ge, and Au should be studied, as these elements could provide clues to the actual conditions during core formation. A recent abstract by *Ringwood et al.* (1989) reports, for example, that large increases in the partition coefficients between metal and silicate are observed for Cr, V, and Mn in experiments relevant to depths greater than 1000 km in the Earth.

Another key question is the physics of metal separation in the proto-Earth. Can extremely small amounts of metal be efficiently segregated into the core, as required by some theories? Can small amounts of metal be left behind? Experimental data is lacking and innovative theories, including core formation from partially molten and totally molten mantles (e.g., *Stevenson,* 1990) need to be investigated.

The siderophile element data may also provide evidence regarding the existence of a terrestrial magma ocean. Several lines of evidence have suggested the existence of a magma ocean, including the large impact theory for the origin of the Moon (*Hartmann,* 1986; *Newsom and Taylor,* 1989; *Benz and Cameron,* 1990; *Melosh,* 1990) and the convergence of the solidus and liquidus at high pressures in the Earth's mantle, which might suggest that the upper mantle is derived from a partial melt (e.g., *Takahashi,* 1986). In contrast, strong evidence against a magma ocean has come from high-pressure partitioning experiments that suggest strong fractionations should be observed among Sc, Zr, and Hf relative to rare earth elements, contrary to the actual abundance data (*Kato et al.,* 1988; *Ringwood,* 1990). One solution would be the solidification of the magma ocean under conditions of rapid convection, as suggested by *Tonks and Melosh* (1990). What are the chemical implications of metal segregation from a largely molten Earth? For the equilibrium S-rich metallic-liquid core formation model, segregation in a molten Earth worsens the calculated fits for P, Co, Ni, Au, and Ir, while significant improvements are seen only for Mo and Re. Assuming a totally molten Earth for the inefficient core formation model results in calculated depletions that are worse for most elements, especially W. No elements have significantly improved fits. The heterogeneous accretion theory is the most compatible with the idea of a molten Earth, because the calculated depletions during the first-stage depletion, involving over 90% of the Earth (Fig. 6a), are essentially unaffected by the degree of partial melting. Only for the second stage of depletion is a low degree of partial melting required to deplete Mo relative to Ge, because the biggest difference in partitioning behavior for these elements is their solid-silicate/liquid-silicate partition coefficients, which are ~ 1 for Ge and ~ 0.01 for Mo.

The accretion of a late veneer on the Earth has been criticized due to the possible lack of evidence for such a veneer of material on the Moon (*Drake,* 1987). The simplest explanation is that the late veneer on the Moon was scavenged into the lunar core by metal from late accreting material if the outer part of the Moon was still molten (e.g., *Newsom,* 1984, 1986). If the late veneer on the Moon is assumed to be represented by the siderophiles in the lunar breccias, then strong constraints are placed on the depth to which the veneer is mixed on the Earth and Moon and on the relative accretion rates for the Earth and Moon. *Morgan et al.* (1981) have shown that the veneer material could be accounted for in the lunar regolith, assuming that the veneer was only mixed to a depth of 670 km on the Earth, and that about 33 times as much material accreted to the Earth as to the Moon. The relative accretion rates of the Earth and Moon have been studied by *Bandermann and Singer* (1973). They have shown that the relative accretion rates vary because of gravitational focusing as a function of the velocity of the accreting material and the distance between the Earth and Moon, which was smaller early in solar system history. For high-velocity material (>30 km/sec) the relative accretion rates or impact rates depend only on the physical cross-sections of the bodies, such that the Earth/Moon ratio is only about 14, and the Earth-Moon distance is not a factor. Gravitational focusing onto the Earth becomes important for material with velocities less than about 10 km/sec, for which an Earth/Moon accretion ratio of 20 to 30 is calculated, the larger value being for greater Earth-Moon distances. For material with velocities less than 5 km/sec, an Earth/Moon accretion ratio of greater than 30 is calculated for all Earth-Moon distances, consistent with the arguments of *Morgan et al.* (1981).

Another possibility is that the present abundances of highly siderophile elements (e.g., late veneer) in the Earth's mantle are due to addition of a small portion of material from the core of an impactor connected with the origin of the Moon (*Newsom and Taylor,* 1989), while the rest of the impactor's core accreted to the Earth's core without significant interaction with the Earth's mantle. The Moon shows little evidence of a late veneer of siderophile elements, supporting a unique event involving the Earth as the explanation for the abundances of the highly siderophile elements in the Earth's mantle. To obtain the observed mantle abundances of the highly siderophile elements requires 3% to 4% of the impactor's core, assuming the total mass of the impactor is between 0.12-0.17 Earth mass respectively. The percentage of impactor core required depends on the size of the impactor, and is independent of the size of the impactor's core because the highly siderophile elements from the impactor are quantitatively concentrated in the metal. The amount of metal containing the siderophiles is 0.2 wt.% of the Earth's mantle for a core size in the impactor of 31 wt.%, or less than 0.2 wt.% for smaller core sizes. Regarding the implications for the moderately siderophile elements, the impactor was probably differentiated into a core and mantle, such that the impactor mantle would be depleted in the moderately siderophile elements to at or below the present terrestrial mantle values. Thus, the addition of impactor mantle amounting to approximately 10% of the Earth's mantle would not significantly change the moderately siderophile element pattern. The amount of moderately siderophile elements from the impactor's core contributed to the Earth's mantle would also be at the same (<1%) level as the highly siderophile elements, which would be insignificant compared to the 5-10% levels of moderately siderophile elements in the impactor- and Earth-mantle material. This model would also rule out the late veneer as a source of water and other volatiles on the Earth, requiring other explanations such as cometary bombardment.

CONCLUSIONS

Revised depletions of siderophile elements in the Earth's mantle have been obtained and their uncertainties estimated. These depletions have been compared with model calculations to test the equilibrium S-rich core formation theory, the inefficient core formation theory, and the heterogeneous accretion theory. For the equilibrium S-rich core formation theory, the best fit was for equilibrium with 50% partially melted mantle. The elements V, W, Ni, Ge, and Au are critical elements that are not successfully modeled by this theory. Antimony and As also are poorly modeled, but their partition coefficients are not well known. For the inefficient core formation theory, the new calculations suggest that the siderophile abundances in the mantle can be approximated if ~ 0.04 wt.% solid metal and 2.5 wt.% S-rich metallic liquid were left behind in the mantle after core formation. Significant problems in matching observed abundances are found for V, Ge, and Au. The predicted S abundance in the mantle is also very high for this model—6000 ppm compared to the actual value of less than 1000 ppm.

For the heterogeneous accretion theory, several aspects have improved in comparison to the analysis of *Jones and Drake* (1986). In particular, the new abundance data for W (*Sims et al.,* 1990) now matches the calculated abundances. The problem with the amount of S added in the late-stage accretion and the veneer has also lessened, with the model predicting roughly 1400 ppm S while the actual abundance is probably less than 1000 ppm. Remaining problems with this theory include discrepancies for the elements Ga, Ge, Sb, and As. The relevant partition coefficients for Sb and As are not well known, however, so some of these problems may be resolved in the future.

Another important conclusion for the heterogeneous accretion theory is that the second-stage accretion episode that established the abundances of the moderately siderophile elements need only involve roughly the last 7% of the accretion of the Earth. This late stage of accretion is therefore almost a veneer in itself.

The understanding of the abundances and partitioning behavior of the siderophile elements in the Earth's mantle is continually improving. A large amount of future work lies ahead in improving our knowledge of these elements and in expanding our studies by more careful quantitative statistical studies of mantle abundances as well as extending the range of pressure conditions for partitioning experiments. Another important area, only briefly mentioned here, is the need for better understanding of the physical mechanisms of core formation (e.g., *Taylor,* 1989; *Stevenson,* 1990). All of these approaches should provide exciting new insights into the earliest history of the Earth.

Acknowledgments. This research was funded by NSF grant EAR 8804070 with additional support from NASA grant NAG-9-30 (to K. Keil). We thank J. Jones for sharing his expertise on inefficient core formation calculations and R. Brett and C. Capobianco for excellent reviews. We also thank S. R. Taylor, G. J. Taylor, K. W. Sims, K. Keil, and P. D. Noll Jr. for discussions and other assistance.

REFERENCES

Ahrens T. J. and Jeanloz R. (1987) Pyrite: shock compression, isentropic release, and composition of the Earth's core. *J. Geophys. Res., 92,* 10363-10375.

Anders E. and Grevesse N. (1989) Abundances of the elements: Meteoritic and solar. *Geochim. Cosmochim. Acta, 53,* 197-214.

Anderson D. L. (1989) Composition of the Earth. *Science, 243,* 367-370.

Arculus R. J. and Delano J. W. (1981) Siderophile element abundances in the upper mantle: Evidence for a sulfide signature and equilibrium with the core. *Geochim. Cosmochim. Acta, 45,* 1331-1343.

Bandermann L. W. and Singer S. F. (1973) Calculation of meteoroid impacts on Moon and Earth. *Icarus, 19,* 108-113.

Basaltic Volcanism Study Project (1981) *Basaltic Volcanism on the Terrestrial Planets.* Pergamon, New York. 1286 pp.

Benz W. and Cameron A. G. W. (1990) Terrestrial effects of the giant impact. In *Origin of the Earth,* this volume.

Brett R. (1984) Chemical equilibration of the Earth's core and upper mantle. *Geochim. Cosmochim. Acta, 48,* 1183-1188.

Chou C.-L. (1978) Fractionation of siderophile elements in the earth's upper mantle. *Proc. Lunar Planet. Sci. Conf. 9th,* pp. 219-230.

Chou C.-L., Shaw D. M., and Crocket J. H. (1983) Siderophile trace elements in the Earth's oceanic crust and upper mantle. *Proc. Lunar Planet. Sci. Conf. 13th,* in *J. Geophys. Res., 88,* A507-A518.

Davies G. F. (1990) Heat and mass transport in the early Earth. In *Origin of the Earth,* this volume.

Drake M. J. (1987) Siderophile elements in planetary mantles and the origin of the Moon. *Proc. Lunar Planet. Sci. Conf. 17th,* in *J. Geophys. Res., 92,* E377-E386.

Drake M. J., Newsom H. E., Reed S. J. B., and Enright M. C. (1984) Experimental determination of the partitioning of gallium between solid iron metal and synthetic basaltic melt: Electron and ion microprobe study. *Geochim. Cosmochim. Acta, 48,* 1609-1615.

Drake M. J., Newsom H. E., and Capobianco J. (1989) V, Cr, and Mn in the Earth, Moon, EPB, and SPB and the origin of the Moon: Experimental studies. *Geochim. Cosmochim. Acta, 53,* 2101-2111.

Hartmann W. K. (1986) Moon origin: The impact trigger hypothesis. In *Origin of the Moon* (W. K. Hartmann, R. J. Phillips, and G. J. Taylor, eds.), pp. 579-608. Lunar and Planetary Institute, Houston.

Hewins R. H. and Newsom H. E. (1988) Igneous activity in the early solar system. In *Meteorites and the Early Solar System* (J. F. Kerridge and M. S. Matthews, eds.), pp. 73-101. Univ. of Arizona, Tucson.

Hofmann A. W. (1988) Chemical differentiation of the Earth: The relationship between mantle, continental crust and oceanic crust. *Earth Planet. Sci. Lett., 90,* 297-314.

Hofmann A. W., Jochum K. P., Seifert M., and White W. M. (1986) Nb and Pb in oceanic basalts: New constraints on mantle evolution. *Earth Planet. Sci. Lett., 79,* 33-45.

Jagoutz E., Palme H., Baddenhausen H., Blum K., Cendales M., Dreibus G., Spettel B., Lorenz V., and Wänke H. (1979) The abundances of major, minor and trace elements in the Earth's mantle as derived from primitive ultramafic nodules. *Proc. Lunar Planet. Sci. Conf. 10th,* pp. 2031-2050.

Jones J. H. and Drake M. J. (1983) Experimental investigations of trace element fractionation in iron meteorites, II. The influence of sulfur. *Geochim. Cosmochim. Acta, 47,* 1199-1209.

Jones J. H. and Drake M. J. (1986) Geochemical constraints on core formation in the Earth. *Nature, 322,* 221-228.

Kato T., Ringwood A. E., and Irifune T. (1988) Experimental determination of element partitioning between silicate perovskites, garnets and liquids: constraints on early differentiation of the mantle. *Earth Planet. Sci. Lett., 89,* 123-145.

Klöck W. and Palme H. (1988) Partitioning of siderophile and chalcophile elements between sulfide, olivine and glass in a naturally reduced basalt from Disko island, Greenland (abstract). In *Lunar and Planetary Science XVIII,* pp. 483-484. Lunar and Planetary Institute, Houston.

Malvin D. J. and Drake M. J. (1987) Experimental determination of crystal/melt partitioning of Ga and Ge in the system forsterite-anorthite-diopside. *Geochim. Cosmochim. Acta, 51,* 2117-2128.

Melosh H. J. (1990) Giant impacts and the thermal state of the early earth. In *Origin of the Earth,* this volume.

Mitchell R. H. and Keays R. R. (1981) Abundance and distribution of gold, palladium and iridium in some spinel and garnet lherzolites, Implication for the nature and origin of precious metal-rich intergranular components in the upper mantle. *Geochim. Cosmochim. Acta, 45,* 2425-2442.

Morgan J. W., Wandless G., Petrie R., and Irving A. (1981) Composition of the Earth's upper mantle—I: Siderophile trace elements in ultramafic nodules. *Tectonophysics, 75,* 47-67.

Newsom H. E. (1984) The lunar core and the origin of the Moon. *Eos Trans. AGU, 65,* 369-370.

Newsom H. E. (1986) Constraints on the origin of the Moon from molybdenum and other siderophile elements. In *Origin of the Moon* (W. K. Hartmann, R. J. Phillips, and G. J. Taylor, eds.), pp. 203-229. Lunar and Planetary Institute, Houston.

Newsom H. E. and Drake M. J. (1982) The metal content of the Eucrite Parent Body, constraints from the partitioning behavior of tungsten. *Geochim. Cosmochim. Acta, 46,* 2483-2489.

Newsom H. E. and Drake M. J. (1983) Experimental investigation of the partitioning of phosphorus between metal and silicate phases: Implications for the Earth, Moon and Eucrite Parent Body. *Geochim. Cosmochim. Acta, 47,* 93-100.

Newsom H. E. and Palme H. (1984) The depletion of siderophile elements in the Earth's mantle: New evidence from molybdenum and tungsten. *Earth Planet. Sci. Lett., 69,* 354-364.

Newsom H. E. and Taylor S. R. (1989) Geochemical implications of the formation of the Moon by a single giant impact. *Nature, 338,* 29-34.

Newsom H. E., White W. M., Jochum K. P., and Hofmann A.W. (1986) Siderophile and chalcophile element abundances in oceanic basalts, lead isotope evolution and growth of the Earth's core. *Earth Planet. Sci. Lett., 80,* 299-313.

Palme H. and Nickel K. G. (1985) Ca/Al ratio and composition of the Earth's upper mantle. *Geochim. Cosmochim. Acta, 49,* 2123-2132.

Rammensee W. (1978) Verteilungsgleichgwichte von Spurenelementen zwischen Metallen und Silikaten. Ph.D. thesis, Univ. of Mainz, F.R. Germany.

Rammensee W. and Wänke H. (1977) On the partition coefficient of tungsten between metal and silicate and its bearing on the origin of the Moon. *Proc. Lunar Sci. Conf. 8th,* pp. 399-409.

Ringwood A. E. (1966a) Chemical evolution of the terrestrial planets. *Geochim. Cosmochim. Acta, 30,* 41-104.

Ringwood A. E. (1966b) Mineralogy of the mantle. In *Advances in Earth Science* (P. Hurley, ed.), pp. 357-398. MIT, Boston.

Ringwood A. E. (1979) *Origin of the Earth and Moon.* Springer-Verlag, New York. 295 pp.

Ringwood A. E. (1990) Earliest history of the Earth-Moon system. In *Origin of the Earth,* this volume.

Ringwood A. E., Hibberson W., Kato T., and Ware N. (1989) High pressure geochemistry of Cr, V and Mn: Implications for origin of the Moon and formation of planetary cores (abstract). *Eos Trans. AGU, 70,* 1419.

Sasaki S. and Nakazawa K. (1986) Metal-silicate fractionation in the growing Earth: Energy source for the terrestrial magma ocean. *J. Geophys. Res., 91,* 9231-9238.

Schmitt W., Palme H., and Wänke H. (1989) Experimental determination of metal/silicate partition coefficients for P, Co, Ni, Cu, Ga, Ge, Mo, and W and some implications for the early evolution of the Earth. *Geochim. Cosmochim. Acta, 53,* 173-185.

Seifert S., O'Neill H. St. C., and Brey G. (1988) The partitioning of Fe, Ni and Co between olivine, metal, and basaltic liquid: An experimental and thermodynamic investigation, with application to the composition of the lunar core. *Geochim. Cosmochim. Acta, 52,* 603-616.

Sims K. W., Newsom H. E., and Gladney E. S. (1990) Chemical fractionation during formation of the Earth's core and continental crust: Clues from As, Sb, W and Mo. In *Origin of the Earth,* this volume.

Stevenson D. J. (1981) Models of the Earth's core. *Science, 214,* 611-619.

Stevenson D. J. (1990) Fluid dynamics of core formation. In *Origin of the Earth,* this volume.

Sun S.-S. (1982) Chemical composition and origin of the earth's primitive mantle. *Geochim. Cosmochim. Acta, 46,* 179-192.

Takahashi E. (1986) Melting of a dry peridotite KLB-1 up to 14 GPa: Implications on the origin of peridotitic upper mantle. *J. Geophys. Res., 91,* 9367-9382.

Taylor G. J. (1989) Metal segregation in asteroids (abstract). In *Lunar and Planetary Science XX,* pp. 1109-1110. Lunar and Planetary Institute, Houston.

Taylor S. R. and McLennan S. M. (1985) *The Continental Crust: Its Composition and Evolution.* Blackwell, Oxford. 312 pp.

Tonks W. B. and Melosh H. J. (1989) The physics of crystal settling and suspension in a turbulent magma ocean. In *Origin of the Earth,* this volume.

Turekian K. K. and Clark S. P. Jr. (1969) Inhomogeneous accumulation of the Earth from the primitive solar nebula. *Earth Planet. Sci. Lett., 6,* 346-348.

Walker R. J., Shirey S. B., and Stecher O. (1988) Comparative Re-Os, Sm-Nd and Rb-Sr isotope and trace element systematics for Archean komatiite flows from Munro Township, Abitibi Belt, Ontario. *Earth Planet. Sci. Lett., 87,* 1-12.

Wänke H. (1981) Constitution of terrestrial planets. *Philos. Trans. R. Soc. London, A303,* 287-302.

Wänke H., Dreibus G., and Jagoutz E. (1984) Mantle chemistry and accretion history of the Earth. In *Archaean Geochemistry* (A. Kroner, G. N. Hanson, and A. M. Goodwin, eds.), pp. 1-24. Springer-Verlag, New York.

Wasson J. T. (1985) *Meteorites, Their Record of Early Solar-System History.* Freeman, New York. 267 pp.

Weckwerth G. (1983) Anwendung der Instrumentellen β-Spektrometrie im Bereich der Kosmochemie Insbesondre zur Messung von Phosphorgehalten. Unpublished thesis, Univ. of Mainz, F.R. Germany.

FORMATION OF THE CONTINENTAL CRUST

The origin of the crust is connected to the main issues of core formation and large-scale melting of the Earth in several ways. The lunar crust is inferred to have largely originated by crystallization of the lunar magma ocean. A similar ancient crust on the Earth does not exist, although evidence of continental crust formed during the first 500 m.y. of the Earth's history is beginning to be uncovered. An important contribution to this story is described in the paper by Bowring and coworkers.

Another connection with the question of the melting of the Earth is the remarkable homogeneity of the mantle in terms of the abundances and ratios for certain elements. Sims and coworkers address this and other questions about the differentiation of the primitive mantle into the reservoirs that now exist.

The two papers that discuss the origin of the continental crust are:

K. W. Sims, H. E. Newsom, and E. S. Gladney: *Chemical Fractionation During Formation of the Earth's Core and Continental Crust: Clues from As, Sb, W and Mo*

S. A. Bowring, T. B. Housh, and C. E. Isachsen: *The Acasta Gneisses: Remnant of Earth's Early Crust*

Sims and coworkers discuss the abundances and geochemical behavior of four siderophile elements (As, Sb, W, and Mo) in different mantle and crustal reservoirs. In the case of Mo, they find a remarkably constant ratio of Mo to the incompatible lithophile elements such as Ce and La in all crustal and mantle-derived rocks. This evidence of a homogeneous mantle is consistent with the previously known homogeneity of Ni and Co abundances in mantle nodules. Through the use of element ratios, they also provide evidence that As, Sb, and W are enriched in the present continental crust to a greater extent than expected from their incompatible behavior during igneous fractionation. In contrast, the enrichment of Mo in the continental crust is consistent with its igneous incompatibility. The extra enrichment observed for As, Sb, and Mo has previously been observed for Pb, and is probably due to a hydrothermal process that has affected at least the upper crust.

The paper by Bowring and coworkers discusses the implications of their discovery of 3.96-b.y.-old rocks in Canada, 160 m.y. older than the previous record ages of 3.8 b.y. for the Isua rocks of West Greenland. More important than their simply being the oldest rocks is the Nd isotopic evidence that they were derived from crustal materials enriched in light rare earth elements, as opposed to most Archean rocks, which were derived from a depleted source. Somewhat older detrital zircon grains (4.0-4.3 b.y.), which also constitute evidence of older continental crustal material, have been found in western Australia. The Canadian rocks therefore provide the first Sm/Nd isotopic evidence for enriched crust older than 3.8 b.y., although the authors indicate that there is abundant evidence for the variable involvement of older crust in other Archean rocks. Furthermore, the Pb isotope data for these rocks indicate a multistage history prior to 3.96 b.y., testifying to the complexity of early crustal events.

Literature about the formation of the Earth's crust and the establishment of the major crustal and mantle reservoirs in the Earth is extensive. For a start, we suggest the following papers: Silver, Carlson, and Olson (1988) *Annual Review of Earth and Planetary Science, vol. 16,* 477-541; also papers in *Earth and Planetary Science Letters, vol. 90,* special issue: *Isotope Geochemistry—The Crafoord Symposium,* especially the paper by Hofmann on pp. 297-314.

CHEMICAL FRACTIONATION DURING FORMATION OF THE EARTH'S CORE AND CONTINENTAL CRUST: CLUES FROM As, Sb, W, AND Mo

K. W. W. Sims[1] and H. E. Newsom

Department of Geology and Institute of Meteoritics, University of New Mexico, Albuquerque, NM 87131

E. S. Gladney

Los Alamos National Laboratory, Health and Environmental Chemistry Group MS K484, Los Alamos, NM 87545

We have analyzed suites of continental crustal rocks of different geologic ages, as well as midocean ridge and ocean island samples for As, Sb, W, and Mo, using radiochemical, thermal, and epithermal neutron activation analysis. The oceanic data show that Mo, As, and Sb have the same incompatible behavior as the light rare earth elements (e.g., Ce) during igneous fractionation, while W behaves as a highly incompatible element and correlates with Ba. The Mo/Ce ratio is the same in oceanic and crustal materials. However, crustal-derived materials, including Archean shales, appear to be enriched in As and Sb relative to Ce, and W relative to Ba, compared to the correlations in the oceanic rocks. A similar enrichment in crustal material has been previously observed for the element Pb. The suite of affected elements (Pb, As, Sb, and W) suggests that the enrichment is probably due to hydrothermal processes during crustal formation. The question of core formation through geologic time has been examined using Mo/Ce ratios; no significant change in this ratio has been observed in samples ranging from recent back to 3.8 Ga in age. The depletions of these siderophile elements relative to CI chondrites in the Earth's primitive mantle (silicate portion of the Earth) provide clues to the accretion and core formation history of the Earth. Our new depletion value for W differs significantly from the previous estimate based only on oceanic samples, and now overlaps the depletion of Co and Ni. The depletion of Mo is confirmed to be greater than that of Co and Ni, but less than that of the highly siderophile elements. The new data therefore provide additional support to the heterogeneous accretion theory, but are also consistent with the inefficient core formation theory.

INTRODUCTION

Present theories of terrestrial differentiation hypothesize that the Earth accreted from essentially chondritic material and that the core (Fe-Ni alloy and 10 wt.% light elements) subsequently segregated toward the Earth's center, while the crust (enriched in incompatible elements) accumulated at the Earth's surface, leaving a mantle deficient in those elements that make up the crust and the core (*Allègre et al.,* 1982; *Sun,* 1982). The chemical composition of the Earth's crust and upper mantle may be estimated through direct analysis, while the chemical composition of the lower mantle and core can only be inferred.

---

[1]Also at Los Alamos National Laboratory, Health and Environmental Chemistry Group MS K484, Los Alamos, NM 875445

Samples from the Earth's silicate portion (i.e., the mantle and crust) show depletions of the siderophile (high iron-affinity) and chalcophile (high sulfur-affinity) elements relative to the lithophile (high oxygen-affinity) elements. These depletions can be used as geochemical "fingerprints" to record the early events of the Earth's history. Information about the Earth's accretion and core formation is recorded in crustal and mantle rocks that postdate core formation because after core formation ceased, the dominant processes responsible for the depletion of siderophile and chalcophile elements relative to the lithophile elements also ceased.

Two important developments have allowed quantitative modeling of the Earth's core formation and accretionary history: (1) improvements in radiochemical techniques for the accurate measurement of siderophile and chalcophile element abundances (e.g., *Gladney,* 1978; *Rammensee and Palme,* 1982; *Newsom and Palme,* 1984) and (2) the determination of metal-silicate partition coefficients (e.g., *Rammensee and Wänke,* 1977; *Newsom and Drake,* 1982, 1983; *Jones and Drake,* 1983, 1986; *Drake et al.,* 1984; *Klöck and Palme,* 1988; *Schmitt et al.,* 1989). Mantle abundances of the highly siderophile noble elements such as Ir, Ru, and Os, and the compatible, moderately siderophile elements Co and Ni, are reasonably well known from their abundances in mantle nodules (*Morgan et al.,* 1980; *Jagoutz et al.,* 1979; *Chou et al.,* 1983). The abundances of several moderately siderophile and chalcophile elements that are incompatible during igneous fractionation such as Mo, W, As, and Sb, have been studied only in mantle nodules or oceanic rocks, and not in the important continental crustal reservoir. Because of uncertainties in the depletions and partition coefficients for the moderately siderophile elements, quantitative modeling has not yet established which of several processes were responsible for establishing the abundances of either the moderately or highly siderophile elements in the Earth's primitive mantle (*Jones and Drake,* 1986; *Newsom,* 1990.)

We have developed a radiochemical epithermal neutron activation analysis technique for measuring low concentrations (sub μg/g) of As, Sb, W, and Mo in silicate rocks. With this technique we have analyzed numerous Archean and Phanerozoic continental crustal and mantle-derived rocks that have been demonstrated by other workers to be representative of these reservoirs. We also report new data for As and Sb abundances in mid-ocean ridge basalts (MORB) and ocean island basalts (OIB), using the samples of *Newsom et al.* (1986). Using these data we have modeled the abundances and depletions of these elements, relative to CI chondrites, in the Earth's depleted mantle, continental crustal, and primitive mantle (mantle + crust) reservoirs. We find that the Mo/Ce ratio is relatively uniform in all the samples we have measured, whereas the As/Ce, Sb/Ce, and W/Ba ratios have continental crustal values that are enriched relative to the uniform values observed for the range of mantle-derived oceanic basalts. In this paper, we discuss these data in context of the Earth's early differentiation including (1) core formation through time, (2) processes of crustal formation, and (3) models proposed to account for the abundances of siderophile elements in the Earth's primitive mantle.

ANALYTICAL METHODS

To determine low-level (sub μg/g) concentrations of As, Sb, W, and Mo in silicate matrices we have developed a radiochemical epithermal neutron activation analysis procedure. This procedure is a modification of a radiochemical thermal neutron activation analysis procedure developed by *Gladney* (1978), which uses an Al_2O_3 inorganic ion exchange technique to separate As, Sb, W, and Mo from the background-producing major elements (e.g., Na, K, Fe, and Mn) in activated rock samples.

Using radiotracers we have reevaluated the chemical parameters of this procedure. Our results are consistent with earlier observations (*Gladney,* 1978) and show that under oxidizing conditions, the oxy-anion complexes of these elements are retained quantitatively (>95%) on the Al_2O_3 column. We have also found that using epithermal neutrons for the irradiations, rather than thermal neutrons, improves the detection limits and reduces the analytical uncertainties of this procedure in two significant ways:

1. Most rock-forming elements have neutron cross-sections that are inversely proportional to the velocity of the bombarding neutrons (*Steinnes,* 1971; *Baedeker et al.,* 1977); however, because As, Sb, W, and Mo have resonances in their excitation functions for epithermal neutrons, the level of specific activity for these elements is enhanced relative to the background activity of the sample. This significantly improves the detection limits for the short-lived radionuclides As ($t_{1/2}$ = 1.10 days) and W ($t_{1/2}$ = 0.996 days) because the overall activity of the sample is low enough that it can be safely processed the same day it is irradiated.

2. The fission of ^{235}U is greatly suppressed with epithermal neutrons. As a result, the $^{235}U(n,f)^{99}Mo$ contribution to the determined Mo concentrations is considerably lower. With the Omega West's epithermal port, the apparent weight of Mo per μg of U is 0.0018 μg, whereas with thermalized neutrons the apparent weight of Mo per μg of U is 1.70 μg (S. Lansberger, unpublished data, 1988). Nonetheless, the corrections from this ^{235}U contribution have been accounted for in the Mo concentrations reported here.

The following is a detailed outline of the radiochemical epithermal neutron activation analysis procedure that we have developed for this study.

1. Prior to any chemical processing the samples are irradiated with epithermal neutrons in a boron-filtered epithermal port at the Los Alamos Omega West Reactor (*Gladney et al.,* 1980).

2. Following 2-4 hours decay, the sample is combined in a nickel crucible with a 5:2 mixture of sodium peroxide and sodium hydroxide. This mixture is then fused in a high-temperature furnace at 575°C for 15 min.

3. The resulting fusion cake is allowed to solidify and is then dissolved, first with 100 ml of H_2O and then with 500 ml of dilute (0.5N) HCl.

4. This solution (<0.5N HCl) is passed through the ion exchange column (max flow rate <8 ml/min), which consists of 5-6 g of acid-washed chromatographic grade 80 mesh Al_2O_3.

5. The Al_2O_3 column is then counted directly on an intrinsic Ge detector. Two counts are employed to take advantage of the relative activities and half lives of the measured isotopes. On the first count As ($t_{1/2}$ = 1.10 days) and W ($t_{1/2}$ = 0.996 days) are determined using the 559 keV and the 686 keV lines, respectively. This first count must be done within 24 hr of irradiation in order to attain sub μg/g detection limits. After allowing the samples to decay for 3 days, Mo ($t_{1/2}$ = 2.75 days) and Sb ($t_{1/2}$ = 2.72 days) are then counted using the 140 keV and 564 keV lines, respectively. This delay allows enough time for secular equilibrium between ^{99}Mo and ^{99m}Tc to be established, which is necessary for accurate Mo determinations since it is actually the Tc 140 keV internal transition that is measured. This also allows approximately 90% of the As activity to decay, substantially reducing the As 561 keV correction to the measured Sb 564 keV peak.

The analytical detection limits for this procedure, based upon 3σ above the background, are matrix dependent. In particular, the amount of P in the sample greatly affects the Mo detection limit. This is because P is also largely retained on the Al_2O_3 column

(*Girardi et al.,* 1970) and the 1.7 MeVβ^- emission from the decay of $^{32}P(t_{1/2}=$ 14.5 days) increases the low energy background of the spectra, making the ^{99m}Tc 140.5 keV peak more difficult to detect.

To demonstrate the accuracy and precision of this technique, we have run replicates of several USGS and NBS standard reference materials (BCR-1, BHVO-1, AGV-1, RGM-1, and NBS 1633A). The values and uncertainties (the mean and standard deviation of the replicate analysis) for these reference materials are compared with accepted values in Table 1. For continuing quality assurance of our analytical accuracy and precision, the NBS Standard Reference Material 1633A Flyash was analyzed as an unknown with every sample batch. Shewart-type quality control charts (*Shewart,* 1931; *Kateman and Pijpers,* 1981) from these analyses are presented in Fig. 1.

ABUNDANCES OF As, Sb, W AND Mo IN THE DEPLETED MANTLE, CONTINENTAL CRUST, AND PRIMITIVE MANTLE

The abundances of compatible siderophile elements in the Earth's primitive mantle, such as Ni and Co, are well known from mantle nodules because they are retained in olivine during partial melting. However, in order to determine the primitive mantle abundances of incompatible siderophile elements, such as As, Sb, W, and Mo, which have been preferentially partitioned into the continental crust, it is necessary to determine their abundances in both the depleted mantle and continental crustal reservoirs. Knowledge of these abundances and the relative masses of these

TABLE 1. *As, Sb, W, and Mo content of USGS and NBS standard reference materials by radiochemical epithermal neutron activation analysis (RENA).*

Sample		No. of Analyses	As (ppm)	Sb (ppm)	W (ppm)	Mo (ppm)
AGV-1	RENA value		0.98 ± 0.03	4.0 ± 0.3	0.7 ± 0.1	3 ± 0.4
	Consensus value (1)	3	0.84 ± 0.27	4.4 ± 0.4	0.53 ± 0.09	3 ± 1
RGM-1	RENA value		3.10 ± 0.08	1.30 ± 0.02	1.49 ± 0.02	2.6 ± 0.2
	Consensus value (2)	4	3.0 ± 0.40	1.26 ± 0.07	1.50 ± 0.18	2.30 ± 0.30
BHVO-1	RENA value		0.54 ± 0.05	0.17 ± 0.01	0.22 ± 0.02	1.04 ± 0.08
	Consensus value (2)	4	0.40 ± 0.22	0.159 ± 0.036	0.27 ± 0.06	1.02 ± 0.10
BCR-1	RENA value		0.57 ± 0.03	0.58 ± 0.04	0.37 ± 0.06	1.6 ± 0.2
	Consensus value (1)	2	0.64 ± 0.14	0.62 ± 0.10	0.40 ± 0.09	1.2±0.2
1633A	RENA value		151 ± 6	7.0 ± 0.3	5.5 ± 0.6	29 ± 2
	Accepted value (3)	26	145 ± 15	7.0 ± 0.5	5.7 ± 0.70	30.0 ± 3.0

Average analyses of As, Sb, W, and Mo in the reference materials BCR-1, AGV-1, RGM-1, BHVO-1, and NBS 1633A analyzed by the radiochemical epithermal neutron activation analysis technique used for this study. Values are quoted as the mean ± one standard deviation of at least three separate analyses, with the exception of BCR-1, which is quoted as the mean and propagated analytical uncertainty of two analyses. The consensus values for the USGS reference materials are taken from *Gladney et al.* (1983) and *Gladney and Roelandts* (1988). The certified (As, Sb) and consensus (W, Mo) values for NBS 1633A are from *Gladney et al.* (1987).

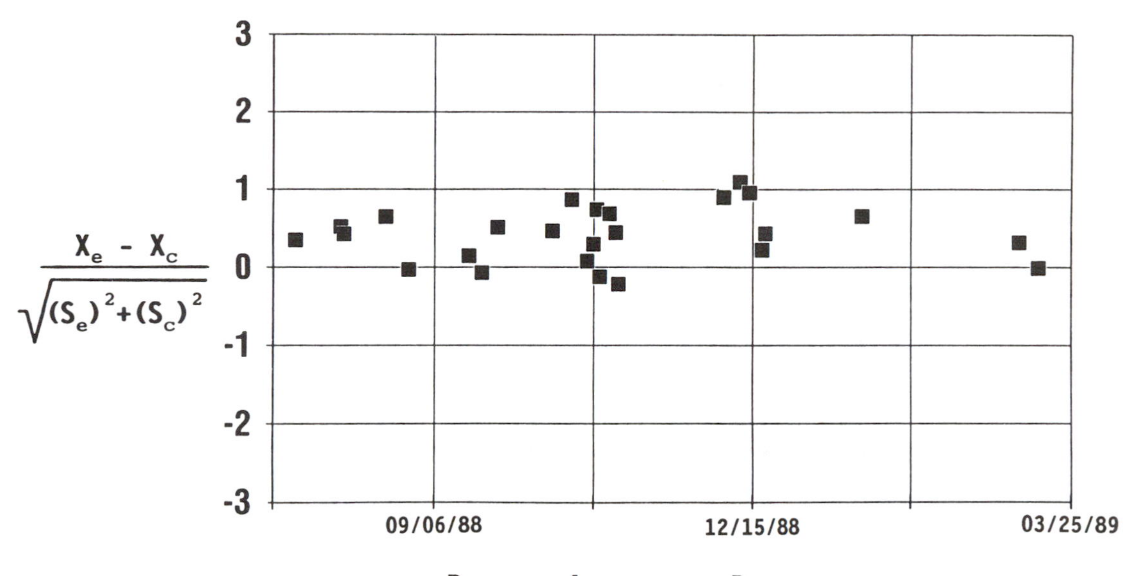

Fig. 1. *Control charts for* **(a)** *As,* **(b)** *Sb,* **(c)** *Mo, and* **(d)** *W. These quality control charts provide an indication of the accuracy and precision of the analytical procedure. During each analytical batch, a similar sized aliquot of NBS-1633A fly ash was analyzed as a quality assurance sample. The analytical batch is considered under control if the value of the difference between our measured result (X_e) and the certified or consensus mean (X_c) is within the propagated standard deviation (1σ) of the experimental uncertainty (S_e) and of the certified mean (S_c) (As, Sb, W, and Mo values of NBS 1633A are given in Table 1).*

reservoirs then allows a reconstruction of the concentrations of these elements in the primitive mantle.

For this purpose, we have chosen to analyze samples that are compositionally and isotopically well characterized and have been demonstrated in other studies to be representative of the continental crustal and depleted mantle reservoirs. In the following sections we discuss the samples, assumptions, and methods we have used to model the abundances of these elements in the Earth's depleted mantle, continental crustal, and primitive mantle reservoirs. The results for specific Archean reservoirs are less complete and are discussed in a later section.

Depleted Mantle Abundances

It is generally accepted that the formation of the continental crust has depleted a significant portion of the Earth's mantle in incompatible trace elements, and that this "depleted mantle" is now the source of oceanic basalts. While As, Sb, W, and Mo are siderophile and/or chalcophile in the presence of metal or sulfides, in silicate systems they behave as incompatible elements such that their concentrations are strongly affected by magmatic processes (i.e., partial melting and fractional crystallization). As a result, the depleted mantle abundances of As, Sb, W, and Mo cannot be determined from their absolute concentrations in oceanic basalts.

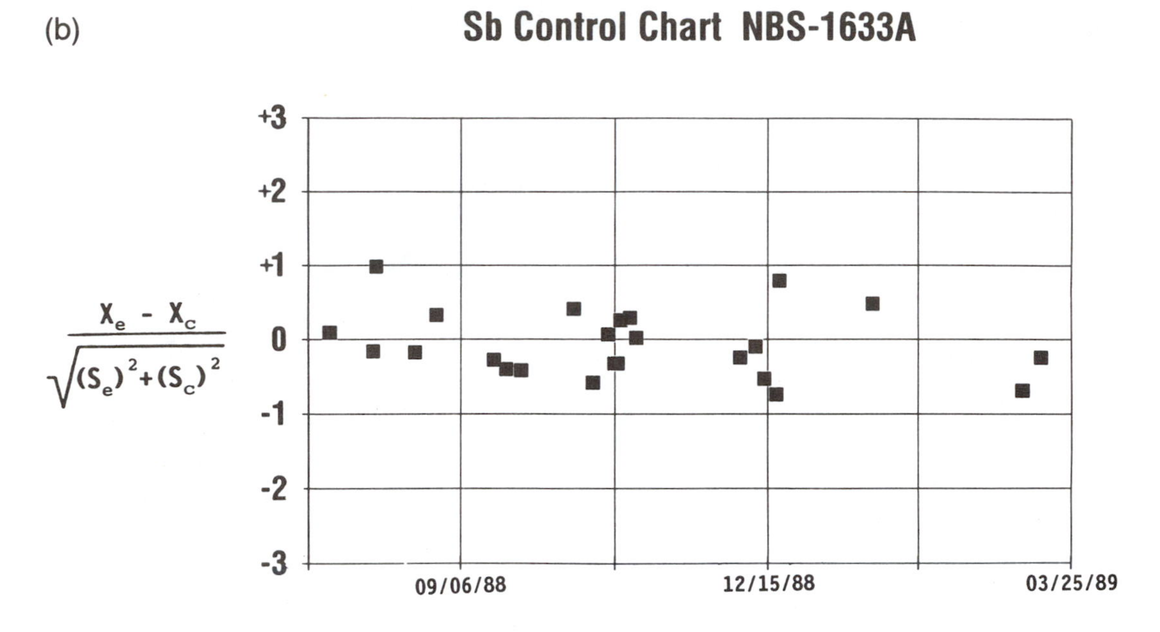

Fig. 1. *(continued).*

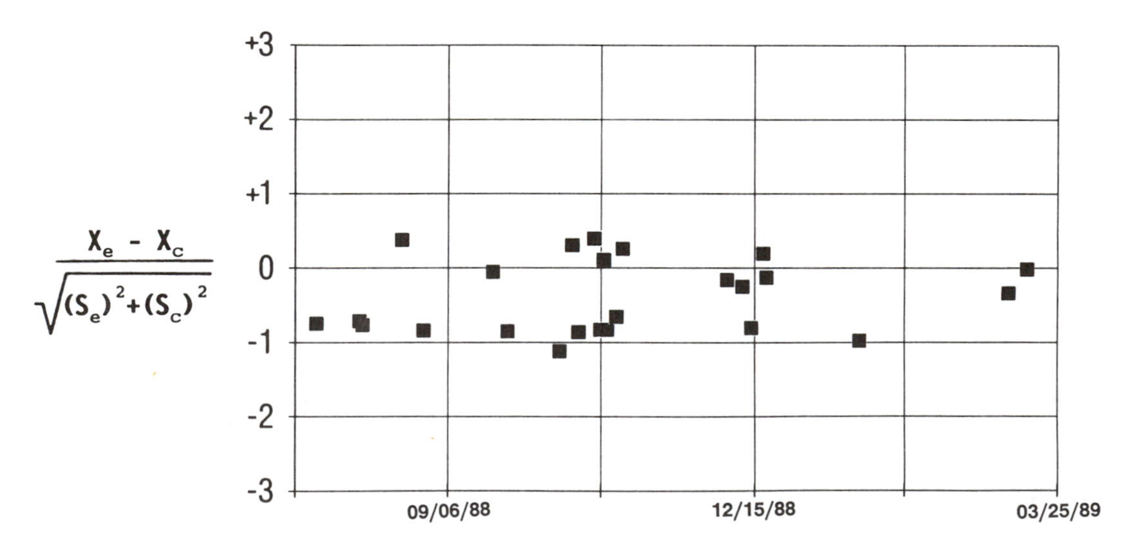

Fig. 1. *(continued).*

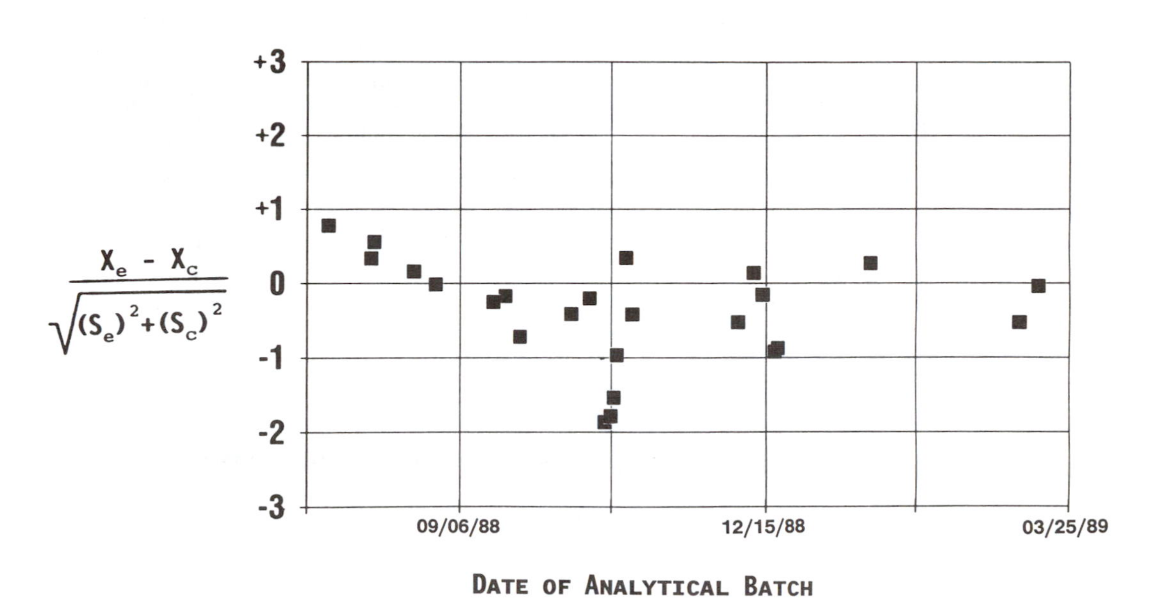

Fig. 1. *(continued).*

From oceanic basalts, however, it is possible to determine the depleted mantle abundances of incompatible siderophile elements relative to equally incompatible lithophile elements, because the ratio of two equally incompatible elements does not change during simple equilibrium partial melting (*Hofmann et al.,* 1986). Therefore, if it can be demonstrated that two incompatible elements, A and B, have a ratio (A/B) that is both constant and independent of those elements concentrations, then A and B are equally incompatible and their ratio (A/B) determined from oceanic basalts is the same as the initial ratio (A/B) in the depleted mantle source. There are several examples of pairs of incompatible trace elements in modern oceanic basalts that have a uniform ratio throughout a wide range of absolute trace element concentrations; these include Rb/Ce, Ba/Rb, K/U, Nb/Th, Pb/Ce, Mo/Pr, and W/Ba (*Hofmann and White,* 1983; *Jochum et al.,* 1983, 1984; *Hofmann et al.,* 1986; *Newsom et al.,* 1986).

To determine the present-day depleted mantle abundances of As, Sb, W, and Mo, we use a series of MORBs and OIBs that encompass almost the entire range of variability of trace element concentrations (Fig. 2) and isotopic ratios observed in modern oceanic basalts. These oceanic basalts were analyzed at the Max-Planck-Institut für Chemie, Mainz, West Germany, by H. Newsom using the metal separation technique of *Rammensee and Palme* (1982), with some additional analyses by K. Sims using the inorganic ion exchange technique described here. The results of these analyses are presented in Table 2. The Ce and Ba literature values we have used for normalization are presented in Table 3.

Within this suite we have found elements that are well correlated with As, Sb, W, and Mo. As indicated in Figs. 3a-d, As, Sb, and Mo are all incompatible and show a good correlation with the light rare Earth element Ce, while W is highly incompatible and is correlated with Ba. The depleted-mantle ratios and

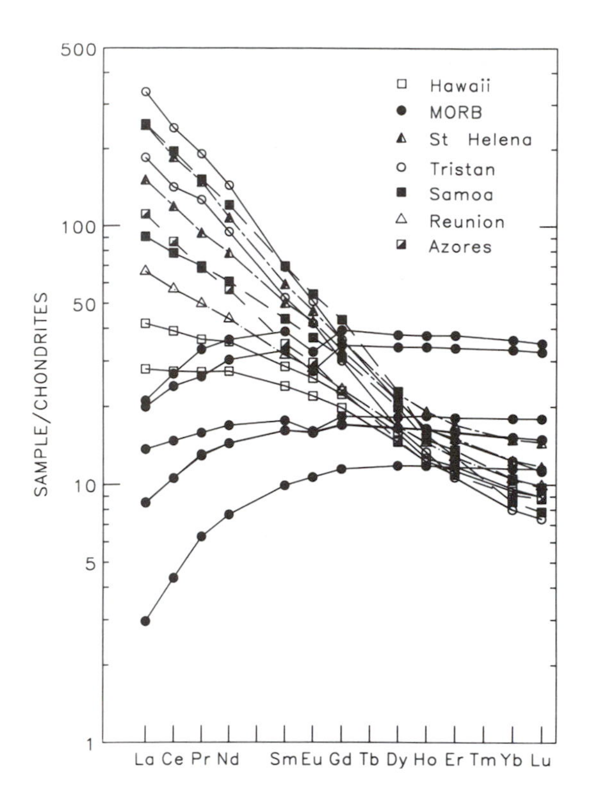

Fig. 2. *REE patterns for the MORB and OIB suites used in this study (after Newsom et al., 1986).*

their uncertainties (1σ) are As/Ce (0.0096 ± 0.0027), Sb/Ce (0.0012 ± 0.0005), Mo/Ce (0.032 ± 0.009), and W/Ba (0.0029 ± 0.0015). Assuming depleted mantle values of 1.41 ppm Ce and 2.40 ppm Ba (*Wänke et al.,* 1984) the absolute abundances in the depleted mantle are 14 ± 4 ppb for As; 1.7 ± 0.7 ppb for Sb; 45 ± 13 ppb for Mo; and 7 ± 4 ppb for W.

Earlier studies have modeled depleted mantle abundances of As, Sb, and W (*Jagoutz et al.,* 1979; *Morgan et al.,* 1980; *Kurat et al.,* 1980; *Wänke et al.,* 1984) by measuring their concentrations in "LIL-element depleted" spinel and garnet lherzolites. Together with our data, we have plotted the As/Ce, Sb/Ce, and W/Ba ratios of several lherzolites from these other studies (Figs. 3a-c). For the W/Ba and Sb/Ce ratios, there is good agreement between the depleted mantle values obtained from oceanic basalts and those obtained from spinel and garnet lherzolites. For As/Ce, however, the lherzolite data are scattered, with As/Ce values significantly higher than the As/Ce values of the oceanic basalts. Although we have not yet resolved this difference, one possibility is that As is substantially enriched in the lherzolites by secondary processes. *Kurat et al.* (1980), showed that leaching of lherzolites (with 0.5N HCl for 10 min) yielded a particularly high concentration of As in the leachable fraction (11 ppm in sample KA-167), while the remaining As in the bulk sample was below detection limits.

Continental Crustal Reservoir

The bulk composition of the Earth's continental crust is difficult to determine because the crust is heterogeneous and only the uppermost parts of the crust are easily sampled. Although we have determined the abundances of As, Sb, W, and Mo in the upper continental crust, our present database is not yet adequate to model the abundances of these elements in the lower continental crust or in the total continental crust. As a result, we have made the assumption that the upper crust values represent the total crust.

Upper continental crust. Although accessible for direct sampling, the upper continental crust is visibly heterogeneous. To determine the upper crust's average composition, a variety of approaches have been employed. These approaches involve either manually trying to average the exposed continental crust (i.e., through composite samples or weighted sums of the exposed rock types) or, more conveniently, allowing the geological processes of erosion and sedimentation to produce an upper crustal average (*Taylor and McLennan,* 1985). The validity of this second approach is indicated by the apparent uniformity of composition of fine-grained sedimentary rocks resulting from a wide range of sampling of the exposed crust and derived through a wide variety of sedimen-

TABLE 2. *Terrestrial abundances of As, Sb, W, and Mo.*

Loess									
Location	N.Z.	N.Z	Germany	U.S.A.					
Sample	BP-2	BP-4	K-2	Cy-4a-c					
No. of analyses	3	3	3	3					
As (ppm)	4.4 ± 0.1	4.3 ± 0.3	6.6 ± 0.2	3.4 ± 0.1					
Sb (ppm)	0.41 ± 0.06	0.25 ± 0.06	0.57 ± 0.06	0.29 ± 0.03					
W (ppm)	1.67 ± 0.06	1.4 ± 0.1	1.23 ± 0.06	1.1 ± 0.1					
Mo (ppm)	0.9 ± 0.3	0.8 ± 0.2	0.93 ± 0.06	0.87 ± 0.06					
Australian Post-Archean Shales									
Location	Mtlsa	Amadeus	Canning	Peth	Canberra				
Sample	M11	AO-7	PL-7	PW-4	SC-5				
No. of analyses	3	3	1	3	1				
As (ppm)	5.4 ± 0.02	7.3 ± 0.3	4.6 ± 0.2	10.8 ± 0.9	5.0 ± 0.3				
Sb (ppm)	0.81 ± 0.06	0.32 ± 0.03	0.43 ± 0.03	0.36 ± 0.02	0.77 ± 0.05				
W (ppm)	1.9 ± 0.2	3.0 ± 0.1	2.8 ± 0.2	2.1 ± 0.3	2.9 ± 0.2				
Mo (ppm)	2.3 ± 0.2	1.2 ± 0.2	0.9 ± 0.2	0.9 ± 0.2	0.8 ± 0.2				
Oceanic Rocks									
Location	P. Fournaise	Faial	P. Galapagos*	Indian*	Indian	Tristan	Kilueaea		
Sample	RE24-1	F-33	K71AD130	A119311	3095	TR-4	KL-2		
No. of analyses	1	1	1	1	1	1	1		
As (ppm)	0.51 ± 0.01	0.90 ± 0.018	0.15 ± 0.01	0.10 ± 0.01	0.134 ± 0.004	0.808 ± 0.016	0.34 ± 0.01		
Sb (ppm)	0.107 ± 0.005	0.064 ± 0.005	0.007 ± 0.002	0.009 ± 0.001	0.021 ± 0.003	0.070 ± 0.004	0.039 ± 0.003		
W (ppm)	0.37	0.694	0.074 ± 0.008	0.015 ± 0.002	0.007	0.786	0.159		
Mo (ppm)	1.3	3.39	0.54 ± 0.05	0.46 ± 0.02	0.48	3.11	0.98		
Oceanic Rocks (continued)									
Location	Mauna Loa	Upolu	Manua	Pacific	P. Galapagos	St. Helena	St. Helena	Tristan	Pacific*
Sample	M1-3B	UPO-7	82MT15	K10AD33A	K62AD143G	StH102	StH2926	TR-1	K73AD123H
No. of analyses	1	1	1	1	1	1	1	1	1
As (ppm)	0.149 ± 0.006	0.738 ± 0.015	0.450 ± 0.014	0.211 ± 0.004	—	2.23 ± 0.05	0.971 ± 0.003	1.76 ± 0.04	0.109 ± 0.009
Sb (ppm)	0.023 ± 0.003	0.050 ± 0.004	0.045 ± 0.004	0.013 ± 0.001	—	0.135 ± 0.005	0.106 ± 0.005	0.125 ± 0.006	0.005 ± 0.002
W (ppm)	0.11	0.938	0.429	0.022	0.065	1.25	0.492	1.71	0.007 ± 0.007
Mo (ppm)	0.743	3.52	1.89	0.689	0.578	5.62	3.25	6.34	0.16 ± 0.07

TABLE 2. *(continued)*

Early-Archean Sedimentary Rocks (High-Grade Terrains)										
Location	Isua	Isua	Malene	Malene	Malene	Malene	Akila	Mt. Narryer	Limpopo	Limpopo
Sample	24 84 84A	24 84 84G	22 1137	22 1136	201 429	201 424	152 769	MN-45	LP-28	LP-30
No. of analyses	1	1	1	2	2	3	2	1	1	3
As (ppm)	0.32 ± 0.02	0.40 ± 0.02	0.46 ± 0.03	0.18 ± 0.02	66 ± 5	0.30 ± 0.02	0.15 ± 0.02	0.044 ± 0.008	0.25 ± 0.02	0.11 ± 0.01
Sb (ppm)	0.09 ± 0.02	0.05 ± 0.01	0.019 ± 0.007	0.042 ± 0.009	0.34 ± 0.02	0.08 ± 0.02	0.08 ± 0.02	0.04 ± 0.01	0.08 ± 0.02	0.19 ± 0.03
W (ppm)	0.76 ± 0.07	1.5 ± 0.1	0.34 ± 0.03	0.19 ± 0.06	4.3 ± 0.3	0.59 ± 0.01	1.5 ± 0.1	0.22 ± 0.05	2.6 ± 0.2	1.3 ± 0.2
Mo (ppm)	0.29 ± 0.09	0.5 ± 0.1	0.84 ± 0.09	1.4 ± 0.1	1.8 ± 0.2	1.4 ± 0.1	1.4 ± 0.2	0.11 ± 0.06	0.6 ± 0.1	2.0 ± 0.2
Early-Archean Sedimentary Rocks (Greenstone Belt Shales)										
Location	Barberton	Barberton	Barberton	Pilbara	Pilbara					
Sample	79 NC 131	79 NC 124	79 NC 118	Pg-2	Pg-6					
No. of analyses	2	1	1	2	3					
As (ppm)	14.3 ± 0.9	11.9 ± 0.6	76 ± 4	26 ± 1	17.4 ± 0.2					
Sb (ppm)	1.20 ± 0.07	0.66 ± 0.04	2.3 ± 0.1	18.1 ± 0.9	17.1 ± 0.4					
W (ppm)	4.9 ± 0.3	1.5 ± 0.1	1.8 ± 0.2	4.2 ± 0.4	3.8 ± 0.1					
Mo (ppm)	2.3 ± 0.2	1.0 ± 0.1	1.4 ± 0.3	1.5 ± 0.3	0.9 ± 0.1					
Komatiites										
Location	Onverwacht	Onverwacht	Onverwacht	Onverwacht	Onverwacht	Onverwacht	Canada	Canada	Canada	
Sample	T5053	Pk5019	Bk5067	Bk5077	Bk5080	T5088	G-47	C-17	Z-2	
No. of analyses	2	2	2	2	2	1	1	1	1	
As (ppm)	0.22 ± 0.02	1.70 ± 0.09	0.52 ± 0.03	0.13 ± 0.02	0.41 ± 0.04	0.98 ± 0.05	0.128 ± 0.009	0.32 ± 0.02	0.40 ± 0.03	
Sb (ppm)	0.06 ± 0.01	0.32 ± 0.04	0.07 ± 0.01	0.08 ± 0.01	0.10 ± 0.02	0.08 ± 0.01	0.014 ± 0.007	0.34 ± 0.03	0.06 ± 0.01	
W (ppm)	0.64 ± 0.07	0.44 ± 0.09	0.29 ± 0.05	0.17 ± 0.06	0.30 ± 0.05	0.20 ± 0.03	0.09 ± 0.02	0.38 ± 0.07	0.12 ± 0.07	
Mo (ppm)	0.29 ± 0.06	0.12 ± 0.09	0.71 ± 0.09	0.14 ± 0.07	0.20 ± 0.09	0.44 ± 0.09	0.16 ± 0.08	0.36 ± 0.09	0.26 ± 0.09	

Analyses of As, Sb, W, and Mo. Depending on the availability of sample, either triplicate (preferred), duplicate, or single analyses were run on each sample. For triplicate analyses, the reported value represents the mean and standard deviation of the three analyses; for duplicates, the reported value and uncertainty represents the mean and propagated analytical uncertainties; and for a single analysis the reported value is the measured concentration and analytical uncertainty. For the oceanic samples, three analyses were done using the RENA procedure described here (indicated by an asterisk); for the rest of the oceanic samples, the As and Sb values are unreported data from Newsom, and the W and Mo values are taken from *Newsom et al.* (1986).

TABLE 3. *Ce and Ba literature values.*

Loess			
Location	N.Z.	N.Z	Germany
Sample	BP-2	BP-4	K-2
Ce (ppm)	74	51	83
Ba (ppm)	590	190	660
Reference	1	1	1

Australian Post-Archean Shales					
U.S.A.	Mtlsa	Amadeus	Canning	Perth	Canberra
Cy-4a-c	M11	AO-7	PL-7	PW-4	SC-5
74.4	52	65	72	115	83
595	450	210	290	620	400
1	2	2	2	2	2

Oceanic Rocks														
Location	P. Fournaise	Faial	P. Galapagos	Indian	Indian	Tristan	Kilueaea	Mauna Loa	Upolu	Manua	Pacific	P. Galapagos	St. Helena	St. Helena
Sample	RE24-1	F-33	K71AD130	A119311	3095	TR-4	KL-2	M1-3B	UPO-7	82MT15	K10AD33A	K62AD143G	StH102	StH2926
Ce (ppm)	49	74.9	20.7	9.12	9.13	122	33.7	23.6	169	67.7	23.1	12.8	160	103
Ba (ppm)	0.291	421	23.5	8.82	8.73	747	124	81.4	533	191	12.3	32	584	369
Reference	7	7	7	7	7	7	7	7	7	7	7	7	7	7

Early-Archean Sedimentary Rocks															
Location	Isua	Isua	Malene	Malene	Malene	Malene	Akila	Mt. Narryer	Limpopo	Limpopo	Barberton	Barbeton	Barberton	Pilbara	Pilbara
Sample	24 84 84A	24 84 84G	22 1137	22 1136	20 1429	201 424	152 769	MN-45	LP-28	LP-30	79 NC 131	79 NC 124	79 NC 118	Pg-2	Pg-6
Ce (ppm)	44.2	66.8	13.7	6.15	36.2	23.3	44.7	20.5	34.1	80.7	54.8	35.7	37.5	65.7	71.5
Ba (ppm)	201	177	416	87.7	452	450	487	460	170	290	458	547	246	1095	998
Reference	3	3	3	3	3	3	3	4	4	4	3	3	3	3	3

Komatiites									
Location	Onverwacht	Onverwacht	Onverwacht	Onverwacht	Onverwacht	Onverwacht	Canada	Canada	Canada
Sample	T5053	Pk5019	Bk5067	Bk5077	Bk5080	T5088	G-47	C-17	Z-2
Ce (ppm)	20.38	—	31.18	10.47	14.09	35.87	5.98	1.85	2.63
Ba (ppm)	—	—	—	—	—	—	—	—	—
Reference	5	5	5	5	5	5	6	6	6

Literature Ce and Ba concentrations for the samples used in this study. References: (1) *Taylor et al.* (1983); (2) *Nance and Taylor* (1976); (3) *Taylor and McLennan* (1985); (4) *Taylor et al.* (1986); (5) *Jahn et al.* (1982); (6) N. Arndt, personal communication, 1989; (7) *Newsom et al.* (1986).

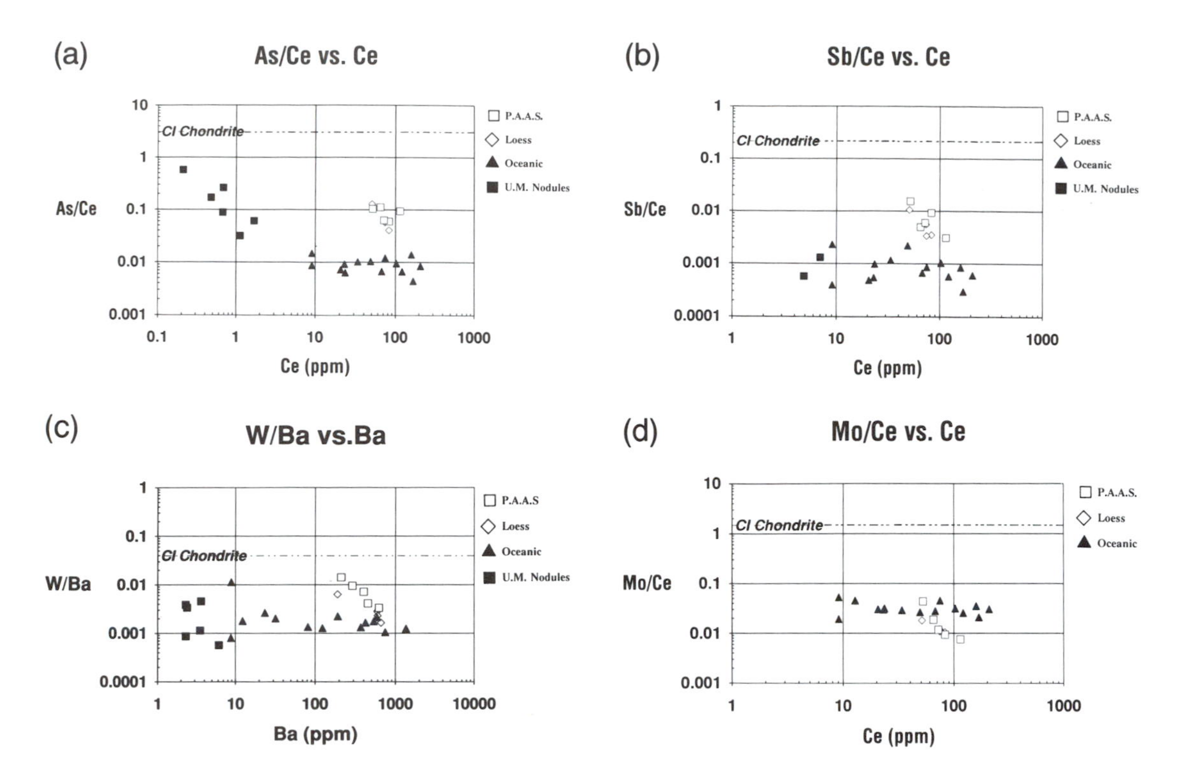

Fig. 3. **(a)** *As/Ce vs. Ce concentration;* **(b)** *Sb/Ce vs. Ce concentration.* **(c)** *W/Ba vs. Ba concentration; and* **(d)** *Mo/Ce vs. Ce concentration. These plots show the Phanerozoic continental crustal and mantle-derived oceanic suites. For W and As the ultramafic mantle nodule data are from Jagoutz et al. (1979) and the Sb data are from Morgan et al. (1980).*

tation processes. This seems to hold true for even the highly mobile elements such as As and Sb, which have similar abundances in the variety of upper crustal sedimentary samples we have measured, despite differences in provenance (three different continents) and differences in the sedimentation processes producing these samples (glacial wind-blown loess and marine shales).

To determine the average present-day upper continental crustal abundances of As, Sb, W, and Mo, we have measured the concentrations of these elements in several Post-Archean Australian shales (PAAS) and glacial loess samples that have previously been demonstrated to be good upper crustal averages (*Nance and Taylor,* 1976; *Taylor et al.,* 1983; *Taylor and McLennan,* 1985). The results of these analyses are presented in Table 2. The Ce and Ba literature values used for normalization are presented in Table 3. The average continental crustal ratios and their uncertainties (1σ) are As/Ce (0.08 ± 0.03); Sb/Ce (0.007 ± 0.002), Mo/Ce (0.018 ± 0.012), and W/Ba (0.006 ± 0.002). Assuming upper continental crustal abundances of 64 ppm Ce and 550 ppm Ba (*Taylor and McLennan,* 1985), the absolute abundances in the continental crust are 5.1 ± 1 ppm for As, 0.45 ± 0.08 ppm for Sb, 1.2 ± 0.4 ppm for Mo, and 3.3 ± 1.1 ppm for W.

The Mo/Ce ratio for upper continental crustal samples and mantle-derived samples are essentially the same, within the range of uncertainty (1σ) (Fig. 3d). However, the As/Ce, Sb/Ce, and W/Ba ratios for upper continental crustal samples are significantly higher than for mantle-derived samples (Figs. 3a-c). As confirmation that the elevated As/Ce, Sb/Ce, and W/Ba continental crustal ratios are a result of As, Sb, and W being relatively enriched, we have normalized As and Sb to Nd and W to U (not shown), with the identical result that the As/Nd, Sb/Nd, and W/U ratios have continental crustal values that are higher than the depleted mantle values. This fractionation between the continental crustal and the depleted mantle ratios has important implications for continental crustal formation as it indicates that crust-mantle differentiation extracted As, Sb, and W from the mantle more efficiently than Ce, Nd, U, and Ba.

Total continental crust. Many estimates for the composition of the continental crust have been proposed (*Wedehpol,* 1975, 1981; *Taylor,* 1964; *Taylor and McLennan,* 1985; *Weaver and Tarney,* 1984). However, due to the lack of data, most models, including our present model, do not distinguish between the *upper* continental crust and the *total* continental crust. A common approach to determine the composition of the total crust has been to combine, in their relative proportions, terrains and samples thought to be representative of the lower and upper crusts (*Weaver and Tarney,* 1984). However, the lower crust is not only difficult to sample, but recent COCORP seismic data indicate that it also extremely heterogeneous (*Oliver,* 1982; *Taylor and McLennan,* 1985). Samples that are thought be representative of the lower crust include lower crustal xenoliths from volcanic pipes, and outcrops of granulite terrains. Of these two types of samples, lower crustal xenoliths are more mafic, thought to be a less biased sampling and show conditions of equilibration more typical of the lowermost crust (*Taylor and McLennan,* 1985; *Bohlen and Mezger,* 1989).

We have measured the abundances of As, Sb, W, and Mo in several Archean high-grade (amphibolite and granulite facies) terrains. However, we feel that our present database is not adequate enough to model the *average* lower crustal abundances of these elements without data from lower crustal xenoliths. The data from the Archean high-grade terrains are important, as they provide us with an indication of how these elements behave under conditions of metamorphism approaching those of the lower crust. The Archean high-grade terrains we have analyzed include the following:

1. The Limpopo Province of South Africa. These are predominantly metasedimentary rocks and have metamorphic pressures and temperatures estimated at about 10 kbar and greater than 800°C (*Taylor et al.,* 1986; *Horracks,* 1983).

2. The Western Gneiss Terrain in the vicinity of Mt. Narryer of Western Australia. Here the metamorphic grade is generally upper amphibolite facies with some local granulite facies development (*Taylor et al.,* 1986).

3. The Isua supracrustal belt and adjacent Akilia association and the younger Malene supracrustals, all from West Greenland. The metamorphic grades for these metasedimentary rocks are amphibolite facies for the Isua and Malene suites and amphibolite-granulite facies for the Akilia suite (*McLennan et al.,* 1984; *Taylor and McLennan,* 1985).

The analytical results for these samples are presented in Tables 2 and 3. Our data indicate that the W/Ba (Fig. 4c) and Mo/Ce (Fig. 4d) ratios for the Archean high-grade terrains and the upper crustal samples are similar. However, for As/Ce (Fig. 4a) and Sb/Ce (Fig. 4b), the Archean high-grade terrains have significantly lower values than the upper crustal samples. A similar depletion of As and Sb (and Au), relative to their crustal abundances, has been observed in the high grade terrains of the Bamble belt in southern Norway (*Cameron,* 1989).

These observed depletions have important implications for mass balance calculations for the total continental crust, as they indicate that the lower crustal abundances of the volatile

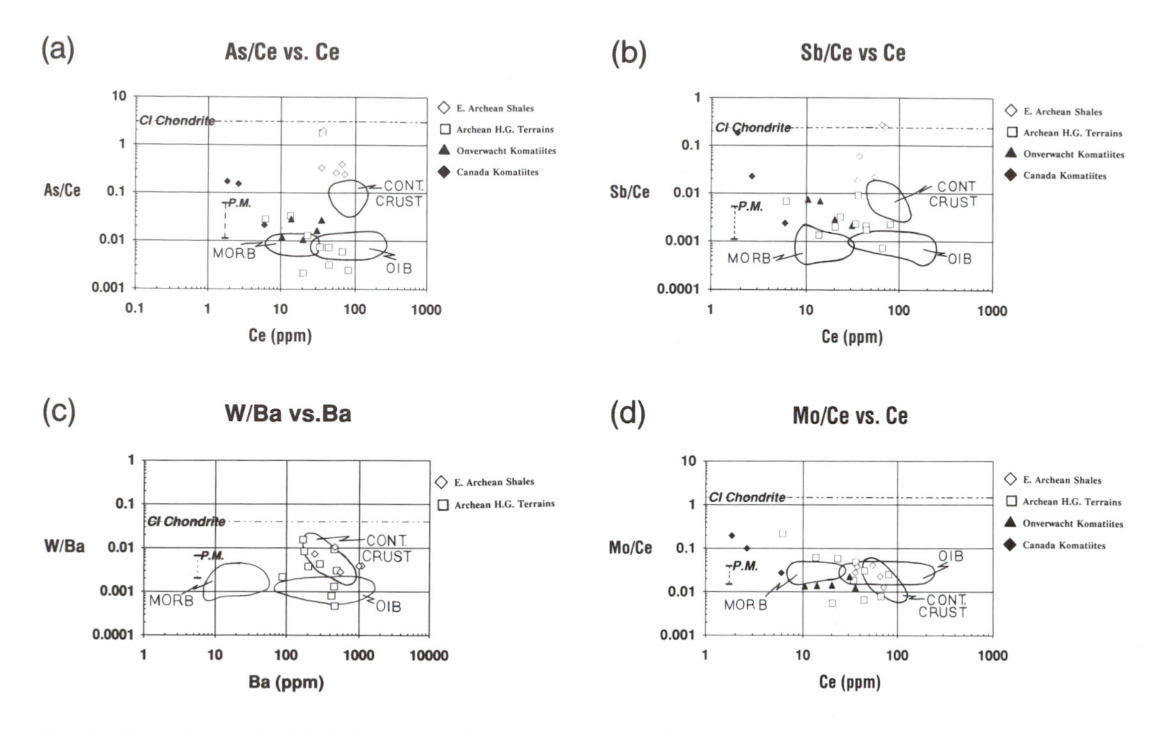

Fig. 4. **(a)** *As/Ce vs. Ce;* **(b)** *Sb/Ce vs. Ce.* **(c)** *W/Ba vs. Ba; and* **(d)** *Mo/Ce vs. Ce. These plots show the relationship between the Archean rocks, the mantle trend defined by OIB and MORB, the present upper continental crust, and the primitive mantle.*

elements As and Sb are probably depleted relative to their upper crustal values. For the refractory elements W and Mo, however, there is, at present, no indication of any *large* crustal reservoirs that have W/Ba and Mo/Ce values that are significantly different than their upper crustal values.

For our total crustal abundances in the primitive mantle mass balance calculations of the following section, we have used the upper crustal "sedimentary average" to represent the total crust for the As/Ce, Sb/Ce, Mo/Ce, and W/Ba ratios. This is a reasonable assumption for W and Mo, but does not reflect the likelihood of a total crust that has lower abundances of As and Sb. In the future, we intend to better characterize the abundances of these elements in the continental crust by analyzing several suites of lower crustal xenoliths, and continental crustal and island-arc volcanics.

Abundances of As, Sb, Mo, and W in the Earth's Primitive Mantle

We assume that the Earth's primitive mantle has been differentiated into the present-day depleted mantle and continental crust and use a mass balance between these two reservoirs to model the primitive mantle abundances of As, Sb, W, and Mo. Alternative models that include other uncharacterized reservoirs, such as the lower mantle (e.g., *Galer and O'Nions,* 1985), are not considered, since they provide no quantitative basis for evaluating primitive mantle abundances of nonradiogenic siderophile trace elements.

Using the depleted mantle $(A/B)_{dm}$ and continental crustal $(A/B)_{cc}$ ratios that we have determined, the mass balance for the primitive mantle ratio, $(A/B)_{pm}$, can be expressed in the following form

$$(A/B)_{pm} = \frac{(A/B)_{dm} \cdot [B_{dm} \cdot (1 - X_{cc})] + (A/B)_{cc} \cdot [B_{cc} \cdot X_{cc}]}{[B_{dm} \cdot (1 - X_{cc})] + [B_{cc} \cdot X_{cc}]} \quad (1)$$

where A is the element of interest (i.e., As, Sb, W, or Mo); B is the "mutually incompatible" normalizing element (i.e., Ce or Ba); the subscripts pm, dm, and cc refer to the primitive mantle, depleted mantle, and the continental crust, respectively; and X_{cc} is the mass fraction of continental crust relative to the amount of mantle that has been differentiated in the formation of the continental crust. Note that the term $(B_{cc} \cdot X_{cc})$, represents the contribution of the continental crust to the bulk Earth concentration of B, while the term $[B_{dm} \cdot (1-X_{cc})]$, represents the contribution of the depleted mantle.

Many of the variables in equation (1) are not well known. For instance, estimates of X_{cc} are model-dependent and vary by a factor of 4, depending upon the amount of mantle that was differentiated in the formation of the crust (*Allègre et al.,* 1983). In addition, the range of uncertainty in the ratios we have determined varies, on average, by a factor of 2. In Appendix A, we have evaluated equation (1) over a range of variables in order to calculate the range of possible primitive mantle As/Ce, Sb/Ce, W/Ba, and Mo/Ce ratios. In all cases, the primitive mantle range is sensitive to uncertainties in the measured ratios. The sensitivity of the primitive mantle estimates to the other variables, X_{cc}, B_{dm}, and B_{cc}, is proportional to the degree of separation between the continental crustal ratio and depleted mantle ratio. As a result, there are two different cases to be considered: (1) when the continental crustal and depleted mantle ratios are similar, and (2) when these ratios are significantly different.

Case 1. In this first and simpler case, which includes Mo/Ce, the continental crustal and depleted mantle reservoirs have essentially the same ratio values, indicating that the trace element ratio has remained unfractionated. The similarity between the Mo/Ce depleted mantle ratio (0.032 ± 0.009) and the Mo/Ce continental crustal ratio (0.018 ± 0.012) results in the $(Mo/Ce)_{pm}$ ratio being *insensitive* to variations in the other parameters of equation (1) (see Appendix A), and is calculated by simply taking the average of the Mo/Ce ratios of all the oceanic and upper crustal samples. The primitive mantle Mo/Ce ratio is 0.027 ± 0.012 (1σ). Assuming a primitive mantle value of 1.73 ppm Ce (*Wänke et al.,* 1984), the absolute abundance of Mo in the primitive mantle is 47 ± 20 ppb.

Case 2. In the second case, which includes As/Ce, Sb/Ce, and W/Ba (Figs. 3a,b, and d), the continental crustal and depleted mantle reservoirs have significantly different ratios. As a result of this fractionation, the primitive mantle ratios calculated with equation (1) are sensitive to (1) uncertainties in the parameters, X_{cc}, B_{cc}, and B_{dm}, (2) uncertainties associated with our assumption that the upper continental crust represents the total continental crust, and (3) uncertainties (1σ) of the ratios $(A/B)_{cc}$ and $(A/B)_{dm}$ themselves (see Appendix A).

Our estimates of the primitive mantle As/Ce, Sb/Ce, and W/Ba ratios using the approach outlined above and in Appendix A are 0.036 ± 0.025, 0.0032 ± 0.0021, and 0.0044 ± 0.002, respectively. The absolute primitive mantle abundances of these elements can be calculated by assuming primitive mantle abundances of 1.73 ppm for Ce and 5.6 ppm for Ba (*Wänke et al.,* 1984), giving As, Sb, and W values of 62.3 ± 43.0 ppb, 5.5 ± 3.7 ppb, and 24.4 ± 11.2 ppb, respectively.

We have observed that the Archean high-grade terrains have lower abundances of As/Ce and Sb/Ce relative to the determined upper crust. Incorporating these data into our total crustal estimate (which we feel would not be justified at present) would bring the total crustal abundances of these elements closer to the depleted mantle abundances and reduce the upper limit of the calculated primitive mantle range, which would, however, still be within the existing limits of the primitive mantle range of As/Ce and Sb/Ce. Therefore, the ranges that we have calculated incorporate the possibility that the lower continental crust (and hence total continental crust) has lower As/Ce and Sb/Ce values.

ABUNDANCES OF As, Sb, W AND Mo IN THE ARCHEAN MANTLE AND ARCHEAN CONTINENTAL CRUST

Archean Mantle

Archean komatiites are generally believed to have been produced through large degrees of partial melting of the upper mantle (*Green,* 1975; *Sun and Nesbitt,* 1978; *Jahn et al.,* 1980, 1982). To address the question of core formation through time, and to compare the abundances of As, Sb, W, and Mo in the Archean upper mantle with the present depleted mantle and our modeled primitive mantle, we have measured the concentrations of these elements in several Archean peridotitic komatiites and basaltic komatiites from the Onverwacht Group, South Africa, 3.56 Ga (*Jahn et al.,* 1982) and the Munro Township, Ontario, Canada, 2.7 Ga (N. Arndt, personal communication, 1989).

The results of our As, Sb, W, and Mo analyses are given in Table 2. Table 3 contains the Ce literature values used for normalization. Although we have measured W concentrations for these komatiites, Ba values have not yet been determined and, consequently, these samples are not plotted on the W/Ba vs. Ba plot (Fig. 4c). The Onverwacht komatiites and one komatiite from the Monro Township have As/Ce, Sb/Ce, and Mo/Ce values that fall within the respective ranges of primitive mantle values determined for these ratios (Figs. 4a,b, and d, respectively). There are, however, two Monro Township komatiites that have low Ce concentrations and high As/Ce, Sb/Ce, and Mo/Ce values relative to the other komatiites. Normalizing As, Sb, and Mo to Nd results in similarly enriched As/Nd, Sb/Nd, and Mo/Nd ratios for these two samples. This indicates that the higher ratios in these two samples are due to the enrichment of As, Sb, and Mo, rather than depletion of Ce.

The fact that *most* of these Archean komatiites have As/Ce and Sb/Ce ratios that fall within the range of their primitive mantle ratios is consistent with the observations of *Jochum et al.* (1987) that show that Archean komatiites also have primitive mantle Nb/Th ratios. According to *Carlson and Shirey* (1988) the primitive mantle Nb/Th ratios support the idea that the Archean mantle did not yet bear the chemical signature of continental crustal extraction. These results should be interpreted cautiously, however, since any crustal contamination could have resulted in elevated As/Ce and Sb/Ce ratios for these mantle-derived samples.

Archean Continental Crust

Taylor and McLennan (1985) have proposed that Archean sedimentary sequences have sampled a complex provenance of volcanic, plutonic, and sedimentary Archean rocks and are thus representative of the Archean upper crust. In order to determine the abundances of these elements in the exposed Archean crust we have measured the concentrations of these elements in several early Archean sedimentary sequences, including early Archean metasedimentary high-grade terrains and early Archean greenstone belts. These samples, provided to us by S. R. Taylor, are the same samples that *Taylor and McLennan* (1985) used in their modeling of the Archean upper crust.

The high-grade Archean metasedimentary terrains analyzed for this study have been discussed previously in reference to the lower crust. Shale samples from the following Archean greenstone belts have also been studied:

1. The Gorge Creek group from the Pilbara Block of Western Australia. This group is bracketed in age between 3.4 Ga and 2.9 Ga

and is derived from a source composed of 70-80% felsic igneous rocks and 20-30% mafic volcanics (*McLennan et al.,* 1983a).

2. The Fig Tree and Moodies Groups from Barberton Mountain Land, South Africa. In modeling the source provenance of these groups, which are approximately 3.3 Ga in age, *McLennan et al.* (1983b) suggests that the Fig Tree group was derived from subequal proportions of mafic and felsic igneous rocks whereas the Moodies Group required a significantly larger component of parental felsic rocks.

The data for the shales from the early-Archean greenstone belts (Tables 2 and 3; Figs. 4a-c) show that the Gorge Creek, Moodies, and Fig Tree Groups have enrichments of As/Ce, Sb/Ce, and W/Ba similar to our average Phanerozoic upper crustal abundances. For Mo/Ce (Fig. 4d) these groups also plot in the area defined by the Phanerozoic continental crustal envelope. This suggests that the same processes responsible for the Phanerozoic crustal enrichments of As, Sb, and W were also occurring in the Archean.

DISCUSSION

Core Formation Through Time

Most oceanic basalts have higher $^{143}Nd/^{144}Nd$ and $^{176}Hf/^{177}Hf$ ratios than those in chondritic meteorites (*White and Hofmann,* 1982; *DePaolo and Wasserburg,* 1976; *Patchett,* 1983). Assuming that the Earth accreted with chondritic Sm/Nd and Lu/Hf ratios, it can be inferred that the mantle has been depleted in the incompatible elements Nd and Hf relative to Sm and Lu. This incompatible element depletion of the oceanic rocks' mantle source is presumed to be the result of the extraction of a partial melt during crustal genesis.

Because of the volatility of Pb, the U/Pb ratio of the bulk Earth cannot be presumed to be chondritic (*Allègre et al.,* 1982; *Newsom et al.,* 1986). However, assuming that (1) the Earth inherited initial Pb isotope ratios equivalent to those of the Canyon Diablo Troilite (CDT) and (2) that calculations indicating an age of 4.55 Ga for the Earth are correct (*Vollmer,* 1977), then the $^{207}Pb/^{204}Pb$ and $^{206}Pb/^{204}Pb$ ratios for the bulk Earth must lie on an isochron passing through the assumed initial CDT Pb isotope ratios (Fig. 5) (*Allègre et al.,* 1982). The trend defined for oceanic island and midocean ridge basalts (*Sun,* 1980) lies distinctly to the right side of this 4.55 Ga isochron, referred to as the geochron (*Stacey and Kramers,* 1975). This implies that the evolution of Pb in the mantle did not occur in a closed system, but that there was an enrichment in U relative to Pb after the accretion of the Earth. However, because U is more incompatible than Pb (*Tatsumoto,* 1978), the Nd, Hf, and Sr systematics would predict that the mantle should be depleted in U relative to Pb due to the extraction of a silicate melt.

One explanation for this anomalous behavior in the U/Pb isotopic systematics, which has been referred to as the "Pb paradox," is that the Earth's core has grown throughout a significant portion of geologic time (*Allègre,* 1969; *Vollmer,* 1977; *Vidal and Dosso,* 1978; *Allègre et al.,* 1982). According to the model of continuous core formation, the extraction of a sulfide-rich phase from the mantle into the Earth's core would deplete the abundances of

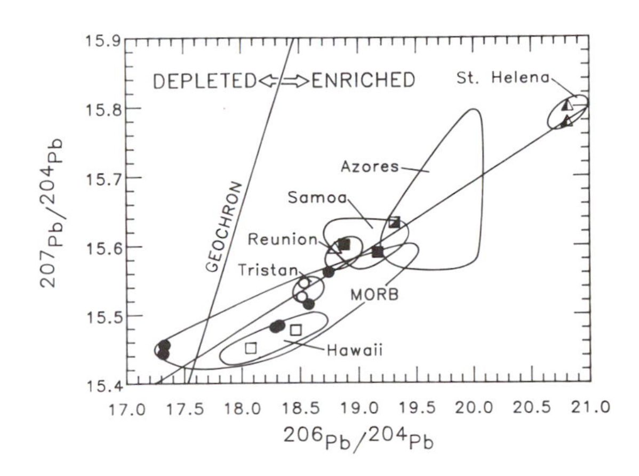

Fig. 5. *Plot of $^{207}Pb/^{204}Pb$ vs. $^{206}Pb/^{204}Pb$ for oceanic samples (after Newsom et al., 1986).*

chalcophile elements in the mantle. Because of the chalcophile affinity of Pb, this process would increase the U/Pb ratio of the mantle.

Newsom et al. (1986) tested this theory by determining the abundances of Mo in young oceanic basalts (OIBs and MORBs) of varying Pb isotopic compositions. Because Mo is more siderophile and chalcophile than Pb, one would expect that if the Pb paradox is a result of the extraction of a sulfide-rich metal phase into the core, then the abundances of Mo relative to similarly incompatible lithophile elements would be linked to the Pb isotopic ratios of the oceanic rocks. The results of *Newsom et al.* (1986) indicated that the ratios of Mo to light rare Earth elements such as Ce or Pr were independent of the initial Pb isotopic ratios, suggesting that the continuous segregation of a sulfide-rich phase into the core was not the cause for the observed Pb paradox in oceanic basalts.

We have extended this study by examining Mo/Ce ratios in mantle-derived and continental crustal rocks from the Archean to the present. With a few exceptions, all the rocks we analyzed have, within the range of uncertainty, similar Mo/Ce ratios (Fig. 4d), indicating that there were no significant temporal variations in the abundances or depletions of Mo in the mantle and continental crustal reservoirs that we have sampled. Based upon this evidence and upon the relative chalcophile affinities of Mo and Pb, we conclude that the observed Pb paradox is not a result of the continued segregation of a sulfide-rich metal phase into the core, reinforcing the conclusions of *Newsom et al.* (1986).

Crustal Formation and Mantle Evolution

Most trace element ratios of equally incompatible elements (e.g., Zr/Hf, Nb/Ta, Ba/Rb, Cs/Rb, K/U, and Y/Ho) are similar to the Mo/Ce ratio (Fig. 4d) in that they have essentially the same values in both the continental crust and upper depleted mantle (*Joron et al.,* 1978; *Bougault et al,* 1980; *Hofmann and White,* 1983; *Jochum et al.,* 1983, 1984). Thus, despite the differentiation of the primitive mantle into the continental crustal and depleted mantle reservoirs, the ratioed pairs of elements have behaved similarly during *all* processes affecting their concentration, such that they have remained unfractionated with respect to their primitive mantle values. However, as has been shown for the Pb/Ce, Nb/Hf, and Ta/Hf ratios, not all trace element ratios with uniform depleted mantle values have remained unfractionated with respect to their primitive mantle values. We have found that the As/Ce, Sb/Ce, and W/Ba ratios (Figs.4a-c) have also been fractionated with respect to their primitive mantle values. These ratios have uniform values in depleted mantle-derived oceanic basalts. However, they have significantly elevated continental crustal values, relative to their depleted mantle values. We have shown that these enrichments in the crust are not due to the choice of normalizing elements (Ce or Nd, Ba or U) and therefore we conclude that during crustal genesis As, Sb, and W were preferentially partitioned into the continental crust.

These "fractionated ratios" provide important information about the complexity of the primitive mantle differentiation, as they indicate that crust-mantle differentiation extracted Pb, As, Sb, and W from the mantle more efficiently than Ce or Ba, yet resulted in uniformly depleted mantle Pb/Ce, As/Ce, Sb/Ce, and W/Ba ratios. To account for this "variable incompatibility" and the uniformly depleted mantle ratio in Pb/Ce, *Hofmann et al.* (1986) have proposed a two stage model of primitive mantle differentiation: In stage 1, a partial melt is extracted from the mantle to form the continental crust during which Pb was more incompatible than Ce. Concurrent with or subsequent to this crustal extraction, the residual mantle is rehomogenized and forms the depleted mantle reservoir; in stage 2, after the initial depletion and rehomogenization, the mantle is differentiated into a relatively more depleted MORB source and a relatively enriched OIB source. During this stage of differentiation Pb and Ce behave similarly and are not fractionated with respect to each other, with the Pb/Ce ratio remaining constant and independent of any enrichments or depletions.

Because our data for As/Ce, Sb/Ce, and W/Ba show enrichment and depletion patterns that are similar to the Pb/Ce data, we conclude that our data are in agreement with the *Hofmann et al.* (1986) model. However, because Pb, As, Sb, and W are siderophile and/or chalcophile and have been affected by core formation processes, the primitive mantle values of their ratios are only indirectly bounded by cosmochemical constraints. As a result, our data cannot preclude the possibility that (1) other unsampled reservoirs were involved in the differentiation of the primitive mantle or (2) the concentrations of these elements, in the reservoirs that we have sampled, have been influenced by fluxes of volatile-rich incompatible material coming from the lower mantle. Furthermore, until there are more data for these elements in the lower crust (particularly the more volatile elements), the extent that these ratios have been fractionated by intracrustal differentiation (as opposed to crust-mantle differentiation) will remain uncertain.

Why have As, Sb, W, and Pb been preferentially partitioned into the upper continental crust relative to their respective normalizing elements Ce and Ba, while during the production of oceanic crust these pairs of elements (As/Ce, Sb/Ce, Pb/Ce, and W/Ba) are similar in their incompatibility? To explain this "variable incompatibility" observed in the Pb/Ce ratio, *Hofmann et al.* (1986) have suggested that either (1) Pb was transported into the continental crust by both igneous and hydrothermal processes or (2) during continental crust formation mantle-melting occurred at depth in the "feldspar absent zone," whereas during the production of oceanic crust, partial melting occurs at a relatively shallow depth and in the presence of feldspar. Because (unlike Pb) As, Sb, and W are not concentrated in plagioclase, our new data provide further evidence that this enrichment is a result of hydrothermal processes acting in conjunction with normal igneous processes. In fact, it has been recently suggested that this particular suite of elements—As, Sb, Pb, and W (also Sn, Au, Ag, Cu, and Zn)—are enriched by "simple" (i.e., low to moderate salinity) subvolcanic hydrothermal systems during the emplacement of large deep-seated granitic batholiths (*Kessler,* 1989). The enrichment of these elements in hydrothermal systems is also corroborated by a literature survey that shows that these elements are extremely enriched, relative to continental crustal values, in hydrothermal deposits (*Gladney et al.,* 1984), in hydrothermal waters (*Helgeson,* 1964; *Ellis,* 1979; *Weissberg et al.,* 1979), and in jasperoids (*Gladney et al.,* 1984).

Depletion of As, Sb, W and Mo in the Earth's Primitive Mantle

In this section we will discuss the implications of our new data to theories of accretion and core formation, while a detailed discussion of these models is presented in the accompanying article by *Newsom* (1990).

Several explanations have been proposed to account for the unexpectedly large siderophile element abundances in the Earth's primitive mantle: (1) partitioning into an FeO-rich iron metal phase at high pressures during core formation (*Ringwood,* 1979) (unfortunately, few partition coefficients for siderophile elements at high pressures are available to test this theory); (2) heterogeneous accretion, involving multiple stages of accretion and core formation (*Ringwood,* 1966; *Chou,* 1978; *Wänke et al.,* 1984); (3) inefficient core formation, involving the retention of Fe-metal and sulfides during core formation (*Mitchell and Keys,* 1981; *Arculus and Delano,* 1981; *Jones and Drake,* 1986); and (4) equilibrium partitioning into a S-rich metal phase during core formation (*Brett,* 1984).

The most important new result in terms of core formation models is the revised depletion for W (Fig. 6). The previous depletion estimates (*Newsom and Palme,* 1984; *Newsom et al.,* 1986) were based on ocean-island and midocean ridge basalts only, and gave a depletion significantly greater than that observed for Co and Ni. As discussed above, we have now shown that the crust is not only a major reservoir of W in the Earth, but the crust has a higher W/Ba (or W/U) ratio compared to the mantle-derived samples. The new primitive mantle depletion for W, based on a mass balance between the crust and depleted mantle, now overlaps the depletion of Co and Ni (Fig. 6). This removes an important objection to the heterogeneous accretion theory, as pointed out by *Newsom and Palme* (1984), and *Jones and Drake* (1986).

The heterogeneous accretion theory, involving multiple stages of accretion and core formation, has been proposed to explain the bimodal pattern of siderophile element depletions in the Earth (*Ringwood,* 1966; *Wänke,* 1981). This theory assumes that the Earth began accreting from a highly reduced component including iron metal. As the proto-Earth grew, core formation depleted the siderophile elements in the mantle to levels significantly below the presently observed abundances. The accreting material later became more oxidizing, such that during the last 5% to 10% of accretion (*Newsom,* 1990) core formation essentially ceased and the moderately siderophile elements, including W, were able to build up in the mantle to the level best defined by Co and Ni. During the last 5% of accretion, just enough Fe-metal or S-rich metallic liquid continued to segregate to keep the highly siderophile elements depleted well below their observed abundances. Finally, the accretion of a late veneer is responsible for the chondritic relative abundances of the highly siderophile elements.

Our new crustal data for Mo confirm the depletion previously estimated from oceanic samples alone (*Newsom and Palme,* 1984; *Newsom et al.,* 1986). Molybdenum is significantly depleted relative to Co and Ni, but this is consistent with segregation of a small amount of a metal or sulfide phase during the last 5% to 10% of the accretion of the Earth, as required by the heterogeneous accretion theory.

We have also obtained new estimates of the primitive mantle depletions of As and Sb (Fig. 6). In contrast to the refractory nature of W and Mo, As and Sb were depleted in the primitive mantle by volatility as well as by core

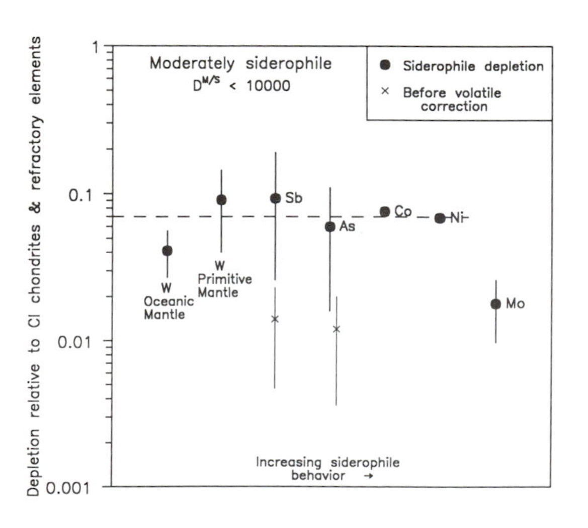

Fig. 6. *Depletion of siderophile elements in the Earth's primitive mantle (silicate portion of the Earth), normalized to mean CI chondrites (Anders and Grevesse, 1989), and refractory elements, such that the initial abundances of refractory elements in the Earth's mantle would fall at a depletion of 1.0. The moderately siderophile elements are arranged roughly in order of increasing siderophile behavior, where the siderophile behavior is measured by the bulk metal/silicate partition coefficients during equilibration between olivine, silicate-melt, and metal at low degrees of partial melting. The new primitive mantle depletion for W clearly overlaps the depletion observed for Co and Ni (Wänke et al., 1984), which is consistent with the heterogeneous accretion theory. In contrast, the old primitive mantle estimate based on mantle-derived oceanic basalts (Newsom et al., 1986), is more depleted than Co and W. The newly determined depletions of Sb and As, corrected for their volatile depletion, are also shown to overlap the depletion of Co and Ni.*

formation. We have corrected the depletions for As and Sb by subtracting the volatile depletions observed in the mantle for lithophile elements with similar volatilities (*Newsom,* 1990). The corrected depletions overlap the depletion of Co and Ni, although the uncertainties are quite large. The calculations reported by *Newsom* (1990) for the heterogeneous accretion theory predict abundances of As and Sb that are at least a factor of 3 too low. However, few of the relevant partition coefficients among solid metal, S-rich metallic liquid, and silicates have been measured.

The depletions of W, Mo, As, and Sb are also consistent with the inefficient core formation theory (*Jones and Drake,* 1986; *Newsom,* 1990). This theory suggests that the siderophile element depletion pattern in the primitive mantle was established by retention of small amounts of Fe-Ni metal and/or sulfur-bearing metallic liquid in the mantle during core formation.

For the equilibrium S-rich core formation theory (*Brett,* 1984), only the calculated depletion of Mo is within a factor of two of the observed value. The strong disagreement for W, which has well-known partition coefficients, is a major problem for this theory, unless the partition coefficients for W are, for example, particularly affected by pressure.

CONCLUSIONS

We have analyzed suites of continental crustal rocks of different geologic ages, as well as midocean ridge and ocean island samples for As, Sb, W, and Mo. The data for As and Sb in oceanic basalts show that these elements have the same incompatible behavior as the light rare Earth elements (e.g., Ce), during igneous fractionation. A similar result was previously obtained for Mo and Pb by *Newsom and Palme* (1984) and *Newsom et al.* (1986). These studies also determined that W behaves as a highly incompatible element and correlates with U or Ba.

An important discovery of this work is that the crustal-derived materials appear to be enriched in As and Sb relative to Ce, while W is enriched relative to Ba. The Mo/Ce ratio, in contrast, is the same in oceanic and crustal materials. The extra enrichment in crustal material has been previously observed for the element Pb (*Newsom et al.,* 1986; *Hofmann et al.,* 1986; *Hofmann,* 1988); and the suite of effected elements (Pb, As, Sb, and W) suggests that the enrichment is probably due to hydrothermal processes during crustal formation.

We have also examined the question of core formation through time, with the result that no obvious changes in the abundances of these elements, especially Mo, have been observed in samples ranging back to 3.8 Ga in age. This supports the arguments of *Newsom et al.* (1986) against core formation through time as the explanation for the increase in the U/Pb ratio observed on oceanic samples, which Pb isotope determinations suggest occurred at a time significantly later than the origin of the Earth.

The concentrations, and depletions relative to CI chondrites, of these siderophile elements in the Earth's primitive mantle (silicate portion of the Earth) have been estimated from our data. The new depletion value for W is significantly different from the previous estimate based only on oceanic samples (*Newsom et al.,* 1986), and now overlaps the depletion observed for Co and Ni (*Wänke et al.,* 1984). The new data therefore provide added support to the heterogeneous accretion theory (*Wänke,* 1981), but are also consistent with the inefficient core formation theory of *Jones and Drake* (1986).

APPENDIX A

Range of Primitive Mantle Abundances for As, Sb, W, and Mo

Using our depleted mantle $(A/B)_{dm}$ and continental crustal $(A/B)_{cc}$ ratios, the mass balance for the primitive mantle ratio, $(A/B)_{pm}$, can be expressed in the following form

$$(A/B)_{pm} = \frac{(A/B)_{dm} \cdot [B_{dm} \cdot (1 - X_{cc})] + (A/B)_{cc} \cdot [B_{cc} \cdot X_{cc}]}{[B_{dm} \cdot (1 - X_{cc})] + [B_{cc} \cdot X_{cc}]} \quad (A1)$$

where A is the element of interest (i.e., As, Sb, W, or Mo); B is the "mutually incompatible" normalizing element (i.e., Ce or Ba); the subscripts pm, dm, and cc refer to the primitive mantle, depleted mantle, and the continental crust, respectively; and X_{cc} is the mass fraction of continental crust relative to the amount of mantle that has been differentiated in the formation of the continental crust.

Estimates of primitive mantle As/Ce, Sb/Ce, W/Ba, and Mo/Ce values are obtained by evaluating equation A1 over a range of the parameters $(A/B)_{cc}$, $(A/B)_{dm}$, X_{cc}, B_{cc}, and B_{dm}. For $(A/B)_{cc}$ and $(A/B)_{dm}$ we use the average and $\pm 1\sigma$ values determined in this study. The other parameters, X_{cc}, B_{cc}, and B_{dm}, are based upon the following literature values:

X_{cc}: Estimates of X_{cc} fall into two categories. (1) Assuming that the whole mantle was involved in the differentiation of the crust, yields values of around 0.5% (i.e., 0.55%, *Jacobson and Wasserburg,* 1979; 0.60%, *Davies,* 1981). (2) Assuming only the mantle above the 670-km discontinuity was involved in the differentiation of the crust yields, approximately 2.0% (*DePaolo,* 1988a,b).

B_{dm}: Concentrations of Ce and Ba in the depleted mantle are from spinel lherzolites that show a range of Ce concentrations from 0.69 ppm to 1.41 ppm and a range of Ba concentrations from 2.3 ppm to 3.5 ppm (*Jagoutz et al.,* 1979; *Weckworth,* 1983).

B_{cc}: Our only estimates of concentrations of Ce and Ba in the total crust are from *Taylor and McLennan* (1985). Ce, 33 ppm; Ba, 250 ppm.

When calculating uncertainties for the primitive mantle As/Ce, Sb/Ce, W/Ba, and Mo/Ce ratios, it is instructive to distinguish between uncertainties resulting from (1) variability in the determined $(A/B)_{cc}$ and $(A/B)_{dm}$ ratios, which are visually apparent in Figs. 3a-d, and (2) the range of values in the parameters X_{cc}, B_{cc}, and B_{dm}. To do this we have divided through equation A1 with $(B_{cc} \times X_{cc})$. This yields

$$(A/B)_{pm} = \frac{(A/B)_{dm} \cdot R + (A/B)_{cc}}{R + 1} \quad (A2)$$

$$\text{where } R = \frac{[B_{dm} \cdot (1 - X_{cc})]}{[B_{cc} \cdot X_{cc}]}$$

R is a unitless parameter representing the fraction of B in the depleted mantle relative to the fraction of B in the continental crust. In Figs. A1a-d, the $(A/B)_{pm}$ is plotted as a function of R for fixed values of $(A/B)_{cc}$ and $(A/B)_{dm}$ (the solid line represents the average ratios and the dashed lines represent the upper and lower 1σ values of these ratios). Much of the observed range of R is a function of the four-fold difference in the estimated fraction of differentiated primitive mantle (whole mantle, $X_{cc} = 0.55$-0.60% vs. upper mantle, $X_{cc} = 2\%$). An additional uncertainty in R comes from the range of estimated values for B_{dm} (we have used a fixed value for B_{cc}, as stated above). The total range of possible R values for each element (1.01 to 7.72 for Ce and 0.45 to 2.53 for Ba) is indicated in Figs. A1a-d.

Mo/Ce (Fig. A1a). The average $(Mo/Ce)_{cc}$ is a factor of 0.7 lower than the average $(Mo/Ce)_{dm}$, although they are statistically indistinguishable. Because the relative separation between these ratios is small, most of the range in the $(Mo/Ce)_{pm}$ value is due to the uncertainties in $(Mo/Ce)_{cc}$ and $(Mo/Ce)_{dm}$. For a fixed R (3.41), the $(Mo/Ce)_{pm}$ varies by a factor of 2.1 as a result of the uncertainties (1σ) in $(Mo/Ce)_{cc}$ and $(Mo/Ce)_{dm}$; whereas, across the entire range of given Rs, the $(Mo/Ce)_{pm}$ only varies by a factor of 0.2, assuming the average $(Mo/Ce)_{cc}$ and $(Mo/Ce)_{dm}$ values.

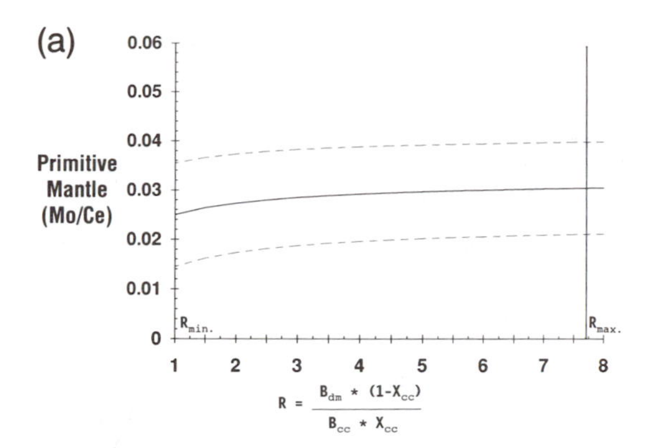

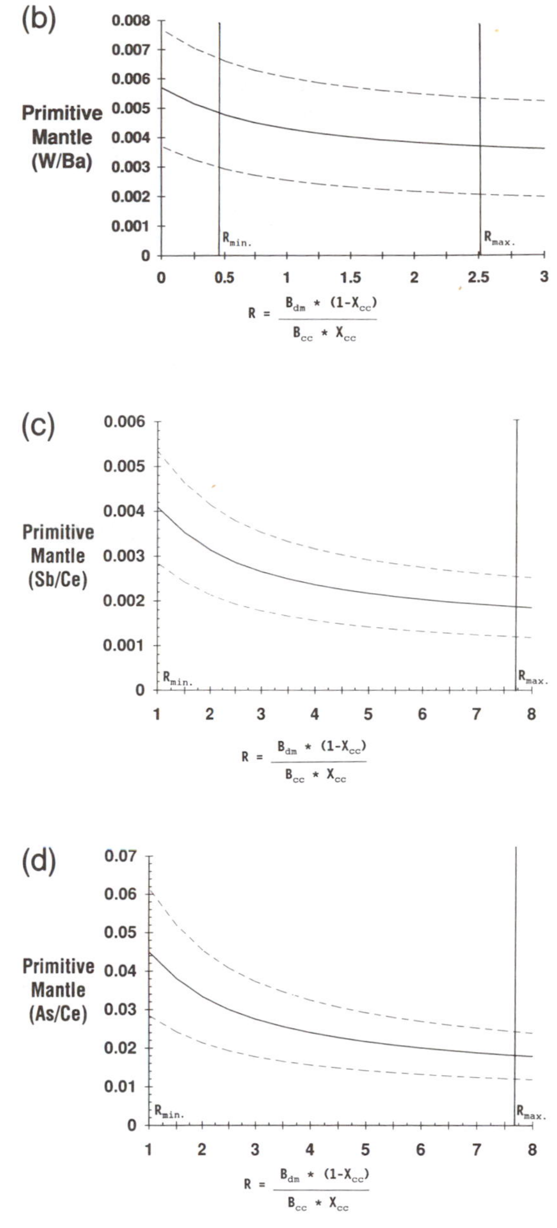

Fig. A1. **(a)** *Mo/Ce;* **(b)** *W/Ba;* **(c)** *Sb/Ce;* **(d)** *As/Ce. These charts plot $(A/B)_{pm}$ as a function of R for fixed values of $(A/B)_{cc}$ and $(A/B)_{dm}$ (the solid line represents the average depleted mantle and continental crustal ratios while the dashed lines represent the upper and lower 1σ ratios).*

Taking into account all possible sources of uncertainty, the Mo/Ce primitive mantle ratio is 0.027 ± 0.012.

W/Ba (Fig. A1b). The average $(W/Ba)_{cc}$ is about a factor of 2 higher than the average $(W/Ba)_{dm}$. Here again the relative separation between these ratios is not very large and most of the uncertainty in the $(W/Ba)_{pm}$ is due to uncertainties in $(W/Ba)_{cc}$ and $(W/Ba)_{dm}$ and not to variations in the parameter R. Taking in account both sources of error, the W/Ba primitive mantle ratio is 0.0044 ± 0.002.

Sb/Ce (Fig. A1c). The average $(Sb/Ce)_{cc}$ is a factor of 6 higher than the average $(Sb/Ce)_{dm}$. Because of this relatively large separation, the $(Sb/Ce)_{pm}$ varies by a factor of 2 across the range of R, for the average $(Sb/Ce)_{cc}$ and $(Sb/Ce)_{dm}$. As a result of the uncertainties (1σ) in $(Sb/Ce)_{cc}$ and $(Sb/Ce)_{dm}$, the $(Sb/Ce)_{pm}$ varies by an additional factor of 2 for a fixed R (3.41). The combined uncertainties result in an Sb/Ce primitive mantle ratio of 0.0033 ± 0.0021.

As/Ce (Fig. A1d). The average $(As/Ce)_{cc}$ is higher than the average $(As/Ce)_{dm}$ by a factor of 8. For the average $(As/Ce)_{cc}$ and $(As/Ce)_{dm}$ values, the $(As/Ce)_{pm}$ ratio varies by a factor of 2.5 across the given range of R. Due to the uncertainties in $(As/Ce)_{cc}$ and $(As/Ce)_{dm}$, the $(As/Ce)_{pm}$ varies by an additional factor of 2.1 for a fixed R (3.41). The combined uncertainties result in a primitive mantle As/Ce ratio of 0.037 ± 0.025.

The obvious conclusion that can be derived from these plots is that when the separation between the continental crustal and depleted mantle ratios is great, the range for the primitive mantle ratio is sensitive to (1) uncertainties in the continental crustal and depleted mantle ratios and (2) variations in the variables of R, X_{cc}, B_{dm}, and B_{cc}. When the relative separation is small, the range of the primitive mantle ratio is mostly a reflection of the uncertainties in the continental crustal and depleted mantle ratios.

Acknowledgments. NSF grant EAR 8804070 (H. Newsom), NASA grant NAG 9-30 (K. Keil), and the Department of Energy. We thank S. R. Taylor, G. Gruau and N. Arndt for generously providing samples, and K. Keil, D. G. Brookins, C. J. Yapp, E. Sklar, J. A. Grambling, and P. D. Noll Jr., for helpful discussions. This paper has been significantly improved by the thoughtful reviews of J. Morgan and C. Capobianco.

REFERENCES

Allègre C. J. (1969) Comportement des systems U-Th-Pb dans le manteau superieur modele d'evolution de ce dernier au cours des temps geologiques. *Earth Planet. Sci. Lett., 5,* 261-269.

Allègre C. J., Dupre B., and Brevart O. (1982) Chemical aspects of the formation of the core. *Philos. Trans. R. Soc. London, A306,* 49-59.

Allègre C. J., Hart S. R., and Minster J.-F. (1983) Chemical structure and evolution of the mantle and continents determined by inversion of Nd and Sr isotopic data, II. Numerical experiments and discussion. *Earth Planet. Sci. Lett., 66,* 191-213.

Anders E. and Grevesse N. (1989) Abundances of the elements: Meteoritic and solar. *Geochim. Cosmochim. Acta, 53,* 197-214.

Arculus R. J. and Delano J. W. (1981) Siderophile element abundances in the upper mantle: Evidence for a sulfide signature and equilibrium with the core. *Geochim. Cosmochim. Acta, 45,* 1331-1343.

Baedeker P. A., Rowe J. J., and Steinnes E. (1977) Application of epithermal neutron activation in multielement analysis of silicate rocks employing both coaxial Ge(Li) and low energy photon detector systems. *J. Radioanal. Chem., 40,* 115-146.

Bohlen S. R. and Mezger K. (1989) Origin of granulite terrains and the formation of the lowermost crust. *Science, 244,* 326-329.

Bougault H., Joron J. L., and Treuil M. (1980) The primordial chondritic nature of large-scale heterogeneities in the mantle: evidence from high and low partition coefficient elements in oceanic basalts. *Philos. Trans. R. Soc. London, A297,* 203-213.

Brett R. (1984) Chemical equilibration of the Earth's core and upper mantle. *Geochim. Cosmochim. Acta, 48,* 1183-1188.

Cameron E. M. (1989) Scouring of gold from the lower crust. *Geology, 17,* 26-29.

Carlson R. W. and Shirey S. B. (1988) Magma oceans, ocean ridges and continental crust: Relative roles in mantle differentiation (abstract). In *Papers Presented to the Conference on Origin of the Earth,* pp. 13-15. Lunar and Planetary Institute, Houston.

Chou C.-L. (1978) Fractionation of siderophile elements in the Earth's upper mantle. *Proc. Lunar Planet. Sci. Conf. 9th,* pp. 219-230.

Chou C.-L., Shaw D. M., and Crocket J. H. (1983) Siderophile trace elements in the Earth's oceanic crust and upper mantle. *Proc. Lunar Planet. Sci. Conf. 13th,* in *J. Geophys. Res., 88,* A507-A518.

Davies G. F. (1981) Earth's neodymium budget and structure and evolution of the mantle. *Nature, 290,* 208-213.

DePaolo D. J. (1988a) *Neodymium Isotope Geochemistry.* Springer-Verlag, New York. 187 pp.

DePaolo D. J. (1988b) Geochemical constraints on the stratification of the mantle. *Eos Trans. AGU, 69,* 486.

DePaolo D. J. and Wasserburg G. J. (1976) Nd isotopic variations and petrogenetic models. *Geophys. Res. Lett., 3,* 249-252.

Drake M. J., Newsom H. E., Reed S. J. B., and Enright M. C. (1984) Experimental determination of the partitioning of gallium between solid iron metal and synthetic basaltic melt: Electron and ion microprobe study. *Geochim. Cosmochim. Acta, 48,* 1609-1615.

Ellis J. A. (1979) Explored geothermal systems. In *Geochemistry of Hydrothermal Ore Deposits* (H. L. Barnes, ed.), pp. 632-684. Wiley, New York.

Galer S. J. G. and O'Nions R. K. (1985) Residence time of thorium, uranium and lead in the mantle with implications for mantle convection. *Nature, 316,* 778-782.

Girardi F., Pietra R., and Sabbioni E. (1970) Radiochemical separations by retention on ionic precipitate adsorption tests on 11 materials. *J. Radioanal. Chem., 5,* 141-147.

Gladney E. S. (1978) Determination of arsenic, antimony, molybdenum, thorium and tungsten in silicates by thermal neutron activation and inorganic ion exchange. *Anal. Lett., All(5),* 429-435.

Gladney E. S. and Roelandts I. (1988) 1987 Compilation of elemental concentration data for USGS BHVO-1, MAG-1, QLO-1, RGM-1, SCo-1, SDC-1, SGR-1 AND STM-1. *Geostandards Newsletter, 12,* 253-367.

Gladney E. S., Burns C. E., and Roelandts I. (1983) 1982 compilation of elemental concentration in eleven United States Geological Survey rock standards. *Geostandards Newsletter, 7,* 3-226.

Gladney E. S., Burns C. E., and Roelandts I. (1984) 1982 compilation of elemental concentration data for the United States Geological Survey's Geochemical exploration reference samples GXR-1 to GXR-6. *Geostandards Newsletter, 8,* 119-154.

Gladney E. S., O'Malley B. T., Roelandts I., and Gills T. E. (1987) Compilation of elemental concentration data for NBS Biological, Geological and Environmental Standard Reference Material. *National Bureau of Standards Special Publication, 260-111.*

Gladney E. S., Perrin D. R., Balagna J. P., and Warner C. L. (1980) Evaluation of a boron-filtered epithermal neutron irradiation facility. *Anal. Chem., 52,* 2128-132.

Green D. H. (1975) Genesis of Archean peridotitic magmas and constraints on Archean geothermal gradients and tectonics. *Geology, 3,* 15-18.

Helgeson H. C.(1964) *Complexing and Hydrothermal Ore Deposition.* Pergamon, New York. 128 pp.

Hofmann A. W. (1988) Chemical differentiation of the Earth: The relationship between mantle, continental crust and oceanic crust. *Earth Planet. Sci. Lett., 90,* 297-314.

Hofmann A. W. and White W. M. (1983) Ba, Rb and Cs in the Earth's mantle. *Z. Naturforsch., 38a,* 256-266.

Hofmann A. W., Jochum K. P., Seufert M., and White W. M. (1986) Nb and Pb in oceanic basalts: New constraints on mantle evolution. *Earth Planet. Sci. Lett., 79,* 33-45.

Horracks P. C. (1983) A corundum and sapphirine paragenesis from the Limpopo Mobile Belt, South Africa. *J. Metamorph. Geol., 1,* 13-23.

Jacobson S. B. and Wasserburg G. J. (1979) The mean age of the crustal and mantle reservoirs. *J. Geophys. Res., 84,* 7411-7427.

Jagoutz E., Palme H., Baddenhausen H., Blum K., Cendales M., Dreibus G., Spettel B., Lorenz V., and Wänke H. (1979) The abundances of major, minor and trace elements in the Earth's mantle as derived from primitive ultramafic nodules. *Proc. Lunar Planet. Sci. Conf. 10th,* pp. 2031-2050.

Jahn B., Auvray B., Blais S., Capdevila R., Vidal F., and Hameurt J. (1980) Trace-element geochemistry and petrogenesis of Finnish Greenstone belts. *J. Petrol., 21,* 201-244.

Jahn B., Gruau G., and Glikson A. Y. (1982) Komatiites of the Onverwacht Group, S. Africa: REE geochemistry, Sm/Nd age and mantle evolution. *Contrib. Mineral. Petrol., 80,* 25-40.

Jochum K. P., Hofmann A. W., and Arndt N. T. (1987) Nb/Th in Precambrian and modern komatiites and basalts (abstract). *Terra Cognita, 7,* 396.

Jochum K. P., Hofmann A. W., Ito E., Seufert H. M., and White W. M. (1983) K, U and Th in mid-ocean ridge basalt glasses and heat production, K/U and K/Rb in the mantle. *Nature, 306,* 431-436.

Jochum K. P., Hofmann A. W., and Seufert H. M. (1984) Global trace element systematics in oceanic basalts (abstract). *27th Int. Geol. Congr., Moscow, Abstr. Vol IX, Part 2,* p. 190.

Joron J. L., Bougault H., Wood D. A., and Treuil M. (1978) Application de la geochimie des elements en traces a l'etude des proprietes et des processes de genese de la croute oceanique et du manteau superier. *Bull. Soc. Geol. Fr., 20,* 521-531.

Jones J. H. and Drake M. J. (1983) Experimental investigations of trace element fractionation in iron meteorites, II. The influence of sulfur. *Geochim. Cosmochim. Acta, 47,* 1199-1209.

Jones J. H. and Drake M. J. (1986) Geochemical constraints on core formation in the Earth. *Nature, 322,* 221-228.

Kateman G. and Pijpers F. W. (1981) Quality control in analytical chemistry. In *Chemical Analysis* (P. J. Elving and J. D. Winerfordner, eds.), pp. 1-133. Wiley, New York.

Kessler S. E. (1989) Subvolcanic hydrothermal systems in the continental crust. *N. M. Bur. Mines Miner. Resour., Bull., 131,* 151.

Klöck W. and Palme H. (1988) Partitioning of siderophile and chalcophile elements between sulfide, olivine and glass in a naturally reduced basalt from Disko Island, Greenland (abstract). In *Lunar and Planetary Science XVIII,* pp. 493-494. Lunar and Planetary Institute, Houston.

Kurat G., Palme H., Spettel B., Baddenhausen H., Hofmeister H., Palme C., and Wänke H. (1980) Geochemistry of ultramafic nodules from Kapfenstein, Austria: Evidence for a variety of upper mantle processes. *Geochim. Cosmochim. Acta, 44,* 45-60.

McLennan S. M., Taylor S. R., and Eriksson K. A. (1983a) Geochemistry of Archean shales from the Pilbara Supergroup, Western Australia. *Geochim. Cosmochim. Acta, 47,* 1211-1222.

McLennan S. M., Taylor S. R., and Kröner A. (1983b) Geochemical evolution of Archean shales from South Africa 1: The Swaziland and Pongola Supergroups. *Precambrian Res., 22,* 93-124.

McLennan S. M., Taylor S. R., and McGregor V. R. (1984) Geochemistry of Archean meta-sedimentary rocks from West Greenland. *Geochim. Cosmochim. Acta, 48,* 1-15.

Mitchell R. H. and Keays R. R. (1981) Abundance and distribution of gold, palladium and iridium in some spinel and garnet lherzolites, Implication for the nature and origin of precious metal-rich intergranular components in the upper mantle. *Geochim. Cosmochim. Acta, 45,* 2425-2442.

Morgan J. W., Wandless G., Petrie R., and Irving A. (1980) Composition of the Earth's upper mantle—II: Volatile trace elements in ultramafic xenoliths. *Proc. Lunar Planet. Sci. Conf. 11th,* pp. 213-233.

Nance W. B. and Taylor S. R. (1976) Rare earth patterns and crustal evolution—I. Australian post-Archean sedimentary rocks. *Geochim. Cosmochim. Acta, 40,* 1539-1551.

Newsom H. E. (1990) Accretion and core formation in the Earth: Evidence for siderophile elements. In *Origin of the Earth,* this volume.

Newsom H. E. and Drake M. J. (1982) The metal content of the Eucrite Parent Body, constraints from the partitioning behavior of tungsten. *Geochim. Cosmochim. Acta, 46,* 2483-2489.

Newsom H. E. and Drake M. J. (1983) Experimental investigation of the partitioning of phosphorus between metal and silicate phases: Implications for the Earth, Moon and Eucrite Parent Body. *Geochim. Cosmochim. Acta, 47,* 93-100.

Newsom H. E. and Palme H. (1984) The depletion of siderophile elements in the Earth's mantle: New evidence from molybdenum and tungsten. *Earth Planet. Sci. Lett., 69,* 354-364.

Newsom H. E., White W. M., Jochum K. P., and Hofmann A. W. (1986) Siderophile and chalcophile element abundances in oceanic basalts, lead isotope evolution and growth of the Earth's core. *Earth Planet. Sci. Lett., 80,* 299-313.

Oliver J. (1982) Probing the structure of the deep continental crust. *Science, 216,* 689.

Patchett P. J. (1983) Hafnium isotope results from mid-ocean ridges and Kerguelen. *Lithos, 16,* 47-51.

Rammensee W. and Palme H. (1982) Metal-silicate extraction technique for the analysis of geological and meteorite samples. *J. Radioanal. Chem., 71,* 401-418.

Rammensee W. and Wänke H. (1977) On the partition coefficient of tungsten between metal and silicate and its bearing on the origin of the moon. *Proc. Lunar Sci. Conf. 8th,* pp. 399-409.

Ringwood A. E. (1966) Mineralogy of the mantle. In *Advances in Earth Sciences* (P. Hurley, ed.), pp. 357-398. MIT, Boston.

Ringwood A. E. (1979) *Origin of the Earth and Moon.* Springer-Verlag, New York. 295 pp.

Schmitt W., Palme H., and Wänke H. (1989) Experimental determination of metal/silicate partition coefficients for P, Co, Ni, Cu, Ga, Ge, Mo, and W and some implications for the early evolution of the Earth. *Geochim. Cosmochim. Acta, 53,* 173-185.

Shewart W. A. (1931) *Economic Control of the Quality of Manufactured Product.* Van Nostrand, New York.

Stacey J. S. and Kramers J. D. (1975) Approximation of terrestrial lead isotope evolution by a two-stage model. *Earth Planet. Sci. Lett., 26,* 207-221.

Steinnes E. (1971) Epithermal neutron activation analysis of geological material. In *Activation Analysis in Geochemistry and Cosmochemistry* (A. Brunfelt and E. Steinnes, eds.), pp. 113-129. Universitetsforlaget, Oslo.

Sun S. S. (1980) Lead isotopic study of young volcanic rocks from mid-ocean ridges, ocean islands and island arcs. *Philos. Trans. R. Soc. London, A297,* 409-445.

Sun S. S. (1982) Chemical composition and origin of the Earth's primitive mantle. *Geochim. Cosmochim. Acta, 46,* 179-192.

Sun S. S. and Nesbitt R. W. (1978) Petrogenesis of Archean ultrabasic and basic volcanics: Evidence from rare earth elements. *Contrib. Mineral. Petrol., 65,* 301-325.

Tatsumoto M. (1978) Isotopic composition of lead in oceanic basalt and its implication to mantle evolution. *Earth Planet. Sci. Lett., 38,* 63-87.

Taylor S. R. (1964) The abundances of chemical elements in the continental crust—a new table. *Geochim. Cosmochim. Acta, 28,* 1273-1285.

Taylor S. R. and McLennan S. M. (1985) *The Continental Crust: Its Composition and Evolution.* Blackwell, Oxford. 312 pp.

Taylor S. R., McLennan S. M., and McCullough M. T. (1983) Geochemistry of loess, continental crustal composition and crustal model ages. *Geochim. Cosmochim. Acta, 47,* 1897-1905.

Taylor S. R., Rudnick R. L., McLennan S. M., and Eriksson K. A. (1986) Rare earth element patterns in Archean high-grade meta-sediments and their tectonic significance. *Geochim. Cosmochim. Acta., 50,* 2267-2279.

Vidal P. and Dosso L. (1978) Core formation: catastrophic or continuous? Sr and Pb isotope geochemistry constraints. *Geophys. Res. Lett., 5,* 169-172.

Vollmer R. (1977) Terrestrial lead isotopic evolution and formation time of the Earth's core. *Nature, 270,* 144-147.

Wänke H. (1981) Constitution of terrestrial planets. *Philos. Trans. R. Soc. London, A303,* 287-302.

Wänke H., Dreibus G., and Jagoutz E. (1984) Mantle chemistry and accretion history of the Earth. In *Archaean Geochemistry* (A. Kröner, G. N. Hanson, and A. M. Goodwin, eds.), pp. 1-24. Springer-Verlag, New York.

Weaver B. L. and Tarney J. (1984) Major and trace element composition of the continental lithosphere. *Phys. Chem. Earth, 15,* 39-68.

Weckworth G. (1983) Anwendung der instrumentellen β-Strektromterie im Bereich der Kosmochemie insbesondre zur Messung von Phosphorgehalten. Unpublished thesis, Universitat Mainz, F. R. Germany.

Wedepohl K. H. (1975) The contribution of chemical data to assumptions about the origins of magma from the mantle. *Fortschr. Mineral., 52,* 141-172.

Wedepohl K. H. (1981) Der primare Erdmantle (Mp) und die durch die Krustenbildung verarmte Mantlezusammensetzung (Md). *Fortschr Miner., 59,* 203-205.

Weissberg B. G., Browne P. R., and Seward T. M. (1979) Ore metals in active geothermal systems. In *Geochemistry of Hydrothermal Ore Deposits* (H. L. Barnes, ed.), pp. 738-780. Wiley, New York.

White W. M. and Hofmann A. W. (1982) Sr and Nd isotope geochemistry of oceanic basalts and mantle evolution. *Nature, 296,* 821-825.

THE ACASTA GNEISSES: REMNANT OF EARTH'S EARLY CRUST

S. A. Bowring, T. B. Housh, and C. E. Isachsen

Department of Earth and Planetary Sciences, Washington University, St. Louis, MO 63130

The 3.96-Ga Acasta gneisses from the Slave craton of northwestern Canada are the oldest rocks yet recognized on Earth. Uranium-lead ion-microprobe analyses of zircons separated from two samples, a banded tonalitic orthogneiss and a granitic orthogneiss, yield the same crystallization age of 3.96 Ga. The U-Pb zircon data from both samples define complex arrays on a concordia diagram and may be interpreted in terms of zircon crystallization at 3.96 Ga followed by an early Pb-loss event and crystallization of new zircon at ca. 3.6 Ga, followed by a Proterozoic or younger Pb-loss event. Neodymium and lead isotopic data suggest that crust older than 3.96 Ga was involved in the generation of these rocks. The tonalitic orthogneiss has a chondritic Nd model age of ca. 4.1 Ga and an ϵ_{Nd}(3.96 Ga) of -1.7. A mafic layer from the tonalitic orthogneiss and the granitic orthogneiss have ϵ_{Nd}(3.96 Ga) of +0.7 and +0.8, respectively. The least radiogenic Pb isotopic compositions of feldspars from the orthogneisses are very radiogenic relative to early Archean model mantle compositions and have model ages considerably younger than their age. The feldspar Pb isotopic data indicate that these rocks were derived from a source region that had experienced a complex multistage history prior to 3.96 Ga and had a high U:Pb ratio during some or all of its evolution. Thus, the Acasta gneisses are the first example of an early Archean isotopically enriched reservoir. The U-Pb geochronology, Nd and Pb isotopic compositions, and geochemistry of the Acasta gneisses are consistent with the involvement of older crust in their formation. By implication, available Nd and Pb isotopic data from other Archean cratons worldwide may be interpreted as reflecting mixtures of older crust and mantle, rather than mantle heterogeneity alone. The 3.96-Ga Acasta gneisses are evidence that early, enriched, sialic crust was preserved, although isotopic and geochemical arguments suggest that it is not the sole enriched complement to the depleted mantle.

INTRODUCTION

Since the recognition of 3.8-Ga supracrustal rocks at Isukasia, Greenland (*Moorbath et al.,* 1972), there has been much debate about Earth's evolution during the first 800 Ma of its history. More recently, the discovery of 4.0-4.3-Ga detrital zircons in western Australia (*Froude et al.,* 1983; *Compston and Pidgeon,* 1986) led to the question of whether intact crust older than the Isua supracrustals would ever be found (e.g., *Moorbath,* 1986). The presence of the 4.0-4.3-Ga detrital grains suggests that differentiated crustal rocks were in existence by 4.3 Ga; nevertheless, the lack of recognition of any rocks much older than 3.8 Ga has been used to infer that the Earth's earliest differentiated crust was either not voluminous (e.g., *Moorbath,* 1986) or was preferentially destroyed (*Jacobsen and Dymek,* 1988). A better understanding of the first billion years of the Earth's history will provide valuable constraints on early planetary accretion, differentiation, and the processes and mechanisms of heat loss and crust generation.

The Acasta gneisses located in northwestern Canada are 3.96 Ga, making them the oldest rocks yet recognized on Earth. The purpose of this paper is to review the geology, U-Pb zircon geochronology, Pb and Nd isotopic composition, and geochemistry of the Acasta gneisses, and to discuss the implications of these data for models of Earth's evolution. Geochemical and isotopic data from the Acasta gneisses require that older crust was involved in their generation. This observation has important implications for the generation of other early Archean tonalites, for the timing of the cessation of catastrophic meteorite bombardment, and for the development and geochemical evolution of Earth's earliest crust.

GEOLOGY OF THE SLAVE CRATON

The Slave craton is an Archean granite-greenstone terrane located in the northwestern part of the Canadian Shield (Fig. 1) with an area of approximately 190,000 km^2. It is bounded on the east by the 2.0-1.9-Ga Thelon orogen, and on the west by the 1.9-1.8-Ga Wopmay orogen (Fig. 1). The geology of the Slave craton is summarized by *McGlynn and Henderson* (1970), *Henderson* (1981, 1985), and *Padgham* (1985), and the regional metamorphism is described by *Thompson* (1978, 1989). Supracrustal rocks collectively termed the Yellowknife Supergroup (*McGlynn and Henderson,* 1970; *Henderson,* 1981) constitute approximately 40% of the outcrop in the Slave craton (Fig. 1) and have been subdivided into separate "belts" (e.g., *Padgham,* 1985). In contrast to other Archean cratons, the supracrustal rocks of the Slave craton contain a higher proportion of metaturbidites relative to metavolcanic rocks (*Padgham,* 1985). The remainder of the craton consists of a variety of granitoid and gneissic rocks, including areas dominated by gneisses older than 2.8 Ga.

Uranium-lead zircon ages from the Slave craton indicate that most of the metavolcanic rocks were erupted between 2.65 and 2.72 Ga and were then intruded by 2.58- to 2.67-Ga granitic to dioritic plutons (*Green and Baadsgaard,* 1971; *Henderson et al.,* 1987; *van Breemen and Henderson,* 1988; *Mortensen et al.,* 1988; and *Isachsen et al.,* 1990). Rocks older than the Yellowknife Supergroup have been found exclusively in the western part of the Slave craton (heterogeneous gneisses of Fig. 1). These older rocks consist of heterogeneous, tonalitic to granitic gneisses, and granitic to dioritic plutons that have yielded ages of 2.82-3.15 Ga [*Easton,* 1985 (PL, Fig. 1); *Frith et al.,* 1986 (GL, Fig. 1); *Krogh and Gibbins,* 1978 (PL, Fig. 1); *Henderson et al.,* 1982 (PL, Fig. 1), 1987 (SD, Fig. 1); C. E. Isachsen and S. A. Bowring, unpublished data, 1989]. Within these regions of old rocks, remnants of supracrustal rocks are found whose ages and relationships to the older rocks are uncertain. Within Wopmay orogen, several exposures of rocks older than 2.9 Ga have been discovered although their lateral extent is not yet known (*Hildebrand et al.,*

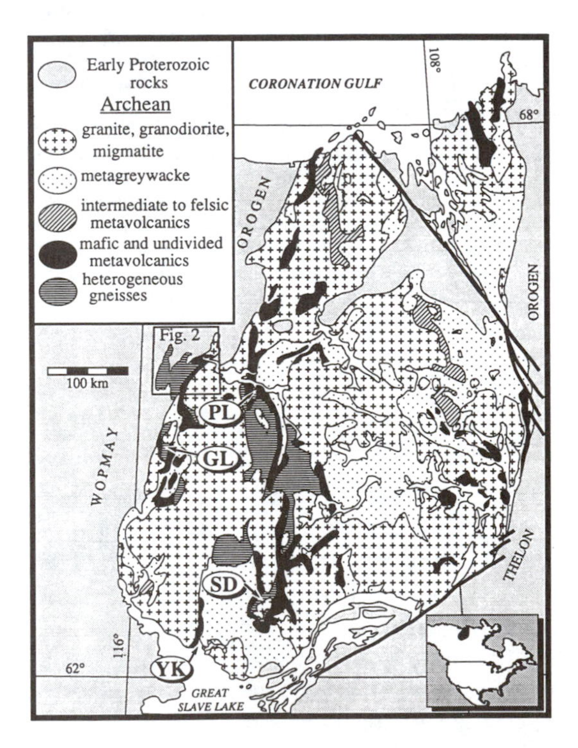

Fig. 1. *Lithologic map of the Slave Province. Box shows location of Fig. 2. Compiled from McGlynn (1977), Henderson (1985), Padgham (1985), and Hoffman (1989). Locations discussed in text: Point Lake (PL); Sleepy Dragon (SD); Grenville Lake (GL); YK indicates the location of the city of Yellowknife.*

1990; *Chamberlain et al.,* 1984). *Nikic et al.* (1980) reported U-Pb zircon minimum ages as old as 3.21 Ga for boulders contained in a diatreme from the Con Mine in Yellowknife (YK, Fig. 1). Furthermore, *Dudas* (1989) has reported Nd isotopic data from the southern Slave craton that suggest the presence of crust as old as 3.3 Ga in the Sleepy Dragon Complex (SD, Fig. 1).

The most detailed study of the gneissic rocks within the Slave craton is that of *Easton* (1985) in the western Point Lake area where he recognized several generations of ortho- and para-gneisses. *Easton* (1985) presented descriptions and representative chemical analyses of the major rock types, described the relationships between greenstones and the gneisses, and speculated that the gneisses underlie much of the western Slave craton. The relationship of these older gneisses to the Yellowknife Supergroup is not well understood. The older rocks have been considered to be stratigraphic basement to the Supergroup (e.g., *Stockwell,* 1933; *Henderson,* 1981; *Easton,* 1985) although unequivocal, unconformable relationships with the Yellowknife Supergroup volcanic rocks have not been observed, and where exposed, contacts are tectonic (*Easton,* 1985; *Kusky,* 1989; *Lambert and van Staal,* 1987). At Point Lake (PL, Fig. 1) an unconformity separating conglomerate and turbidites from 3.15-Ga granite has been invoked as evidence that volcanic and sedimentary rocks were deposited in response to rifting of the basement (*Henderson,* 1981, 1985; *Easton,* 1985). Subsequent work, however, has shown that the volcanic rocks are below this unconformity, and their contact with the 3.15-Ga granite is a shear zone (*Kusky,* 1989). *Kusky* (1987) and *Hoffman* (1989) interpreted these relationships as indicating that the conglomerates were deposited during a thrusting event that occurred after structural juxtaposition of the volcanic rocks and basement.

The role of older crust is important in models for the growth and assembly of the Slave craton. Models for the granite-greenstone belts proposed to date range from collapsed ensialic rifts (*Henderson,* 1981; *Easton,* 1985) to accreted island arc complexes (*Folinsbee et al.,* 1968; *Hoffman,* 1986; *Fyson and Helmstaedt,* 1988; *Kusky,* 1989). *Kusky* (1989) interpreted the remnants of pre-Yellowknife Supergroup rocks in the western Slave craton as a microcontinent that collided with a complex arc terrane ca. 2.6 Ga, followed by an arc-polarity reversal.

Initial sampling of the region containing the very old rocks described in this paper occurred in 1983 during regional geological mapping and geochronological studies by members of the Geological Survey of Canada (GSC) and the senior author, respectively, in early Proterozoic Wopmay orogen (*St-Onge et al.,* 1984, 1988; *Bowring and Van Schmus,* 1984). *Fraser* (1960) first recognized and mapped the structural culminations shown in Figs. 1 and 2 that expose the ancient gneisses, although their age was not then known. The first recognition that the rocks of this region were continuous with the rocks of Slave craton and underlay the early Proterozoic rocks of Wopmay orogen was the result of regional mapping by M. St-Onge, J. King, and A. Lalonde of the GSC (*St-Onge et al.,* 1984, 1988). The rocks were first dated as a test of the hypothesis that an early Proterozoic (2.0-2.3-Ga) terrane had overthrust the western edge of Slave craton prior to development of the early Proterozoic Coronation Supergroup. In this model, the Acasta gneisses would have been the leading edge of the overthrust terrane. J. King of the GSC collected a sample of tonalitic gneiss (BG 83 A) from the Acasta River region (Fig. 3) as part of a regional geochronologic study to test this model, and it yielded an age of 3.48 Ga (*Bowring and Van Schmus,* 1984; *Bowring,* 1985; *Bowring et al.,* 1989a). Additional sampling and mapping in this region by the authors led to the recognition of rocks with complex zircon systematics that had Pb-Pb ages as old as 3.84 Ga with even older Nd model ages (*Bowring et al.,*

1989a). Resolution of the zircon systematics required the use of the ion-microprobe (SHRIMP) at the Australian National University, which indicated that the two samples had crystallization ages as old as 3.96 Ga (*Bowring et al.,* 1989b). The gneisses were informally termed the Acasta gneisses (*Bowring et al.,* 1989b) for their exposures along the Acasta River. The river was named after a vessel of the British Merchant Marine that was sunk in the first year of World War II.

THE ACASTA GNEISSES

Rocks currently known to be older than 2.8 Ga in the Slave craton occur exclusively in its westernmost part. The Slave craton extends at least 100 km beneath the deformed and metamorphosed supracrustal rocks of early Proterozoic Wopmay orogen (Figs. 1 and 2). Supracrustal rocks of Wopmay orogen were detached and thrust eastward over the western edge of Slave craton during the ca. 1.9-Ga Calderian Orogeny (*Hoffman and Bowring,* 1984; *King,* 1986; *Hoffman et al.,* 1988). The structural culminations that expose Archean gneisses beneath Proterozoic cover are the result of interference between two thick-skinned, basement-involved fold sets that probably formed between 1.89 and 1.84 Ga (*Bowring,* 1985; *King,* 1986). The first set of basement involved folds are north-trending high-amplitude folds having a ca. 35-km wavelength and an apparent structural relief, in the region shown in Fig. 2, of approximately 6 km (*King,* 1986; *Hoffman et al.,* 1988). The second set of folds trend northeast and have wavelengths of 80-140 km and amplitudes of up to 15 km (*King,* 1986; *Hoffman et al.,* 1988). Metamorphic conditions in the autochthonous Proterozoic cover are estimated to have been up to 9.5 kbar and 620°C (*St-Onge and King,* 1987a,b). The autochthonous Proterozoic cover is characterized by inverted metamorphic isograds caused by thrusting of hot allochthonous sheets onto the relatively cool autochthon followed by rapid (1.5-2.7 mm/yr) unroofing and relaxation of isotherms into the basement (*St-Onge and King,* 1987b). The Archean gneisses in the culminations have a recognizable Proterozoic fabric only near the basement-cover interface (*St-Onge et al.,* 1984; *King et al.,* 1987). The fabric is a schistosity defined by alignment of retrograde chlorite grains and a lineation defined by recrystallized quartz, feldspar, and hornblende (*St-Onge et al.,* 1984; *King et al.,* 1987). The samples discussed in this paper are from the structural saddle between the two westernmost culminations (Exmouth and Scotstoun culminations, Fig. 2). Here the Proterozoic metamorphism of the basement is interpreted to be biotite-grade or less (*King,* 1986; *Hoffman et al.,* 1988). This suggests that the relatively high temperature and pressure metamorphism recorded in the Proterozoic cover sequence did not greatly affect the Acasta gneisses described here.

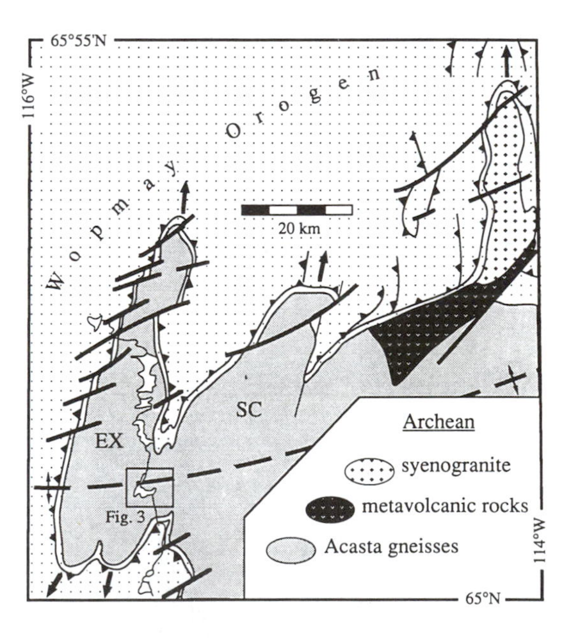

Fig. 2. *Generalized geologic map of the Exmouth and Scotstoun culminations, showing the location of the samples discussed in the paper and the location of Fig. 3 (modified after St-Onge et al., 1988). Barbed lines are Proterozoic-aged thrust faults; solid dark lines are Proterozoic-aged transcurrent faults; arrows and heavy dashed line represent Proterozoic fold axes.*

The Acasta gneisses are a heterogeneous assemblage composed predominantly of strongly foliated to mylonitic, biotite-hornblende tonalitic to granitic orthogneisses commonly interlayered on a centimeter to meter scale with amphibolitic and chloritic schlieren, boudins, and layers. Relatively large areas composed predominantly of amphibolite also occur (Fig. 3). Less common lithologies include calc-silicate gneisses, quartzites, biotite schists, and ultramafic schists (tremolite-chlorite and serpentine-tremolite-talc ± forsterite). The ultramafic schists are the only rocks yet found that contain mineral assemblages constraining metamorphic grade. The rare occurrence of the assemblage tremolite-serpentine-talc-forsterite indicates that temperatures did not exceed approximately 650°C, yet were in excess of 400°C (*Winkler,* 1979; *Turner,* 1981). All the rocks are intruded by weakly foliated, gabbroic to dioritic dikes and pods and weakly to strongly foliated to mylonitic biotite-bearing granites. The gneisses are intruded by north-trending diabase dikes

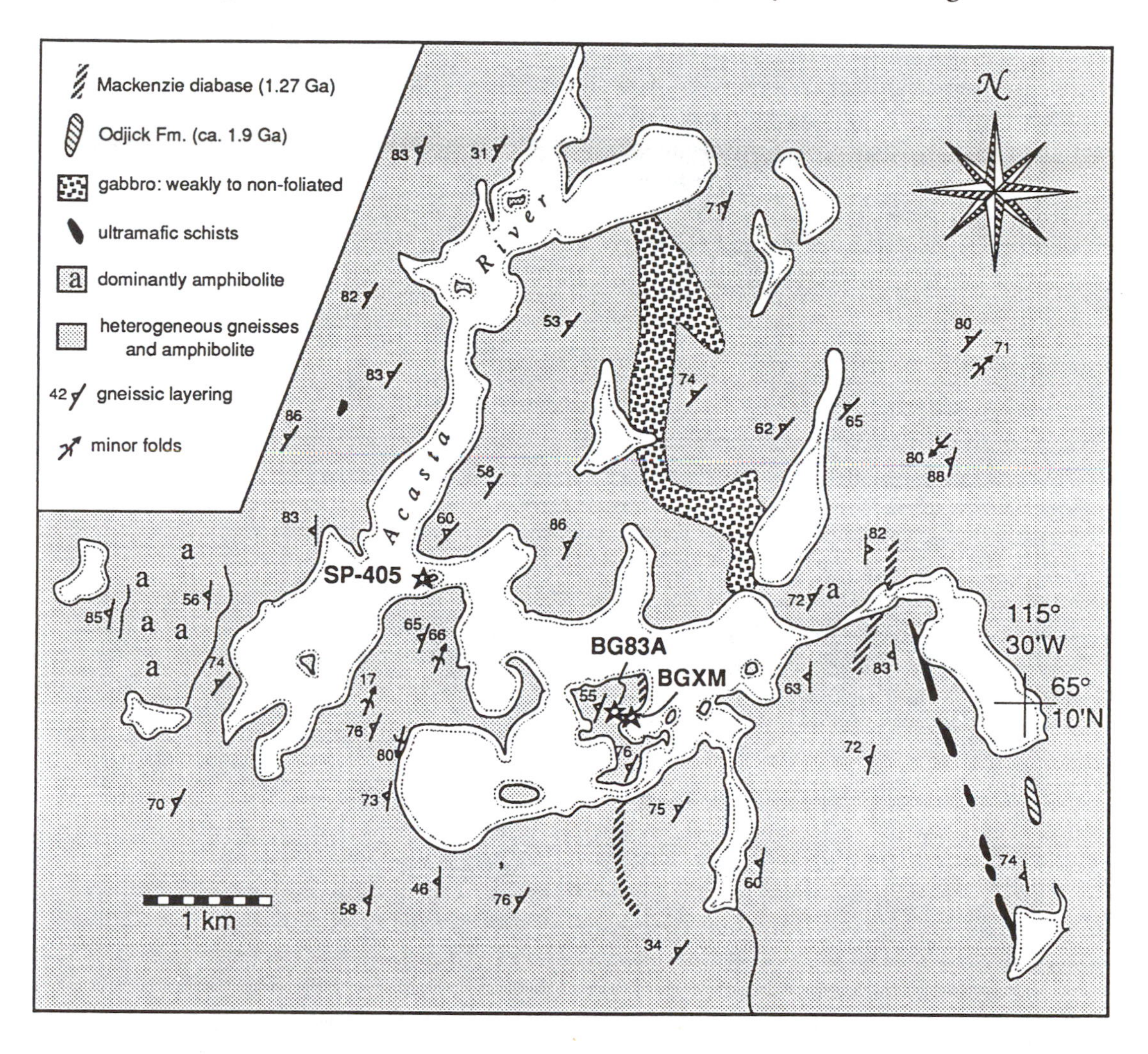

Fig. 3. *Reconnaissance geologic map of a portion of the Acasta gneisses showing locations of the samples discussed in the text.*

tentatively correlated with the ca. 1.267-Ga Mackenzie dike swarm (*LeCheminant and Heaman,* 1989).

Two samples of Acasta gneisses, SP-405 and BGXM, have been dated at 3.96 Ga (*Bowring et al.,* 1989b). BGXM is a layered amphibolitic to tonalitic orthogneiss in which the amphibolitic and tonalitic layers alternate on a centimeter scale. The tonalitic layers consist of granoblastic plagioclase and quartz with plagioclase porphyroclasts and accessory allanite, sphene, and zircon. The amphibolitic layers consist of subhedral biotite and hornblende with variable amounts of anhedral plagioclase and abundant sphene and zircon. SP-405 is a granitic orthogneiss consisting of granoblastic plagioclase, alkali feldspar, and quartz with larger (ca. 10 mm) alkali feldspar porphyroclasts. Mafic minerals within SP-405 are concentrated in thin layers and consist of subhedral biotite, tremolite, and poikiloblastic garnet.

Geochemistry

Chemical analyses of BGXM and SP-405 are given in Table 1. BGXM is similar in composition to other early Archean tonalites (see *Martin,* 1986; *Condie,* 1981; *O'Nions and Pankhurst,* 1978) in that it has rather low K_2O/Na_2O (0.67), low Yb_N (3.73), moderate $(La/Yb)_N$ (11.6), and high Sr/Ba (0.55). The rare earth element pattern for BGXM (Fig. 4) is similar to those for the Group B Amitsoq tonalitic to granodioritic gneisses (*O'Nions and Pankhurst,* 1974, 1978), including the presence of a significant positive Eu anomaly ($Eu/Eu^* = 1.78$). The fractionated rare earth patterns and low heavy rare earth element abundances of BGXM and the Group B Amitsoq gneisses are characteristic of other early Archean tonalites (*Martin,* 1986). Also shown in Fig. 4 are the extreme compositions of basement granodiorite and orthogneisses from the Point Lake area (PL, Fig. 1; *Easton,* 1985). BGXM is most similar to a 3155-Ma granodiorite (*Krogh and Gibbins,* 1978) that defines the low end of the Point Lake array.

TABLE 1. *Chemical analyses of BGXM and SP-405.*

	BGXM	SP-405
SiO_2	65.52	71.95
TiO_2	0.34	0.15
Al_2O_3	13.84	14.92
Fe_2O_3*	5.37	1.63
MnO	0.09	0.04
MgO	4.19	0.63
CaO	4.10	1.82
Na_2O	3.52	4.65
K_2O	2.10	3.26
P_2O_5	0.06	0.04
LOI	0.88	0.14
TOTAL	100.01	99.22
Cr	278	
Co	17.61	
Sc	11.62	
Ni	75	
La	14.22	
Ce	23.6	
Nd	8.8	
Sm	1.58	
Eu	0.892	
Tb	0.204	
Yb	0.746	
Lu	0.117	
Rb	47.7	
Sr	332	
Ba	602	
Zr	166	
Hf	4.44	
Ta	0.300	
Th	1.55	
U	1.52	

* Total Fe reported as Fe_2O_3.

Major element analyses determined by XRF at Washington University; trace element analyses determined by INAA at Washington University.

SP-405, the granitic orthogneiss, is distinctly more felsic than BGXM; nevertheless, SP-405 has nearly the same K_2O/Na_2O ratio (0.70) as does the tonalitic orthogneiss, BGXM. The modal expression of the K_2O is different between the two samples: In BGXM the K_2O reflects the abundance of biotite, while in SP-405 it reflects the presence of alkali feldspar and some biotite.

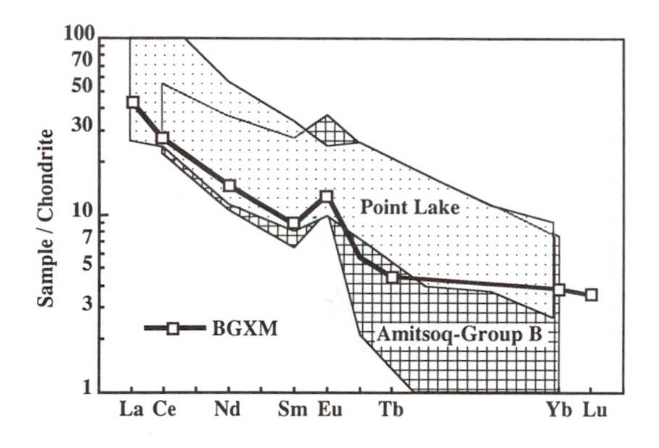

Fig. 4. *Chondrite-normalized rare earth element diagram for sample BGXM. Also shown in grid pattern are the extreme values for for the Amitsoq Group B gneisses (tonalitic to granodioritic; O'Nions and Pankhurst, 1974, 1978) and in dotted pattern the basement granodiorites and orthogneisses from the Point Lake area (Easton, 1985).*

U-Pb Zircon Geochronology

Both conventional U-Pb analyses and SHRIMP (Sensitive High-mass Resolution Ion Micro-Probe) analyses of zircons separated from the tonalitic and granitic orthogneisses were published by *Bowring et al.* (1989a,b). For details of the analyses as well as the table of isotopic data the reader is referred to *Bowring et al.* (1989a,b). In addition, new conventional analyses from the granitic orthogneiss (SP-405) are presented here (Table 2). The previously published ion-probe and conventional analyses are shown in Figs. 5a-d as well as the new analytical data for SP-405. The conventional U-Pb zircon data for sample BGXM were originally interpreted as indicating a mixture between cores whose ages were older than 3.84 Ga and younger overgrowths, thought to be 3.7 Ga (*Bowring et al.,* 1989a). However, subsequent analyses using the SHRIMP facility at the Australian National University (*Bowring et al.,* 1989b), have shown that most of the zircons from both BGXM and SP-405 crystallized 3.96 Ga.

BGXM. Zircon occurs throughout the sample as large (up to 1 mm long, ca. 250 μg), euhedral to subhedral grains. There are several distinct morphological types of zircon present in the sample, and there exists a relationship between morphology and age. The structurally

TABLE 2. *U-Pb isotopic data for SP-405.*

	Fractions		Concentrations		Measured	Ratios corrected for blank and common Pb				Age (Ma)
No.	Properties	Weight (mg)	U (ppm)	Pb (ppm)	206 Pb / 204 Pb	208 Pb / 206 Pb	206 Pb / 238 U	207 Pb / 235 U	207 Pb / 206 Pb	207 Pb / 206 Pb
1.	2 clear grains	0.061	226	166	2200	0.1496	0.55513	28.5665	0.373216	3804.5
2.	1 clear grain	0.057	242	204	3827	0.2040	0.61781	32.5774	0.382438	3841.4
3.	2 clear grains	0.038	172	125	1598	0.1353	0.56696	27.1317	0.347076	3694.2
4.	m2aa sm cl eu	0.116	672	518	5875	0.0810	0.63995	27.9220	0.316446	3552.6
5.	3 clear grains	0.051	288	225	474	0.1151	0.56932	28.5142	0.363250	3763.5
6.	3 clear grains	0.020	202	160	900	0.1335	0.61022	28.8818	0.343269	3677.4
7.	2 clear grains	0.051	1767	824	1616	0.0391	0.40411	14.996	0.269205	3301.4
8.	5 clear grains	0.125	622	309	257	0.1056	0.34584	15.7348	0.329981	3617.0

Notes: m, magnetic; number following refers to degrees of tilt on the Frantz isodynamic separator; aa, air-abraded; sm, small; cl, clear; eu, euhedral.

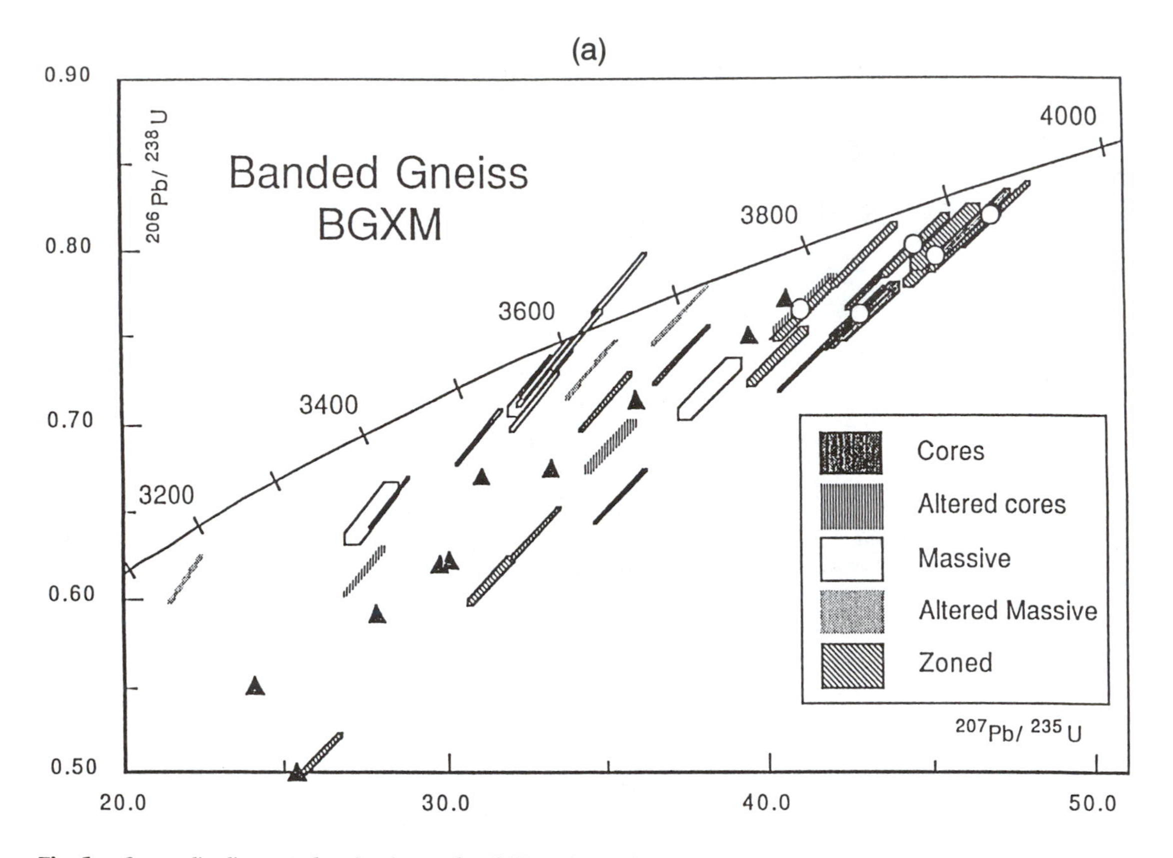

Fig. 5. *Concordia diagram showing ion probe Pb-U analyses (from Bowring et al., 1989b) of two samples of Acasta gneisses. Shading identifies recognized structural types discussed in text and Bowring et al. (1989b). Analytical uncertainties are 1 sigma.* **(a)** *Tonalitic orthogneiss, BGXM; most concordant analyses; circles represent analyses of different areas of a single grain (grain 11 of Bowring et al., 1989b). Triangles are conventional small sample and single grain analyses of zircons from BGXM (Bowring et al., 1989a).*

oldest zircons form rounded cores within other grains. The cores are overgrown by thick layers of euhedral, finely zoned zircon that are overgrown in turn by massive, structureless zircon that also occurs as separate, equant grains. Euhedral, zoned zircon with or without thin massive overgrowths is the most abundant structural type present in the sample. The cores are clear and unzoned, although portions of them are commonly altered.

Zircons from both samples yield complex data arrays on concordia diagrams. For BGXM there are two distinct clusters of data near concordia, one exceeding 3.9 Ga and the other at ca. 3.6 Ga (Fig. 5a). The oldest components in the zircons are the cores and the zoned zircon, which are, within analytical uncertainty, the same age. Both the zoned zircon and the cores show a significant range in radiogenic $^{207}Pb/^{206}Pb$. Of the 23 analyses of zoned zircon (*Bowring et al.,* 1989b), the six highest $^{207}Pb/^{206}Pb$ are equal within error and correspond to an average age of 3964 ± 4 Ma.

The massive zircon, with one exception, is distinctly younger than the zoned zircon, with an average of the five most concordant analyses corresponding to a $^{207}Pb/^{206}Pb$ age of 3621 ± 6 Ma. The massive grains are interpreted as representing a second generation of zircon growth. We do not interpret the array of U-Pb data that is subparallel to concordia between 3.96 and 3.6 Ga as mixing between

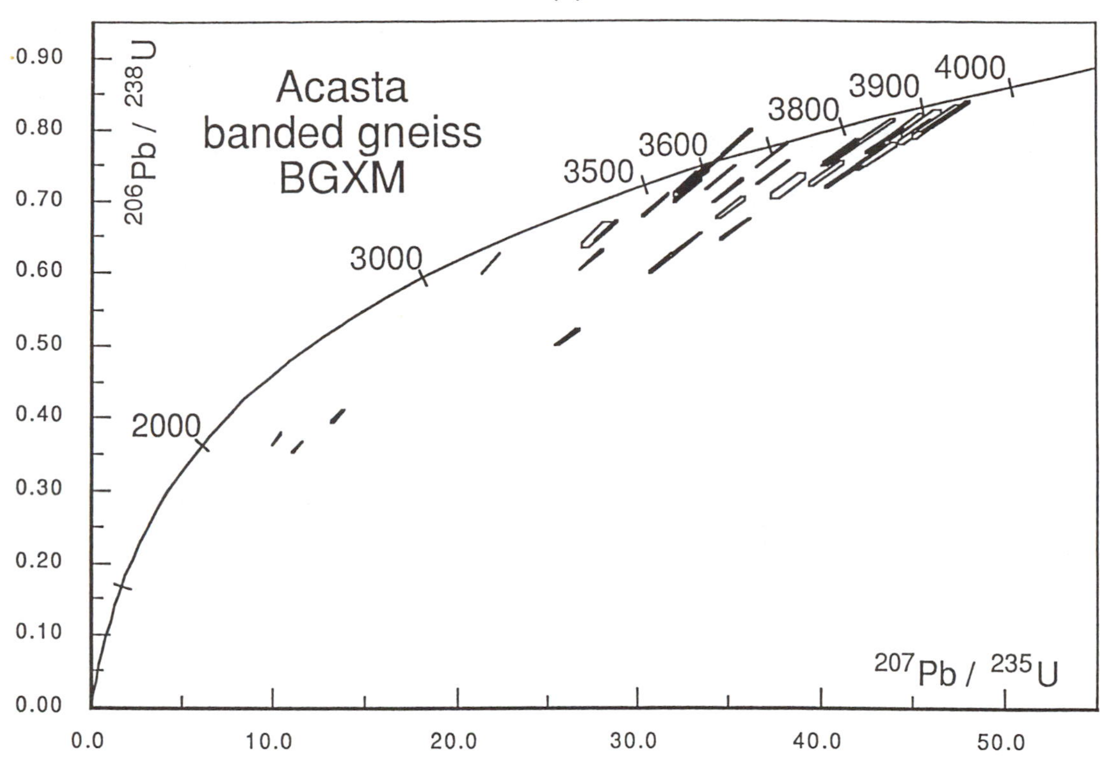

Fig. 5. *(continued)* **(b)** *BGXM; full-scale concordia diagram showing the complete dataset.*

the younger component and older cores but rather as the effect of an early Pb-loss event, perhaps synchronous with the time of new zircon growth (ca. 3.6 Ga). This interpretation is supported by multiple analyses from grain 11 (Fig. 5a), which show the same pattern of discordance as the entire dataset. Whether the new zircon growth is metamorphic or represents small amounts of tonalitic melt intruded into the gneiss protoliths prior to or during metamorphism is not known. As previously mentioned, a sample of tonalitic gneiss (BG-83-A), collected less than 50 m away from BGXM (Fig. 3), has an age of 3.48 Ga with cores that are slightly older (*Bowring et al.,* 1989a). Thus, the 3.48-Ga intrusion, the Pb isotopic data presented below, and unpublished U-Pb sphene and rutile ages from other gneisses in the region are consistent with periods of intrusion, deformation, and metamorphism ca. 3.4-3.6 Ga.

The full-scale concordia diagram for BGXM (Fig. 5b) clearly shows that the dataset as a whole disperses both subparallel to concordia and toward a Proterozoic lower intercept, suggesting that a younger Pb-loss episode(s) is superimposed on an earlier episode (*Bowring et al.,* 1989b). Also shown on Fig. 5a are conventional single and multigrain analyses from *Bowring et al.* (1989a). The complex history of Pb loss within different domains of single crystals makes even single grain conventional zircon data difficult to interpret.

SP-405. The zircons from SP-405 are distinctly different from those of BGXM and consist of four morphologies. The majority of the grains are prismatic and show well-developed fine euhedral zoning. Other euhedral grains lack well-developed zoning and are clear. A few of these euhedral prismatic grains contain structureless cores. About 10% of the zircons are equidimensional, massive to weakly

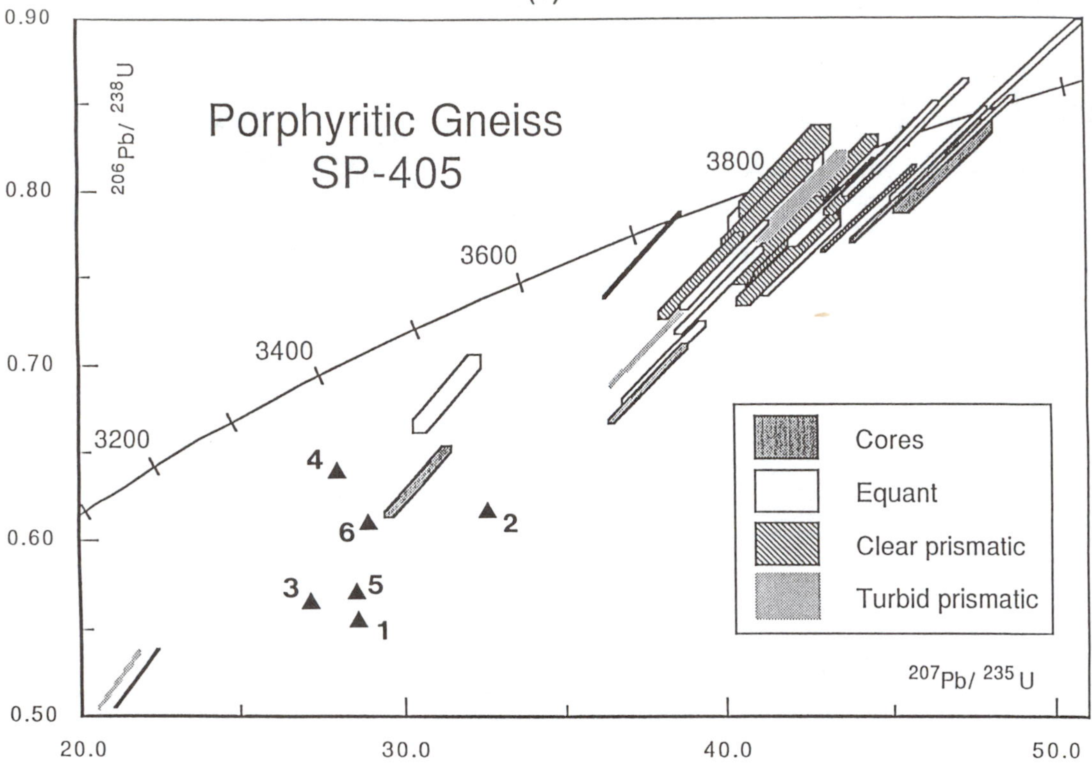

Fig. 5. *(continued)* **(c)** *SP-405, granitic orthogneiss; most concordant analyses; triangles are conventional analyses from Table 2.*

zoned, and in many cases overgrown by a thin zoned mantle.

Zircons from SP-405 show a history similar to, but simpler than, those from BGXM. The four oldest analyses yield an average $^{207}Pb/^{206}Pb$ age of 3958 ± 8 Ma. This is indistinguishable from the oldest group of ages from BGXM. The zircons as a whole show the same dispersion subparallel to concordia as seen in BGXM (Figs. 5c,d). The cores range widely in $^{207}Pb/^{206}Pb$, even within a single core, implying variable Pb loss. One core contains considerable unsupported radiogenic Pb and plots above concordia (Fig. 5d). The unsupported radiogenic Pb may be a result of the low uranium core taking up radiogenic Pb from the high uranium zircon surrounding it. The cores do not yield older ages than their overgrowths, and hence the 3958-Ma age is interpreted as the best estimate of their age of crystallization. The dispersion of analyses subparallel to concordia is interpreted to be the result of early Pb loss. A few grains appear to have distinctly younger ages that may be related to new zircon growth ca. 3.6 Ga, as observed in BGXM.

New conventional zircon data for SP-405 is presented in Table 2 and plotted on Figs. 5c and 5d. As with BGXM, the conventional analyses are distinctly more discordant than the ion-probe analyses. This reflects the complex internal variation in U-Pb systematics within single grains. Accordingly, analyses of single grains average the disparate isotopic compositions produced by variable Pb loss from the different domains within each crystal. A cluster of three highly discordant ion-probe analyses with $^{207}Pb/^{206}Pb$ ages of ca. 1.6 Ga

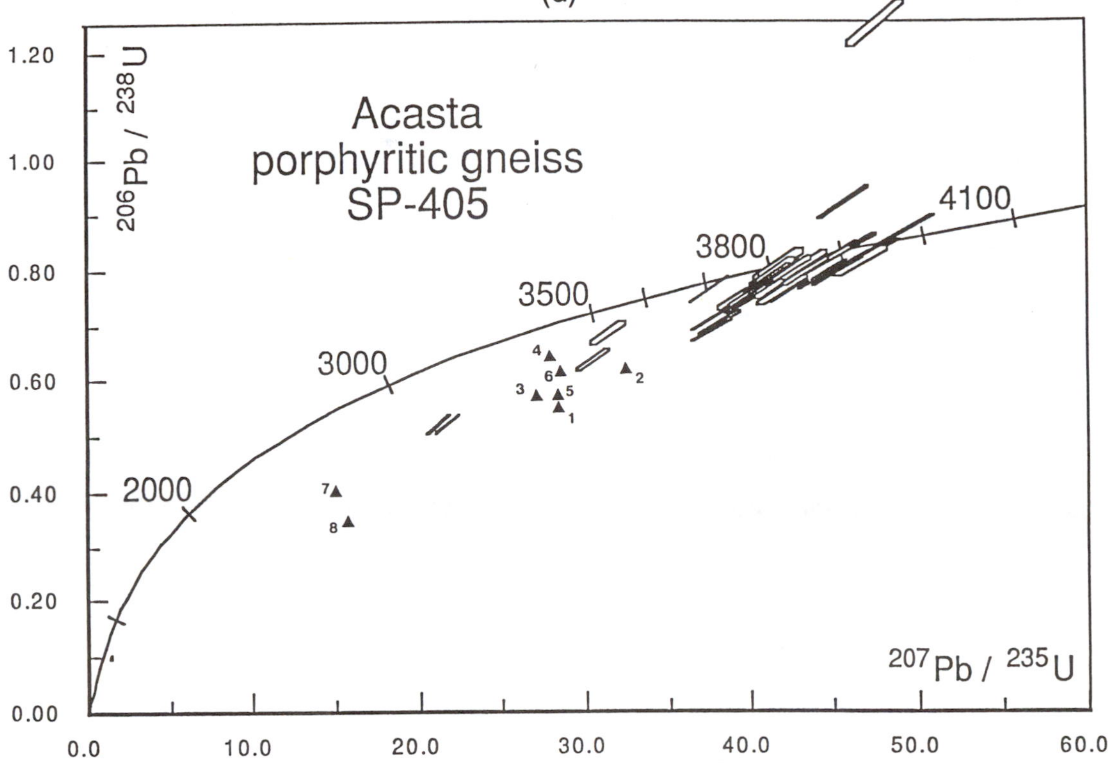

Fig. 5. *(continued)* **(d)** *SP-405; full-scale concordia diagram showing the complete dataset.*

and $^{206}Pb/^{238}U$ ages of ca. 0.6 Ga (Fig. 5d) are from altered core regions containing 2500-4500 ppm U. These highly discordant cores have suffered extreme Pb loss, perhaps related to Proterozoic tectonothermal activity.

The simplest interpretation of the data from both samples is that they crystallized as igneous rocks at 3.96 Ga. There is no textural or geochronologic evidence that either rock is a metamorphosed sedimentary rock. The zircons from both rocks are internally complex and have undergone at least two episodes of Pb loss. The first episode is perhaps related to metamorphism shortly after formation, and a second episode to a younger (Proterozoic?) event(s). The early Pb-loss event is responsible for the dispersion in the data subparallel to concordia and the younger Pb-loss event(s) for the dispersion toward the origin.

Most of the zircon cores from both samples yield $^{207}Pb/^{206}Pb$ ages indistinguishable from the overgrowths, although this does not preclude them from being significantly older. The zircons from BGXM are all slightly discordant, and therefore the average $^{207}Pb/^{206}Pb$ age must be viewed as a minimum. Structurally older cores may have lost much of their Pb during the event that produced the gneisses. Additional analyses of the core regions may yield even older ages consistent with the Nd model ages discussed below.

Nd Isotopic Data

Neodymium isotopic analyses were performed on whole-rock powders of BGXM and SP-405, as well as on an amphibolitic layer from BGXM. An aliquot of BGXM whole-rock powder has an initial $\epsilon_{Nd}(3.962)$ of -1.7 and a chondritic model age of 4.10 Ga (Table 3). From the BGXM hand-specimen, an amphibolitic layer (BGXM-M) was separated and powdered. An aliquot of BGXM-M yields an ϵ_{Nd}(3.962) of +0.7. SP-405 yields an ϵ_{Nd}(3.962) of +0.8. These data are given in Table 3 and shown graphically in Fig. 6. The main conclusions from the Nd isotopic data are that first, BGXM is heterogeneous in its Nd isotopic composition, with the amphibolitic component having a slightly more depleted signature than the whole rock. Second, another component(s) in BGXM was enriched at the time of its formation at 3.96 Ga, giving rise to a whole-rock ϵ_{Nd}(3.962) of -1.7. It is generally assumed that Sm and Nd are not significantly fractionated during metamorphism (e.g., *McCulloch and Wasserburg,* 1978). In detail, however, the Sm-Nd isotopic systematics of polymetamorphic high-grade rocks often show evidence of disturbance (e.g., *Collerson et al.,* 1989; *Black and McCulloch,* 1987). The disturbance of Sm-Nd isotopic systematics during metamorphism is not well understood; therefore, the extrapolation of the Nd isotopic data beyond ca. 3.6 Ga (the time of metamorphism recorded by zircon overgrowths and feldspar Pb systematics) must be viewed with some caution. Nevertheless, the Nd isotopic compositions of BGXM and SP-405 at 3.6 Ga are still the most

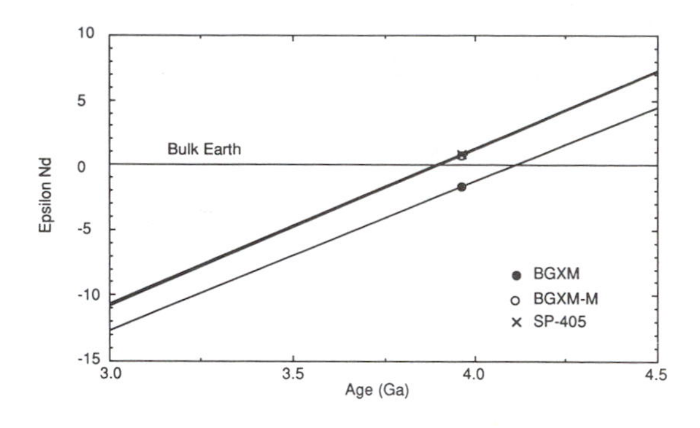

Fig. 6. *Neodymium isotopic evolution diagram for the three samples discussed in text with bulk earth ϵ_{Nd} = 0 for reference.*

enriched compositions [ϵ_{Nd}(3.6) = -6 and -3.5, respectively) reported to date (Fig. 6), consistent with chondritic model ages similar to or in excess of their crystallization ages.

Due to the generally low metamorphic grade of the Acasta gneisses, we interpret the observed difference as reflecting original protolith heterogeneity as opposed to metamorphic differentiation. The whole-rock data require that BGXM was derived from, or interacted with, a reservoir that is considerably older than the zircon crystallization age. The chondritic model age of 4.1 Ga is the oldest model age yet reported for terrestrial rocks. The slightly more depleted nature of the amphibolitic portion of BGXM (BGXM-M) at

TABLE 3. *Neodymium isotopic data.*

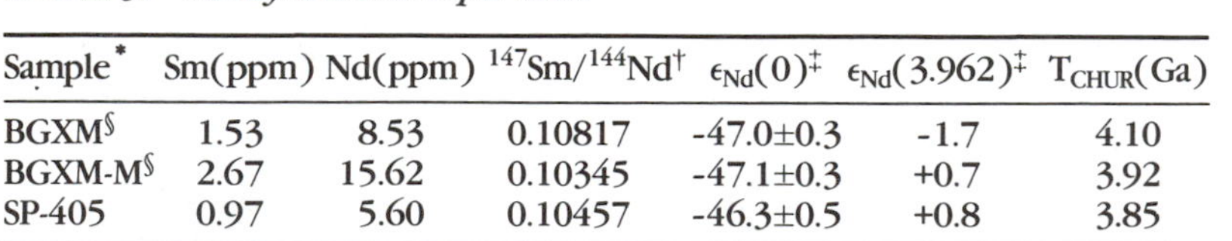

Sample*	Sm(ppm)	Nd(ppm)	$^{147}Sm/^{144}Nd$†	$\epsilon_{Nd}(0)$‡	$\epsilon_{Nd}(3.962)$‡	T_{CHUR}(Ga)
BGXM§	1.53	8.53	0.10817	-47.0±0.3	-1.7	4.10
BGXM-M§	2.67	15.62	0.10345	-47.1±0.3	+0.7	3.92
SP-405	0.97	5.60	0.10457	-46.3±0.5	+0.8	3.85

* Powdered samples were totally spiked and dissolved in a HF-HNO_3 mixture in a pressurized teflon capsule at 180°C for five days; procedures for chemical processing and isotope analysis are described in *Bowring and Podosek* (1989).
† Determined by mixed ^{149}Sm and ^{150}Nd spike calibrated against CIT mixed normal; uncertainty 0.1%.
‡ $^{143}Nd/^{144}Nd$ cited as parts in 10^4 deviation from CHUR, defined by present $^{143}Nd/^{144}Nd$ = 0.512638 (for discrimination corrections by $^{146}Nd/^{144}Nd$ = 0.7219) and $^{147}Sm/^{144}Nd$ = 0.1966; measured values are adjusted to be consistent with normal ϵ_{Nd} = -15.2 for La Jolla standard Nd; errors (2σ) are reproducibility of replicate analyses.
§ These analyses were reported in *Bowring et al.* (1989a).

3.96 Ga suggests that there is an even more enriched component present in the whole rock that when mixed with the mafic portion yields an intermediate value for the whole rock. A depleted mantle model age for the whole rock using existing approximations for the evolution of the mantle (e.g., *Galer et al.,* 1989) would be about 0.1 Ga older than the chondritic age, although constraints on the Nd isotopic composition of the mantle at 4.0 Ga do not exist. The $\epsilon_{Nd}(3.962)$ of +0.83 from the granitic orthogneiss (SP-405) suggests that it was derived from either a younger or more depleted source than the tonalitic orthogneiss (BGXM). This is somewhat surprising considering the relatively evolved character of the granitic orthogneiss relative to the tonalitic orthogneiss, but nevertheless, points to a complex evolution for the Acasta gneisses.

Pb Isotope Systematics

Lead isotopic compositions were determined for feldspars from two samples of the Acasta gneisses. The feldspars (alkali feldspar from SP-405; plagioclase from BGXM) were leached in 5% HF in a series of four to five steps to obtain the least radiogenic Pb isotopic composition as the best estimate of the initial Pb isotopic composition of the sample [see *Housh et al.* (1989) for details of the procedure]; the results are presented in Table 4. In a plot of $^{206}Pb/^{204}Pb$ vs. $^{207}Pb/^{204}Pb$ (Fig. 7), the isotopic compositions of the BGXM plagioclase

TABLE 4. *Lead isostopic compositions of feldspar leaches.*

Sample/ Leach #	$^{206}Pb/^{204}Pb$	$^{207}Pb/^{204}Pb$	$^{208}Pb/^{204}Pb$
BGXM,L2	15.202	15.369	33.657
BGXM,L3	14.957	15.290	33.534
BGXM,L4	15.033	15.311	33.652
SP-405, L1	14.297	15.137	34.215
SP-405, L2	14.167	15.118	34.105
SP-405, L3	14.131	15.097	34.043
SP-405, L4	14.128	15.097	34.039
SP-405, L5	14.122	15.083	34.006

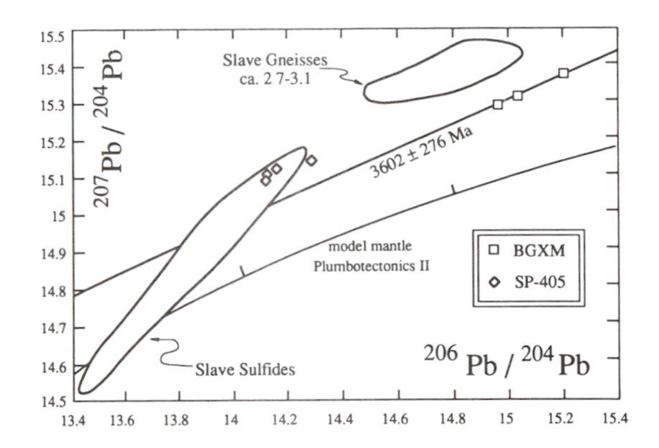

Fig. 7. *Lead isotopic compositions of leaches of feldspars from BGXM and SP-405, of ca. 2.7-Ga volcanogenic massive sulfides for the Slave Craton (Franklin and Thorpe, 1982), and of feldspars from younger (ca. 2.7–3.1 Ga) gneisses of the Slave craton (unpublished data, 1988). The mantle growth curve is from Zartman and Doe (1981). The isochron was calculated for the three leaches of plagioclase from BGXM and corresponds to an age of 3602 ± 276 Ma (m.s.w.d. = 0.225). Analytical uncertainties (0.01%) are smaller than the symbols used for SP-405 and BGXM.*

leaches define a linear array, while the isotopic compositions of the alkali feldspar leaches from SP-405 show more scatter. In general, the isotopic compositions of the leaches become less radiogenic with progressive leaching. The leaches of the plagioclase from BGXM define an isochron with an age of 3602 ± 276 Ma (m.s.w.d. = 0.225). This age overlaps with the 3621 ± 5 Ma age of the younger, metamorphic (or magmatic) zircon component within BGXM (*Bowring et al.,* 1989b), suggesting that the isotopic compositions of Pb reservoirs within plagioclase from BGXM were homogenized at this time.

The Pb isotopic compositions of leaches of alkali feldspar from SP-405 do not define a linear array. The most radiogenic and least radiogenic leaches from SP-405 lie on an "isochron" with a slope corresponding to an age of 3.5 Ga, while the last four leaches lie along a trend with a steeper slope, consistent with an older age. Thus, the observed dispersion in the Pb isotopic composition of alkali feldspar leaches from SP-405 may be the result of incomplete homogenization of the Pb isotopic reservoirs within the alkali feldspar during the ca. 3.6-Ga event.

The least radiogenic Pb isotopic composition of leaches of the alkali feldspar from SP-405 is distinctly less radiogenic than those of feldspars from other early (ca. 2.7-3.1 Ga) Archean gneisses from the western portion of the Slave craton. The least radiogenic Pb isotopic composition of the leaches of the plagioclase from BGXM is more comparable to, but has lower $^{207}Pb/^{204}Pb$ than, the younger (2.7-3.1 Ga) gneisses. The Acasta gneisses are unique in that they have elevated $^{207}Pb/^{204}Pb$ isotopic compositions and young model ages relative to terrestrial growth models. For example, the least radiogenic Pb isotopic composition of the alkali feldspar from SP-405 has a Stacey-Kramers model age of 2789 Ma (*Stacey and Kramers,* 1975), while the plagioclase from BGXM has a Stacey-Kramers model age of 2319 Ma. The elevated $^{207}Pb/^{204}Pb$ and young model ages of the feldspars require derivation from material that had experienced growth in an environment characterized by a high U:Pb ratio prior to 3.96 Ga. The least radiogenic leach of the alkali feldspar from SP-405 lies to the right of a 4.57- to 3.96-Ga isochron from primordial Pb, and hence cannot be modeled by single-stage growth from primordial Pb. Similarly, the radiogenic composition precludes a two-stage evolution from primordial Pb. This indicates that the crustal reservoirs from which these rocks were derived had undergone a complex multistage history prior to 3.96 Ga.

The least radiogenic Pb isotopic compositions of plagioclase from BGXM and other early Archean gneisses (ca. 2.7-3.1 Ga) from the Slave craton are distinctly more radiogenic than the isotopic compositions of sulfides from 2.7-Ga volcanogenic massive sulfide deposits from the Slave craton. The least radiogenic Pb isotopic compositions of the alkali feldspars from SP-405, however, overlap the field of the Pb isotopic compositions of the ca. 1-b.y. younger volcanogenic massive sulfides. The Pb isotopic compositions of these massive sulfides define a steep array that has been interpreted as a secondary isochron reflecting mineralization at 2.7 Ga and a source age of ca. 4.0 Ga (*Robertson and Cumming,* 1968; *Franklin and Thorpe,* 1982; *Thorpe,* 1982). Growth paths for the Pb isotopic compositions of feldspars from the Acasta gneisses and other early Archean gneisses from the Slave craton, calculated from their age of formation until 2.7 Ga, lie along the radiogenic extension of the array defined by the volcanogenic massive sulfides, suggesting that this array might reflect mixing of nonradiogenic, juvenile Pb (at 2.7 Ga) and older, crustal Pb derived from the early Archean gneisses within Slave craton. It is interesting to note that the most radiogenic sulfides in the array are from volcanic belts occurring in the western Slave craton where older basement occurs, whereas the least radiogenic sulfides are from the eastern half of the craton where older basement is not recognized.

DISCUSSION

The volume, composition, and likelihood of survival of Earth's earliest crust and lithosphere must constrain models for the evolution of Earth's crust and mantle. Recognized remnants of Earth's crust older than 3.8 Ga, however, are extremely rare. Since the discovery of ca. 3.8-Ga rocks from Greenland, several other indications of crust older than 3.8 Ga have been recognized. The most spectacular example is the 4.0-4.3-Ga detrital zircon grains from the Jack Hills and Mt. Narrayer regions of western Australia (*Froude et al.,* 1983; *Compston and Pidgeon,* 1986). Other examples are the 3.87-Ga orthogneisses from Mount Sones, Antarctica (*Black et al.,* 1986), 3.86-Ga cores in zircons from the Uivak gneisses in Labrador (*Schiøtte et al.,* 1989), 3.82-Ga gneissic rocks from Greenland (*Kinney,* 1986), and 3.88-Ga detrital zircon grains from quartzites in Greenland (*Kinney et al.,* 1988).

An important question for models of crustal evolution is how much pre-3.8-Ga crust is preserved. The recent discovery of the Acasta gneisses in the Slave craton of northwestern Canada point out that much is still unknown about the distribution of ages in the Archean cratons of the world. It is likely, as geologic mapping and isotopic studies continue, that more crust older than 3.8 Ga will be found. The methods used in searching for old crust in Archean cratons are a key factor, and the Slave craton is an excellent example of how old gneisses have gone unrecognized. Although ca. 2.9-3.15-Ga gneisses have been known for some time in the Slave craton, older crust had only been suspected on the basis of regional isotopic studies of ore deposits (*Franklin and Thorpe,* 1982). Furthermore, a number of studies, including Nd isotopic analyses of Archean sedimentary rocks (*Miller and O'Nions,* 1986) and composite gneiss samples (*McCulloch and Wasserburg,* 1978), U-Pb zircon geochronology of detrital zircons from greywackes (*Schärer and Allègre,* 1982), and Hf isotopic studies of detrital zircons in greywackes (*Stevenson and Patchett,* 1989), failed to detect a large component of older crust. *Schärer and Allègre* (1982), in a single grain dating study, did recognize several grains estimated to be as old as 3.3 and 3.5 Ga, although the majority were ca. 2.7 Ga.

The lack of evidence for the old rocks of the Slave craton is perhaps due in large part to tectonic juxtaposition of the older gneisses with the supracrustal rocks. This could explain why the sedimentary rocks of the Slave craton studied to date (*Schärer and Allègre,* 1982; *Stevenson and Patchett,* 1989; *Miller and O'Nions,* 1986) do not contain a large component of older detritus. Neodymium and Pb isotopic studies of post-tectonic granites, which have largely been derived from melting of the lower crust, may be a better way to recognize ancient crust. In this way, examination of late granitoids from different geographic localities across a craton could facilitate recognition of areas underlain by older crust.

It is unknown whether the present-day distribution of crust 3.8 Ga and older is representative of the original volume of crust formed or whether large amounts of early-formed continental crust were destroyed. Destruction of early-formed crust could have been accomplished by meteorite bombardment and/or tectonic processes. Another possibility is that large volumes of pre-3.8-Ga crust never formed. This possibility and the prospect of massive crustal recycling early in the Earth's history can be evaluated through the isotopic record of early Archean rocks and consideration of the development and evolution of geochemical reservoirs within the Earth.

Meteorite bombardment on the Moon and, inferentially, the Earth is believed to have declined approximately exponentially between 3.9 and 3.3 Ga (*Neukum et al.,* 1975), a period that has been referred to as the "late heavy bombardment." Based on the exponential decay between 3.9 and 3.3 Ga, most of the preserved large impact basins on the Moon were probably formed between 4.2 and 3.9 Ga (*Shoemaker,* 1984). It is interesting to note that the oldest rocks on Earth (until now, 3.8 Ga) correspond to the age of major impact-related lunar magmatism. If the Earth received a similar flux during this period, it is important to understand whether these impacts were a factor in crustal recycling. If sialic crust did not form in large amounts prior to 3.8 Ga, most impacts during 4.6-3.8 Ga would take place on oceanic crust. Thus, the absence of large volumes of crust older than 3.8 Ga does not require recycling by impacts. Similarly, the absence of evidence for impacts in the Earth's oldest crust does not mean that recycling by intensive bombardment was unimportant. The Acasta gneisses place a younger limit of 3.96 Ga on when catastrophic impacts capable of completely destroying the terrestrial crust were no longer occurring.

The isotopic systematics of the oldest preserved rocks are an important way to evaluate crustal growth-destruction models. Samarium-neodymium isotopic systematics have received a lot of attention in this regard due to the observation that the rare earth elements are not strongly fractionated by most crustal processes. Accordingly, the Sm-Nd system has been used to estimate the time at which crust "separates" from the mantle (e.g., *DePaolo,* 1988) and the relative roles of crust and mantle in the growth of continents (*Patchett and Arndt,* 1986). Compilations of Nd isotopic data for both felsic and mafic Archean rocks (e.g., *Shirey and Hanson,* 1986; *Chase and Patchett,* 1989; *Galer et al.,* 1989) indicate that, until now, the oldest rocks analyzed have depleted Nd isotopic compositions by 3.8 Ga, reflecting derivation from a depleted source (Fig. 8). The development of a depleted mantle has often been linked to the growth of continental crust (e.g., *DePaolo,* 1981a,b; *Hofmann,* 1988). If, as most agree, the depleted mantle was in existence by 3.8 Ga, then what is the nature of the enriched complement to this mantle? In addition to continental crust, there are several other possible enriched reservoirs that may be complementary to the depleted mantle. These include continental lithospheric mantle, oceanic crust, and cryptic enriched mantle (*Silver et al.,* 1988; *Jordan,* 1981; *Chase and Patchett,* 1989; *Galer and Goldstein,* 1988; *Carlson and Shirey,* 1988). If the depletion of the mantle is related to continental crust formation, the positive ϵ_{Nd} of the oldest "mantle-derived" rocks requires substantial separation of enriched crust prior to 3.8 Ga (*Silver et al.,* 1988). Until now, the lack of evidence for enriched pre-3.8 Ga crust, even as a reworked component, has been used to argue against a complementary relationship between crust and mantle (*Silver et al.,* 1988).

Until the discovery of the Acasta gneisses, there has been no pre-3.8 Ga Archean crust with an enriched Nd isotopic signature. The tonalitic gneiss BGXM has an initial ϵ_{Nd} (3.96 Ga) of -1.7 and would have an $\epsilon_{Nd}(3.8)$ of -4, whereas "mafic" rocks worldwide at 3.8 Ga have ϵ_{Nd} as high as +4. Thus, there is a range of at least 8 epsilon units by 3.8 Ga (Fig. 8). The evolution trajectory for the Acasta

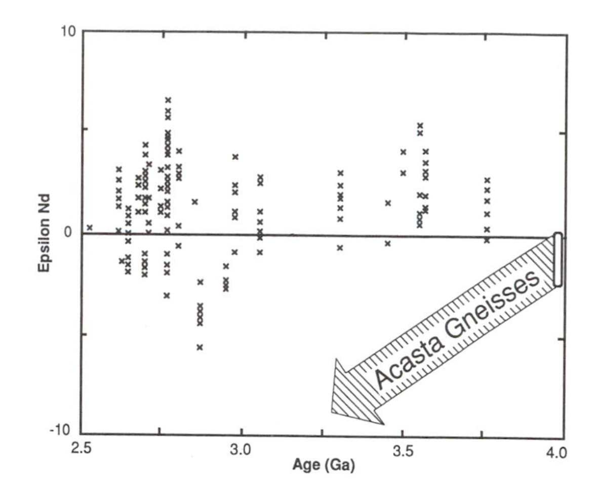

Fig. 8. *Initial ϵ_{Nd} values for Archean "mantle-derived" rocks compared with bulk earth ($\epsilon_{Nd}=0$). Evolution trajectory for Acasta gneisses is also shown for reference. (After Shirey and Hanson, 1986.)*

gneisses is shown in Fig. 8 along with initial values for a variety of Archean rocks. One of the striking aspects of the dataset is its spread at 3.8 Ga. This spread in data has been variously attributed to mantle heterogeneity and/or crustal contamination [see *Shirey and Hanson* (1986) for a review]. Given that most Archean rocks are preserved today because they were erupted onto, intruded into, or structurally incorporated into older continental lithosphere, it is likely that most will have some isotopic imprint of continental crust.

It is becoming increasingly clear that intermediate composition rocks, such as those found in continental magmatic arcs, can be explained as mixtures of a depleted mantle component and enriched crust (e.g., *DePaolo,* 1981c; *Housh et al.,* 1989; *Grove and Donnelly-Nolan,* 1986; *McMillan et al.,* 1989; *Hildreth and Moorbath,* 1988). The isotopic composition of the resulting mixture is in large part determined by the age of the overriding arc crust. In regions where the arc crust is old the effect of crustal interaction is more pronounced, while where it is young, its isotopic composition may be very similar to that of the mantle. This effect is well demonstrated in the Andes. In the Peruvian Andes the arc is underlain by Precambrian crust, and the isotopic effect of this crust on the arc magmas is quite pronounced (e.g., *Harmon et al.,* 1984; *Barreiro,* 1984; *Barreiro and Clark,* 1984). In the southern Andes, however, where the arc is underlain by Paleozoic and Mesozoic rocks, the isotopic effects of assimilation are less obvious (e.g., *Davidson et al.,* 1987; *McMillan et al.,* 1989). Therefore, the generally depleted but highly variable Nd isotopic signature of Archean rocks cannot be used to rule out the involvement of older crust.

The initial Nd isotopic values measured for Archean rocks, which until 3.0 Ga are broadly positive and range between 0 and +4 (Fig. 8), could be interpreted as mixtures of more enriched crust and depleted mantle, which is rarely sampled directly. Thus, the extreme initial values are a more reliable indicator of crust-forming processes. The significance attributed to the range of depleted signatures for early Archean rocks should be reevaluated in light of the evidence from the Slave craton that indicates very old crust with a distinctly enriched Nd isotopic signature at 3.96 Ga is preserved.

The Pb isotopic system is a particularly sensitive indicator of crustal interaction in magmatic rocks. This arises for two reasons. First, the short half-life of ^{235}U relative to ^{238}U and ^{232}Th means that in the early history of the Earth, rocks with elevated U:Pb ratios, such as the upper crust, will develop highly elevated $^{207}Pb/^{204}Pb$. Accordingly, the involvement of early Archean upper crustal rocks in later magmatism will be manifested in elevated $^{207}Pb/^{204}Pb$. Second, unlike Nd, there is a large difference in Pb concentration between crustal and mantle rocks. The average concentration of Pb in mantle rocks is 0.1-0.5 ppm vs. 10-20 ppm in crustal rocks. The Pb isotopic signature of rocks produced by interaction between mantle-derived melts and crustal rocks will therefore be dominated by the crustal component. Hence, the Pb isotopic systematics of feldspar and galena are ideally suited for detecting the involvement of small amounts of early Archean crust.

The Pb isotopic compositions of feldspars from Archean provinces worldwide are shown in Fig. 9. There is a distinct provinciality of the Pb isotopic compositions of feldspars ranging from unradiogenic compositions, such as the Amitsoq gneisses, to radiogenic compositions, such as the early Archean gneisses of the Slave craton. Comparison of feldspar Pb isotopic data from similar age cratons shows that this variability is not merely a function of age. For example, the ca. 3.8-Ga Amitsoq gneisses of Greenland and the early Archean gneisses of the Slave craton nearly account for most of the observed diversity in initial Pb isotopic compositions of Archean rocks. This diversity must be in part related to the prehistory of these cratons.

The feldspar Pb isotopic data from ca. 2.7-Ga granites and gneisses of the Superior craton and from the ca. 3.8-Ga Amitsoq gneisses are only slightly more radiogenic than the Pb isotopic compositions of sulfides from associated volcanogenic massive sulfide deposits. This similarity suggests that their Pb isotopic compositions were ultimately dominated by a depleted reservoir or were influenced by only slightly older rocks. The feldspar Pb isotopic data from the early Archean gneisses of the Slave craton and of the Yilgarn and Pilbara cratons, on the other hand, are significantly more radiogenic than the isotopic compositions of coeval volcanogenic massive sulfides (Yilgarn and Pilbara), or even significantly younger massive sulfide deposits (Slave). The Wyoming craton and the Karnataka craton of South India are also characterized by radiogenic feldspar Pb ratios. Overall, the radiogenic nature of these feldspar Pb ratios indicates that the felsic rocks from these cratons were derived from significantly older crust that had not lost U or Th. For example, the Pb isotopic data of SP-405 requires that its source region had undergone a complex multistage history prior to 3.962 Ga. Interest-

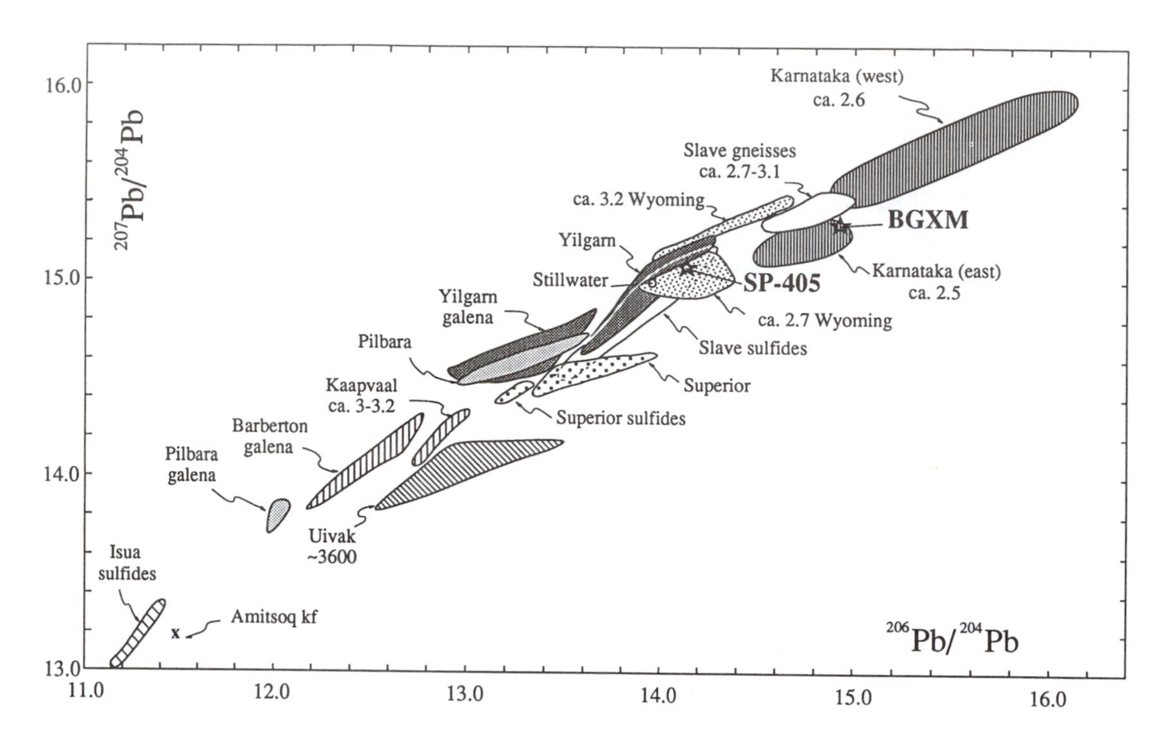

Fig. 9. *Lead isotopic compositions of alkali feldspars and sulfides from volcanogenic massive sulfides from Archean provinces worldwide. All of the fields correspond to analyses of alkali feldspars unless specified as sulfide or galena. Sources of data: Slave and Superior: Franklin and Thorpe, 1982; Gariépy and Allègre, 1985; Brévart et al., 1986. Australia: Claoué-Long et al., 1988; Hill et al., 1989; Browning et al., 1987; Bickle et al., 1983; Richards et al., 1981; Oversby, 1976; Brévart et al., 1986; Oversby, 1975; McNaughton and Bickle, 1987. S. Africa: Sinha and Tilton, 1973; Stacey and Kramers, 1975; Ulrych et al., 1967; Saager and Köppel, 1976. Uivak (Labrador): Baadsgaard et al., 1979. Greenland: Gancarz and Wasserburg, 1977; Appel et al., 1981. Wyoming: Rosholt et al., 1973; Stuckless and Nkomo, 1978; Stuckless et al., 1985; Naylor et al., 1970; Wooden and Mueller, 1988; Fischer and Stacey, 1986; Czamanske et al., 1986.*

ingly, there is evidence from all of these cratons for the presence at some time of older crust. This evidence includes the presence of 4.0-4.3-Ga detrital zircons from the Jack Hills and Mount Narrayer gneisses within the Yilgarn Craton (*Froude et al.,* 1983; *Compston and Pidgeon,* 1986), Nd-depleted mantle model ages older than 3.6 Ga (*De Laeter et al.,* 1981), the presence of 3.1- to 3.45-Ga xenocrystic zircons in the 2.7-Ga volcanic and and plutonic rocks of the Yilgarn craton (*Claoué-Long et al.,* 1988; *Hill et al.,* 1989), >3.5-Ga Nd crustal residence ages for gneisses and supracrustal rocks from the Wyoming craton (*Peterman and Futa,* 1988; *Koesterer et al.,* 1987), zircon cores >3.17 Ga in gneissic enclaves in the western Karnataka craton (*Krogstad et al.,* 1989), and the 3.96-Ga zircon ages and 4.1-Ga chrondritic Nd model ages from the Acasta gneisses.

It is clear that there is a continuum of Pb isotopic signatures for Archean rocks that can be explained most simply by variable involvement of older crust. Some have appealed to enriched mantle or mantle heterogeneity to explain the range in both Nd and Pb isotopic values in the Archean, but the relationship between enriched signatures and the presence of old crust does not support this interpretation. The large range in values and the difference between cratons suggest that small remnants of old crust are scattered about the present-day continents and that they have participated in subsequent crustal growth.

The geochronologic and isotopic database for Archean cratons around the world suggest that small amounts of pre-3.8-Ga crust are present and must be considered in models of crust and mantle evolution. Nevertheless, as discussed above, it is not likely that these fragments of crust are the sole complement to the depleted mantle already in existence by 3.8 Ga. A number of workers have suggested that Earth's earliest crust might have been enriched basaltic crust (*Chase and Patchett,* 1989; *Galer and Goldstein,* 1988; *Carlson and Shirey,* 1988). Thickening and melting of an enriched oceanic crust by subduction or collision could lead to tonalitic rocks such as those from the Slave craton. This would be compatible with an early depletion of light rare earth elements from the Archean mantle. In this manner, small sialic nuclei would form and participate in subsequent crust-forming processes. The general unsubductable nature of continental crust, its paucity in the geologic record prior to 3.8 Ga, and the observation that it cannot be the sole enriched complement to the depleted mantle all suggest that it was not until 3.8 Ga that large cratons began to form and stabilize.

The geochemistry of the 3.96-Ga tonalite gneiss (BGXM) has some important implications for the petrogenesis of other early Archean tonalites. Compositionally, BGXM is very similar to typical early Archean tonalites (see Fig. 4). The low heavy rare earth element abundances and strongly fractionated rare earth patterns of these rocks are generally interpreted to reflect a two-stage petrogenetic process: (1) formation of a mafic crust by partial melting of the mantle, and (2) partial melting of this crust after being transformed into garnet-bearing amphibolite or quartz eclogite (*Arth and Hanson,* 1975; *Hanson,* 1978; *Condie,* 1981; *Martin,* 1986). Some have suggested that this mafic crust melted as a subducted slab, or at least at mantle depths, to produce early Archean tonalites (*Arth and Hanson,* 1975; *Hanson,* 1978; *Condie,* 1981; *Martin,* 1986). While the trace-element characteristics are consistent with BGXM being related to melting of a garnet-bearing amphibolite, the isotopic data presented here are not consistent with melting of this mafic material at mantle depths (particularly the slab), as the sample preserves evidence for a substantial crustal prehistory. Hence, either early formed crust foundered and was assimilated back into the mantle (*Arndt and Goldstein,* 1988), or mantle-derived melts interacted with preexisting crust or lower crustal rocks that melted to produce tonalitic liquids. This is consistent with a model in which tonalitic magmas are formed in magmatic arc settings, some of which were superimposed on older continental crust, similar to the model for granite-greenstone terranes suggested by *Tarney et al.* (1976).

CONCLUSIONS

The Slave craton, located in the northwest Canadian shield, contains remnants of the oldest rocks yet discovered on Earth. Both a tonalitic orthogneiss and a granitic orthogneiss yield U-Pb zircon ages of 3.96 Ga. Both samples have a Nd isotopic composition consistent with involvement of older crust, and the tonalite gneiss has a chondritic model age of 4.1 Ga. Feldspars from the two gneisses have elevated initial Pb isotopic ratios indicative of derivation from or interaction with a source region that had a complex multistage history prior to 3.96 Ga. The major- and trace-element geochemistry of the tonalite sample is very similar to typical tonalites from other Archean cratons. The tonalite from the Slave craton is different from the others in having isotopic evidence for involvement of older crust, a conclusion that is at odds with models that invoke derivation of tonalitic magmas from partial melting of the mantle.

Available Nd and Pb isotopic data from the major Archean cratons suggest that small amounts of 3.8 Ga and older crust are present, and its involvement in younger crust-forming events is responsible for much of the observed isotopic variability. The isotopic signature of Archean intermediate magmatism is probably dominated by the age and isotopic character of preexisting crust rather than mantle heterogeneity.

The geologic, geochemical, isotopic, and geochronologic constraints on Earth's earliest crust are consistent with a model in which sialic crust developed slowly over the first 700 to 800 Ma of the Earth's history and could not be stabilized into large cratons until ca. 3.8 Ga. The 3.96-Ga differentiated rocks of the Slave craton are evidence that early, enriched, sialic crust was preserved, although isotopic and geochemical arguments suggest that it is not the sole enriched complement to the depleted mantle.

Acknowledgments. This work has been supported by NASA grant NAG 9-223, NSF grant EAR-8618612, and Indian and Northern Affairs, Yellowknife. W. A. Padgham has been a constant source of encouragement and logistical support and has made much of this work possible. The paper benefitted from reviews by P. F. Hoffman, R. S. Hildebrand, W. R. Van Schmus, R. F. Dymek, L. A. Haskin, and two anonymous reviewers. The two editors of this volume, J. Jones and H. Newsom, are thanked for their constructive comments and patience. Finally, B. Fegley is thanked for encouraging us to submit a paper to this volume.

REFERENCES

Appel P. W. U., Moorbath S., and Taylor P. N. (1981) Least radiogenic terrestrial lead from Isua, West Greenland. *Nature, 272,* 524-526.

Arndt N. T. and Goldstein S. L. (1988) Recycling of lower continental crust through foundering of cumulates from contaminated mafic intrusions (abstract). In *Workshop on the Growth of Continental Crust* (L. D. Ashwal, ed.), pp. 35-37. LPI Tech. Rpt. 88-02, Lunar and Planetary Institute, Houston.

Arth J. G. and Hanson G. N. (1975) Geochemistry and origin of the early Precambrian crust of northeastern Minnesota. *Geochim. Cosmochim. Acta, 39,* 325-362.

Baadsgaard H., Collerson K. D., and Bridgewater D. (1979) The Archean gneiss complex of northern Labrador. 1. Preliminary U-Th-Pb geochronology. *Can. J. Earth Sci., 16,* 951-961.

Barreiro B. A. (1984) Lead isotopes and Andean magmagenesis. In *Andean Magmatism: Chemical and Isotopic Constraints* (R. S. Harmon and B. A. Barreiro, eds.), pp. 21-30. Shiva, Orpington, England.

Barreiro B. A. and Clark A. H. (1984) Lead isotopic evidence for evolutionary changes in magma-crust interaction, central Andes, southern Peru. *Earth Planet. Sci. Lett., 69,* 30-42.

Bickle M. J., Bettenay L. F., Barley M. E., Chapman H. J., Groves D. I., Campbell I. H., and De Laeter J. R. (1983) A 3500 Ma plutonic and volcanic calc-alkaline province in the Archaean east Pilbara block. *Contrib. Mineral. Petrol., 84,* 25-35.

Black L. P. and McCulloch M. T. (1987) Evidence for isotopic equilibrium of Sm-Nd whole-rock systems in early Archean crust of Enderby Land, Antarctica. *Earth Planet. Sci. Lett., 82,* 15-24.

Black L. P., Williams I. S., and Compston W. (1986) Four zircon ages from one rock: The history of a 3930 Ma-old granulite from Mount Sones, Enderby Land, Antarctica. *Contrib. Mineral. Petrol., 94,* 427-437.

Bowring S. A. (1985) U-Pb zircon geochronology of early Proterozoic Wopmay Orogen, NWT, Canada: An example of rapid crustal evolution. Unpublished Ph.D. dissertation, Univ. of Kansas, Lawrence.

Bowring S. A. and Podosek F. A. (1989) Nd isotopic evidence from Wopmay Orogen for 2.0-2.4 Ga crust in western North America. *Earth Planet. Sci. Lett., 94,* 217-230.

Bowring S. A. and Van Schmus W. R. (1984) U-Pb zircon constraints on evolution of Wopmay Orogen, N.W.T. (abstract). In *Geol. Assoc. Canada/ Miner. Assoc. Canada, Programs with Abstracts, 9,* 47.

Bowring S. A., King J. E., Housh T. B., Isachsen C. E., and Podosek F. A. (1989a) Neodymium and lead isotope evidence for enriched early Archaean crust in North America. *Nature, 340,* 222-225.

Bowring S. A., Williams I. S., and Compston W. (1989b) 3.96 Ga gneisses from the Slave Province, N.W.T. Canada. *Geology, 17,* 971-975.

Brévart O., Dupré B., and Allègre C. (1986) Lead-lead age of komatiitic lavas and limitations on the structure and evolution of the Precambrian mantle. *Earth Planet. Sci. Lett., 77,* 292-302.

Browning P., Groves D. I., Blockley J. G., and Rosman K. J. R. (1987) Lead isotope constraints on the age and source of gold mineralization in the Archean Yilgarn Block, western Australia. *Econ. Geol., 82,* 971-986.

Carlson R. W. and Shirey S. B. (1988) Magma oceans, ocean ridges, and continental crust: Relative roles in mantle differentiation (abstract). In *Papers Presented to the Conference on Origin of the Earth,* pp. 50-51. Lunar and Planetary Institute, Houston.

Chamberlain V. E., Lambert R. St. J., Bradley M., and Martin B. E. (1984) U-Pb ages on tonalitic gneisses from the Big Bend area of the Coppermine River, District of Mackenzie, N.W.T. (abstract). In *Geol. Assoc. Canada/Miner. Assoc. Canada, Programs with Abstract, 9,* 51.

Chase C. G. and Patchett P. J. (1989) Stored mafic/ultramafic crust and early Archean mantle depletion. *Earth Planet. Sci. Lett., 91,* 66-72.

Claoué-Long J. C., Compston W., and Cowden A. (1988) The age of the Kambalda greenstones resolved by ion-microprobe: Implications for Archaean dating methods. *Earth Planet. Sci. Lett., 89,* 239-259.

Collerson K. D., McCulloch M. T., and Nutman A. P. (1989) Sr and Nd isotope systematics of polymetamorphic Archean gneisses from southern West Greenland and northern Labrador. *Can. J. Earth Sci., 26,* 446-466.

Compston W. and Pidgeon R. T. (1986) Jack Hills, evidence of more very old detrital zircons in Western Australia. *Nature, 321,* 766-769.

Condie K. C. (1981) *Archean Greenstone Belts.* Elsevier, Amsterdam. 434 pp.

Czamanske G. K., Wooden J. L., and Zientek M. L. (1986) Pb isotopic data for plagioclase from the Stillwater Complex (abstract). *Eos Trans AGU, 67,* 1251.

Davidson J. P., Dungan M. A., Ferguson K. M., and Colucci M. T. (1987) Crust-magma interactions and the evolution of arc magmas; the San Pedro—Pellado volcanic complex, southern Chilean Andes. *Geology, 15,* 443-446.

De Laeter J. R., Fletcher I. R., and Rosman K. J. R. (1981) Early Archaean gneisses from the Yilgarn block, Western Australia. *Nature, 242,* 322-324.

DePaolo D. J. (1981a) Nd isotopic studies: Some new perspectives on earth structure and evolution. *Eos Trans. AGU, 62,* 137-140.

DePaolo D. J. (1981b) Neodymium isotopes in the Colorado Front Range and crust-mantle evolution in the Proterozoic. *Nature, 291,* 193-196.

DePaolo D. J. (1981c) A neodymium and strontium isotopic study of the Mesozoic calc-alkaline batholiths of the Sierra Nevada and the Peninsular Ranges, California. *J. Geophys. Res., 86,* 10470-10488.

DePaolo D. J. (1988) *Neodymium Isotope Geochemistry; An Introduction.* Springer-Verlag, Berlin. 187 pp.

Dudas F. Ö. (1989) Nd isotopic compositions from the Slave craton: The case of the missing mantle. In *Geol. Assoc. Canada/Miner. Assoc. Canada, Programs with Abstracts, 14,* A24.

Easton R. M. (1985) The nature and significance of pre-Yellowknife Supergroup rocks in the Point Lake area, Slave Structural Province, Canada. In *Evolution of Archean Supracrustal Sequences* (L. D. Ayres, P. C. Thurston, K. D. Card, and W. Weber, eds.), pp. 156-167. Geol. Assoc. Canada Spec. Pap. 28.

Fischer L. B. and Stacey J. S. (1986) Uranium lead zircon ages and common lead measurements for the Archean gneisses of the Granite Mountains, Wyoming. *U.S. Geol. Surv. Bull., 1622,* 13-23.

Folinsbee R. E., Baadsgaard H., Cumming G. L., and Green D. C. (1968) A very ancient island arc. *AGU Geophysical Monograph 12,* pp. 441-448. AGU, Washington, DC.

Franklin J. M. and Thorpe R. I. (1982) Comparative metallogeny of the Superior, Slave and Churchill provinces. In *H.S. Robinson Memorial Volume, Precambrian Sulphide Deposits* (R. W. Hutchinson, C. D. Spence, and J. M. Franklin, eds.), pp. 3-90. Geol. Assoc. Canada Spec. Pap. 25.

Fraser J. A. (1960) North-central District of Mackenzie, N.W.T. *Geol. Surv. Canada Map 18-1960.*

Frith R. A., Loveridge W. D., and van Breemen O. (1986) U-Pb ages on zircon from basement granitoids of the western Slave Province, northwestern Canadian Shield. In *Current Research, Part C*, pp. 113-119. Geol. Surv. Canada Pap. 86-1C.

Froude D. O., Ireland T. R., Kinney P. D., Williams R. S., Compston W., Williams A. R., and Myers J. S. (1983) Ion microprobe identification of 4,100-4,200 Myr-old detrital zircons. *Nature, 304,* 616-618.

Fyson W. K. and Helmstaedt H. (1988) Structural patterns and tectonic evolution of supracrustal domains in the Archean Slave Province, Canada. *Can. J. Earth Sci., 25,* 301-315.

Galer S. J. G. and Goldstein S. L. (1988) Some chemical and thermal consequences of early mantle differentiation (abstract). In *Papers Presented to the Conference on Origin of the Earth,* p. 22. Lunar and Planetary Institute, Houston.

Galer S. J. G., Goldstein S. L., and O'Nions R. K. (1989) Limits on chemical and convective isolation in the Earth's interior. *Chem. Geol., 75,* 257-290.

Gancarz A. J. and Wasserburg G. J. (1977) Initial Pb of the Amitsoq gneiss, West Greenland, and implications for the age of the Earth. *Geochim. Cosmochim. Acta, 41,* 1283-1301.

Gariépy C. and Allègre C. J. (1985) The lead isotope geochemistry and geochronology of late-kinematic intrusives from the Abitibi greenstone belt, and the implications for late Archaean crustal evolution. *Geochim. Cosmochim. Acta, 49,* 2371-2383.

Green D. C. and Baadsgaard H. (1971) Temporal evolution and petrogenesis of an Archean crustal segment at Yellowknife, N.W.T., Canada. *J. Petrol., 12,* 177-217.

Grove T. L. and Donnelly-Nolan J. M. (1986) The evolution of young silicic lavas at Medicine Lake volcano, California: Implications for the origin of compositional gaps in calc-alkaline series lavas. *Contrib. Mineral. Petrol., 92,* 281-302.

Hanson G. N. (1978) The application of trace elements to the petrogenesis of igneous rocks of granitic composition. *Earth Planet. Sci. Lett., 38,* 26-43.

Harmon R. S., Barreiro B. A., Moorbath S., Hoefs J., Francis P. W., Thorpe R. S., Déruelle B., McHugh J., and Viglino J. A. (1984) Regional O-, Sr-, and Pb-isotope relationships in late Cenozoic calc-alkaline lavas of the Andean Cordillera. *J. Geol. Soc. London, 141,* 803-822.

Henderson J. B. (1981) Archean basin evolution in the Slave Province, Canada. In *Precambrian Plate Tectonics* (A. Kroner, ed.), pp. 213-235. Elsevier, Amsterdam.

Henderson J. B. (1985) Geology of the Yellowknife-Hearne Lake Area, District of Mackenzie: Segment across an Archean Basin. *Geol. Surv. Canada Mem. 414.* 135 pp.

Henderson J. B., Loveridge W. D., and Sullivan R. W. (1982) A U-Pb study of zircon from granitic basement beneath the Yellowknife Supergroup, Point Lake, District of Mackenzie. In *Current Research, Part C,* pp. 173-178. Geol. Surv. Canada Pap. 82-1C.

Henderson J. B., van Breemen O., and Loveridge W. D. (1987) Some U-Pb zircon ages from Archean basement, supracrustal and intrusive rocks, Yellowknife-Hearne Lake area, District of Mackenzie. In *Radiogenic Age and Isotopic Studies: Report 1,* pp. 111-121. Geol. Survey Canada Pap. 87-2.

Hildebrand R. S., Bowring S. A., and Housh T. (1990) The Wopmay line, central Wopmay Orogen, District of Mackenzie. In *Current Research.* Geol. Surv. Canada Paper 90-1C, in press.

Hildreth W. and Moorbath S. (1988) Crustal contribution to arc magmatism in the Andes of central Chile. *Contrib. Mineral. Petrol., 98,* 455-488.

Hill R. I., Campbell I. H., and Compston W. (1989) Age and origin of granitic rocks in the Kalgoorlie-Norseman region of Western Australia: Implications for the origin of Archaean crust. *Geochim. Cosmochim. Acta, 53,* 1259-1275.

Hoffman P. F. (1986) Crustal accretion in a 2.7-2.5 Ga "Granite Greenstone" terrane, Slave Province, NWT: A prograding trench-arc system? *Geol. Assoc. Canada/Miner. Assoc. Canada, Programs with Abstracts, 11,* 82.

Hoffman P. F. (1989) Precambrian geology and tectonic history of North America: an overview. *Geol. Soc. Am., Decade of North American Geology, vol. A,* 447-512.

Hoffman P. F. and Bowring S. A. (1984) A short-lived 1.9 Ga continental margin and its destruction, Wopmay Orogen, northwest Canada. *Geology, 12,* 68-72.

Hoffman P. F., Tirrul R., King J. E., St-Onge M. R., and Lucas S. B. (1988) Axial projections and modes of crustal thickening, eastern Wopmay orogen, northwest Canadian shield. In *Processes in Continental Lithospheric Deformation* (S. P. Clark, B. C. Burchfiel, and J. Suppe, eds.), pp. 1-29. Geol. Soc. Am. Spec. Pap. 218.

Hofmann A. W. (1988) Chemical differentiation of the earth: The relationship between mantle, continental crust, and oceanic crust. *Earth Planet. Sci. Lett., 90,* 297-314.

Housh T., Bowring S. A., and Villeneuve M. (1989) Lead isotopic study of early Proterozoic Wopmay Orogen, NW Canada: Role of continental crust in arc magmatism. *J. Geol., 97,* 735-747.

Isachsen C. E., Bowring S. A., and Padgham W. A. (1990) U-Pb zircon geochronology of the Yellowknife volcanic belt, N.W.T., Canada: New constraints on the timing and duration of greenstone belt magmatism. *J. Geol.*, in press.

Jacobsen S. B. and Dymek R. F. (1988) Nd and Sr isotope systematics of clastic metasediments from Isua, West Greenland; Identification of pre-3.8 Ga differentiated crustal components. *J. Geophys. Res., 93,* 338-354.

Jordan T. H. (1981) Continents as a chemical boundary layer. *Philos. Trans. R. Soc. London, A301,* 359-373.

King J. E. (1986) The metamorphic-internal zone of Wopmay Orogen (Early Proterozoic), Canada: 30 km of structural relief in a composite section based on plunge projection. *Tectonics, 5,* 973-994.

King J. E., Barrette P. D., and Relf C. D. (1987) Contrasting styles of basement deformation and longitudinal extension in the metamorphic-internal zone of Wopmay Orogen, N.W.T. In *Current Research, Part A,* pp. 515-531. Geol. Surv. Canada Pap. 87-1A.

Kinney P. D. (1986) 3820 Ma zircons from a tonalitic Amitsoq gneiss in the Godthåb district of southern West Greenland. *Earth Planet. Sci. Lett., 79,* 337-347.

Kinney P. D., Compston W., and McGregor V. R. (1988) The early Archaean crustal history of West Greenland as recorded by detrital zircons (abstract). In *Workshop on the Growth of Continental Crust* (L. D. Ashwal, ed.), pp. 79-81. LPI Tech. Rpt. 88-02, Lunar and Planetary Institute, Houston.

Koesterer M. E., Frost C. D., Frost B. R., Hulsebosch T. P., Bridgewater D., and Worl R. G. (1987) Development of the Archean crust in the Medina Mountain area, Wind River Range, Wyoming (U.S.A.). *Precambrian Res., 37,* 287-304.

Krogh T. E. and Gibbins W. A. (1978) U-Pb isotopic ages of basement and supracrustal rocks in the Point Lake area of the Slave Province, Canada (abstract). In *Geol. Assoc. Canada/Miner. Assoc. Canada, Programs with Abstracts, 3,* p. 438.

Krogstad E. J., Hanson G. N., and Rajamani V. (1988) U-Pb ages and Sr, Pb, and Nd isotope data for gneisses near the Kolar Schist Belt: Evidence for the juxtaposition of discrete Archean terranes. In *Workshop on the Deep Crust of South India* (L. D. Ashwal, ed.), pp. 60-61. LPI Tech. Rpt. 88-06, Lunar and Planetary Institute, Houston.

Kusky T. M. (1987) An Archean fold and thrust belt at Point Lake, NWT. In *Geol. Assoc. Canada Summer Meeting: Yellowknife, Canada, Indian Affairs and Northern Development, Abstracts with Programs,* p. 22.

Kusky T. M. (1989) Accretion of the Archean Slave Province. *Geology, 17,* 63-67.

Lambert M. B. and van Staal C. R. (1987) Archean granite-greenstone boundary relationship in the Beaulieu River volcanic belt, Slave Province, N.W.T. In *Current Research, Part A,* pp. 605-618. Geol. Surv. Canada Pap. 87-1A.

LeCheminant A. N. and Heaman L. M. (1989) Mackenzie igneous events, Canada: Middle Proterozoic hotspot magmatism associated with ocean opening. *Earth Planet. Sci. Lett., 96,* 38-48.

Martin H. (1986) Effect of steeper Archean geothermal gradient on geochemistry of subduction-zone magmas. *Geology, 14,* 753-756.

McCulloch M. T. and Wasserburg G. J. (1978) Sm-Nd and Rb-Sr chronology of continental crust formation. *Science, 200,* 1003-1011.

McGlynn J. C. (1977) Geology of Bear-Slave structural provinces, District of Mackenzie. *Geol. Surv. Canada Open-File 445,* scale 1:1,000,000.

McGlynn J. C. and Henderson J. B. (1970) Archean volcanism and sedimentation in the Slave Structural Province. In *Symposium on Basins and Geosynclines of the Canadian Shield* (A. J. Baer, ed.), pp. 31-44. Geol. Surv. Canada Pap. 70-40.

McMillan N. J., Harmon R. S., Moorbath S., Lopez-Escobar L., and Strong D. F. (1989) Crustal sources involved in continental arc magmatism: A case study of volcan Mocho-Choshuenco, southern Chile. *Geology, 17,* 1152-1156.

McNaughton N. J. and Bickle M. J. (1987) K-feldspar Pb-Pb isotope systematics of Archaean post-kinematic granitoid intrusions of the Diemals area, central Yilgarn Block, Western Australia. *Chem. Geol., 66,* 193-208.

Miller R. G. and O'Nions R. K. (1986) Source of Precambrian chemical and clastic sediments. *Nature, 314,* 325-330.

Moorbath S. (1986) The most ancient rocks revisited. *Nature, 321,* 725.

Moorbath S., O'Nions R. K., Pankhurst R. J., Gale N. H., and McGregor V. R. (1972) Further rubidium-strontium age determinations on the very early Precambrian rocks of the Godthåb district, West Greenland. *Nature Phys. Sci., 240,* 78-82.

Mortensen J. K., Thorpe R. I., Padgham W. A., King J. E., and Davis W. J. (1988) U-Pb zircon ages for felsic volcanism in Slave Province, N.W.T. In *Radiogenic Age and Isotopic Studies: Report 2,* pp. 85-86. Geol. Surv. Canada Pap. 88-2.

Naylor R. S., Steiger R. H., and Wasserburg G. J. (1970) U-Th-Pb and Rb-Sr systematics in 2700×10^6-year old plutons from the southern Wind River Range, Wyoming. *Geochim. Cosmochim. Acta, 34,* 1133-1159.

Neukum G., König B., Fechtig H., and Storzer D. (1975) Cratering in the Earth-Moon system: Consequences for age determination by crater counting. *Proc. Lunar Sci. Conf. 6th,* pp. 2597-2620.

Nikic Z., Baadsgaard H., Folinsbee R. E., Krupicka J., Payne-Leech A., and Saasaki A. (1980) Boulders from the basement, the trace of ancient crust? *Geol. Soc. Am. Spec. Pap. 182,* pp. 169-175.

O'Nions R. K. and Pankhurst R. J. (1974) Rare-earth element distribution in Archaean gneisses and anorthosite, Godthåb area, west Greenland. *Earth Planet. Sci. Lett., 22,* 328-338.

O'Nions R. K. and Pankhurst R. J. (1978) Early Archaean rocks and geochemical evolution of the Earth's crust. *Earth Planet. Sci. Lett., 38,* 211-236.

Oversby V. M. (1975) Lead isotopic systematics and ages of Archaean acid intrusives in the Kalgoorlie-Norseman area, Western Australia. *Geochim. Cosmochim. Acta, 39,* 1107-1125.

Oversby V. M. (1976) Isotopic ages and geochemistry of Archean acid igneous rocks from the Pilbara, Western Australia. *Geochim. Cosmochim. Acta, 40,* 817-829.

Padgham W. A. (1985) Observations and speculations on supracrustal successions in the Slave Structural Province. In *Evolution of Archean Supracrustal Sequences* (L. D. Ayres, P. C. Thurston, K. D. Card, and W. Weber, eds.), pp. 156-167. Geol. Assoc. Canada Spec. Pap. 28.

Patchett P. J. and Arndt N. T. (1986) Nd isotopes and tectonics of 1.9-1.7 Ga crustal genesis. *Earth Planet. Sci. Lett., 78,* 329-338.

Peterman Z. E. and Futa K. (1988) Contrasts in Nd crustal residence ages between the Superior and Wyoming cratons (abstract). In *Geol. Soc. Am. Abstracts with Programs, 20,* A137.

Richards J. R., Fletcher I. R., and Blockley, J. G. (1981) Pilbara galenas: Precise isotopic assay of the oldest Australian leads; model ages and growth-curve implications. *Miner. Deposita, 16,* 7-30.

Robertson D. K. and Cumming G. L. (1968) Lead and sulphur isotope ratios from the Great Slave Lake area, Canada. *Can. J. Earth Sci., 5,* 1269-1276.

Rosholt J. N., Zartman R. E., and Nkomo I. T. (1973) Lead isotope systematics and uranium depletion in the Granite Mountains, Wyoming. *Geol. Soc. Am. Bull., 84,* 989-1002.

Saager R. and Köppel V. (1976) Lead isotopes and trace elements from sulfides of Archean greenstone belts in South Africa—A contribution to the knowledge of the oldest known mineralizations. *Econ. Geol., 71,* 44-57.

Schärer U. and Allègre C. J. (1982) Investigation of the Archean crust by single-grain dating of detrital zircon: a graywacke of the Slave Province, Canada. *Can. J. Earth Sci., 19,* 1910-1918.

Schiotte L., Compston W., and Bridgewater D. (1989) Ion probe U-Th-Pb zircon dating of polymetamorphic orthogneisses from northern Labrador, Canada. *Can. J. Earth Sci., 26,* 1533-1556.

Shirey S. B. and Hanson G. N. (1986) Mantle heterogeneity and crustal recycling in Archean granite-greenstone belts: Evidence from Nd isotopes and trace elements in the Rainy Lake area, Superior Province, Ontario, Canada. *Geochim. Cosmochim. Acta, 50,* 2631-2651.

Shoemaker E. M. (1984) Large body impacts through geologic time. In *Patterns of Change in Earth Evolution* (H. D. Holland and A. F. Trendall, eds.), pp. 15-40. Springer-Verlag, Berlin.

Silver P. G., Carlson R. W., and Olsen P. (1988) Deep slabs, geochemical heterogeneity and the large scale structure of mantle convection. *Annu. Rev. Earth Planet. Sci., 16,* 477-541.

Sinha A. K. and Tilton G. R. (1973) Isotopic evolution of common lead. *Geochim. Cosmochim. Acta, 37,* 1823-1849.

Stacey J. S. and Kramers J. D. (1975) Approximation of terrestrial lead isotope evolution by a two-stage model. *Earth Planet. Sci. Lett., 26,* 207-221.

Stevenson R. K. and Patchett P. J. (1989) Constraints on the evolution of Archean continental crust: Implications from Hf isotopes in detrital zircons (abstract). In *Geol. Assoc. Canada/Miner. Assoc. Canada, Programs wilth Abstracts, 14,* A74.

Stockwell C. H. (1933) Great Slave Lake—Coppermine River area, N.W.T. *Geol. Surv. Canada Summary Report 1932, Part C,* pp. 37-63.

St-Onge M. R. and King J. E. (1987a) Evolution of regional metamorphism during back-arc stretching and subsequent crustal shortening in the 1.9 Ga Wopmay orogen, Canada. *Philos. Trans. R. Soc. London, A321,* 199-218.

St-Onge M. R. and King J. E. (1987b) Thermo-tectonic evolution of a metamorphic internal zone documented by axial projections and petrological P-T paths, Wopmay orogen, north-west Canada. *Geology, 15,* 155-158.

St-Onge M. R., King J. E., and Lalonde A. E. (1984) Deformation and metamorphism of the Coronation Supergroup and its basement in the Hepburn Metamorphic-Plutonic Zone of Wopmay Orogen: Redrock Lake and the eastern portion of Calder River map areas, District of Mackenzie. In *Current Research, Part A,* pp. 171-180. Geol. Surv. Canada Pap. 84-1A.

St-Onge M. R., King J. E., and Lalonde A. E. (1988) Geology, east-central Wopmay Orogen, District of Mackenzie, Northwest Territories. *Geol. Surv. Can. Open-File Rept. 1923,* 3 sheets, scale 1:125,000.

Stuckless J. S. and Nkomo I. T. (1978) Uranium-lead isotope systematics in uraniferous alkali-rich granites from the Granite Mountains, Wyoming: Implications for uranium source rocks. *Econ. Geol., 73,* 427-441.

Stuckless J. S., Hedge C. E., Worl R. G., Simmons K. R., Nkomo I. T., and Wenner D. B. (1985) Isotopic studies of the late Archean plutonic rocks of the Wind River Range, Wyoming. *Geol. Soc. Am. Bull., 96,* 850-860.

Tarney J., Dalziel I. W. D., and de Wit M. J. (1976) Marginal basin 'Rocas Verdes' Complex from S. Chile; a model for Archaean greenstone belt formation. In *The Early History of the Earth* (B. F. Windley, ed.), pp. 131-146. Wiley, New York.

Thompson P. H. (1978) Archean regional metamorphism in the Slave Structural Province—a new perspective on some old rocks. In *Metamorphism in the Canadian Shield* (J. A. Fraser and W. W. Heywood, eds.), pp. 85-102. Geol. Surv. Canada Pap. 78-10.

Thompson P. H. (1989) Moderate overthickening of thinned sialic crust and the origin of granitic magmatism and regional metamorphism in low-P-high-T terranes. *Geology, 17,* 520-523.

Thorpe R. I. (1982) Lead isotope evidence regarding Archean and Proterozoic metallogeny in Canada. *Rev. Bras. Geocienc., 12,* 510-521.

Turner F. J. (1981) *Metamorphic Petrology,* 2nd edition. McGraw-Hill, New York. 524 pp.

Ulrych T. J., Burger A., and Nicolaysen L. O. (1967) Least radiogenic terrestrial leads. *Earth Planet. Sci. Lett., 2,* 179-184.

van Breemen O. and Henderson J. B.(1988) U-Pb zircon and monazite ages from the eastern Slave Province and Thelon Tectonic Zone, Artillery Lake area, N.W.T. In *Radiogenic Age and Isotopic Studies: Report 2,* pp. 73-84. Geol. Surv. Canada Pap. 88-2.

Winkler H. G. F. (1979) *Petrogenesis of Metamorphic Rocks,* 5th edition. Springer-Verlag, New York. 334 pp.

Wooden J. L. and Mueller P. A. (1988) Pb, Sr, and Nd isotopic compositions of a suite of Late Archean igneous rocks, eastern Beartooth Mountains: Implications for crust-mantle evolution. *Earth Planet. Sci. Lett., 87,* 59-72.

Zartman R. E. and Doe B. R. (1981) Plumbotectonics—the model. *Tectonophysics, 75,* 135-162.

EVOLUTION OF THE HYDROSPHERE AND THE ORIGIN OF LIFE

The origin of the present hydrosphere and atmosphere of the Earth has important implications for understanding the accretion process. The present surface of the Earth is volatile rich and oxidized, in contrast to the Earth's core. In order to understand how this hydrosphere and atmosphere originated, we need to understand the effects of life and the clues to early environments and bombardment history provided by the existence of life early in the planet's history.

The two papers in this section are:

E. Tajika and T. Matsui: *The Evolution of the Terrestrial Environment*
J. C. G. Walker: *Origin of an Inhabited Planet*

The paper by Tajika and Matsui discusses the temperature of the surface environment of the Earth over geologic history. During the early history of the Earth, the sun provided about 30% less heat than at present, which would result in a frozen Earth. The solution to this problem is to have large amounts of the greenhouse gas CO_2 in the early atmosphere. Tajika and Matsui's paper discusses a quantitative model of the distribution of carbon (among different geochemical reservoirs through time) that is required to approximately maintain the present surface temperatures and to establish the present inventories of carbon in the different reservoirs. An important conclusion of this work is the necessity of continental growth in order for the terrestrial environment to evolve to its present state.

The paper by Walker discusses the effects of biologic activity on the evidence of the early state of the Earth. Walker concludes that the sketchy evidence for early life on the Earth does not constrain theories of the Earth's origin, and even sterilizing events are possible through much of the Earth's history. Another interesting conclusion is that the chemical unmixing activities of life have not greatly affected the oxidation state of the bulk Earth, and may have preserved it by preventing the loss of reducing species from the Earth's atmosphere.

The timing and environment of the origin of life are interesting questions with implications not only for the Earth, but also for planets such as Mars, and even for the mystery of life in the universe. While not specifically addressing these issues, many of the papers in this book indirectly address the question of the terrestrial environment, such as the existence of a magma ocean, the possibility of a dense early atmosphere, and the existence of processes such as bombardment that might have destroyed the early continental crust. Of particular interest, both Sasaki and Ahrens (section 3) discuss the possibility of a hot, dense early atmosphere due to the late accretion of large planetesimals or the accretion of nebular gas. The discovery of 3.96-b.y.-old rocks reported by Bowring and coworkers (section 5) holds out the hope that more direct evidence regarding these questions will be forthcoming.

For additional information and general reviews of these topics see articles in *Origin and Evolution of Planetary and Satellite Atmospheres* (S. K. Atreya, J. B. Pollack, and M. S. Matthews, eds., Univ. of Arizona, 1989), and in *Earth's Earliest Biosphere* (J. W. Schopf, ed., Princeton Univ., 1983).

THE EVOLUTION OF THE TERRESTRIAL ENVIRONMENT

E. Tajika and T. Matsui

Geophysical Institute, Faculty of Science, University of Tokyo, Tokyo, Japan

The evolution of the Earth's environment during its 4.6-b.y. history is investigated using a carbon cycle model in which the carbon is assumed to circulate among five reservoirs (the atmosphere, ocean, continents, seafloor, and mantle). In this model, we consider continental weathering, carbonate precipitation in the ocean, carbonate accretion to the continents, metamorphism of carbonates following CO_2 degassing through arc volcanism, carbon regassing into the mantle, and CO_2 degassing from the mantle. We also take into account changes in external conditions such as an increase in the solar luminosity, continental growth, and a decrease in tectonic activity with time, which obviously affect the carbon cycle. We numerically calculate the temporal variation of the carbon content of each reservoir under varying external conditions over the entire history of the Earth. We find that continental growth is required for the terrestrial environment to evolve to the present state, and that the carbon cycle has had an important role in stabilizing the surface temperature of the Earth throughout its entire history. The distribution of carbon at the surface is mainly controlled by one parameter, an accretion ratio, which represents the fraction of the seafloor carbonates accreted to the continents. When this value is equal to 0.7, correct values for the present distribution of carbon at the surface of the Earth can be obtained from the model.

INTRODUCTION

Recently Matsui and Abe (*Matsui and Abe,* 1986a; *Abe and Matsui,* 1985, 1986) have proposed that the surface of the accreting Earth was covered with a magma ocean due to the blanketing effect of an impact-induced steam atmosphere. This steam atmosphere became unstable with the decrease in the impact energy flux at the end of the accretionary period, and H_2O in the proto-atmosphere condensed to form the proto-ocean (*Matsui and Abe,* 1986b; *Abe and Matsui,* 1988). The composition of the atmosphere just after the formation of the ocean was thought to be mainly CO_2 because it is the second most abundant volatile on the surface of the Earth at the present time (*Holland,* 1978; *Kitano,* 1984). Before the continents formed, atmospheric CO_2 was controlled by the dissolution equilibrium between the atmosphere and the ocean. The partial pressure of CO_2 would not have decreased to a level lower than about 10 bar at this aquaplanet stage (*Walker,* 1985).

Because CO_2 gas has a greenhouse effect, the evolution of atmospheric CO_2 should have had a dominant effect on the evolution of the Earth's environment. This evolutionary scenario, from an impact-induced steam atmosphere to CO_2 atmosphere after oceanic formation, is consistent with the solution for the so-called "Faint Young Sun Paradox" (*Sagan and Mullen,* 1972). It has been known that the luminosity of the sun was initially about 30% lower than it is today (*Gough,* 1981) and that it has been increasing continually during the history of the Earth. So if the atmospheric composition had always been the same as it is today, the surface temperature of the Earth would have been below 0°C up until 2 b.y. ago (*Sagan and Mullen,* 1972). However, there is no evidence indicating that the Earth was cold at that time, so this creates a paradox. If the atmosphere had more CO_2 in the past than it does today, the temperature could have been warmer even if the solar luminosity was lower (*Owen et al.,* 1979; *Walker,* 1982; *Matsui and Abe,* 1986b; *Kasting,* 1987).

Then the remaining problem is how such a CO_2-rich atmosphere has evolved to the present N_2-rich atmosphere. The changes in atmospheric CO_2 level are related to the carbon cycle among the various surface reservoirs and the interior of the Earth. The present carbonate-silicate geochemical cycle between the atmosphere, the ocean, and the continents has been studied by *Berner et al.* (1983) and *Lasaga et al.* (1985). Their results suggest that tectonic activity (such as the spreading rate of the seafloor) and the surface area of the land play an important role in surface temperature change. But these factors have changed a great deal throughout the entire history of the Earth. Also, the luminosity of the sun, which, as stated earlier, has increased by 30% of the present value, should have had an important effect on the surface temperature of the Earth.

Therefore, we will conduct a numerical simulation, starting from an impact-induced atmosphere model proposed by Matsui and Abe, and examine the response of the Earth's environment to change in external conditions such as the solar luminosity, continental growth, and tectonic activity, using the carbon cycle model.

MODEL

Present-Day Amount of Carbon

The amount of carbon at the surface of the Earth at the present time has been estimated by a number of geochemists (*Rubey,* 1951; *Poldervaart,* 1955; *Ronov and Yaroshevsky,* 1967, 1976; *Ronov,* 1968; *Hunt,* 1972; *Javoy et al.,* 1982; etc.). It is possible to estimate this amount by studying the chemical composition of the crust because almost all of the carbon exists as carbonate rocks on the continents and the seafloor. In addition to the amount, we need to know the distribution of the carbon in our study. Among these estimates, *Ronov and Yaroshevsky* (1976) studied the composition of three different types of crust (continental, subcontinental, and oceanic). Therefore, we will adopt their estimate of 9.27×10^{21} mol as the present-day amount of carbon at the surface.

The distribution of surface carbon estimated by Ronov and Yaroshevsky is shown in Table 1.

TABLE 1. *Distribution of carbon at the surface (in mol).*

Continental crust	6.07×10^{21}
Subcontinental crust	1.34×10^{21}
Oceanic crust	1.86×10^{21}
Atmosphere	5.75×10^{16}
Ocean	3.33×10^{18}

The amounts of carbon in the three types of crust are estimated by *Ronov and Yaroshevsky* (1976); the others are by *Holland* (1978).

The amounts of carbon in the atmosphere and ocean are also shown. As shown in this table, 65.5% of the total carbon is in the continental crust (land area), 14.5% is in the subcontinental crust (continental shelf), and 20.0% is in the oceanic crust.

The amount of carbon in the mantle is much more difficult to determine. *Javoy et al.* (1982) estimated it as 3.33×10^{22} mol by multiplying the concentration of carbon in peridotites by the total mass of the upper mantle. However, such a method is not yet accurate enough.

It should be noted that carbon in the crust exists in volcanic, granitic, metamorphic, and sedimentary rocks. According to *Hunt* (1972), three-fourths of it is in the form of carbonate and over half of the total carbon is in the form of sedimentary carbonates. Organic carbon comprises only 20% of the total and it is mainly in the form of clays and shales. Living organic matter is 2.5×10^{16} mol, which is less than the amount of carbon in the atmosphere. For this reason we will not consider a cycle related to organic carbon in this model.

Carbon Cycle

We will consider five reservoirs of carbon in our model. They are the atmosphere, the ocean, the continents, the seafloor, and the mantle. In this case, the treatment of the continental shelf presents a small problem. The continental shelf should play a different role as a carbon reservoir than the continents or the seafloor. But here, for simplicity, we will assume it is a part of the continents, since the carbonates on both reservoirs never subduct into the interior.

The model is shown in Fig. 1. Each box represents a carbon reservoir and each arrow represents a flux produced by some process. These processes comprise the carbon cycle as follows: Atmospheric CO_2 dissolves in raindrops and becomes carbonic acid (H_2CO_3). When the rain falls to the surface, minerals such as silicates and carbonates are weathered. Cations and bicarbonate ions produced by this weathering are carried to the oceans by rivers. There, these ions react with each other to form carbonates. These carbonates are deposited on the seafloor where they move with the motion of the plates, and are eventually subducted under the continental crust. They are then metamorphosed under certain conditions of temperature and pressure, and CO_2 is reproduced. Finally, this CO_2 returns to the atmosphere by volcanism. This is known as "the Carbonate-Silicate Geochemical Cycle" (*Walker et al.,* 1981; *Berner et al.,* 1983; *Lasaga et al.,* 1985).

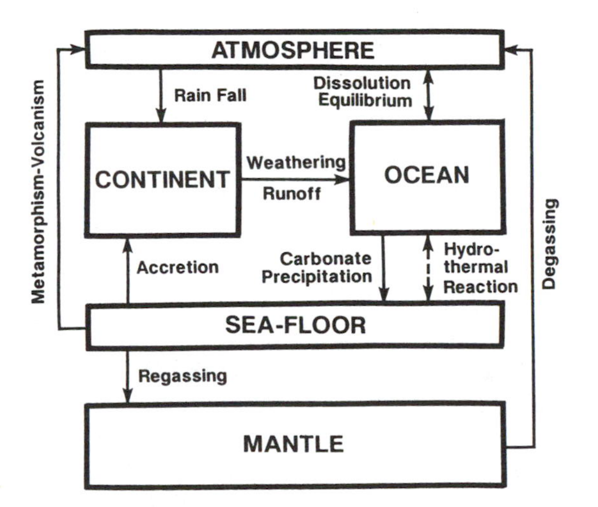

Fig. 1. *Carbon cycle model in which each box represents a carbon reservoir and each arrow represents a carbon flux between reservoirs due to some process. The broken arrow relates to cation flux. See text for details about each process.*

This cycle is affected by the evolution of the tectonic environment. The changes in flux due to tectonic evolution such as continental growth and tectonic activity will influence the flow of all components of the system. For example, sediments, including carbonates, that are accreted to the continental crust result in continental growth. On the other hand, sediments that are subducted under the continental crust may be reabsorbed into the mantle, which eventually allows CO_2 to be released from the mantle at the midocean ridge. Therefore, we must take into account all of these processes in our model.

Processes in the Carbon Cycle

Weathering. Silicates and carbonates are weathered by rainfall. We consider the weathering of $CaSiO_3$, $MgSiO_3$, $CaCO_3$, and $MgCO_3$. The reactions are

$$2CO_2 + CaSiO_3 + H_2O \rightarrow Ca^{2+} + 2HCO_3^- + SiO_2$$
$$2CO_2 + MgSiO_3 + H_2O \rightarrow Mg^{2+} + 2HCO_3^- + SiO_2$$
$$CO_2 + CaCO_3 + H_2O \rightarrow Ca^{2+} + 2HCO_3^-$$
$$CO_2 + MgCO_3 + H_2O \rightarrow Mg^{2+} + 2HCO_3^-$$

These reactions are only symbolic for silicate weathering and, in fact, we will use the weathering function for feldspar. For carbonates, we will consider calcite and magnesite, but not dolomite. The reason for this will be discussed later. The present-day fluxes of cations and bicarbonate ions due to weathering are estimated by multiplying the concentrations of these ions in world-average river water by the total amount of runoff in the world (*Berner et al.,* 1983).

$F_W^{cs\bullet} = 2.8 \times 10^{18}$ (mol/Ma):Ca^{2+} derived from Ca-silicate weathering

$F_W^{ms\bullet} = 3.1 \times 10^{18}$ (mol/Ma):Mg^{2+} derived from Mg-silicate weathering

$F_W^{calcite\bullet} = 8.3 \times 10^{18}$ (mol/Ma):Ca^{2+} derived from calcite weathering

$F_W^{dolomite\bullet} = 2.1 \times 10^{18}$ (mol/Ma):Ca^{2+} derived from dolomite weathering

$F_W^{dolomite\bullet} = 2.1 \times 10^{18}$ (mol/Ma):Mg^{2+} derived from dolomite weathering

where • represents the present value.

Instead of considering dolomite weathering here, we use the cation fluxes due to calcite and magnesite weathering. Then

$F_W^{cc\bullet} = 10.4 \times 10^{18}$ (mol/Ma):Ca^{2+} derived from Ca-carbonate weathering

$F_W^{mc\bullet} = 2.1 \times 10^{18}$ (mol/Ma):Mg^{2+} derived from Mg-carbonate weathering

We use the following weathering function

$$F_W = K_W \cdot g(T_S, P_{CO_2}) \cdot L$$

where K_W is a constant, L is the reservoir size (in mol), and g is

$$g(T_s, P_{CO_2}) = \left(\frac{P_{CO_2}}{P^{\bullet}_{CO_2}}\right)^{0.3} \cdot \exp\left(\frac{T_s - 285}{13.7}\right) \cdot a \cdot b$$

which is derived from modification of the weathering function by *Walker et al.* (1981); a and b are correction factors determined from the condition that $g \rightarrow 0$ ($T_S \rightarrow 273$ K)

$$a = \begin{cases} 1 & (T_s \geq 285\text{K}) \\ \ln\left(\frac{T_s}{273}\right) / \ln\left(\frac{285}{273}\right) & (273\text{K} \leq T_s \leq 285\text{K}) \\ 0 & (T_s \leq 273\text{K}) \end{cases}$$

and that $g \rightarrow 1$ ($T_S \rightarrow 288$ K)

$$b = \exp\left\{\left(\frac{288 - 285}{13.7}\right)\right\}^{-1} = 0.80$$

To estimate the constants, we assume that all carbon in the continental crust is in carbonate form and is divided into Ca-carbonate and Mg-carbonate by using the ratio $CaCO_3$:$MgCO_3$ = 8:2 (*Berner et al.,* 1983). Then

$$K_W^{cc} = F_W^{cc\bullet}/g^{\bullet} \cdot L^{\bullet}_{cc} = 10.4 \times 10^{18}/(7.41 \times 10^{21} \times 0.8) = 1.75 \times 10^{-3}\ \text{Ma}^{-1}$$

$$K_W^{mc} = F_W^{mc\bullet}/g^{\bullet} \cdot L^{\bullet}_{mc} = 2.1 \times 10^{18}/(7.41 \times 10^{21} \times 0.2) = 1.42 \times 10^{-3}\ \text{Ma}^{-1}$$

where $g^{\bullet} = 1$.

For the constants in the equation for silicate weathering, we assume that the size of the silicate reservoirs are proportional to the continental surface area, i.e., $L \propto S_C \propto f_{S_C}$, where S_C is the continental area and f_{S_C} is the ratio of continental area at a given time to the present value ($=S_C/S_C^{\bullet}$). We define the silicate weathering function as follows

$$F_W = K_W \cdot g(T_S, P_{CO_2}) \cdot f_{S_C}$$

Then

$$K_W^{cs} = F_W^{cs*}/g^* \cdot f_{S_C}^* = 2.8 \times 10^{18} \text{ mol/Ma}$$

$$K_W^{ms} = F_W^{ms*}/g^* \cdot f_{S_C}^* = 3.1 \times 10^{18} \text{ mol/Ma}$$

where $g^* = f_{S_C}^* = 1$.

It is noted that this function includes the effects of an increase in rainfall and runoff due to an increase in surface temperature (see *Walker et al.,* 1981).

Carbonate precipitation. Cations and bicarbonate ions are brought to the oceans through rivers. In the oceans, they react with each other and form carbonates. The reactions are

$$Ca^{2+} + CO_3^{2-} \rightarrow CaCO_3$$

$$Mg^{2+} + CO_3^{2-} \rightarrow MgCO_3$$

$$Ca^{2+} + Mg^{2+} + 2CO_3^{2-} \rightarrow CaMg(CO_3)_2$$

At the present day, about 40% of carbonates are dolomites. It is known from several experiments that dolomites are hard to produce by direct precipitation even under supersaturation conditions. They can be produced through secondary metamorphism by substituting magnesium ions for calcium ions in calcite at high temperature ($\geq 150°C$) (see *Kitano,* 1984; *Holland,* 1978). Magnesite has an even larger solubility product than dolomite as is shown by the lack of evidence for magnesite formation in the present oceans. Because of this, the production of magnesite gives us an upper limit to the concentration of magnesium ions in the oceans. Therefore, we will consider magnesite as the only product of magnesium precipitation and calcite as the only product of calcium precipitation.

Almost all of the present carbonates are produced by organisms. Inorganic carbonates such as oolite are hardly seen. However, in the absence of organic processes, carbonates would be produced inorganically if the ion activity product of carbonate in the ocean exceeds some critical value. The difference between the organic and inorganic processes is mainly the level for the equilibrium and possibly the place where the carbonates are precipitated. For simplicity, we will only consider the inorganic production of carbonates. This flux F_P is estimated so as to maintain chemical equilibrium between the atmosphere and the oceans; carbonates are precipitated when the ion activity product of carbonate exceeds the thermodynamic solubility product (see Table 2; details will be shown in the next section).

TABLE 2. *Equilibrium constants and activity coefficients.*

K_0	3.48×10^{-2}	α_{H_2O}	0.967
K_1	4.45×10^{-7}	$\gamma_{H_2CO_3}$	1.130
K_2	4.69×10^{-11}	$\gamma_{HCO_3^-}$	0.550
K_{sp}^{cc}	3.60×10^{-9}	$\gamma_{CO_3^{2-}}$	0.021
K_{sp}^{mc}	1.00×10^{-5}	$\gamma_{Ca^{2+}}$	0.203
		$\gamma_{Mg^{2+}}$	0.260

Accretion. Some fraction of the carbonates deposited on the seafloor are considered to accrete onto the continental crust at the subduction zones. This idea is based on the accretion model in which an accretionary prism is formed by the accretion of seafloor sediments onto the continental crust (*Sleely et al.,* 1974; *Karig and Sharman,* 1975). Seamounts also accrete onto the continents. Examples of this are the west coast of North America and East Asia. To take into account this process, we define the accretion ratio as a parameter describing the amount of carbonates that accrete onto the continents (Fig. 2).

This flux is equal to the ratio of the accreting carbonates multiplied by the flux of the subducting carbonates and is expressed as

$$F_A = A \cdot t_r^{-1} \cdot P$$

where A is the accretion ratio, t_r is the residence time of carbon on the seafloor, and P is the amount of carbon in the seafloor reservoir (in mol).

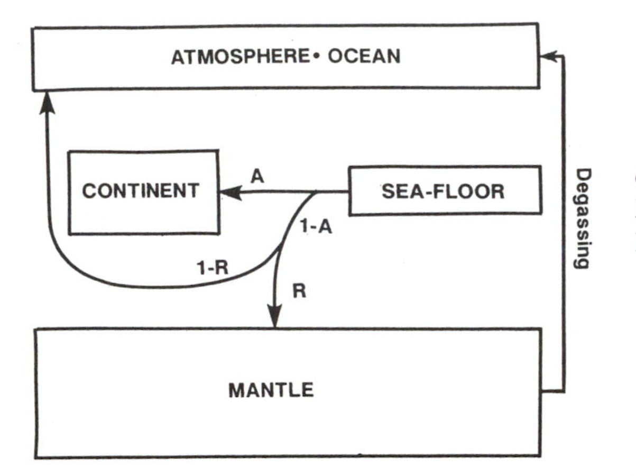

Fig. 2. *Definition of parameters A (accretion ratio) and R (regassing ratio). Parameter A represents the ratio of seafloor carbonates that accrete onto the continents to the total amount of precipitated seafloor carbonates; R represents the ratio of seafloor carbonates that regas into the mantle to the subducted carbonates. Note that the actual fraction of carbon regassed into the mantle is (1-A) · R.*

Regassing. Some fraction of carbonates subducted under the continental crust may be reabsorbed into the mantle. Actually, we do not know how much carbon is regassed into the mantle, but it should be related to the geotherm and the carbonate metamorphism conditions. According to *DesMarais* (1985), carbonates could not subduct into the mantle in the Archean because of the higher geotherm then. But at present-day values, 3-50% of the carbonates reaching subduction zones should be able to subduct into the mantle (*DesMarais,* 1985).

Here we define the regassing ratio as a second parameter (Fig. 2). The flux of regassing is expressed as

$$F_R = (1 - A) \cdot R \cdot t_r^{-1} \cdot P$$

where R is the regassing ratio.

Metamorphism-volcanism. CO_2 is produced by metamorphism of carbonates subducted under the continental crust. This CO_2 can be released back into the atmosphere by volcanic activity.

The metamorphic reactions are as follows:

$$CaCO_3 + SiO_2 \rightarrow CaSiO_3 + CO_2$$
$$MgCO_3 + SiO_2 \rightarrow MgSiO_3 + CO_2$$

This flux is also difficult to estimate. *Berner et al.* (1983) estimated it at 5.9×10^{18} mol/Ma by explaining the present CO_2 budget. Here we express this flux as

$$F_M = (1 - A) \cdot (1 - R) \cdot t_r^{-1} \cdot P$$

Degassing. CO_2 is also degassed from midocean ridges and hot spots. It should be distinguished from the degassing of CO_2 as a volcanic gas at island arcs because the former comes from the mantle but the latter is considered to be a recycled gas.

This flux is estimated by measuring the C/^{3}He or $^{13}C/^{12}C$ in the hydrothermal fluid and in the basaltic glasses at the midocean ridge. Some recent estimates are $10\text{-}20 \times 10^{18}$ mol/Ma (*Javoy et al.,* 1982), $1\text{-}8 \times 10^{18}$ mol/Ma (*DesMarais,* 1985), or 2.2×10^{18} mol/Ma (*Marty and Jambon,* 1987). Among these, the value of Javoy et al. is considered to be too large when compared with more recent estimates. Therefore, we will use a flux of 4×10^{18}, which is the average of the latter two estimates.

Primordial degassing of CO_2 from the mantle is thought to have been small. This is because the solubility of CO_2 in a magma ocean is very low, so most of the CO_2 produced by impact degassing during accretion (e.g., *Matsui and Abe,* 1986a) would have existed in the proto-atmosphere. We assume as an extreme case that the initial content of CO_2 in the mantle was nearly zero, and if regassing of CO_2 into the mantle had not occurred, CO_2 degassing would be zero. But if regassing has occurred, the current degassing would be higher. We also assume that the amount of degassing is proportional to the carbon content in the mantle and the degassing rate is proportional to the seafloor production rate, which is, in turn, proportional to heat flow. The expression of this flux is

$$F_D = K_D \cdot f_Q \cdot C_{mantle}$$

where K_D is a constant, f_Q is the ratio of the heat production rate at a given time to the present value, and C_{mantle} is the carbon content in the mantle. K_D is determined by giving the initial amount of carbon, C_{total}, which is taken as a parameter, then

$$K_D = \frac{F_D^\bullet}{f_Q^\bullet \cdot C_{mantle}^\bullet} = \frac{F_D^\bullet}{C_{total}^0 - C_{surface}^\bullet}$$

where

$$f_Q^\bullet = 1.0$$

$$C_{total}^0 = C_{surface}(t) + C_{mantle}(t)$$

Volcanic-seawater reaction (hydrothermal reaction). Magnesium ions in the ocean react with seafloor basalts and are exchanged for stoichiometrically equivalent calcium ions (*Holland,* 1978; *Wolery and Sleep,* 1976). This occurs mainly at the midocean ridges, but it also occurs at other lower temperature regions of the seafloor.

The reaction is

$$Mg^{2+} + CaSiO_3 \rightarrow MgSiO_3 + Ca^{2+}$$

where $CaSiO_3$ represents, for example, volcanic glasses, plagioclases, pyroxenes, and olivines, and $MgSiO_3$ represents smectites, amphiboles, chlorites, and other hydrous phases.

We will assume that this flux is proportional to the amount of magnesium in the ocean and to the spreading rate of the seafloor, which is, in turn, proportional to the heat flow. The flux is

$$F_V = K_V \cdot f_Q \cdot M_{Mg}$$

where K_V is a constant and M_{Mg} is the amount of magnesium ions in the ocean (in mol).

This reaction controls the amount of both the magnesium and calcium ions in the ocean. Their present values are $M_{Ca}^\bullet = 1.37 \times 10^{19}$ mol and $M_{Mg}^\bullet = 7.26 \times 10^{19}$ mol (*Broecker,* 1974).

To estimate the constant K_V, we assume that magnesium ions added to the ocean by weathering are taken up by this reaction today (see *Berner et al.,* 1983). Then

$$K_V = \frac{F_V^\bullet}{f_Q^\bullet \cdot M_{Mg}^\bullet} = \frac{F_W^{ms\bullet} + F_W^{mc\bullet}}{M_{Mg}^\bullet} = 0.0716$$

Other processes. All carbon-related species in the atmosphere, oceans, and seafloor are in chemical equilibrium. This equilibrium depends on the pH, temperature, pressure, and salinity of the ocean. Among these, pH is set as a free parameter, but our simple box model cannot take into account the other factors. Therefore, for all equilibrium values, we will use constants, as will be shown in the next section.

Of course, the seafloor is weathered by seawater. At the present time, however, this process is very limited compared to continental weathering because of its low temperature, so we do not treat this process here. It might be of more importance, however, in the cases where the bottom of the ocean is rather hot or the pH in the ocean is very low.

External Conditions

There are some factors that have a strong influence on the Earth's environment. These are solar luminosity, continental surface area, and tectonic activity such as plate motion. These factors have been changing throughout the entire history of the Earth. We will attempt to analyze the response of the box model to variations in these external conditions. ("External" means that they are not constrained by the system of the carbonate-silicate geochemical cycle.)

Increase in solar luminosity. According to the theory of stellar evolution, the luminosity of the sun has been increasing since the Earth's formation. There are several evolutional models, but here we adopt the model proposed by *Gough* (1981). According to his model, the luminosity of the sun at 4.6×10^9 years ago was about 30% less than that of today. The luminosity change given by him is

$$L(t) = \left[1 + \frac{2}{5}\left(1 - \frac{t}{t^*}\right)\right]^{-1} \cdot L^*$$

where L is the luminosity of the sun, L^* is its present value, and t^* is the present age of the sun ($=4.6 \times 10^9$ yr). The solar constant at the Earth's orbit is

$$S(t) = \frac{L(t)}{4\pi D_E^2}$$

where D_E is the distance of the Earth from the sun. So we have

$$S(t) = \left[1 + \frac{2}{5}\left(1 - \frac{t}{t^*}\right)\right]^{-1} \cdot S^*$$

where S^* is the present value of the solar constant.

This change (shown in Fig. 3) affects the surface temperature of the Earth, as suggested by *Sagan and Mullen* (1972).

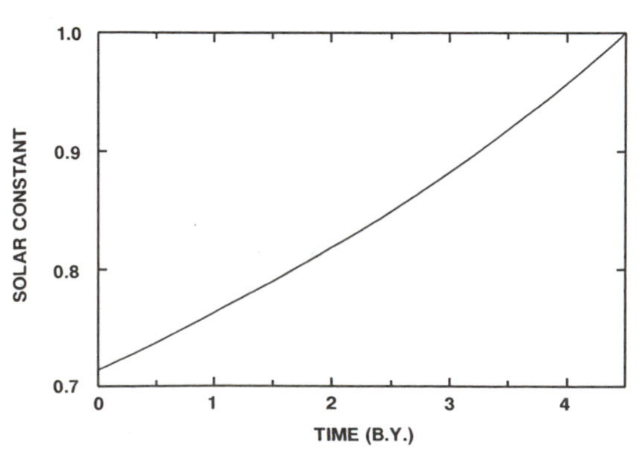

Fig. 3. *Temporal variation of the solar constant (Gough, 1981).*

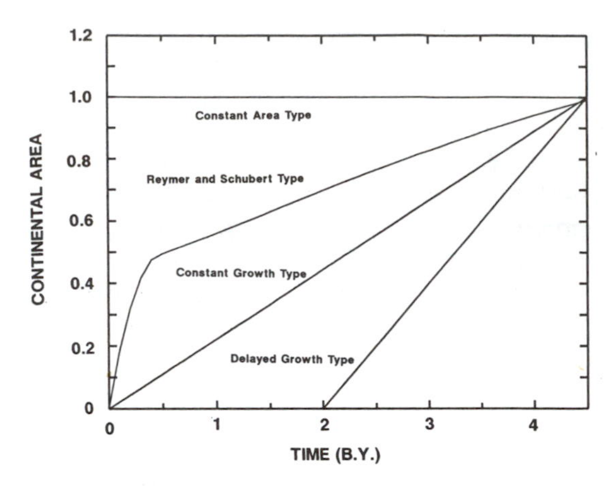

Fig. 4. *Continental growth models used in our simulation. The vertical axis represents the ratio of the continental area at a given time to that at the present time.*

Continental growth. Although many continental growth models have been proposed, we still have very little understanding of the process, especially in the Hadean and early Archean periods. For the present purpose, we do not need a detailed model. We will use several specific models: no continent model ($S_C = 0$), constant area model ($S_C = S_C^*$), constant growth model ($S_C \propto t$), delayed growth model ($S_C = 0$ for $t \leq t_C$ and $S_C \propto t$ for $t \geq t_C$), and the growth model proposed by *Reymer and Schubert* (1984). These models are shown in Fig. 4. (We should note here that, since we do not consider the continental shelves separately in this study, S_C means the area of the entire continental crust, not just the actual land area.)

We define the ratio of continental area at a given time to that at the present as

$$f_{S_C}(t) = \frac{S_C(t)}{S_C^*}$$

Decrease in tectonic activity. Tectonic activity is thought to be proportional to the heat flow from the interior of the Earth. To know its true temporal variation, we need to calculate the thermal history of the Earth,

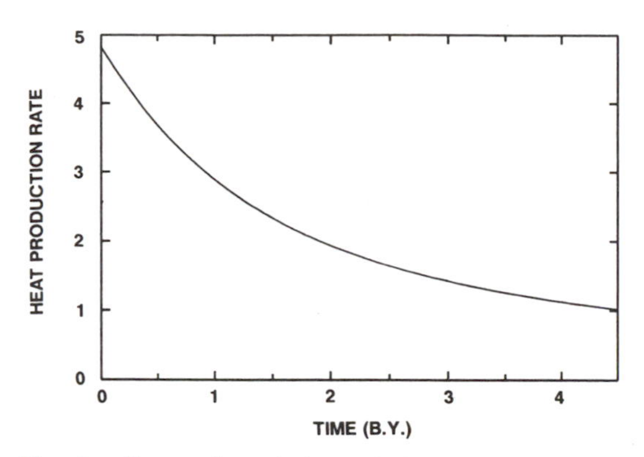

Fig. 5. *Temporal variation of the radiogenic heat production rate that is representative of tectonic activity. The vertical axis represents the ratio of the heat production rate at a given time to that at the present time.*

taking into account the energy of accretion and differentiation, and radiogenic heat production. However, for simplicity, we assume that the heat flow is proportional only to the radiogenic heat production rate (*Turcotte and Schubert,* 1982) as shown in Fig. 5. We define the ratio of the heat production rate at a given time to the present one as

$$f_Q(t) = \frac{Q(t)}{Q^*}$$

Then, the residence time of the seafloor is defined by using this heat production rate

$$t_r(t) = \frac{Q^*}{Q(t)} \frac{\sqrt{S_O}}{\sqrt{S_O^*}} \cdot tr^*$$

where t_r^* is the present seafloor residence time, equal to 60 Ma (*Sclater and Parsons,* 1981; *Sprague and Pollack,* 1980).

Surface Temperature

We need to calculate the surface temperature of the Earth for a given atmospheric CO_2 level and solar constant. For this purpose, we will use the results of a numerical calculation by *Kasting and Ackerman* (1986). They calculated the greenhouse effect of various CO_2 levels for two different solar constants ($S/S^* = 0.7, 1.0$) using a radiative-convective equilibrium model. In their calculation, the tropospheric lapse rate was set equal to its moist adiabatic value and the effect of clouds was not included explicitly. As a consequence of their relative humidity parameterization, a sharp increase in the surface temperature is seen in the present-day Earth model. However, this feature has only a small influence on our simulation, as will be seen later.

Their results are given for only two values of the solar constant. Therefore, we interpolate their results to calculate the surface temperature for the intermediate values of the solar constant.

METHOD

Basic Equations

In order to determine the evolution of each reservoir, we solve the following mass balance equations

$$\frac{dL_{cc}}{dt} = F_A^{cc} - F_W^{cc}$$

$$\frac{dL_{mc}}{dt} = F_A^{mc} - F_W^{mc}$$

$$\frac{dP_{cc}}{dt} = F_P^{cc} - F_A^{cc} - F_M^{cc} - F_R^{cc}$$

$$\frac{dP_{mc}}{dt} = F_P^{mc} - F_A^{mc} - F_M^{mc} - F_R^{mc}$$

$$\frac{dC_{mantle}}{dt} = F_R^{cc} + F_R^{mc} - F_D$$

$$\frac{dM_{Ca}}{dt} = F_W^{cc} + F_W^{cs} + F_V - F_P^{cc}$$

$$\frac{dM_{Mg}}{dt} = F_W^{mc} + F_W^{ms} - F_V - F_P^{mc}$$

where cc = calcium carbonate, mc = magnesium carbonate, cs = calcium silicate, and ms = magnesium silicate. L represents the continental reservoir, P represents the seafloor reservoir, and M represents the oceanic reservoir. F_W, F_P, F_A, F_R, F_M, F_D, and F_V are the fluxes already explained in the previous section.

In the above equations, F_P^{cc} and F_P^{mc} are the carbonate precipitation fluxes. They are determined so as to sustain a chemical equilibrium among the atmosphere, the oceans, and the carbonate on the seafloor. To estimate these fluxes, we need to know the concentrations of the cations in the ocean. Therefore, we consider two equations for the concentrations of the two species of cations Ca^{2+} and Mg^{2+}.

We solve the mass balance equations under the constraints of the mass conservation of carbon and the condition of chemical equilibrium between the carbon-related species in the atmosphere, the ocean, and the seafloor reservoir. Under these constraints and by varying the external conditions described in the previous section, we can solve the mass balance equations.

Mass conservation of carbon. We set the total amount of carbon in the system (surface + mantle) as a free parameter. This amount is assumed to be conserved throughout the entire history of the Earth.

$$C_{total} = C_{surface} + C_{mantle}$$

In this equation

$$C_{surface} = L_C + P_C + A_C + M_C$$

where A_C is the atmospheric carbon content in mol, and

$$M_C = M_{H_2CO_3} + M_{HCO_3} + M_{CO_3}$$
$$L_C = L_{cc} + L_{mc}$$
$$P_C = P_{cc} + P_{mc}$$

Chemical equilibrium. As stated earlier, we assume that chemical equilibrium is always maintained between the carbon-related species in the atmosphere, the oceans, and the seafloor reservoir. These are expressed as

$$CO_2(g) + H_2O \longleftrightarrow H_2CO_3$$
$$H_2CO_3 \longleftrightarrow HCO_3^- + H^+$$
$$HCO_3^- \longleftrightarrow CO_3^{2-} + H^+$$
$$Ca^{2+} + CO_3^{2-} \longleftrightarrow CaCO_3$$
$$Mg^{2+} + CO_3^{2-} \longleftrightarrow MgCO_3$$

We can rewrite these relations by using the law of mass action

$$K_0 = \frac{a_{H_2CO_3}}{P_{CO_2} \cdot a_{H_2O}}$$

$$K_1 = \frac{a_{HCO_3^-} \cdot a_{H^+}}{a_{H_2CO_3}}$$

$$K_2 = \frac{a_{CO_3^{2-}} \cdot a_{H^+}}{a_{HCO_3^-}}$$

$$K_{sp}^{cc} = a_{Ca^{2+}} \cdot a_{CO_3^{2-}}$$

$$K_{sp}^{mc} = a_{Mg^{2+}} \cdot a_{CO_3^{2-}}$$

where the K_i are the equilibrium constants and the a_i are the activities. The activities are defined by

$$a_i = \gamma_i \cdot m_i$$

where γ_i is an activity coefficient and m_i is the concentration of ions of the i-th component in the seawater in mol/l. The equilibrium constants and the activity coefficients depend on the pH, temperature, pressure, and salinity of the ocean. However, as stated earlier, we use the constant values for them at 25°C and 1 atm total pressure as shown in Table 2 (*Kramer,* 1965; *Berner,* 1965; *Kitano,* 1984).

We use four parameters in our simulation. They are C_{total}, pH, A, and R. C_{total} is assumed to be equal to the initial amount of carbon at the surface, $C^0_{surface}$, for the reason we stated earlier. There are lower and upper limits to this value. The lower limit is the present-day amount of carbon at the surface, and the upper limit is determined by the conditions required for liquidization of the impact-induced steam atmosphere. According to *Abe* (1988a), the latter corresponds to 1 kbar of CO_2 partial pressure. Then

$$9.27 \times 10^{21} \text{ mol} \leq C_{total} \leq 1.2 \times 10^{23} \text{ mol}$$

Numerical Scheme

We use the finite difference scheme to solve the mass balance equations numerically. In the mass balance equations, however, we do not know the carbonate precipitation fluxes (F_p^{cc}, F_p^{mc}). Therefore, we cannot solve them by using a simple forward difference scheme.

In this study, we consider that calcite precipitates in the ocean when the seawater is supersaturated with respect to calcite, which means that the ion activity product of calcite ($IAP = a_{Ca^{2+}} \cdot a_{CO_3^{2-}}$) exceeds the thermodynamic solubility product ($= K_{sp}^{cc}$; shown in Table 2). On the other hand, calcite does not precipitate when the seawater is undersaturated with calcite ($IAP \leq K_{sp}^{cc}$). We consider a similar condition for magnesite precipitation. To calculate the ion activity products, however, we first need to know the amounts of the cations and carbonate ions carried to the oceans in each time step. Then we can estimate the carbonate precipitation fluxes by comparing IAP with K_{sp}.

For this purpose, we use the Euler-backward scheme. This is one of the difference schemes that is expressed as

$$\begin{cases} X^{n+1'} = X^n + \Delta t \cdot f^n \\ X^{n+1} = X^n + \Delta t \cdot f^{n+1'} \end{cases}$$

where $f^{n+1'} = f(X^{n+1'}, (n+1)\Delta t)$. The first approximation $X^{n+1'}$ is determined from the Euler scheme and X^{n+1} is derived from the backward scheme (but in this case is not the implicit scheme). However, our scheme is different from this in the sense that $X^{n+1'}$ is not just an approximation but has a physical meaning: It is the amount of cations carried into the ocean during one time step.

The carbonates on the seafloor will dissolve in the seawater when the ocean is undersaturated with the carbonates. It is therefore noted that the carbonate precipitation fluxes can even have a negative value until the carbonates on the seafloor have been exhausted.

The time step is initially taken to be 1.0 Ma as the standard. We choose a different, appropriate time step if the numerical calculation becomes unstable.

Initial Conditions

As described before, we assume that all the carbon initially existed in the atmosphere because of the low solubility of CO_2 in a magma ocean during planetary accretion. Therefore, as an initial condition for the amount of carbon

$$C_{surface}^0 = C_{total}$$

$$C_{mantle}^0 = 0$$

The distribution of carbon in each reservoir is determined by the initial amount of the cations. When the initial amount of the cations was nearly zero, the carbon distributed between the atmosphere and the ocean depends on the value of the pH.

In the proto-atmosphere, about 340×10^{17} kg of HCl and 48×10^{17} kg of SO_2 are considered to have existed. Because these gases can easily dissolve into water, the proto-ocean just after formation should have been a strong acid solution with a molality $\approx$ 0.5 mol/ℓ (*Kitano,* 1984). This solution immediately eroded the seafloor basalts, and cations were supplied. Then the ocean would have been neutralized. Dissolving experiments of basalts in 0.5 mol/ℓ HCl solution and also measurements of the composition of HCl acid hot springs give us information on the composition of such a proto-ocean. One of these estimates is shown in Table 3 (*Ronov,* 1964).

TABLE 3. *The composition of proto- and present ocean (%) (Ronov, 1964).*

	Ca^{2+}	Mg^{2+}	Na^{+}	K^{+}	Total
Proto-ocean	29	24	30	17	100
Present ocean	3.2	10.7	83.1	3.0	100

We will use this estimate as an initial condition for the cations. We assume that the total charge of the cations in the ocean has never changed. (This is the same as assuming the amounts of Cl^- and SO_4^{2-} have not changed.) Then we can determine the concentration of cations in the proto-ocean as shown in Table 4.

We can estimate the initial amount of cations by multiplying the concentration by the volume of the ocean V_o (= $1.36 \times 10^{18}\ m^3$)

$$M^{0'}_{Ca} = 115 \times V_o = 1.56 \times 10^{20}\ \text{mol}$$

$$M^{0'}_{Mg} = 95 \times V_o = 1.29 \times 10^{20}\ \text{mol}$$

Therefore, we can determine A^0_C, M^0_C, L^0_C, P^0_C, M^0_{Ca}, and M^0_{Mg} by using the method described in the previous section.

Actually, the later evolution of each reservoir is not very sensitive to the initial amount of the cations except for the case in which too many cations exist compared to carbon ($\approx 10^{22}$ mol).

TABLE 4. *The concentration (upper row) and charge (lower row) of cations (mol/m³).*

	Ca^{2+}	Mg^{2+}	Na^{+}	K^{+}	Total
Proto-ocean	115	95	119	67	—
	230	190	119	67	606
Present ocean	10	53	470	10	—
	20	106	470	10	606

STATIONARY SOLUTIONS

Present-Day Values of A and R

At the present time, the carbon cycle system may be in a stationary state. If so, we may be able to estimate A and R based on the present fluxes between each reservoir as shown in Fig. 6. In this figure, we number each reservoir (1 = the atmosphere and the ocean, 2 = the seafloor, 3 = the continents, and 4 = the mantle), and represent a flux from reservoir i to j as F_{ij}. In a stationary state, the flux of carbonate precipitation in the ocean will be equivalent to that of the weathering from the continents to the oceans to adjust the oceanic cation budget. Therefore

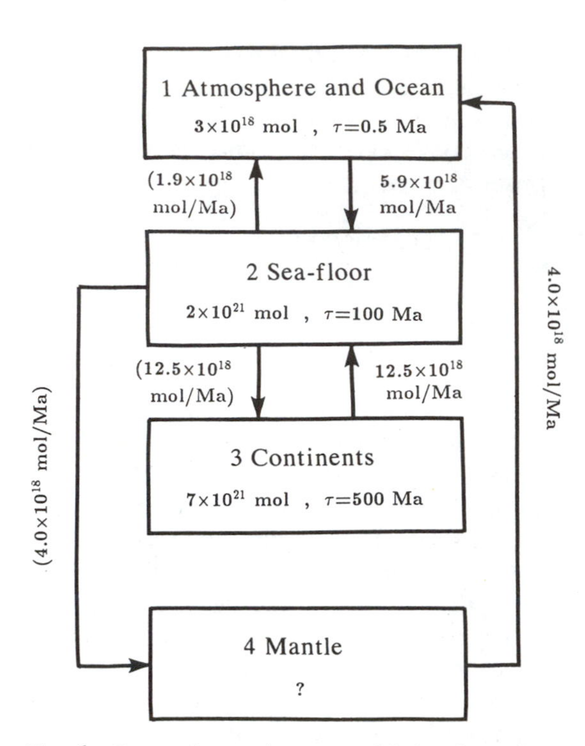

Fig. 6. *Present-day stationary model for the carbon cycle. Each of the values represents the carbon fluxes between the reservoirs in mol/Ma. (Those in parentheses are the adjusted values for the stationary state of this sytem; the others are observed values.) The reservoir sizes (in mol) and the residence times (in Ma) are also shown in each reservoir.*

$$F^*_{41} = F^*_D = 4.0 \times 10^{18}\ \text{mol/Ma}$$

$$F^*_{12} = F^{cs*}_W + F^{ms*}_W = 5.9 \times 10^{18}\ \text{mol/Ma}$$

$$F^*_{32} = F^{cc*}_W + F^{mc*}_W = 12.5 \times 10^{18}\ \text{mol/Ma}$$

To adjust the other fluxes to the above observed values so as to satisfy a stationary state

$$F^*_{21} = F^*_{12} - F^*_{41} = 1.9 \times 10^{18} \text{ mol/Ma}$$
$$F^*_{23} = F^*_{32} = 12.5 \times 10^{18} \text{ mol/Ma}$$
$$F^*_{24} = F^*_{41} = 4.0 \times 10^{18} \text{ mol/Ma}$$

Using these fluxes, we can calculate the values of A and R as

$$A = \frac{F^*_{23}}{F^*_{21} + F^*_{23} + F^*_{24}} = 0.68 \simeq 0.7$$
$$R = \frac{F^*_{24}}{F^*_{21} + F^*_{24}} = 0.68 \simeq 0.7$$

These values are required for the present-day stationary state.

The ratio of the carbonates that subduct into the mantle to the total subducted carbonates is

$$(1 - A) \cdot R = 0.21$$

which is in the range of the recent estimate of 3-50% (*DesMarais,* 1985).

Present-Day Amount of Surface Carbon

We can derive the stationary solutions for the given system by setting the left sides of the mass balance equations equal to zero. The amount of carbon at the surface is written as

$$\begin{aligned} C_{surface} &= L_C + P_C + A_C + M_C \\ &\approx L_C + P_C \\ &= \frac{A}{1 - A} \frac{K^{cs}_W + K^{ms}_W}{K^{cc}_W} \cdot f_{S_C} + \frac{t_r}{(1 - A) \cdot R} \cdot F_D \end{aligned}$$

It is shown from this solution that the amount of carbon at the surface is a function of the tectonic parameters that characterize the model. The existence of this stationary solution suggests that the stable amounts of carbon on the continents and the seafloor are determined by a given tectonic condition.

To examine a controlling factor that determines the present-day amount of carbon at the surface, we investigate the effect of A and R on the stationary solution by using the present observed values for other parameters ($f^*_{S_C} = 1.0$, $t^*_r = 60$ Ma, and $F^*_D = 4.0 \times 10^{18}$ mol/Ma). The result is shown in Fig. 7. It is shown that the amount of carbon at the surface does not depend greatly on the value of the regassing ratio R, but does depend on the value of the accretion ratio A. The recent estimate for the amount of carbon at the surface by *Ronov and Yaroshevsky* (1976) can be explained when the accretion ratio is equal to 0.7.

This is consistent with the value obtained by the previous section on the present-day stationary flux. On the other hand, the value of the regassing ratio cannot be obtained from the above discussion.

The values of A and R would have been changed through the Earth's history. (For example, the regassing ratio depends on the geotherm, which should have been changed with time.) However, we will use the values A = 0.7 and R = 0.7, which are required for

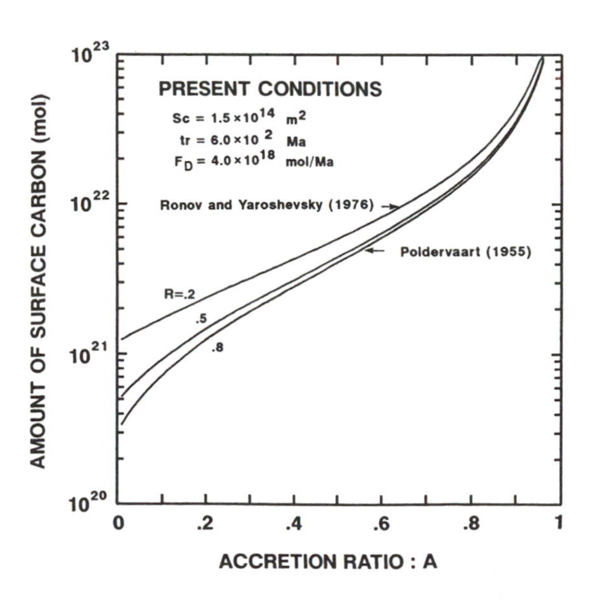

Fig. 7. *Amount of surface carbon under the present conditions as a function of the accretion ratio and the regassing ratio. The arrows show the estimates for the amount of surface carbon by Poldarvaart (1955) and Ronov and Yaroshevsky (1976), respectively.*

the present-day stationary state, as the standard in our calculations. The geophysical meaning of parameter A will be discussed later.

RESULTS

Evolution of Carbon Reservoirs

We consider the case in which the carbonates on the seafloor subduct into the mantle (regassing ratio $\neq 0$). In this case, the amount of CO_2 degassing from the mantle will be proportional to the increase in the carbon content of the mantle. At first the total amount of carbon is assumed to be 1×10^{22} mol, which is nearly equal to the present surface value ($C_{total} = 1 \times 10^{22}$ mol $\approx C^{\bullet}_{surface}$), and the pH is taken to be equal to 8 (the present value).

No continental growth model. If there were no continents on the Earth, the level of the CO_2 in the atmosphere-ocean system would never decrease throughout the Earth's history in Fig. 8. The atmospheric CO_2 level

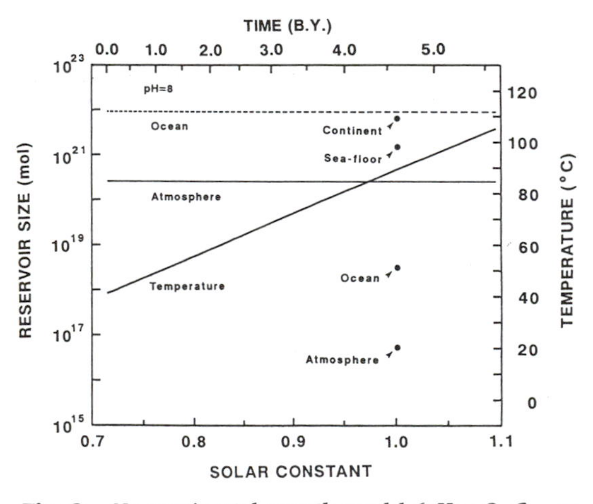

Fig. 8. *No continental growth model (pH = 8, $C_{total} = 1 \times 10^{22}$ mol). Variation of carbon levels (in mol) in each reservoir and surface temperature (°C) of the Earth are shown as a function of the solar constant and time. The heavy solid line represents the surface temperature, the light solid line represents the atmospheric carbon reservoir, and the dashed line represents the oceanic carbon reservoir. The solid circles represent the present values of each reservoir. It is noted that surface temperature increases monotonically with an increase in the solar constant in this case.*

corresponds to about 2 bar in this model. This is because the lack of a supply of cations from the continents to the ocean prohibits precipitation of CO_2 in the ocean as carbonates, resulting in no regassing. The ratio of the amount of carbon in the atmosphere to that in the ocean is determined only by the value of the pH in the ocean. This is because we considered the initial amount of cations in the ocean to be equal to zero, as well as assuming that there was no supply of cations from the continents. Therefore, there is no carbonate in the seafloor reservoir. Even if we assume appropriate values for the initial amount of cations in the ocean (as stated in the previous section), the result is almost the same as in this model. A large amount of carbonates precipitate at first, but they soon subduct and return to the atmosphere through volcanism. After that, the situation becomes the same as mentioned above. It is noted here, however, that we neglect the effect of seafloor weathering that may also supply the cations to the oceans. Therefore, this estimate corresponds to the maximum limit for the atmospheric CO_2 level at a given ocean pH. However, as will be discussed later, even if we consider the effect of seafloor weathering, the partial pressure of atmospheric CO_2 would never decrease to a certain level. In any case, it is shown that the surface temperature of the Earth increases with an increase in the solar constant. In other words, the Earth's environment would not have evolved to the present state if the continents were not formed.

A continental growth model. There are continents on the Earth at present. Thus, we should turn our attention to the more realistic models that include continental growth. Here we study the delayed continental growth model in which the continents begin to grow 2 b.y. after the formation of the Earth (Fig. 9). Both the accretion and regassing ratios are 0.7. Before the formation of the continents, the situation is the same as mentioned in the first case. Once the continents begin to form, cations are supplied to the ocean and thus

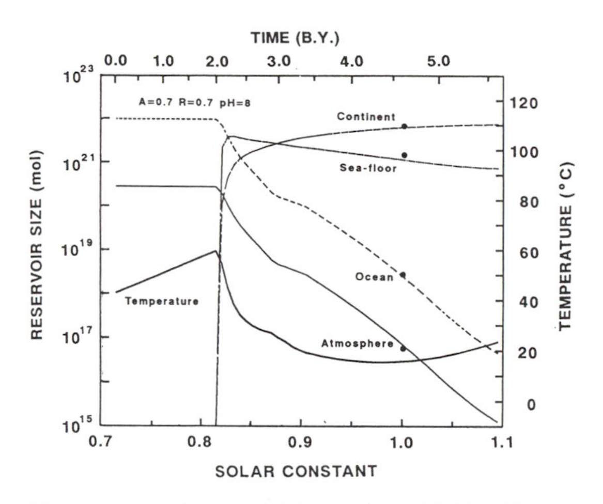

Fig. 9. *Delayed continental growth model ($A = 0.7$, $R = 0.7$, $pH = 8$, $C_{total} = 1 \times 10^{22}$ mol). Variation of carbon levels in each reservoir and surface temperature of the Earth as a function of the solar constant and time. The meaning of the curves and symbols are the same as in Fig. 8.*

carbonates can precipitate onto the seafloor. Some fraction of these carbonates subducts somewhere, but the rest accretes to the continents. Subducted carbonates are disintegrated into carbon dioxide and calcium silicate, and CO_2 returns to the atmosphere through volcanism. The accretion of seafloor carbonates onto continents results in a decrease of CO_2 in the atmosphere-ocean system. The surface temperature therefore decreases. This suggests that the surface temperature could be stabilized in spite of the increase in the solar constant. The sharp decreases in the temperature and the reduction rate of the atmospheric and oceanic reservoirs shown in Fig. 9 are due to the sharp increase in the temperature profile of *Kasting and Ackerman* (1986) that we used in this study.

The continents are a very stable carbon reservoir and provide a longer residence time for carbonates than the other reservoirs do. This is another important role of continental formation in addition to that of being a cation source for the ocean. As shown in Fig. 9, this model explains well the present distribution of carbon.

An alternative continental growth model. Various continental growth models have been proposed. Here we study the effect of the continental growth curve on the numerical results. In this model, the continents are assumed to grow in a manner proposed by *Reymer and Schubert* (1984) (Fig. 10). The accretion and regassing ratios are the same as in the previous model. The evolution of each reservoir and the surface temperature are almost determined at the beginning of the Earth's history, since the continents grow rapidly. The basic features of the evolution are, however, almost the same as in the previous model. It is suggested that the evolution of each reservoir does not depend greatly on the continental growth model after the continents grow.

This model also explains well the present distribution of carbon in each surface reservoir. It is shown that the distribution of carbon depends mainly on the accretion ratio. In order to explain the present distribution, the accretion ratio A needs to be 0.7. For other values, the distribution of carbon is quite

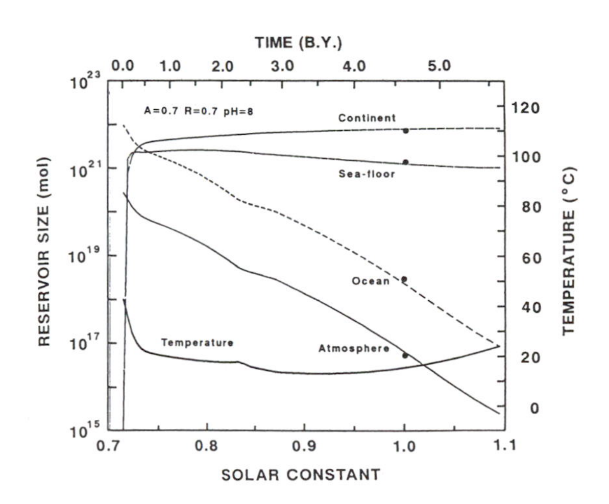

Fig. 10. *Reymer and Schubert (1984) growth model ($A = 0.7$, $R = 0.7$, $pH = 8$, $C_{total} = 1 \times 10^{22}$ mol). Variation of carbon levels in each reservoir and surface temperature of the Earth as a function of the solar constant and time. The meaning of the curves and symbols are the same as in Fig. 8.*

different from the present one. For example, a lower value of A will result in a lower carbon content on the continents than the seafloor.

The distribution of the carbon also depends on a regassing ratio that is equal to 0.7 in this model. The distribution changes when we assume other values for the regassing ratio. However, the effect of R on the distribution is much weaker than that of the accretion ratio, A. We can conclude that the distribution of carbon is mainly determined by A, and that the present distribution is well explained when A is equal to 0.7.

Effect of pH. Even when we use 7 as the value of the pH instead of 8, the result does not change significantly except for the ratio of carbon in the atmosphere to that in the ocean (Fig. 11). It is noted that the level of atmospheric CO_2 does not depend on the pH as long as the pH is larger than 6. Hence, the surface temperature stays at the same level as when the pH is equal to 8. We can conclude that the pH of the oceans does not greatly affect the surface temperature as long as the continents exist.

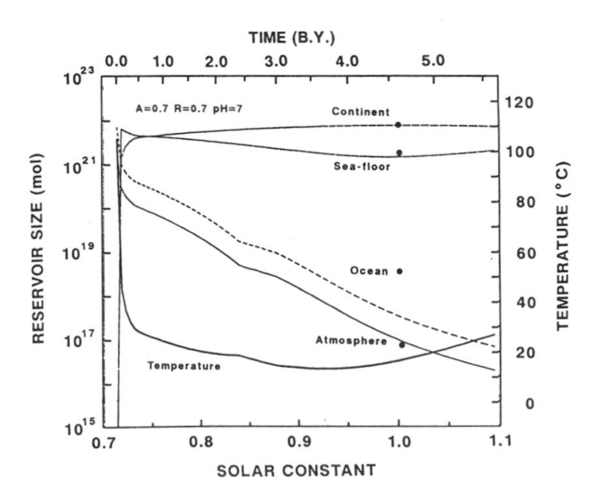

Fig. 11. *Reymer and Schubert (1984) growth model with different pH value (A = 0.7, R = 0.7, pH = 7, C_{total} = 1 × 10^{22} mol). Variation of carbon levels in each reservoir and surface temperature of the Earth as a function of the solar constant and time. The meaning of the curves and symbols are the same as in Fig. 8.*

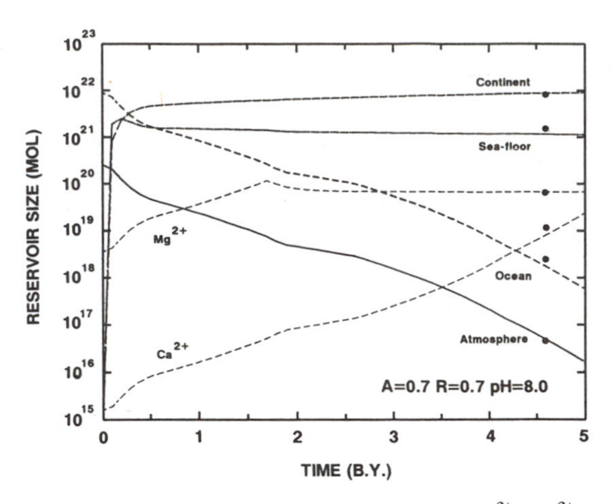

Fig. 12. *Temporal variation in the cation (Ca^{2+}, Mg^{2+}) reservoirs in addition to the carbon reservoirs for the Reymer and Schubert (1984) growth model (A = 0.7, R = 0.7, pH = 8, C_{total} = 1 × 10^{22} mol). The meaning of the curves and symbols are the same as in Fig. 8.*

Evolution of cation reservoirs. The evolution of the cation reservoirs is shown in Fig. 12. Basically, the cations (Ca^{2+}, Mg^{2+}) are in chemical equilibrium with the carbonate ions (CO_3^{2-}), so they are inversely proportional to the amount of carbon in the ocean. Calcium ions are shown to be in equilibrium with carbonate ions throughout the entire history of the Earth. This may be supported by the geological evidence that calcite has been produced continuously since the time of the oldest carbonate formation.

On the other hand, magnesium ions are shown to be in equilibrium with carbonate ions only during the early Archean period, but not after that. This is because the equilibrium level of magnesium in the ocean increases with decreasing atmospheric CO_2 level. This means that the supply of magnesium ions from the continents becomes too low to sustain their consumption by hydrothermal reaction after the magnesite on the seafloor reservoir is exhausted. Once the equilibrium is broken, the concentration of the magnesium ions in the ocean is determined only by the balance of the supply of those ions from the continents and the consumption of them by the hydrothermal reaction. Because this estimate gives the upper limit for the concentration of magnesium ions, we can say that magnesites might have formed only in the early Archean period, but not after that. This is consistent with the lack of geological evidence that magnesite was formed recently.

When we consider chemical equilibrium and assume the ocean pH to be near the present value, the concentrations of calcium and magnesium ions in the oceans should be less in the past than today because of the higher atmospheric CO_2 level in the past. Also, judging from the values of the solubility products, we know that the concentration of magnesium ions should always be higher than that of calcium ions. We can also explain the present value of these ions by assuming an accretion ratio equal to 0.7.

Closed system model. Finally, we consider the case where the seafloor carbonates do not subduct into the mantle. In this case, CO_2 does not degas from the mantle because of the assumption that almost all of the carbon existed as CO_2 in the proto-atmosphere initially. Therefore, we name this case the closed system.

We set the total amount of carbon equal to the present-day amount at the surface ($C_{total} = C^{*}_{surface}$), the pH equal to 8 (the present value), and the regassing ratio equal to zero ($R = 0$).

As shown in Fig. 13 the result is almost the same as for that of the open system models (in the case of $R \neq 0$). Before the formation of the continents, CO_2 in the atmosphere-ocean system cannot decrease at all and hence the surface temperature increases with an increase in the solar constant. Once continental growth begins, however, CO_2 in the atmosphere-ocean system decreases and the surface temperature becomes stabilized. There seems to be no difference in the results between the closed and open system models. The distribution of carbon depends on an accretion ratio that, to explain the present distribution, again needs to be equal to 0.7.

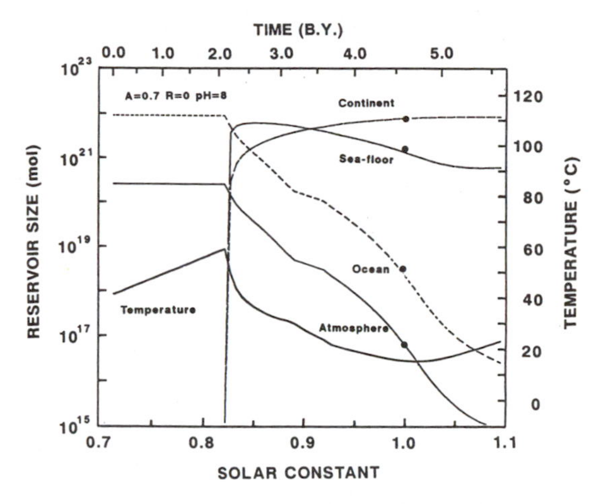

Fig. 13. *Delayed continental growth model ($A = 0.7$, $R = 0$, $pH = 8$, $C_{total} = C_{surface}{}^{*}$). Variation of carbon levels in each reservoir and surface temperature of the Earth as a function of the solar constant and time. The meaning of the curves and symbols are the same as in Fig. 8.*

Effect of total carbon amount. Because we do not know the present carbon content of the mantle, the total amount of carbon cannot be determined for the open system model. However, each reservoir evolves to the present-day environment irrespective of the assumed total amount of carbon. This might be because we used the CO_2 degassing function from the mantle that produces the present flux at the present condition irrespective of the carbon contents in the mantle. In every case, the amount of carbon at the surface reaches a stationary level, which is determined by some tectonic factors such as the CO_2 degassing flux (as stated in the section on stationary solutions).

One example of the result of regassing and degassing fluxes is shown in Fig. 14. The "extra" amount of carbon at the surface precipitates to the seafloor as carbonates that eventually subduct into the mantle, and thus the amount of carbon at the surface decreases. ("Extra" refers to the extra value above a stationary level.) However, once it decreases to a certain level, the fluxes of the CO_2 regassing into the mantle and the CO_2 degas-

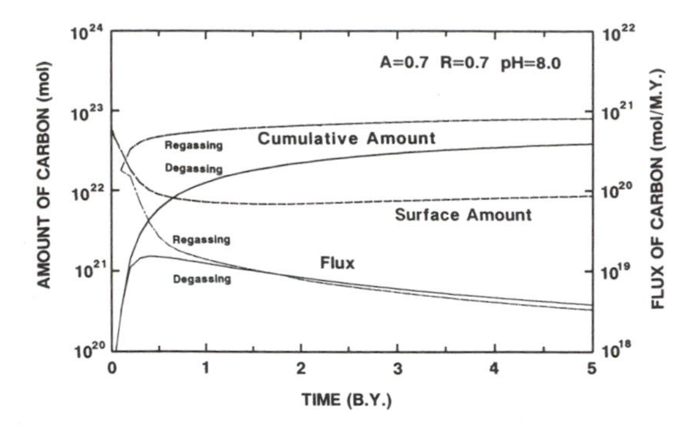

Fig. 14. *Variation in the fluxes of regassing carbon (the surface to the mantle) and degassing CO_2 (the mantle to the surface) in mol/m.y. The cumulative amount represents the integrated amount of the regassing and degassing carbon fluxes since the beginning of the Earth's history (mol). The variation in the surface amount of carbon is also shown. It is noted that since these two fluxes become almost balanced, the amount of surface carbon decreases to a certain level. [Reymer and Schubert (1984) growth model; $A = 0.7$, $R = 0.7$, $pH = 8$, $C_{total} = 5 \times 10^{22}$ mol.]*

sing from the mantle become balanced and reach a stationary state for the amount of carbon at the surface.

It is noted here that the timescale for establishing an equilibrium between surface and mantle carbon during the Archean was roughly 500 m.y., and that, whatever the amount of the carbon in the mantle at present, it is in a steady state with the surface.

A Mechanism to Stabilize Surface Temperature

The evolution of the surface temperature for various models is shown in Fig. 15. In spite of the large differences in the parameters A, R, and pH, surface temperatures stay within a fairly narrow range. This is because a decrease in atmospheric CO_2 results in a decreasing greenhouse effect, which compensates for the effect of the increase in solar luminosity. This means that the negative feedback mechanism proposed by *Walker et al.* (1981) plays a key role in stabilizing surface temperature even when continental growth and tectonic evolution are taken into account. Here, the negative feedback mechanism is described as follows. When the surface temperature increases, weathering becomes more efficient because of the increase in evaporation of H_2O from the ocean, that is, the increase in rainfall. This results in a decrease of CO_2 in the atmosphere, if we assume that the supply of CO_2 through volcanism remains the same. Then the surface temperature will decrease because of the decrease in the greenhouse effect of CO_2. On the other hand, when the surface temperature decreases, weathering becomes less efficient, and CO_2 will accumulate in the atmosphere. Then the surface temperature will begin to increase. In this way, such a negative feedback mechanism will stabilize the surface temperature (*Walker et al.,* 1981, 1983; *Kasting and Toon,* 1988).

In modeling this mechanism, they estimated separately the effects of changes in the effective temperature (which relates to the solar constant) and the rate of CO_2 release through volcanism (which relates to tectonic activity) from the present values. However, all factors (solar constant, tectonic activity, and continental area) are greatly changed throughout the entire history of the Earth. The

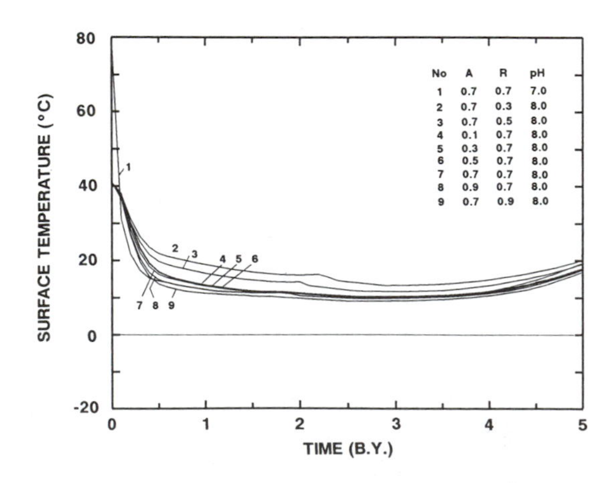

Fig. 15. *Temporal variation of the surface temperature of the Earth for various values of the parameters. It is noted that the surface temperature is stabilized over the entire history of the Earth except for the early stage where the continents had not yet grown large. [Reymer and Schubert (1984) growth model; $C_{total} = 1 \times 10^{22}$ mol.]*

coupled changes of all these factors would have affected the carbon cycle. As shown in the numerical results, the surface temperature is relatively high until the continents grow to a certain degree. But after that, the temperature becomes stabilized even if these conditions are greatly changed. The level at which the temperature is stabilized depends mainly on the tectonic factors such as f_{S_C}, tr, F_D, A, and R. Therefore, the negative feedback mechanism would have a stabilizing effect on the surface temperature of the Earth throughout most of its history.

This mechanism does not work when there are no continents and no tectonic activity (no plate motion and no CO_2 degassing from the mantle). Since the carbonate-silicate geochemical cycle does not work in these cases, the surface temperature cannot be stabilized.

DISCUSSION

Accretion Ratio

We showed that the accretion ratio plays an important role in our carbon cycle model. As shown in the section on stationary solutions, the accretion ratio is 0.7 at the present time. We also showed that this value results in the present distribution of carbon at the surface. Then we discussed what the physical meaning of the accretion ratio is.

There is some geological evidence that the carbonate sediments on seamounts accreted onto the continental crust in the past (*Ozawa,* 1986). At present, however, evidence of the accretion of seafloor sediments onto the continents has not yet been found. Although deep seafloor drilling experiments have not been widely done, all the experiments have revealed that not the accretion type but the tectonic erosion type has occurred (*Nagumo,* 1987).

It may be better, therefore, to define the present accretion ratio as the ratio of the amount of carbonates that will not subduct under the continents to the total precipitated carbonates. In this respect, we need to take note of the carbonates deposited on the continental crust directly (Fig. 16). The present ocean has a CCD (calcium carbonate compensation depth), so carbonates can precipitate only at the continental shelves and the shallower depth regions around the midocean ridges. In fact, at present, very large amounts of carbonates are formed at coral reefs. Therefore, the present accretion ratio is estimated from the carbonate precipitation rate at the continental shelves and at the midocean ridges, and also the growth rate of the worldwide coral reefs.

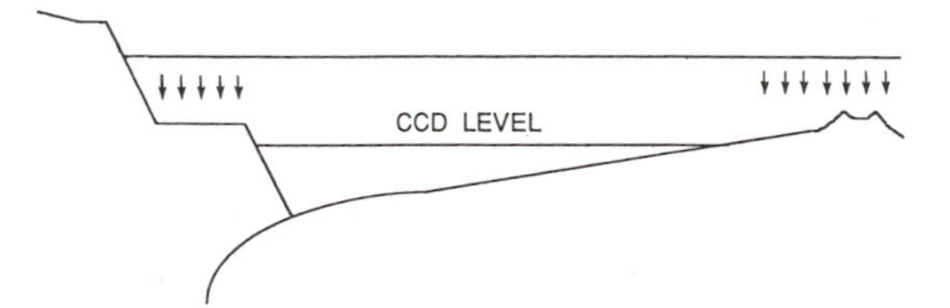

Fig. 16. *Schematic diagram of the carbonate precipitation in the present ocean. Because there exists a CCD (calcium carbonate compensation depth) level in the present ocean, carbonates can precipitate only in the shallower regions such as around the midocean ridges and the continental shelves.*

In this study, we considered the continental shelves as a part of the continents. But the features of the continental shelves as carbon reservoirs are different from both the continents and the seafloor. It will be better to treat them as an independent reservoir in future studies.

Weathering of the Seafloor

The weathering of the seafloor is not as important as that of the continent today. However, when the water at the bottom of the ocean is hot, or the pH of the oceans is low, it becomes more effective. This suggests that the seafloor weathering effect may be important during the "aquaplanet" stage (no continent period).

The effect of the weathering of the seafloor was discussed by *Walker* (1985) for the case where R = 0. In his paper he considered the carbon cycle in the atmosphere, the ocean, and the seafloor at the aquaplanet stage. He assumed a continuous supply of cations from the midocean ridges that precipitates as carbonates onto the seafloor and is carried back into the mantle with the plate motion. In this respect, this estimate corresponds to the upper limit for the effect of the seafloor weathering. Based on the comparison of the timescale of seawater circulation and the residence time of carbon on the seafloor, he concluded that the partial pressure of atmospheric CO_2 would never decrease to a level lower than about 0.2 bar when the pH is equal to 8. (However, in his paper he considered that the pH of the early ocean was much lower than today. When pH was lower than 6, the CO_2 partial pressure is about 10 bar.) Our results show that the CO_2 partial pressure would never decrease to a level lower than about 2 bar. This is because we do not take into account the weathering of the seafloor. In this respect, our treatment corresponds to the lower limit for the effect of seafloor weathering. For both cases, however, the partial pressure of atmospheric CO_2 is constant during the aquaplanet stage. Therefore, the surface temperature increases with an increase in the solar constant.

Even if we consider the effect of seafloor weathering, our main conclusions do not change. For the case of no continents, the Earth's environment would never evolve to the present state.

Depletion of Carbon at the Surface

It is well known that the amount of carbon at the Earth's surface is depleted (1/10 ~ 1/100) compared to other volatiles judging from the value inferred from C/^{36}Ar and C/H for the various kinds of meteorites (*Otting and Zahringer,* 1967). If we assume that an impact degassing occurred during the formation of the Earth, almost all of the carbon-related gas species should have been degassed at that time and constituted a major atmospheric component. The upper limit of the initial amount of CO_2 is given by the conditions necessary for the formation of the ocean. According to *Abe* (1988a), the upper limit is almost 1 kbar. If the Earth formed by the accretion of planetesimals with a chondritic C/H ratio, we need to consider a sink for the decrease of such a large amount of CO_2 to the present level. The only sink so far proposed is the core. CO_2 dissolves into the magma ocean and reacts with metallic iron, and is removed from the magma ocean by core formation (*Abe,* 1988b).

However, there is a possibility that the mantle becomes a sink for carbon through the carbon cycle between the surface reservoirs and mantle. Characteristic timescales for the carbon cycle may determine the distribution of carbon between the surface and the mantle. For example, the surface carbon would be determined by the residence time of carbon in the mantle, which is obviously related to the tectonic (and thermal) evolution of the Earth.

The amount of H_2O, which is the most abundant volatile at the surface, was determined basically at the time of the Earth's accretion by the dissolution equilibrium between a steam atmosphere and a magma ocean (*Matsui and Abe,* 1986a,b). On the other hand, the amount of surface CO_2, which is the second most abundant volatile at the surface, might have been determined during the tectonic evolution of the Earth after ocean formation by the carbon cycle between the surface reservoirs and the mantle.

Problem of Organic Carbon

At the present time carbonates are produced mainly by organisms such as *Coccolithophores* (algae), *Planktonic foraminifera, Pteropoda,* or coral as their hard material in the ocean (*Broecker,* 1974). Of course, through either organic or inorganic processes or both, carbonates are formed in the ocean as long as the seawater is saturated. The organic carbon (CH_2O) is also produced as the soft parts of such organisms, and it also precipitates to the seafloor.

The ratio of carbonate carbon to organic carbon may have been constant at 4:1 during the last 3.8 b.y. according to the isotopic record (*Schidlowski,* 1988; *Junge et al.,* 1975; *Broecker,* 1970). This constancy may be explained by using the carbon cycle model in which the ratio of the flux of organic carbon precipitation to the flux of carbonate precipitation ($= F_P^{C_{org}}/F_P^{C_{carb}}$, where C_{org} represents the organic carbon and C_{carb} represents the carbonate carbon) and the ratio of the weathering constant of organic carbon to that of carbonate ($= K_W^{C_{org}}/K_W^{C_{carb}}$) are assumed to be constant (*Junge et al.,* 1975).

In this study we did not take into account the organic carbon-related processes. However, judging from the ratio (4:1), we consider that our results are not affected significantly by taking into account the organic processes. However, the organic carbon-related processes play a key role in the evolution of oxygen in the atmosphere, because oxygen has been accumulated due to biogenic photosynthesis

$$CO_2 + H_2O \rightarrow CH_2O + O_2$$

The incorporation of the organic processes into our model is a future problem.

Future of the Earth's Environment

In this paper we showed the change in surface temperature from the beginning of the Earth's history to a time in the near future. It is obvious from geological evidence that there were several glaciation periods in the past, but our results cannot predict them. This is because the Earth's environment is perturbed by many factors that we did not take into account in this study. Such factors may cause local and even global changes in surface temperature for very short periods compared to the Earth's history. Such factors are, for example, the distribution and topography of the continents (*Barron and Washington,* 1982, 1984), changes in the spreading rate of the seafloor (*Berner et al.,* 1983; *Owen and Rea,* 1985), changes in the orbital elements (*Berger,* 1988), and, possibly, giant impacts (*Kasting et al.,* 1986). It is noted, however, that characteristic timescales related to these factors are very short in comparison with the age of the Earth. Therefore, we can consider that the basic trend of the evolution of the Earth's environment has been regulated mainly by the CO_2 cycle.

As seen in Fig. 15, however, the long-range variation of the surface temperature seems to be at a turning point at present, that is, from decreasing to increasing. This is because the present greenhouse effect of CO_2 is at its lowest level of the Earth's history. At present, the greenhouse effect due to H_2O is comparable with that due to CO_2 (*Walker,* 1982). Therefore, from now on, the Earth may not be able to maintain the surface temperature by decreasing the CO_2 partial pressure in response to increasing solar luminosity. In addition, as pointed out by *Lovelock and Whitfield* (1982), green plants may not be able to photosynthesize once the atmospheric CO_2 level drops below 150 ppm. From these points of view, the period in which the Earth's environment was regulated by the CO_2 cycle may have already ended. As a rough trend, a long-term increase in the surface temperature of the Earth would be expected. Someday the Earth will become like Venus, unless some unknown mechanism such as cloud feedback plays a role in reducing the surface temperature.

SUMMARY

The main results are summarized as follows:

1. Continental growth is required for the terrestrial environment to evolve to the present state.

a. If there were no continents on the Earth, the atmospheric CO_2 could not decrease to a certain level. This is because the cations that react to form carbonates would not be supplied much to the oceans. The partial pressure of CO_2 would therefore be around 2 bar (when the pH is equal to 8), resulting in an increase in the surface temperature with an increase in the solar luminosity.

b. The formation of the continents makes possible the supply of various cations to the oceans, so carbonates can precipitate to the seafloor. If these carbonates accrete to the continents, the continents act as a sort of storage for carbon in the CO_2 cycle because the continents have a very long residence time for the carbon cycle. In this case, the Earth's environment will evolve to the present state.

c. Even if we take into account the effects of regassing into and degassing from the mantle, continental growth plays the key role in controlling the environment.

d. The distribution of carbon at the surface is determined by a parameter called the accretion ratio, and the present distribution can be explained when this value is equal to 0.7.

2. The carbon cycle, including the carbonate-silicate geochemical cycle, has an important role in controlling the Earth's environment. Throughout the entire history of the Earth, the surface temperature has been stabilized by this cycle against changes in external conditions such as the solar constant, continental growth, and tectonic activity. The surface temperature cannot be stabilized when this cycle cannot work (for example, no oceans, no continents, or no tectonic activity).

3. Cations in the oceans are basically in chemical equilibrium with carbonate ions. Calcium ions have been in equilibrium with carbonate ions throughout the history of the Earth. In the case of magnesium ions, however, the equilibrium has been broken ever since the supply of magnesium ions from the continents was less than enough to balance their consumption, perhaps since sometime during the Archean period.

Acknowledgments. The authors express their thanks to E. Duxbury for assistance in manuscript preparation. One of the authors (E.T.) would like to thank Y. Abe and M. Kumazawa for their critical comments. This paper is a part of his Master's thesis. Calculations were done in the computer center of the University of Tokyo. This research was partially supported by the grants-in-aid for Scientific Research (No. 63611004) of the Ministry of Education of Japan.

REFERENCES

Abe Y. (1988a) Conditions required for formation of water ocean on an Earth-sized planet (abstract). In *Lunar and Planetary Science XIX*, pp. 1-2. Lunar and Planetary Institute, Houston.

Abe Y. (1988b) Abundance of carbon in an impact-induced proto-atmosphere. *Proc. 21st ISAS Lunar Planet. Symp.*, pp. 238-244.

Abe Y. and Matsui T. (1985) The formation of an impact-induced H_2O atmosphere and its implications for the early thermal history of the Earth. *Proc. Lunar Planet. Sci. Conf. 15th*, in *J. Geophys. Res., 90*, C545-C559.

Abe Y. and Matsui T. (1986) Early evolution of the Earth: Accretion, atmosphere formation and thermal history. *Proc. Lunar Planet. Sci. Conf. 17th*, in *J. Geophys. Res., 91*, E291-E302.

Abe Y. and Matsui T. (1988) Evolution of an impact-generated H_2O-CO_2 atmosphere and formation of a hot proto-ocean on Earth. *J. Atmos. Sci., 45*, 3081-3101.

Barron E. J. and Washington W. M. (1982) Atmospheric circulation during warm geologic periods; Is the equator-to-pole surface-temperature gradient the controlling factor? *Geology, 10*, 633-636.

Barron E. J. and Washington W. M. (1984) The role of geographic variables in explaining paleoclimates: Results from Cretaceous climate model sensitivity studies. *J. Geophys. Res., 89*, 1267-1279.

Berger A. (1988) Milankovitch theory of climate. *Rev. Geophys., 26*, 624-657.

Berner R. A. (1965) Activity coefficients of bicarbonate, carbonate and calcium ions in sea water. *Geochim. Cosmochim. Acta, 29*, 947-965.

Berner R. A., Lasaga A. C., and Garrels R. M. (1983) The carbonate-silicate geochemical cycle and its effect on atmospheric carbon dioxide over the past 100 million years. *Am. J. Sci., 283*, 641-683.

Broecker W. S. (1970) A boundary condition on the evolution of atmospheric oxygen. *J. Geophys. Res., 75*, 3553-3557.

Broecker W. S. (1974) *Chemical Oceanography*. Harcourt Brace Jovanovich, New York. 214 pp.

DesMarais D. J. (1985) Carbon exchange between the mantle and crust, and its effect upon the atmosphere: Today compared to Archean time. In *The Carbon Cycle and Atmospheric CO_2: Natural Variations Archean to Present* (E. T. Sundquist and W. S. Broecker, eds.), pp. 602-611. AGU, Washington, DC.

Gough D. O. (1981) Solar interior structure and luminosity variations. *Sol. Phys., 74,* 21-34.

Holland H. D. (1978) *The Chemistry of the Atmosphere and Oceans.* Wiley, New York. 351 pp.

Hunt J. M. (1972) Distribution of carbon in crust of the Earth. *Bull. Am. Assoc. Petrol. Geol., 56,* 2273-2277.

Javoy M., Pineau F., and Allégre C. J. (1982) Carbon geodynamical cycle. *Nature, 300,* 171-173.

Junge C. E., Schidlowski M., Eichmann R., and Pietrek H. (1975) Model calculations for the terrestrial carbon cycle: Carbon isotope geochemistry and evolution of photosynthetic oxygen. *J. Geophys. Res., 80,* 4542-4552.

Karig D. E. and Sharman G. F. (1975) Subduction and accretion in trenches. *Bull. Geol. Soc. Am., 86,* 377-389.

Kasting J. F. (1987) Theoretical constraints on oxygen and carbon dioxide concentrations in the Precambrian atmosphere. *Precambrian Res., 34,* 205-229.

Kasting J. F. and Ackerman T. P. (1986) Climatic consequences of very high carbon dioxide level in the Earth's early atmosphere. *Science, 234,* 1383-1385.

Kasting J. F. and Toon O. B. (1989) Climate evolution on the terrestrial planets. In *Origin and Evolution of Planetary and Satellite Atmospheres* (S. K. Atreya, J. B. Pollack, and M. S. Matthews, eds.), pp. 423-449. Univ. of Arizona, Tucson.

Kasting J. F., Richardson S. M., Pollack J. B., and Toon O. B. (1986) A hybrid model of the CO_2 geochemical cycle and its application to large impact events. *Am. J. Sci., 286,* 361-389.

Kitano Y. (1984) *Environmental Chemistry of the Earth.* Syokabo, Tokyo. 237 pp. (In Japanese.)

Kramer J. R. (1965) History of sea water: Constant temperature-pressure equilibrium models compared to liquid inclusion analyses. *Geochim. Cosmochim. Acta, 29,* 921-945.

Lasaga A. C., Berner R. A., and Garrels R. M. (1985) An improved geochemical model of atmospheric CO_2 fluctuations over the past 100 million years. In *The Carbon Cycle and Atmospheric CO_2: Natural Variations Archean to Present* (E. T. Sundquist and W. S. Broecker, eds.), pp. 397-411. AGU, Washington, DC.

Lovelock J. E. and Whitfield M. (1982) Life span of the biosphere. *Nature, 296,* 561-563.

Marty B. and Jambon A. (1987) $C/^3He$ in volatile fluxes from the solid Earth: Implications for carbon geodynamics. *Earth Planet. Sci. Lett., 83,* 16-26.

Matsui T. and Abe Y. (1986a) Evolution of an impact-induced atmosphere and magma ocean on the accreting Earth. *Nature, 319,* 303-305.

Matsui T. and Abe Y. (1986b) Impact-induced atmospheres and oceans on Earth and Venus. *Nature, 322,* 526-528.

Nagumo S. (1987) About the accretionary prism model. *Chikyu Monthly, 9,* 294-303. (In Japanese.)

Otting W. and Zahringer J. (1967) Total carbon content and primordial rare gases in chondrites. *Geochim. Cosmochim. Acta, 31,* 1949-1960.

Owen R. M. and Rea D. K. (1985) Sea-floor hydrothermal activity links climate to tectonics: The Eocene carbon dioxide greenhouse. *Science, 227,* 166-169.

Owen T., Cess R. D., and Ramanathan V. (1979) Early Earth: An enhanced carbon dioxide greenhouse to compensate for reduced solar luminosity. *Nature, 277,* 640-642.

Ozawa T. (1986) The process of the formation of East Asia from the point of view of paleobiogeography. *Kagaku, 56,* 303-311. (In Japanese.)

Poldervaart A. (1955) Chemistry of the Earth's crust. *Geol. Soc. Am. Spec. Pap. No. 62,* 119-144.

Reymer A. and Schubert G. (1984) Phanerozoic addition rates to the continental crust and crustal growth. *Tectonics, 3,* 63-77.

Ronov A. B. (1964) Common tendencies in the chemical evolution of the Earth's crust, ocean and atmosphere. *Geokhimiya, 8,* 715-743.

Ronov A. B. (1968) Probable change in the composition of sea water during the course of geological time. *Sedimentology, 10,* 25-43.

Ronov A. B. and Yaroshevsky A. A. (1967) Chemical structure of the Earth's crust. *Geokhimiya, 11,* 1285-1309.

Ronov A. B. and Yaroshevsky A. A. (1976) A new model for the chemical structure of the Earth's crust. *Geokhimiya, 12,* 1761-1795.

Rubey W. (1951) Geologic history of sea water, an attempt to state the problem. *Bull. Geol. Soc. Am., 62,* 1111-1148.

Sagan C. and Mullen G. (1972) Earth and Mars: Evolution of atmospheres and surface temperatures. *Science, 177,* 52-56.

Schidlowski M. (1988) A 3,800-billion-year isotopic record of life from carbon in sedimentary rocks. *Nature, 333,* 313-318.

Sclater J. G. and Parsons B. (1981) Oceans and continents: Similarities and differences in the mechanisms of heat loss. *J. Geophys. Res., 86,* 535-552.

Sleely D. R., Vail P. R., and Walton G. G. (1974) Trench slope model. In *The Geology of Continental Margins* (C. A. Burk and C. L. Drake, eds.), pp. 249-260. Springer-Verlag, Berlin.

Sprague D. and Pollack H. N. (1980) Heat flow in the Mesozoic and Cenozoic. *Nature, 285,* 393-395.

Turcotte D. R. and Schubert G. (1982) *Geodynamics.* Wiley, New York. 450 pp.

Walker J. C. G. (1982) Climatic factors on the Archean Earth. *Paleogeogr. Paleoclimat. Paleoecol., 40,* 1-11.

Walker J. C. G. (1985) Carbon dioxide on the early earth. *Origins of Life, 16,* 117-127.

Walker J. C. G., Hays P. B., and Kasting J. F. (1981) A negative feedback mechanism for the long-term stabilization of Earth's surface temperature. *J. Geophys. Res., 86,* 9776-9782.

Walker J. C. G., Klein C., Schidlowski M., Schopf J. W., Stevenson D. J., and Walter M. R. (1983) Environmental evolution of the Archean-early Proterozoic Earth. In *Earth's Earliest Biosphere, Its Origin and Evolution* (J. W. Schopf, ed.), pp. 260-290. Princeton Univ., Princeton, New Jersey.

Wolery T. J. and Sleep N. H. (1976) Hydrothermal circulation and geochemical flux at mid-ocean ridges. *J. Geol., 84,* 249-275.

ORIGIN OF AN INHABITED PLANET

J. C. G. Walker

Space Physics Research Laboratory,
Department of Atmospheric, Oceanic, and Space Science, and Department of Geological Sciences,
The University of Michigan, Ann Arbor, MI 48109

How do the existence and history of life on Earth constrain theories of planetary origin? Something can be learned from the unambiguous record of the antiquity and continuity of life. The oldest clear evidence of life dates back to 3.5 b.y. ago; by that time the processes of planetary growth had ameliorated enough to permit a habitable environment. It is not clear from the paleontological record, however, that the environment was permanently habitable as long ago as 3.5 b.y. A clearly continuous paleontological record of Archean life does not exist, and therefore whole-Earth sterilizing impact events cannot be ruled out on biological grounds until early in the Proterozoic period. Organisms are uniquely able to separate the constituents of the environment. In particular, photosynthetic organisms can generate reduced organic compounds even in a strongly oxidizing environment. Their activities have resulted in accumulations of reduced compounds at the surface of the Earth, throughout the upper crust, and possibly at deeper levels. Equivalent amounts of strongly oxidized compounds have also accumulated. The resultant heterogeneity makes it hard to determine the original oxidation state of Earth's outer layers. Although life has oxidized the ocean and atmosphere, it probably has not contributed to overall oxidation of the Earth. If anything, organisms' abilities to break down environmental compounds have preserved the oxidation state of the mantle from larger changes that would have occurred on an uninhabited planet.

INTRODUCTION

The original focus of this paper was to review what we can learn about the origin of the Earth from the existence of Earth life, and to examine how life has affected or does affect our efforts to understand the origin of the Earth. After conscientiously giving this assignment my best effort, I have come up with very little. My paper will therefore focus on three aspects of my topic: What does the origin of life tell us about the origin of the Earth? What does the history of life tell us about the origin of the Earth? And what has life done to the observable record of Earth origin?

ORIGIN OF LIFE

At this stage of our understanding of the origin of life, I believe that we can learn remarkably little about the origin of the Earth from the fact that Earth is inhabited. Since 1957 an army of emulators of Stanley Miller have synthesized biological molecules in a wide range of environments using every conceivable energy source (*Miller and Orgel,* 1974; *Chang et al.,* 1983). At the same time, organic molecules have been discovered in space, in the interstellar medium, and in meteorites. It is not at all clear where chemical evolution occurred, by what process, or under what conditions. I believe that chemical evolution, the formation of biological monomers, tells us at this stage nothing about the origin of the Earth. The key question now in understanding the origin of life is how to build biological macromolecules and, ultimately, living cells. There has been little progress on these aspects of the origin of life study. While life obviously did originate somewhere, sometime, and probably not very far away from the Earth, we simply do not know where, when, or how, so the origin of life sets few constraints on our ideas concerning the origin of the Earth.

HISTORY OF LIFE

There is undoubted evidence of life on Earth 3.5 b.y. ago, both from the Pilbara Block of Western Australia and from the Barberton Mountain Land of South Africa (*Walsh and Lowe,* 1985; *Awramik,* 1986; *Byerly et al.,* 1986; *Schopf and Packer,* 1987). The evidence is in the form of stromatolites and microfossils and is no longer controversial. Therefore the Earth was obviously habitable 3.5 b.y. ago. It was neither too hot nor too cold, there was a stable interface between solid and fluid, and the fluid was probably water. This is very useful environmental information, for it tells us that Earth origin was complete by that time, which I think most of us believe anyway (*Maher and Stevenson,* 1988; *Oberbeck and Fogleman,* 1989).

Actually, the previous statement is too strong an interpretation of the evidence. Let me review the paleontological record of early life. Earth was indeed habitable at 3.5 b.y. ago, but there is no evidence that it has been continuously habitable since that time. Figure 1 shows the fossil record of Archean life, based on a summary by *Walter* (1983). The record consists mostly of stromatolites but includes some microfossils (*Schopf and Walter,* 1983). Only reliable occurrences are considered in this summary, including the ages of the rocks and, parenthetically, the uncertainties in the ages. The episodes in which life demonstrably did exist on Earth are few and far between. The point of this demonstration is that for most of the Archean there is in fact no evidence of life on Earth, nor is there any paleontological evidence that Earth was habitable. Conversely, of course, there is no evidence that Earth was not habitable. There simply is no evidence one way or the other.

The Archean record is in marked contrast to the record beginning in the early Proterozoic, for which there is an abundance of fossil evidence that the Earth was continuously habitable (*Hofmann and Schopf,* 1983). However, the late Archean and early Proterozoic biota are not demonstrably descended from the older biota. There is no evidence, therefore, for continuity of life. Thus, strictly speaking, the fossil record does not set a date for the end of Earth origin. From the paleontological point of view, it is entirely possible that there were, for example, sterilizing impacts on the Earth in the late Archean. I am not aware of anybody advocating massive impacts that late in Earth history, but such a postulate could not be ruled out on paleontological grounds.

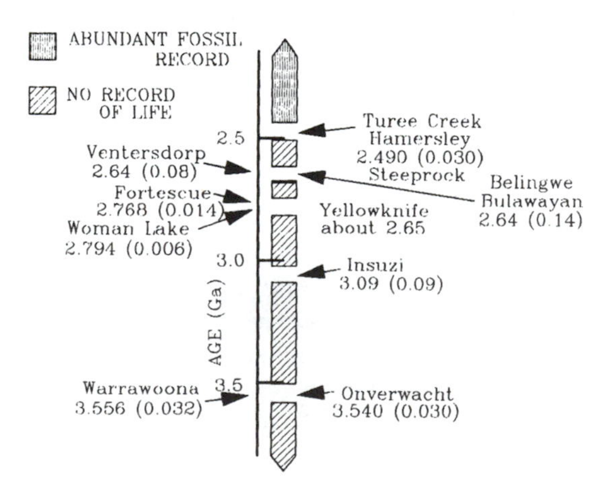

Fig. 1. *The reliable fossil record of Archean life is clearly sparse (Walter, 1983).*

INFLUENCE OF LIFE ON THE RECORD OF EARTH ORIGIN

As we all know, life is uniquely able to divide chemical constituents, to separate reduced species from oxidized species, and to alter isotope ratios. This activity of life has indeed left a mark on the record, particularly for carbon and sulfur isotopes (*Schidlowski,* 1988). The biological influence sometimes makes it hard to determine whether a particular anomalous isotope ratio, for example, is biogenic or identifies interesting primitive material. There is a similar influence of life on the oxidation state of Earth materials. In particular, it is possible that the upper mantle has been significantly oxidized during the course of Earth history by the incorporation of oxidized material from the surface. Are we faced today with uncertainty about the oxygen fugacity of the upper mantle that would not exist if life had not been active?

First, the exogenic system of ocean, atmosphere, and sediments has undoubtedly exported oxidant to the mantle. Figure 2 shows a budget of oxidants and reductants in the exogenic system, based on data on the average composition of sediments accumulated by *Ronov* (1983). The reductants are mostly organic carbon in sediments. To arrive at an upper limit for the oxidants, I have assumed that all sulfate in ocean and sediments was originally sulfide and that all ferric iron in the sediments was produced by the oxidation of ferrous iron. The data are not of the highest quality, but each individual sediment type in Ronov's tabulation with the exception of evaporites, which are quantitatively insignificant, shows the same excess of reductant over oxidant, which in my opinion is an observed fact.

The reductant was all formed from carbon dioxide by photosynthetic organisms, so the production of reductant involved a corresponding production of oxidant. This oxidant

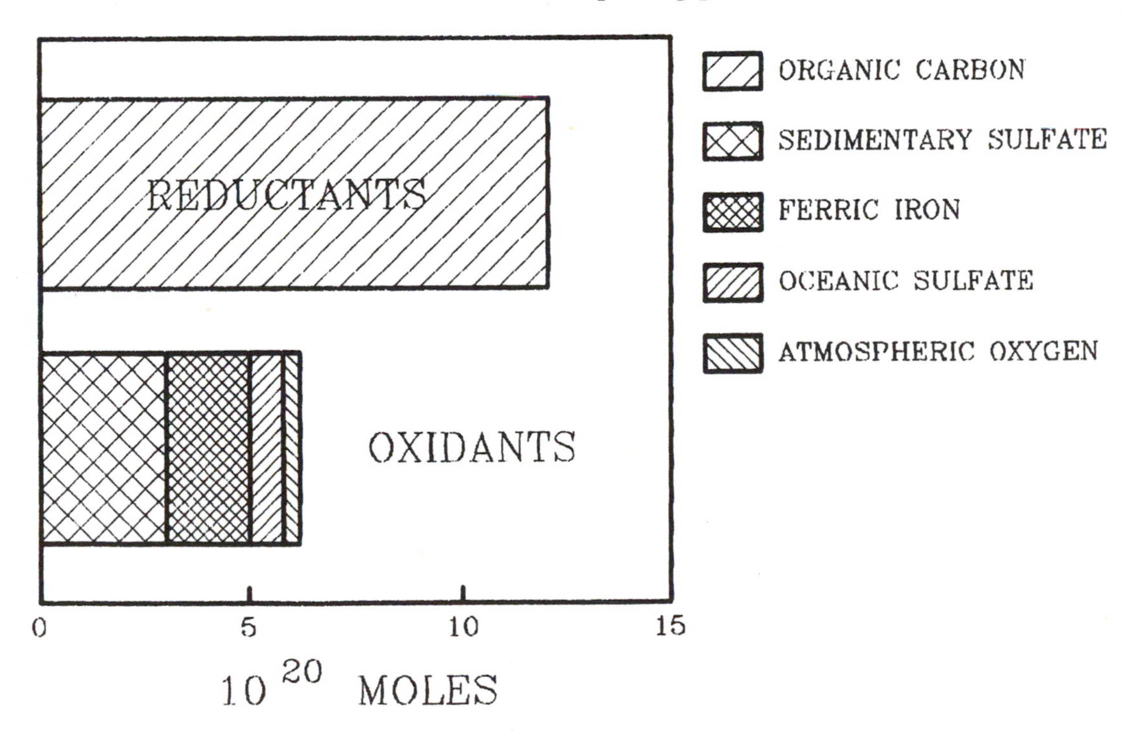

Fig. 2. *Oxidation-reduction balance of the exogenic system, based on data of Ronov (1983).*

is now missing; the exogenic system has lost oxygen equivalents. Oxygen does not escape from the atmosphere to space. The only place it can have gone is into the mantle. The amount identifiable in the budget of Fig. 2 is not much in mantle terms, only enough to oxidize perhaps a tenth of a percent of mantle iron. The significance is that this observation establishes the possibility of the oxidation of the mantle by the exogenic system.

The only way the Earth as a whole can become more oxidized is by escape of reduced gas to space. All other terrestrial processes simply move oxygen equivalents around. Continuing this line of thought, the only way the mantle and core can become oxidized is either by escape of hydrogen to space or by storage of reduced matter in sedimentary rocks. The mass of atmosphere and ocean is too small to be a significant long-term reservoir for oxygen equivalents. Storage of organic carbon in sediments is the component I have already examined using the data of *Ronov* (1983). Escape of hydrogen to space is the component I propose now to explore.

Earth may have lost large amounts of hydrogen during the course of accretion by the process of hydrodynamic escape (*Hunten et al.,* 1989). This is not the process with which I am concerned here. The questions I examine are whether continuing escape during the course of Earth history may have significantly changed Earth's oxidation state and whether life may have influenced this continuing escape.

The rate of escape of hydrogen from a terrestrial atmosphere that is not composed principally of hydrogen is proportional to the concentration of all chemical forms of hydrogen in the middle atmosphere (*Hunten,* 1973; *Walker,* 1977). Today the escape rate is negligibly small, and the most important sources of hydrogen in the middle atmosphere are water vapor, methane, and molecular hydrogen. Water vapor is much more abundant in the troposphere, but it is trapped at low altitudes by condensation resulting from low temperatures at the tropopause. The tropopause temperature is related to the flux of infrared energy leaving the planet (*Goody and Walker,* 1972). It is not likely to have been larger in the past. Escape rates large enough to change Earth's oxidation state would have required much larger concentrations of hydrogen compounds other than water vapor. Today these compounds, methane and hydrogen, are products of life. This is the basis of the suggestion that life may have influenced Earth's oxidation state.

However, because the capacity of ocean and atmosphere to store oxygen equivalents is small, the flux of hydrogen to space cannot, in the long term, exceed the flux of reducing equivalents out of the mantle. Mantle hydrogen, for example, can in principle flow directly through the exogenic system to space. Mantle-reduced iron, manganese, or sulfur can be oxidized by organisms or by inorganic processes, in principle releasing equivalent fluxes of hydrogen that can escape. But the exogenic system, including life, cannot create reducing equivalents. The present-day flux of reduced constituents out of the mantle is estimated at about 10^{12} mol/yr. This rate, if continued for a billion years, would oxidize only a few tenths of a percent of the iron in the mantle, and that amount only if there were no flux of reductants from the exogenic system back into the mantle.

The flux of escaping hydrogen today is only about 10^{10} mol/yr (*Walker,* 1977), much smaller than the flux of reduced constituents in midocean ridge hydrothermal fluids. Reducing equivalents must be returned to the mantle, presumably by subduction of organic carbon and reduced minerals, many of them the products of diagenetic reaction of organic carbon. Life has produced an oxygenic atmosphere, depressing the concentrations of reduced gases and the rate of escape of hydrogen to space. Paradoxically, therefore, by oxidizing the atmosphere, life has diminished the rate of oxidation of the mantle, preserving the original oxidation state.

CONCLUSION

My conclusions are that (1) the origin of life does not constrain theories of Earth origin; (2) the history of life does not usefully constrain theories of Earth origin; and (3) life's unmixing activities have not greatly affected the oxidation state of the bulk Earth and may indeed have preserved it.

Acknowledgments. This research was supported in part by the National Aeronautics and Space Administration under Grant NAGW-176 to The University of Michigan. I am grateful to an anonymous referee for suggesting that life has helped to preserve the original oxidation state of the mantle.

REFERENCES

Awramik S. M. (1986) New fossil finds in old rocks. *Nature, 319,* 446-447.

Byerly G. R., Lowe D. R., and Walsh M. M. (1986) Stromatolites from the 3,300-3,500-Myr Swaziland Supergroup, Barberton Mountain Land, South Africa. *Nature, 319,* 489-491.

Chang S., Des Marais D., Mack R., Miller S. L., and Strathearn G. E. (1983) Prebiotic organic synthesis and the origin of life. In *Earth's Earliest Biosphere* (J. W. Schopf, ed.), pp. 53-92. Princeton Univ., Princeton, New Jersey.

Goody R. M. and Walker J. C. G. (1972) *Atmospheres.* Prentice-Hall, Englewood Cliffs, New Jersey. 150 pp.

Hofmann H. J. and Schopf J. W. (1983) Early Proterozoic microfossils. In *Earth's Earliest Biosphere* (J. W. Schopf, ed.), pp. 321-360. Princeton Univ., Princeton, New Jersey.

Hunten D. M. (1973) The escape of light gases from planetary atmospheres. *J. Atmos. Sci., 30,* 1481-1494.

Hunten D. M., Donahue T. M., Kasting J. F., and Walker J. C. G. (1989) Escape of atmospheres and loss of water. In *Origin and Evolution of Planetary and Satellite Atmospheres* (S. K. Atreya, J. B. Pollack, and M. S. Matthews, eds.), pp. 386-422. Univ. of Arizona, Tucson.

Maher K. A. and Stevenson D. J. (1988) Impact frustration of the origin of life. *Nature, 331,* 612-614.

Miller S. L. and Orgel L. E. (1974) *The Origins of Life on Earth.* Prentice-Hall, Englewood Cliffs, New Jersey. 229 pp.

Oberbeck V. R. and Fogleman G. (1989) Impacts and the origin of life. *Nature, 339,* 434.

Ronov A. B. (1983) *The Earth's Sedimentary Shell.* AGI Reprint Series Vol. V, American Geological Institute, Falls Church, Virginia. 80 pp.

Schidlowski M. (1988) A 3,800-million-year isotopic record of life from carbon in sedimentary rocks. *Nature, 333,* 313-318.

Schopf J. W. and Packer B. M. (1987) Early Archean (3.3-billion to 3.5-billion-year-old) microfossils from Warrawoona Group, Australia. *Science, 237,* 70-73.

Schopf J. W. and Walter M. R. (1983) Archean microfossils: New evidence of ancient microbes. In *Earth's Earliest Biosphere* (J. W. Schopf, ed.), pp. 214-239. Princeton Univ., Princeton, New Jersey.

Walker J. C. G. (1977) *Evolution of the Atmosphere.* Macmillan, New York. 318 pp.

Walsh M. M. and Lowe D. R. (1985) Filamentous microfossils from the 3,500-Myr-old Onverwacht Group, Barberton Mountain Land, South Africa. *Nature, 314,* 530-532.

Walter M. R. (1983) Archean stromatolites: Evidence of the Earth's earliest benthos. In *Earth's Earliest Biosphere* (J. W. Schopf, ed.), pp. 187-213. Princeton Univ., Princeton, New Jersey.

INDEX

* Page number refers to the first page of an article in which a term is discussed.